Thin Layer Chromatography in Chiral Separations and Analysis

CHEM

CHROMATOGRAPHIC SCIENCE SERIES

A Series of Textbooks and Reference Books

Editor: JACK CAZES

Thin Layer Chromatography in Chiral Separations and Analysis

Teresa Kowalska
University of Silesia
Katowice, Poland

Joseph Sherma
Lafayette College
Easton, Pennsylvania, U.S.A.

CRC Press
Taylor & Francis Group
Boca Raton   London   New York

CRC Press is an imprint of the
Taylor & Francis Group, an **informa** business

Chemistry Library

CRC Press
Taylor & Francis Group
6000 Broken Sound Parkway NW, Suite 300
Boca Raton, FL 33487-2742

© 2007 by Taylor & Francis Group, LLC
CRC Press is an imprint of Taylor & Francis Group, an Informa business

No claim to original U.S. Government works
Printed in the United States of America on acid-free paper
10 9 8 7 6 5 4 3 2 1

International Standard Book Number-10: 0-8493-4369-0 (Hardcover)
International Standard Book Number-13: 978-0-8493-4369-8 (Hardcover)

Library of Congress Cataloging-in-Publication Data

Thin layer chromatography in chiral separations and analysis / edited by Teresa
  Kowalska and Joseph Sherma.
       p. cm. -- (Chromatographic science series ; 98)
  Includes bibliographical references and index.
  ISBN 978-0-8493-4369-8
  1. Thin layer chromatography. 2. Enantiomers--Separation. 3. Chirality--Industrial
applications. I. Kowalska, Teresa. II. Sherma, Joseph. III. Title. IV. Series.

QD79.C454T55 2007
543'.84--dc22   2007000600

Visit the Taylor & Francis Web site at
http://www.taylorandfrancis.com

and the CRC Press Web site at
http://www.crcpress.com

Contents

Preface

The purpose of this book is to present practical, comprehensive information on the field of chiral thin layer chromatography (TLC). As with the past book we edited for CRC/Taylor & Francis (*Preparative Layer Chromatography*, 2006), this is the first book on the title topic to become available. The book's coverage of state-of-the-art chiral TLC is divided into two main sections: theory and procedures (Chapter 1 to Chapter 9) and applications (Chapter 10 to Chapter 15). The book will be of great benefit to scientists with diverse interests for better understanding and wider use of chiral TLC. It will be a critical resource for researchers, analysts, and teachers with limited to broad experience in TLC and chiral separations because of its blend of introductory, background, and detailed, advanced experimental material.

We believe that chiral TLC is currently undervalued as a tool for enantiomer separations and analyses. Advantages include simplicity; relatively low cost; ability to run multiple samples at the same time under essentially identical conditions, thereby reducing time per sample; and the storage of chromatograms on the layer with the possibility of applying multiple detection, documentation, and *in situ* quantification procedures for the separated zones. It is our hope that this book will highlight these and other virtues of enantioseparations and analyses by TLC, and encourage commercial manufacturers to produce a greater variety of stationary phases for chiral separations, such as additional impregnated layers and derivatives of cellulose.

The contributors to this book are internationally recognized experts in the topics about which they wrote. They were encouraged to adopt an outline of topics that would suit the presentation of their material in the most useful and understandable manner, including tables, figures, and up-to-date references along with the text. All authors exhibited great enthusiasm for publication of the book, and their varied countries of origin ensure that it will have an international viewpoint.

We are indebted to Dr. Jack Cazes, Barbara Glunn, and the other CRC/Taylor & Francis staff members involved in the approval and production of our book, and for their great cooperation and commitment to its success.

Teresa Kowalska and Joseph Sherma

Editors

Teresa Kowalska is currently a professor in the Department of General Chemistry and Chromatography at the University of Silesia (Katowice, Poland). Her scientific interests include the physicochemical foundations of liquid chromatography and gas chromatography. Dr. Kowalska is the author of more than 200 scientific papers, more than 300 scientific conference papers, and a vast number of the book chapters and encyclopedia entries in the field of chromatography. It is perhaps noteworthy that she has authored (and then updated) the chapter entitled "Theory and Mechanism of Thin-Layer Chromatography" for all three editions of the *Handbook of Thin-Layer Chromatography*, edited by J. Sherma and B. Fried, and published by Marcel Dekker. In addition, she coedited (with J. Sherma) *Preparative Layer Chromatography*, published by CRC Press/Taylor & Francis Group in 2006.

Dr. Kowalska has served as editor of *Acta Chromatographica*, the annual periodical published by the University of Silesia and devoted to all chromatographic and hyphenated techniques, its establishment in 1992. *Acta Chromatographica* appears as a hard copy journal and also online. Its contributors originate from an international academic community, and it is meant to promote the development in separation sciences. It apparently serves its purpose well, as can be judged from its wide readership, abundant citations throughout the professional literature, and also from the ISI ranking quota.

In the course of the past 30 years, Dr. Kowalska has been active as organizer (and in recent years as a cochairperson) of the annual all-Polish chromatographic symposia with international participation, held each year (since 1977) in the small mountain resort of Szczyrk in Southern Poland. Integration of an international community of chromatographers through these meetings has been regarded by Dr. Kowalska as a specific yet important contribution to chromatography.

Joseph Sherma is John D. and Frances H. Larkin Professor Emeritus of Chemistry at Lafayette College, Easton, Pennsylvania. He is author or coauthor of over 525 scientific papers and editor or coeditor of over 50 books and manuals in the areas of analytical chemistry and chromatography. Dr. Sherma is coauthor, with Bernard Fried, Kreider Professor Emeritus of Biology at Lafayette College, of *Thin Layer Chromatography* (editions 1 to 4) and coeditor with Professor Fried of the *Handbook of Thin Layer Chromatography* (editions 1 to 3), both published by Marcel Dekker, Inc. He served for 23 years as the editor for residues and trace elements of the *Journal of AOAC International* and serves currently on the editorial advisory boards of the *Journal of Liquid Chromatography and Related Technologies*, the *Journal of Environmental Science and Health (Part B)*, the *Journal of Planar Chromatography — Modern TLC, Acta Chromatographica,* and *Acta Universitatis Cibiniensis, Seria F. Chemia.* Dr. Sherma received his Ph.D. degree (1958) from Rutgers State University, New Brunswick, New Jersey. He received the 1995 American Chemical Society (ACS) Award for Research at an Undergraduate Institution, sponsored by Research Corporation.

Contributors

Danica Agbaba
Faculty of Pharmacy
University of Belgrade
Belgrade, Serbia

Ravi Bhushan
Department of Chemistry
Indian Institute of Technology
  (Roorkee)
Uttaranchal, India

Jacek Bojarski
Faculty of Pharmacy
Collegium Medicum
Jagiellonian University
Kraków, Poland

Hans Brückner
Interdisciplinary Research
  Center (IFZ)
University of Giessen
Giessen, Germany

Alessandra Cincinelli
Department of Chemistry
University of Florence
Florence, Italy

Virginia Coman
Raluca Ripan Institute for
  Research in Chemistry
Cluj-Napoca, Romania

Massimo Del Bubba
Department of Chemistry
University of Florence
Florence, Italy

Władysław Gołkiewicz
Faculty of Pharmacy
Medical Academy of Lublin
Lublin, Poland

Urszula Hubicka
Faculty of Pharmacy
Collegium Medicum
Jagiellonian University
Kraków, Poland

Branka Ivković
Faculty of Pharmacy
University of Belgrade
Belgrade, Serbia

Teresa Kowalska
Institute of Chemistry
University of Silesia
Katowice, Poland

Jan Krzek
Faculty of Pharmacy
Collegium Medicum
Jagiellonian University
Kraków, Poland

Piotr Kuś
Institute of Chemistry
University of Silesia
Katowice, Poland

Anna Kwiecień
Faculty of Pharmacy
Collegium Medicum
Jagiellonian University
Kraków, Poland

Ewa Leciejewicz-Ziemecka
Office for Registration of Medicinal
   Products, Medical Devices and
   Biocidal Products
Warsaw, Poland

Luciano Lepri
Department of Chemistry
University of Florence
Florence, Italy

Jürgen Martens
Institute of Pure and Applied
   Chemistry
University of Oldenburg
Oldenburg, Germany

Patrick Piras
UMR 6180, CNRS Laboratory of
   Chirotechnologies, Catalysis and
   Biocatalysis
Universite Paul Cézanne
Marseille, France

Irma Podolak
Faculty of Pharmacy
Collegium Medicum
Jagiellonian University
Kraków, Poland

Beata Polak
Faculty of Pharmacy
Medical Academy of Lublin
Lublin, Poland

Mieczysław Sajewicz
Institute of Chemistry
The University of Silesia
Katowice, Poland

Joseph Sherma
Department of Chemistry
Lafayette College
Easton, Pennsylvania

Antoine-Michel Siouffi
Faculte des Sciences de Saint Jerome
Universite Aix-Marseille III
Marseille, France

Aleksander Sochanik
Comprehensive Cancer Centre
Maria Sklodowska-Curie Memorial
   Institute
Gliwice, Poland

1 Overview of the Field of Chiral TLC and Organization of the Book

Teresa Kowalska and Joseph Sherma

CONTENTS

1.1 CHIRALITY BACKGROUND

Chirality is a form of stereoisomerism that embraces the relationship between two or more isomers having the same structure (i.e., the same linkages between atoms) but with different configurations (spatial arrangements). Chiral compounds are further classified into enantiomers, that is, molecules that are mirror images of each other, and diastereoisomers, which are stereoisomers that are not mirror images.

An *enantiomer* is one of a pair of nonsuperimposable mirror image molecules. Two molecules are enantiomers if they are mirror images of each other that cannot be superimposed by any rotation or translation. Physical and chemical properties of the two enantiomers making a pair (also known as two antimers or antipodes) are almost identical, except for their optical property of rotating polarized light in opposite directions. The almost identical physical and chemical properties of the two antimers pose a very difficult task for all who, for one reason or another, need to separate them as two different species. In other words, enantioseparation is among the most difficult tasks in analytical chemistry.

Diastereoisomers are stereoisomers that have more than one center of asymmetry in their structure and are not enantiomers or mirror images of each other.

1

Contrary to enantiomers, diastereoisomers can have different physical properties and reactivity. Because of greater differences in physical and chemical properties among diastereomers, separation and isolation of these isomeric compounds having the same structure but different spatial arrangement are relatively easier than with the enantiomers, but still a difficult task.

The phenomenon of chirality is omnipresent in nature, and its presence in humans, animals, and plants determines their chemical structure and also the majority of their living functions. One of the greatest, and so far unexplained, mysteries of biophysics is the predominance of homochirality among living organisms. According to the definition adopted by the International Union of Pure and Applied Chemistry (IUPAC), only a sample that contains all molecules of the same chirality type can be considered as homochiral (of course, within the limits of the available detection sensitivity). The most striking manifestation of homochirality in nature is that human and animal organisms are built exclusively of the L-amino acids (i.e., the left-handed form) and of the D-carbohydrates (i.e., the right-handed form).

This biophysical puzzle spans many different areas of the natural and life sciences, but in a certain sense it also poses an important question for philosophy and even religion. Why do we not encounter on earth "the life reflected in the mirror," and why do we not encounter our own antimeric "quasi-twins?" The simplest scientific and also philosophical question related to homochirality can be formulated in this way: Is homochirality an inevitable precursor of the organic life on our planet, or, to the contrary, can we consider homochirality as a free play of the forces of nature and a purely random phenomenon? At this point, one starts raising questions about the origin of life on earth and comes very close not only to theoretical organic geochemistry and astrobiology but also to the domain of religion.

Due to the omnipresence of chirality in nature, the majority of metabolic processes occurring in living organisms are stereospecific. It is a well-recognized fact that the biological catalysts known as *enzymes* have asymmetric active centers to properly fit in biological receptors that are also asymmetric. In this sense, homochirality, even if we do not understand its origin, seems to reflect the wisdom of nature and its great economy. It undoubtedly leads to a massive gain in time and energy, because living organisms do not need to select the properly handed substrates from a pool of the left- and right-handed ones at each individual metabolic step.

1.2 BASIC STRATEGIES OF CHROMATOGRAPHIC ENANTIOSEPARATIONS

Chromatographic separation of two antimers, most often referred to as *enantioseparation*, can be carried out following either a direct or an indirect strategy. It is perhaps noteworthy that this duality of alternative options is characteristic not only of enantioseparations by means of thin layer chromatography (TLC), which is the primary subject of this book, but also of high-performance liquid chromatography (HPLC) and gas chromatography (GC).

The first strategy, known as *direct separation*, consists of introducing to the given chromatographic system a mixture of the two enantiomers without any pre-processing, that is, without derivatization, prior to the chromatographic run. It is a widely assumed conviction (and in most cases a true one) that direct separation can only be obtained in a chiral chromatographic system. Each chromatographic system is composed of a stationary phase and a mobile phase. Chiral chromatographic systems are most frequently composed of either a chiral stationary phase (CSP) with a nonchiral mobile phase or vice versa. Chromatographic systems composed of two chiral phases are avoided, one reason being that the optically pure stereoisomers are costly and especially because one chiral phase is sufficient. In HPLC, chromatographic systems with CSPs are more frequently applied than those with chiral mobile phases. An important reason is that chiral columns can be reused for many consecutive analyses, whereas the expensive chiral mobile phase modifiers often cannot be purified and reused. Chiral separations by means of TLC in many cases follow in the conceptual footsteps of successful, practical solutions elaborated earlier for HPLC, and this is one reason why applications of CSPs are more frequent than those of mobile phases containing chiral modifiers, even though both TLC plates and mobile phases are used only once.

The second strategy for separating enantiomer pairs is indirect separation. In this case, a mixture of the antipodes is derivatized with a chiral agent prior to chromatography in an appropriate system to give a respective mixture of diastereoisomers. Chemical structures of diastereoisomers and their resulting physical and chemical properties are much more differentiated than in the case of a pair of original enantiomers, and the separation of diastereoisomers is, therefore, a considerably easier experimental task than the separation of the corresponding enantiomers. In fact, chromatographic separation of diastereoisomers can be obtained in a chromatographic system composed of an achiral stationary phase and an achiral mobile phase. From a purely chromatographic point of view, indirect separation of the enantiomers can be obtained in much simpler and also less expensive chromatographic systems than those required for direct separation. On the other hand, preliminary derivatization of an enantiomer mixture can often prove to be a relatively complex and time-consuming step that many analysts would rather prefer to avoid. This book covers all the important aspects of both direct and indirect enantioseparations by means of TLC.

1.3 THE AREAS OF PRIMARY DEMAND FOR ENANTIOSEPARATIONS

The first strong impulse to seriously consider the development of small- and large-scale separations of enantiomers as an important and urgent research task came from the pharmaceutical field, initiated by the infamous case of introducing to the market the racemic sedative drug known by the name *thalidomide*. Thalidomide was first introduced in Germany (and later in other countries, mostly Western European) on October 1, 1957, by the German pharmaceutical firm, Chemie Grünenthal. The drug was particularly advertised as an effective and safe sedative to pregnant women. At that time, nobody was aware of the fact that one drug enantiomer from a given pair can exert a positive therapeutic effect and its antipode can prove to be ineffective or even dangerous and harmful. This was precisely the case with thalidomide. One enantiomer acted as a very efficient sedative, whereas its antipode was strongly teratogenic. As a result, thousands of seriously crippled babies with underdeveloped limbs were born in many countries, since then known as generation of "thalidomide babies" or "flipper babies."

The dramatic effect of thalidomide on human fetuses is still considered as one of the greatest mistakes ever committed in the history of modern pharmacy. However, it also served as the starting point of a massive research effort in the field of chromatographic enantioseparations. Perhaps, it is also worth mentioning that this once feared drug is now being rediscovered and is again in use [1]. In 1964, a physician treating leprosy patients in Israel for a painful condition known as *erythema nodosum leprosum* (ENL) prescribed thalidomide as a sedative. Surprisingly, the drug alleviated the symptoms of this painful condition. From that point onward, thalidomide was the therapy of choice for this application, including designation by the World Health Organization (WHO).

Moreover, thalidomide's effects on the rapidly dividing cells of embryos suggested that it might destroy cancer cells. What was unique about thalidomide was its powerful teratogenicity that could be related to anticancer effects. Now, it is considered that thalidomide's potential applications are essentially limitless. It can be used, under an emergency U.S. Food and Drug Administration (FDA) approval, to treat more than 70 forms of cancer and various skin, digestive, and immunological diseases. It may also be useful in treating the autoimmune deficiency syndrome (AIDS)-associated cachexia (wasting) and tuberculosis.

Even from this single example of thalidomide and the diverse biological effects and curing potentials of its two antimers, it can easily be deduced that enantioseparations are crucial in human (and also veterinary) pharmacy and medicine. In the present period of dynamic development of advanced computer-aided strategies for molecular drug design, the number of newly devised molecular structures with the anticipated curing potential is growing quickly. Many of these compounds have an asymmetric structure and, hence, they can appear in both left- and right-handed forms. From the pharmacokinetic point of view, each of the two antipodes has to be treated as a completely different compound that is able to exert a unique effect on living organisms, which has to be assessed at a very early stage of investigation of the prospective drug. To carry out the preliminary biological and other tests with each individual enantiomer, efficient working tools that enable the separation and evaluation of the quantitative proportions between the two antimers in a reaction mixture, followed by their ultimate preparative isolation, are needed.

The search for drugs that are safe from a toxicological point of view makes investigations of the analysis of enantiomeric antipodes and their biological activity the most important tasks in a long and meticulous sequence of steps leading from the computer-aided molecular drug design to the implementation of a successful pharmaceutical product on the drug market. However, there is one field of scientific activity, in a superficial way rather similar to computer-aided molecular drug design, which is focused on selection of the compounds with a properly tailored toxic potential. The chemometric strategies elaborated for drug design have been widely and successfully adopted in the search for efficient candidate pesticide compounds [2,3]. Also, in this case, many chemometrically devised and then synthesized substances are asymmetric, and biological activity of each individual antimer from a given pair has to be carefully examined. The principal objective is selection of the enantiomers with well-balanced biochemical properties, combining a possibly low-toxic potential toward humans with a possibly high-toxic potential toward the target organisms to be destroyed. It is a well-known fact that such target organisms (plants, when we consider herbicides, insects for insecticides, etc.) most often play a significant role in the interspecies nutrition chain, and through this chain they can relatively easily be transported by ingestion to humans. Another route for the dangerous migration of pesticides to humans is through natural precipitation (i.e., rain and snow) that is able to carry them into reservoirs of potable water (i.e., underground water, rivers, lakes, artificial reservoirs, etc.).

Chirality studies also proved very important for the dating of the organic fossils, and, hence, fortified archeology, geoarcheology, paleontobiology, and the related fields of knowledge with a very well-performing diagnostic tool based on naturally occurring homochirality. The principle is that all amino acids except glycine (the simplest amino acid that lacks an asymmetric carbon atom in its structure) are chiral, and in living human and animal bodies, they appear exclusively in the left-handed form. At the moment of biological death of the organism, the spontaneous racemization of the L-amino acids commences, with continuously growing content of the right-handed antimer and corresponding diminishing content of the left-handed one. It has not been difficult to experimentally determine the kinetic parameters (i.e., half-times and rate constants) of the spontaneous racemization with selected amino acids, depending on the external physical parameters of running this process (e.g., temperature, pH, pressure, reaction medium, etc.). This knowledge enabled elaboration of the method of dating the archeological findings, based on the degree of the spontaneous racemization of the amino acids contained in the organic fossils. The first amino acid selected to serve the purpose of a natural age marker was L-aspartic acid, and the first article on this subject was published in 1980 in the journal *Nature* [4]. It originated from the leading research group investigation on chirality and homochirality, and it is still considered as a momentous breakthrough in the field of dating archeological findings. The article [4] deals with the dating of the Dead Sea scroll parchments, based on unfolding of the collagen molecules and racemization of L-aspartic acid contained therein. It is well established now that dating of the organic fossils often proves more accurate when quantification is carried out based on the spontaneous racemization of amino acids than when using radiocarbon dating based on measuring the contents of the ^{14}C-radioisotope in the investigated samples.

Last, but not least, chirality studies are also of vital interest for astrochemistry and astrobiology. In these fields, scientists are interested in tracing evidence of possible extraterrestrial life in the meteorites and other material samples of extraterrestrial origin. Because the most important evidence for past life can be inscribed in trace amounts of amino acids, astrochemists and astrobiologists are also quasi-automatically confronted with the issue of extraterrestrial chirality. For example, from the numerical D/L ratios of amino acids extracted from the meteorites of Martian or cometary origin, theories are devised as to the biotic or abiotic origin of the respective compounds.

In summary, the omnipresence of chirality in the structure of the living organic matter and its importance for the natural life processes are today out of any question. Chirality is involved in the metabolic processes, most of which appear to be stereoselective. The understanding of a vital role played in our world by chirality is relatively new, and the subject still needs further extensive studies. One important precondition of success with such studies is an easy availability of convenient and well-performing analytical tools that are powerful enough to separate the enantiomer mixtures and to isolate individual antimers for further investigations.

1.4 THE ROLE OF TLC

From the preceding sections, it is clear that the separation of enantiomer pairs is of vital importance for the general well being of humans. It is also a well-established fact that the most efficient tool to separate various mixtures of compounds is chromatography. On the other hand, it is equally well known that the separation of enantiomers still remains among the most challenging tasks in the separation sciences, even today handled with a considerable effort and only limited success. Polarimetry is used in many laboratories for the determination and control of optical purity, but this method has well-known drawbacks (see Chapter 2). The first reported enantiomer separations were performed by paper chromatography, a planar chromatography method preceding TLC; amino acid enantiomers were separated by paper chromatography as early as 1951 [5]. Historically, the first instrumental chromatographic technique used for this purpose and reasonably well performing with certain compound classes was GC. In 1966, the first GC enantioseparation was reported [6]. Later, with the gradual development of methods and improved instrumentation for HPLC and capillary electrophoresis (CE), these two techniques also became involved in chiral separations. It is no surprise then that the subject of chiral separations by means of fully instrumental chromatographic techniques has already been covered in earlier books [7–11].

Although HPLC and TLC use similar stationary and mobile phases, which have closely related separation mechanisms, and can provide equally reliable quantitative data, the latter method is still less frequently utilized for enantioseparations at the present time. There is no good reason for this except lack of adequate knowledge of the principles, techniques, and potential of TLC, which this book is designed to provide.

TLC is known for having a generally lower separation efficiency than the fully instrumental methods (e.g., GC, HPLC, and CE), which can be expressed in terms of the lower number of theoretical plates (N) in the TLC systems. However, it can be stated with absolute certainty that the available resolution in TLC systems, that includes the combination of center-to-center zone separation (selectivity) and compactness of the zones in the direction of development (N), is sufficient to separate any pair of enantiomeric species. TLC is the most versatile and flexible chromatographic technique. It is applicable to virtually all compound classes, and it has advantages compared with GC, HPLC, and CE for enantioseparations and analysis. For example, GC has been proved to be most suitable only for the separation of chiral amino acids and their derivatives. Although HPLC can be applied to separate a wider spectrum of compounds compared with GC, it often makes use of the specially esterified polysaccharides as stationary phases. Polysaccharide solids are relatively soft, with a marked and unwelcome tendency to swell in contact with mobile phases. This property often hampers proper (i.e., an undisturbed and low pressure) flow of mobile phase through a column and negatively affects the overall outcome of enantioseparations. Enantiomers separated by means of TLC can be visualized directly on layers after evaporation of mobile phase, which alleviates detection problems caused by ultraviolet (UV) absorbance of chiral additives in HPLC mobile phases. A greater variety of stationary and mobile phases is available in TLC, and detection methods for analytes are more flexible and varied. Sample throughput is higher in TLC (analysis per sample is faster), and the possibility of using different types of interactions in two-dimensional development can lead to very high-resolution separations. TLC coupled with *in situ* densitometry provides an excellent means of quantitative analysis of the separated enantiomers. High throughput is especially important when large numbers of analyses are needed, and the relative simplicity and low cost of TLC makes it ideal for simple reaction control of a synthesis, on the spot by laboratory personnel.

Because of these advantages, the importance of TLC for successful enantiomer separation and analysis is growing steadily. It has been applied not only to amino acids, but also to a very wide spectrum of different drug classes, ranging from plant medications to antibiotics, and also to other chiral compounds. It can certainly be recommended for such diverse separation tasks as rapid control of enantiocomposition with selected pharmaceuticals and for micropreparative enantioseparations. The experience of chromatographers in the field of enantioseparation and analysis with the aid of TLC is also growing steadily, and the same can be said about the output of scientific reports on this subject, dispersed in the form of original research papers and review articles in chemical, biochemical, pharmaceutical, and other life sciences literature. At this point, there is an evident need to organize and expand all of this dispersed information on the relevant procedures, materials and instruments, and applications in an authoritative monograph. So far, no such comprehensive monograph on TLC applied to chiral separations and analysis is available; therefore, our book will fill this void and hopefully encourage wider use of the method up to the level deserved by its advantages.

1.5   ORGANIZATION OF THE BOOK

The first section of this book (Chapter 1 through 9) includes chapters on the principles and practice of TLC chromatographic enantioseparations and analysis.

After this introductory Chapter 1, Chapter 2 provides a proper chemical introduction to the phenomenon of chirality and to the polarimetric and spectroscopic methods of its assessment. First, the authors discuss structural asymmetry of the compounds in general terms and also a variety of structural preconditions for its appearance. Then they focus in greater detail on different classes of organic compounds, such as alkanes and cycloalkanes, alkenes, aromatic compounds, heterocyclic compounds, other compounds containing heteroatoms, and complexes.

Chapter 3 through 6 deal with the commercial and noncommercial stationary phases used for the direct and indirect enantioseparations by means of TLC and with the chiral modifiers of mobile phases, which are used exclusively in direct separations. Chapter 3 describes the commercial chiral and nonchiral sorbent materials and commercial precoated layers used in chiral separations. Thus, it deals with silica gel; native and esterified cellulose; chiral plates (reversed phase plates impregnated with a chiral selector); and C-18, C-18W, diol, diphenyl, and C-2 chemically bonded silica gel. At the end of this chapter, the author discusses the quantification of enantiomers by using densitometry, depending on the type of the stationary phase employed.

In Chapter 4, the authors discuss the noncommercial CSPs, their preparation, and the in-laboratory preparation (home coating) of the chromatographic plates by using these sorbents. Among the stationary phases discussed are cellulose, cellulose triacetate, cellulose tribenzoate, cellulose tricarbamate molecular imprinting polymers (MIPs), and β-cyclodextrin (β-CD) bonded to the silica gel matrix.

Chapter 5 presents another group of noncommercial stationary phases employed in chiral separations by means of TLC, known as chiral complexation phases (CCPs). This group of stationary phases makes use of commercial layers that are home impregnated with a complexation agent (e.g., with transition metal ions or with acidic and basic compounds that can participate in ion-pair formation with electrolytically dissociated chiral antimers).

In Chapter 6, an overview is given with respect to chiral mobile phase modifiers used for indirect enantioseparations. The authors discuss β-CD and its water-soluble derivatives, selected macrocyclic antibiotics, selected chiral counter ions (also known as chiral heterons or chiral selectors), and bovine serum albumin (BSA).

Chapter 7 is devoted to important physicochemical — basically mechanistic — aspects of the direct enantioseparations, carried out by using either CSP or mobile phase. In such cases, the diversity of the involved separation mechanisms is much greater than the most of other chromatographic modes (and, particularly, when compared with the relatively simple physicochemical rules governing adsorption or partition liquid chromatography). Thus, the author of this chapter discusses enantioseparation in terms of the solute-chiral selector complexation constants, stoichiometry and selectivity of complexation, the nature of the binding sites on the stationary phase surface, and, finally, the supramolecular mechanisms of complexation.

Chapter 8 provides a discussion of the indirect separation of enantiomer pairs, which, as described above, can be obtained with the help of a nonchiral chromatographic system. The author addresses such crucial issues as the principle of derivatization, structural demands imposed on derivatizing agents, reasons for the choice of a given agent, and, finally, present the compounds most frequently used for derivatization. The chapter ends with an overview of the separations of diastereoisomers performed with the aid of TLC.

Chapter 9 is the last chapter in the first section of this book. Its intent is to draw the readers' attention to the possible occurrence of specific cases when, in spite of the availability of most efficient tools to perform a successful enantioseparation, the goal cannot be obtained. Such bottlenecks occur, for example, in cases when one enantiomer, when dissolved in an ampholytic solvent, can easily and spontaneously change its steric configuration and undergo oscillatory transenantiomerization in the course of its storage. In the first instance, observation of such behavior was made with selected chiral 2-arylpropionic acids (APAs) that are nowadays very common and in a wide use as nonsteroidal anti-inflammatory drugs (NSAIDs; e.g., ibuprofen and naproxen). In the case of such compounds, both their enantioseparation and reliable densitometric quantification are questionable.

The second section of the book comprises Chapter 10 through 15, covering a considerable spectrum of practical TLC enantioseparations within a variety of important compound classes. Chapter 10 deals in a general way with the chirality of pharmaceutical product racemates, viewing them both from a perspective of their use as medicines and also as chiral selectors in the TLC separation procedures. This chapter also provides important information on the methods of controlling the racemate purity.

In Chapter 11, the authors introduce both the direct approaches to enantioseparation of selected adrenergic drugs (carried out using MIPs, impregnated CSPs, or chiral mobile phase modifiers) and the indirect ones. They also discuss selected physicochemical aspects of the approaches discussed (e.g., effects of the concentration of the stationary phase impregnants and chiral mobile phase modifiers, temperature, and pH on the enantioseparation result).

Chapter 12 covers the important field of amino acid enantioseparation, in historical terms the first class of chiral compounds that were successfully separated by means of GC. The authors discuss direct enantioseparations with commercial and the noncommercial CSPs, the use of chiral modifiers of mobile phases, and derivatization of the amino acid antipodes followed by indirect separation.

In Chapter 13, the authors present the most efficient TLC leading to successful enantioseparations of selected NSAIDs. This group of compounds embraces popular drugs such as ibuprofen, naproxen, ketoprofen, and fenoprofen, most of which are widely sold over the counter almost everywhere in the world. Even if they are presently considered to be entirely harmless with both antimers exerting a similar curative effect, occasional reports have been released with respect to the negative medical results of prolonged treatment by using these compounds. In these circumstances, efficient analytical tools for enantioseparation and assessment of the NSAIDs are important for our safety and well-being.

Chapter 14 is devoted to enantioseparation of chiral antibiotics from the groups of cephalosporins and quinolones, and also of certain other antimer pairs having significant pharmaceutical importance. Especially, in view of the fact that antibiotics exert a vigorous pharmacokinetic effect and are usually dispensed in serious medical states, separation and isolation of their respective antimers, followed by a thorough laboratory scrutiny for the possible negative health effects of each individual antimer separately, can be regarded as a vital priority in pharmaceutical research. In these circumstances, the ability to effectively enantioseparate and enantioisolate these compounds prior to their further examination is analytically important.

The final chapter, Chapter 15, is devoted to chiral separations using Marfey's reagent (1-fluoro-2,4-dinitrophenyl-5-L-alanine amide; FDAA) and its analogs, their synthesis, and applications, mostly in advanced biochemical research with the different amino acids and peptides.

1.6 EPILOGUE

Thin Layer Chromatography is a fast, simple, versatile, and inexpensive method with adequate efficiency and resolution for the successful separation and analysis of enantiomers from all classes of chiral compounds. The purpose of this first book on chiral TLC is to present information from experienced, international expert practitioners in the field to illustrate its advantages, guide users in the principles and proper techniques, and encourage wider future analytical applications.

REFERENCES

1. Lewis, R., *The Scientist*, 15, 1, 2001.
2. Reynolds, C.H., Cox, H.K., and Holloway, M.K. (Eds.), *Computer-Aided Molecular Design: Applications in Agrochemicals, Materials, and Pharmaceuticals*, ACS Symposium No. 589, American Chemical Society, Washington, DC, 1995.
3. Voss, G. and Ramos, G. (Eds.), *Chemistry of Crop Protection: Progress and Prospects in Science and Regulation*, C.H.I.P., Weimar, TX, 2003.
4. Weiner, S., Kustanovich, Z., Gil-Av, E., and Traub, W., *Nature*, 287, 820–823, 1980.
5. Kotake, M., Sakan, T., Nakamura, N., and Senoh, J., *J. Am. Chem. Soc.*, 73, 2973–2974, 1951.
6. Charles-Sigler, R. and Gil-Av, E., *Tetrahedron. Lett.*, 35, 4231–4238, 1966.
7. Gubitz, G. and Schmid, M.G., *Chiral Separations, Methods and Protocols*, Vol. 243, Humana Press, Totowa, NJ, 2004.
8. Beesley, T.E. and Scott, R.P.W., *Chiral Chromatography*, John Wiley & Sons, New York, NY, 1999.
9. Chankvetadze, B., *Capillary Electrophoresis in Chiral Analysis*, John Wiley & Sons, New York, NY, 1997.
10. Ahuja, S., *Chiral Separations by Chromatography*, American Chemical Society, Washington, DC, 2000.
11. Aboul-Enein, H.Y., *Chiral Separations by Liquid Chromatography*, CRC Press, Boca Raton, FL, 2003.

2 Chirality

Piotr Kuś and Aleksander Sochanik

CONTENTS

2.1   INTRODUCTION

Symmetry appears as one of the most prevalent features of living beings [1]. Looking at the majority of them, we notice elements of symmetry in their external appearance, despite awareness that in their internal structure often there are no equivalents of such elements. If we take an even closer look at the smallest constituents forming living organisms, that is, we go down to molecular level, then we notice that there are often no objects (molecules) having any symmetry whatsoever. Most frequently, from among two mutually symmetrical molecules only one of them is found in a living organism. The study of mutual arrangement of atoms making up molecules is the subject of stereochemistry.

Stereochemistry is one of the most important branches of chemistry. It investigates spatial makeup of chemical compounds, mainly organic, as well as of complexes and coordination compounds forming the borderland between organic and inorganic chemistry. Stereochemical considerations may apply to small molecules, large polymers obtained in the laboratory, or naturally occurring macromolecular compounds (e.g., proteins, nucleic acids, and polysaccharides). The surroundings of a single carbon, nitrogen, phosphorus, or silicon atom may not be of much importance for a single isolated molecule. However, in the event such a molecule is present at the site of catalyst- or enzyme-controlled chemical reactions, such ambiguity could lead to serious disturbances in chemical processes controlled by these factors. This sort of relationships has long ago forced chemists and technologists to develop methods that allowed obtaining products of precisely determined stereochemical properties. Contemporary technologies of obtaining compounds that find application in biochemistry, molecular biology, medicine, pharmacology, or nanotechnology cannot be imagined without methods controlling stereochemistry of products formed.

Nature has been providing many examples regarding the importance of stereochemistry in case of not only single molecules but also in case of their mutual interactions. Such interactions are the subject of supramolecular stereochemistry. It is difficult to classify the latter into any one branch of chemistry. Its subject concerns investigating interactions between polymers, crystals, dendrimers, as well as of those found in nanochemistry, bioorganic chemistry, etc. It also deals with conformational analysis and molecular recognition chemistry. Supramolecular stereochemistry is among the most rapidly developing areas of stereochemistry [2,3].

Stereochemistry of chemical compounds is associated with their chirality. Formerly, the latter was linked with asymmetry of compounds, which in turn was associated with a carbon atom having four different substituents (for nitrogen or phosphorus (III), three substituents, respectively; see Figure 2.1). There is no element of symmetry present (center, axis, or plane of symmetry) in such a molecule.

Compounds of this type show optical activity, that was previously associated with their asymmetry. It turned out, however, that the asymmetry of a compound is not a prerequisite for optical activity to be present. A molecule of tartaric acid

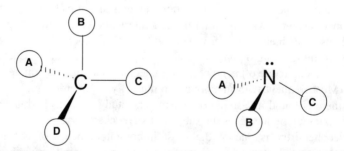

FIGURE 2.1 Carbon atom linked to four different substituents and nitrogen atom linked to three different substituents, as examples of molecules with asymmetric structure.

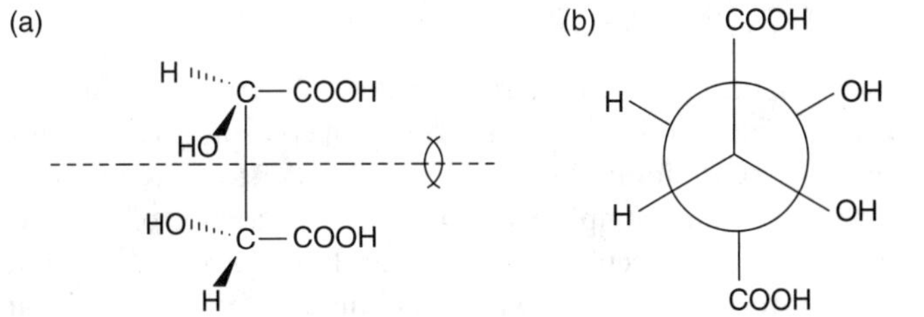

FIGURE 2.2 Chemical formula of D-(+) tartaric acid showing (a) double axis of symmetry and (b) one of its possible conformations shown here as Newman projection.

provides a classic example here; it is optically active despite the fact that it has a double axis of symmetry (see Figure 2.2).

As we can see, optically active molecules can have a symmetry axis, but they cannot have a symmetry center or a symmetry plane. Such molecules were labeled chiral (from a Greek word *cheir*, meaning hand). The name applies to those molecules that cannot be superimposed on their mirror image (much the same as our left hand cannot be superimposed on our right hand) and that show optical activity, that is, they have the property of rotating the plane of polarized light.

Carbon atom linked to four different substituents is called *chiral*. Occurrence of this type of carbon atoms in a molecule is not a necessary condition for a given chemical compound to show optical activity. Compounds differing in the configuration of substituents at the chiral carbon atom are called *enantiomers*. In case there are a greater number of chiral carbon atoms in a molecule, the compounds that are mutual mirror images are termed enantiomers. Enantiomeric compounds, besides differing in rotation of the plane of polarized light, do not differ in physical or chemical properties. All other isomers that possess chiral atoms but are not enantiomers, are called *diastereoisomers*. Configuration of substituents in a chiral molecule does not have direct impact on the direction of rotation of polarized light. It often happens that chemically very similar molecules show opposite signs of optical rotation activity. There is no simple method for determining optical rotation based on the chemical formula. For pure isomers of optically active compounds, this quantity can be determined experimentally.

Process of crystallization of achiral organic compounds can also generate chirality. Compounds that crystallize in chiral form belong to various classes of organic compounds, including substituted benzene derivatives, polyaryl and polycyclic derivatives, heterocyclic compounds, etc. These compounds were reviewed in short in literature [4]. It is evident that chirality of this type occurs only in crystalline form and disappears upon dissolution.

Web page of the International Union of Pure and Applied Chemistry (IUPAC) (www.chem.qmul.ac.uk/iupac/stereo): IUPAC Recommendations 1996 provide basic terminology used in stereochemistry. As stereochemistry and chirality of organic compounds have been the subject of numerous books, readers interested in these subjects should refer to the references provided in [5].

2.2   CHIRALITY IN NATURE

Chirality is one of the greatest mysteries of nature. We encounter this phenomenon in all molecules found in living forms. This property must have been important even for the earliest life forms, but it is not well understood, why, even at present [6]. Can one clearly and completely explain the occurrence of only one kind of isomers of biological molecules? Attempts to do so have been continuing and, almost certainly, this phenomenon will be the subject of various conjectures in the future [7]. The importance of chirality may be appreciated when we realize that substitution of one isomer with its mirror image could have grave consequences for the functioning of whole organisms. Ladik and Szekeres [8] demonstrated, with the help of molecular dynamic simulation, that the chirality change in a few chiral centers of an enzyme totally alters its secondary structure, that in consequence leads to the disappearance of catalytic activity. Let us have a look (Figure 2.3) how even minor changes in the molecule's structure can pose a serious danger to the basic vital activities of living creatures.

One of the simplest, optically active amino acids forming proteins that occur in living beings, alanine, can exist as two distinct isomers: L and D. In the animate world, only the L-isomer of this amino acid is found. Just think! A switch from this isomer to its mirror image (D-isomer) might have fatal consequences for a living being. Likewise, for the rest of the amino acids (except for glycine, which is not optically active) a change of configuration at one (α) carbon atom can alter the properly functioning metabolism of the organism, leading to disease or even

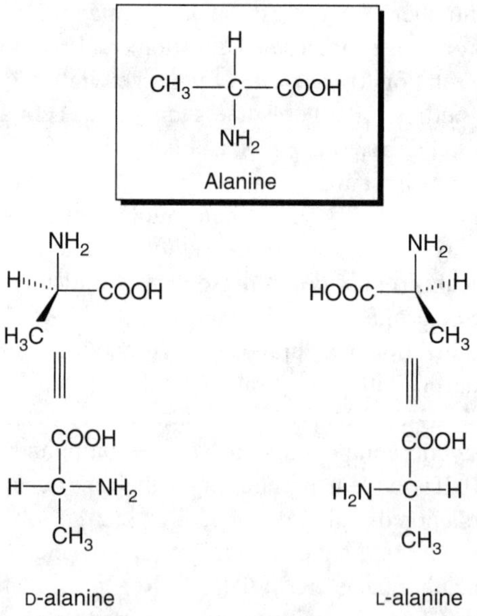

FIGURE 2.3   Structural formula of alanine and various ways of depicting its two enantiomers.

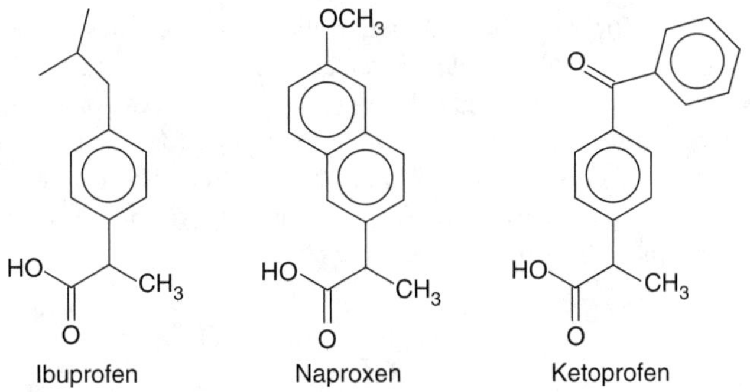

FIGURE 2.4 Structural racemic formulas of ibuprofen, naproxen, and ketoprofen.

death. Stereoselective and regioselective reactions are basic for proper functioning of all living beings.

Pharmacological interference with metabolism of living creatures must take into account stereochemistry of administered medicines (of course, if the therapeutically active substance occurs as enantiomers). Many medications in use are actually mixtures of enantiomers and manufacturers have not always been prone to provide data concerning biological activity of both isomers in case when only one of them exerts positive pharmacological action [9]. In the event the other "inactive" isomer alleviates or weakens the effect exerted by the "active" isomer, it can be the cause of undesirable side effects.

As an example, let us have a look at some derivatives of propionic acid, such as ketoprofen, ibuprofen, and naproxen, that are commonly used as nonsteroidal anti-inflammatory drugs (NSAIDs) and, hence, frequently subjected to chromatographic studies. These compounds, the formulas of which are shown in Figure 2.4, have one asymmetrical carbon atom in the molecule. It has been known that only *S*-isomers of all three compounds have therapeutic utility, whereas from among *R*-enantiomers only ketoprofen of such configuration finds specific application [10]. It is not, however, the same as application as that of the *S*-enantiomer.

It should be recognized that racemic compounds generated every day by chemical industry for use in innumerable applications, besides affecting living organisms directly, are also a source of additional, often harmful, environmental ballast in the form of biologically inactive isomers. This is a serious problem that is being solved, thanks to novel biological and chemical synthetic methods [11].

2.3 METHODS FOR DETERMINING COMPOUNDS' STEREOCHEMISTRY

2.3.1 POLARIMETRIC METHODS

The measurement of optical rotation of organic compounds is among the most frequently used methods for determining their optical activity. It is also one of the oldest methods for investigating physicochemical properties of organic compounds. An instrument called *polarimeter* accomplishes the measurement. The angle of rotation of polarized light depends on the thickness of the layer of an optically active substance. When the measurement is performed for a solution, the angle value depends on the concentration of the solute. The result of measurement is reported as specific rotation. The latter quantity for liquid substances was calculated using the following formula:

$$[\alpha] = \frac{\alpha}{l \times d},$$

where α is observed rotation (in degrees) of the plane of polarized light of a given wavelength by the examined substance of density, d, and thickness layer, l.

In case of solutions, the above formula becomes

$$[\alpha] = \frac{\alpha \times X}{l \times m \times d},$$

where X is the mass of the solution, m is the mass of an optically active compound, α, l, and d are as described in the previous formula. Specific rotation of solutions depends on the solvent used and on the solute concentration.

Quite often the so-called molar rotation is used, which can be calculated from the following formula:

$$[\alpha_{mol}] = \frac{M \times [\alpha]}{100},$$

where M is the molar mass of the investigated optically active compound.

Spectropolarimetry is the second currently used method for investigating optically active compounds [12]. In this method, one measures the rotation dependence on the wavelength of polarized light, changing over a wide range. The results of these measurements are presented as optical rotatory dispersion (ORD) curves. For the wavelength range, where optically active compounds do not absorb, these curves are monotonic, whereas near absorption maxima they abruptly change sign and have abnormal shape due to the so-called Cotton effect. Figure 2.5 shows an example of ORD curves for two enantiomers. The shape of the curves is affected by various factors associated with the molecule of the examined compound, such as its conformation and configuration, as well as the number and kind of chromophores present. The shape of the curve often depends on the solvent used and temperature during measurement.

Yet another method used for studying optically active compounds is circular dichroism (CD). CD curves that are obtained with the help of a dichrograph, represent wavelength dependence on differential absorption of the circularly (levorotatory and dextrorotatory) polarized light.

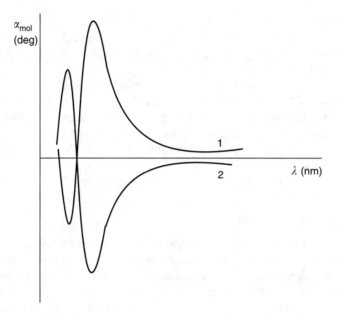

FIGURE 2.5 Schematic representation of ORD curves of two enantiomers: (+) enantiomer (curve 1) and (−) enantiomer (curve 2).

This method uses two kinds of units.

Differential absorption (difference of absorption coefficients):

$$\Delta\varepsilon = \varepsilon_d - \varepsilon_l$$

and molecular ellipticity, $[\Theta]$, which is determined indirectly, based on the value of CD, $\Delta\varepsilon$, and which are related as follows:

$$[\Theta] = 3300\Delta\varepsilon.$$

Applications of this method are similar to those of the ORD method since they both are based on the same effect, namely interaction between matter and light. CD allows determining the chirality of a chemical compound as well as enables studying the influence of various effects (solvent effect, temperature effects, and optical activity induced effect by a chiral medium) on its optical activity.

CD is very sensitive to the detection of small Cotton effects and is, therefore, more practical to use. Both methods can be used in the absence of Cotton effect or when it occurs at very short wavelengths.

Determining absolute configuration of chiral compounds devoid of any chromophore groups can be accomplished with the help of a method based on vibrational CD (VCD) spectroscopy as well as its subsequent modifications [13].

These methods supply more information that is used to create theoretical models of the studied compounds. Comparing spectroscopic data gathered using this method with theoretical calculations allows determining precisely the absolute configuration of the investigated compound.

Chiral optical methods are not devoid of certain shortcomings that result from the purity of the studied compounds. Each contamination of sample with an optically active compound, especially with the one having high coefficient of light refraction (high rotation), or having opposite rotation can lead to large errors in determining enantiomeric purity of the studied sample. Therefore, besides the methods described here, spectroscopic and chromatographic approaches are also used to determine the purity of the compound in question.

2.3.2 SPECTROSCOPIC METHODS

Besides chiral optical methods used to differentiate between optical isomers, spectroscopic methods are also an important tool in studying chiral compounds. Especially, magnetic resonance spectroscopy allows differentiation between various isomers of the same optically active chemical compound. NMR spectra of enantiomers as well as spectra of racemic mixtures are identical if measurements are performed in standard NMR solvents. However, shifts of certain proton groups occurring in such compounds can be discerned if these compounds are transformed into diastereoisomers [14]. Classic methods used to accomplish this goal involve:

- Derivatization with enantiomerically pure compounds
- Use of chiral shift reagents
- Use of chiral solvating agents

These methods are based on the association of chiral compounds with chiral reagents; as a result, new diastereoisomeric compounds (in case of derivatization) or molecular complexes (in case of using chiral solvents or shift reagents) are formed.

Derivatization of enantiomers using enantiomerically pure compounds is the most frequently used method for determining optical purity of studied compounds. Although simple in principle, it requires using reagents of great optical purity. Similar to separating a racemic mixture of amino acids into individual enantiomers, where an optically pure basic or acidic reagent is used depending on the method applied; in spectroscopic investigations too, this type of reagent is used. Initially, they were most frequent derivatives of mandelic acid or of phenylethylamine (Figure 2.6) [15].

Compounds that contain other atoms (besides carbon and hydrogen), the isotopes of which can be observed using magnetic resonance (P, N, Si, and Se), have been tried as derivatization agents. This method has certain limitations resulting from the conditions of derivatization reaction as well as procedures used in purification of reaction products. The reaction may not lead to racemization of the starting

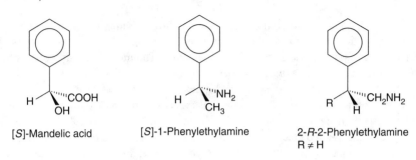

FIGURE 2.6 Mandelic acid and two phenylethylamine derivatives used as chiral reagents.

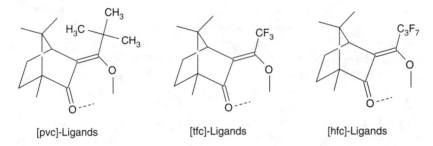

FIGURE 2.7 Examples of camphorates used as ligands in paramagnetic shift reagents.

compound. Purification may not lead to increased yield of one of the diastereoisomers obtained. Most often, chromatographic methods are employed for this purpose since they permit a fast and easy selection of conditions for purifying the reaction mixture.

Chiral shift reagents belong to the group of compounds capable of forming additional complexes with numerous organic compounds [16]. The central ion of chiral shift reagents is that of a lanthanide (Eu, Pr, and Yb), which forms a six-coordinated complex. A compound of this type added to a sample of the studied compound forms with it an unstable associate. As a result, the chemical surroundings of protons and carbons in the analyte become altered. This, in turn, causes chemical shifts of these atoms, depending on the distance from the central metal ion. This effect can be used as an advantage in determining stereochemical structure of the newly attached chiral ligand.

Most frequently used (albeit not unique) lanthanide ligands serving as shift reagents are pivaloyl-D-camphorate (pvc), trifluorohydroxymethylene-D-camphorate (tfc) and heptafluorohydroxymethane-D-camphorate (hfc), dipivaloyl methane (dpm), 2,2,6,6-tetramethyl-3,5-heptanedione (thd), and 1,1,1,2,2,3,3-heptafluoro-7,7-dimethyl-4,6-octanedione (fod) (Figure 2.7).

Chiral solvating agents, enantiomerically pure compounds that are added to the solution of enantiomeric mixtures directly into the NMR test tubes, form diastereomeric solvation complexes with the investigated compounds. Parker [15] presented the theory underlying the observed shifts. Compounds used as chiral solvating agents belong to various groups of organic compounds (amines, alcohols, derivatives of phosphoric acid, and organic acids). The most frequently observed interactions occurring between studied species and chiral solvating agents are classic hydrogen bonds and weak hydrogen bonds of O—H$\cdots\pi$ or C—H$\cdots\pi$ type. By observing shifts of various protons present in the analytes and comparing these shifts with theoretical models of such compounds, it is possible to determine the type of the studied enantiomer or enantiomeric purity of the studied sample.

The aforementioned methods are not free from certain inconveniences related to possible racemization of the investigated compounds during derivatization as well as frequent difficulties in choosing suitable chiral solvating agents for various optically active compounds. In the recent decades, several other NMR-based methods of enantiomer analysis have been described. As an example, one of them is based on NMR spectra of enantiomers dissolved in a binary mixture of nematic and cholesteric thermotropic liquid crystals [17]. In such mixtures, R- and S-isomers possess differing ordering properties, which results in different shifts of the proton signals from these isomers. A later modification of this method involved enantiomer analysis in polypeptide lyotropic liquid crystals [18]. In the cited example, a solution of poly-γ-benzyl-L-glutamate in dichloromethane was used.

Proton NMR spectra can be used for determining enantiomeric purity of the investigated compounds, although more useful are (deuterium) ^{2}H NMR spectra. They require partially deuterated compounds for study. This method was successfully used for determining enantiomeric purity of several compounds belonging to alcohols, amines, carboxylic acids, esters, ethers, epoxides, tosylates, as well as halogen derivatives and hydrocarbons, which could not have been studied before using NMR.

Fukui et al. described a method for determining absolute configuration of alcohols [19] and diols [20] based on nuclear Overhauser effect (NOE), that is, assessment of distance between spatially interacting protons in the molecules. The Overhauser effect has been used for determining stereochemistry of various groups of compounds, for example, ethylenic derivatives, and cyclic and bicyclic molecules [14]. For derivatization, the following axially chiral reagents were used: 2-methoxy-1,1'-binaphthalene-8-carboxylic acid (MBCA) for alcohols and 2-methoxy-1,1'-binaphthalene-8-carbaldehyde (MBC) for diols (Figure 2.8).

Alcohols were derivatized to MBCA esters and diols to suitable MBC acetals. On the basis of the possible correlations present in NOE spectra, absolute configuration of the examined enantiomers could be determined. It appears that NMR techniques using Overhauser effect, together with novel chiral reagents, should soon become leading methods for determining absolute configurations of optically active compounds.

Stereochemical studies of molecules have also been observed using mass spectrometry. Most frequently, mass spectra of various isomers have been compared [21]. The problem with this technique is that both enantiomeric species possess the same mass as well as similar physical characteristics important from the perspective of ionization process taking place in a mass spectrometer. One can use

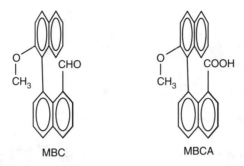

MBC   MBCA

FIGURE 2.8 Formulas of MBC and MBCA used as "chiral matrices" for determining absolute configuration of diols and alcohols.

to the advantage, however, of different interactions of enantiomers with chiral compounds (selectors). Formation of diastereomeric ion-molecule aggregates in the gaseous phase has been the subject of intense studies aimed at finding fast and sensitive methods for enantiomer studies, methods that would provide a wealth of structural information [22–24]. Electron impact (EI) mass spectrometry is a method rarely used in studying enantiomers, albeit there are reports of use of this technique for gas phase enantiomeric distinction [25].

Using techniques employing milder ionization techniques, such as fast atom bombardment (FAB MS), electrospray ionization (ESI MS), and matrix-assisted laser desorption ionization (MALDI), supramolecular constructs have been employed. Chiral compounds used as selectors form "host–guest" or inclusion-type conjugates with the analyzed molecules. The compounds of this type should possess a suitable cavity that can recognize appropriate stereoisomers (molecular recognition). "Recognition" consists of forming multiple noncovalent bonds between host and guest. The number of these bonds for a given host is different for different enantiomers; therefore, stability of this type of conjugates is variable; and this effect has been considered as an advantage in mass spectrometry studies.

One group of the compounds most frequently investigated by various mass spectrometry techniques is that including amino acids. By mixing two different chiral amino acids, one obtains a mixture of homochiral and heterochiral dimers and trimers of varying stability. This stability is directly associated with ion abundance in FAB MS. Absolute chirality of amino acids can be determined by mixing the species in question with another amino acid of known absolute configuration [26]. This method can be applied to investigate "host–guest" type interactions or to molecular recognition, even at very small differences in ion stability (0.1–0.2 kJ/mol).

Sawada et al. [27, 28] have proposed a high-accuracy method ($R^2 > .999$) of determining enantiomeric excess for amino acid derivatives. The methodology based on FAB MS and "host–guest" type interactions requires, however, determining deuterated and nondeuterated pairs of compound acting as "host" and a deuterated compound acting as "guest." As "host" molecules, chiral crown

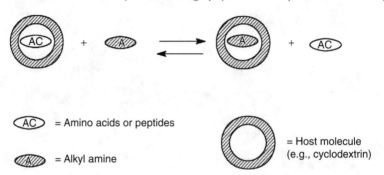

AC = Amino acids or peptides

A = Alkyl amine

= Host molecule
(e.g., cyclodextrin)

FIGURE 2.9 Schematic representation of reactions taking place during exchange of amino acid, forming a "host–guest" type compound with cyclodextrin, for gaseous phase amine.

ethers or chiral-fragment-containing podands have been used. The advantage of FAB MS method is that it enables determining enantiomeric excess without calibration curve.

ESI MS is a technique more often used in the studies of chiral amino acids. All chiral amino acids commonly occurring in proteins were determined using *N*-*tert*-butoxycarbonyl derivatives of proline, phenylalanine, and *O*-benzylserine [29] as chiral selectors. Formation of dimers, trimers, and tetramers was studied using collision-induced dissociations (CIDs) method. The observed dissociation efficiency was strongly dependent on the chirality of the amino acids and the type of chiral selectors.

Cyclodextrins and their derivatives have been frequently employed as chiral selectors in determining amino acids' chirality. Protonated complexes forming directly in the gaseous phase are subjected to reactions with gaseous amines. The exchange reactions taking place during the investigation are enantioselective. This allows determining enantiomeric excess in a mixture of enantiomers [30,31] (Figure 2.9).

Enantiomeric composition of ibuprofen was determined using a kinetic method based on the formation of trimeric complexes of this compound with chiral reference compounds (D-glucose, D-mannose, D-tartaric acid, D-galactose, and D-ribose) in electrosprayed solutions in the presence of divalent metal cations [32]. The described method can be applied to investigate interconversion of ibuprofen enantiomers under various conditions of the drug administration. This is of importance in view of ibuprofen standing in the pharmaceutical industry.

Metal cations with chiral reference were used as binding metal ions for the evaluation of stereoselective discrimination and chiral recognition for few optical isomers. A high degree of chiral recognition ability was observed for di-*O*-benzoyl-tartaric acid dibutyl ester isomers when two cations, zinc (II) and copper (II), were used as binding metal ions, and L-tryptophan was used as chiral reference [33]. In a similar manner, calcium (II) ions were used for stereoselective discrimination and quantification of arginine and *N*-blocked arginine derivatives. As a reference, pure arginine enantiomer was used in studying *N*-blocked arginine derivatives and, conversely, a pure enantiomer of N-blocked arginine derivative was utilized as a reference for investigating purity of enantiomeric arginine [34]. In addition, copper (II) complexes [35] as well as cobalt (III) complexes [36] were used for chiral recognition of amino acids and dipeptides [37]. In each of these methods of chiral recognition of the studied amino acids, molecular modeling and kinetic methods were used; the type of binding ion played an important role in the process. Different enantiomers form complexes in various ways and this results in different dissociation pathways and unique chiral recognition characteristics.

The separation of amino acids into particular enantiomers was the subject of mass spectrometric studies but also ESI MS was applied to determine by kinetic resolution the enantiomeric excess of optically active alcohols and amines in nanoscale by diastereoselective derivatization with optically active acids [38]. This method has several distinctive features: among others, easily available chiral acids can be used (the authors used N-benzoyl proline derivatives), no chromatographic separations are required, it is insensitive to certain impurities, it is fast and requires only small amount of substrate (10 nmol or less). The method can take an advantage when accuracy of enantiomeric excess measurement is sufficient within ±10% limits.

With the help of ESI MS, one can also determine anomeric configuration of glycosides [39]. Such studies were performed on sugar phosphate derivatives of glucopyranose, fucopyranose, galactopyranose, mannopyranose, ribofuranose, and xylopyranose. Collision studies (tandem mass spectrometry) allowed detailed identification of compounds including information on their anomeric configuration.

X-ray diffraction methods have been successfully employed in determining stereochemical relationships in organic and metalloorganic compounds, where chiral centers may also occur. This method finds application in studies of crystals forming solids. Fortunately, a number of compounds form crystals while for many particularly "refractive" ones (e.g., nucleic acids and proteins) and methods were elaborated that allow induced crystallization. By means of crystallography, it was estimated that all naturally occurring amino acids have the same configuration at α atom. Crystallographic methods have been used to determine the absolute configuration of compounds, but these methods are often hampered by the fact that if racemic mixtures were used for investigation then crystals formed would contain mixed enantiomers. Examples of determining absolute or relative configurations of numerous compounds belonging to various classes of organic species can be found in literature [40].

2.4   ON THE METHODS OF OBTAINING STEREOISOMERS

Nucleophilic substitution, electrophilic addition to the double bond, nucleophilic addition to the double bond in which atoms differ in their electronegativity, intramolecular conversions, and reduction are the most important reactions that may result in the appearance of asymmetric carbon atoms [41]. Unfortunately, usually racemic mixtures form. If pure optically active stereoisomers are desired, either the obtained racemic mixtures have to be separated or stereospecific reactions need to be carried out leading to pure stereoisomers.

Several methods can be employed for separating racemic mixtures:

1. Manual separation of crystals, in case the separated compound crystallizes in the form of species having different optical rotation values. This method, used for the first time by Pasteur in 1848, has been seldom used although spontaneous generation of chirality in crystals occurs quite often [4].
2. Chemical separation of enantiomers via diastereoisomers, which can be formed in reactions with optically pure reagents or by using nonspecific bonds in complexes or inclusive conjugates [42].
3. Enzymatic, biochemical, or microbiological methods based on selective "transformation" of one of the isomers into the desired product, or, removal of one of the enantiomers from the mixture. Such methods are currently used in biotechnological processes leading to the procurement of various compounds, including optically active ones [43].
4. Methods taking advantage of differing ability of enantiomers to adsorb on the surface of a carrier, that is, chromatographic methods [44].

All of the aforementioned methods of obtaining pure enantiomers are based on racemic mixtures.

Products obtained by these separation procedures can be characterized by optical purity or enantiomeric excess.

Optical purity is a measure of effectiveness of an accomplished process of separating a racemic mixture into particular enantiomers. It is expressed as follows:

$$\text{Optical purity} = \frac{[\alpha]}{[\alpha]_{abs}} \times 100\%,$$

where $[\alpha]$ denotes specific rotation of sample, $[\alpha]_{abs}$ is specific rotation of a pure enantiomer. It is expressed as a percentage ratio and corresponds to an excess of one of enantiomers in the mixture.

Enantiomeric excess (ee) is defined by the following formula:

$$ee = 100 \times \left[\frac{(x_R - x_S)}{(x_R + x_S)} \right]$$

in case where $x_R > x_S$, x being the molar fraction and R and S denoting the respective R- and S-enantiomers [45].

Chemical industry, especially pharmaceutical companies, cannot afford producing large quantities of compounds that do not meet requirements (most often inactive isomer is a by-product). High costs of separating chiral mixtures have been for many years driving the research of novel chemical methods yielding compounds that are optically pure or have large enantiomeric excess of the desired species. Asymmetric synthesis, also called stereoselective synthesis, is a branch of chemistry dealing with making optically active products, and is heavily based on many recently introduced organic synthetic methods. Asymmetric synthesis can make use of optically active substances obtained from natural sources; they can be transformed further, yielding novel optically active species. Asymmetric synthesis may use, however, substances from sources other than nature, and starting compounds need not be optically active. In such cases, where one of the enantiomers is obtained from an optically inactive compound, the reaction is called *enantioselective* and the whole process is termed *asymmetric induction*. In asymmetric synthesis, one can distinguish four different synthesis methods based on [41]:

- Active substrates
- Active reagents
- Active catalysts or solvents
- Reactions carried out in the presence of circularly polarized light

The method of asymmetric synthesis based on active catalysts is of great importance at present. The Japanese researcher Akabori [46], who obtained amino acids by hydrogenating suitable substrates in the presence of a palladium catalyst, first carried out such a synthesis. In turn, Izumi noticed that far better results could be obtained when the catalyst has suitable optically active ligands attached to it [47]. Contemporary methods of asymmetric synthesis, especially those used worldwide by pharmaceutical industry, employ catalysts with chiral ligands that allow far more efficient and cost-effective synthesis of chiral chemicals. Such catalysts control the course of asymmetric hydrogenation, oxidation, or pericyclic reactions [48].

2.5 STEREOCHEMISTRY OF ORGANIC COMPOUNDS

This section deals with stereochemistry principles applied to basic classes of organic compounds. Particular emphasis is placed on conditions under which members of such classes exhibit (or may exhibit) optical activity.

2.5.1 ALKANES AND CYCLOALKANES

Alkanes are compounds in which stereochemical differentiation results from their various possible conformations. A number of these compounds, especially with different substituents present in the chain, may exist as enantiomers or diastereoisomers, that is, compounds that are optically active. Figure 2.10 shows Newman projections of six different conformations for a molecular fragment consisting of two adjacent carbon atoms with substituents.

Different conformations of a molecule vary in energy. In the case of alkanes, synperiplanar and synclinal conformations have the highest energies. Under given

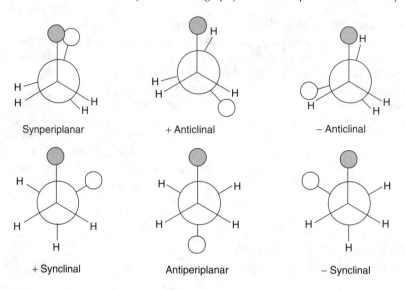

FIGURE 2.10 Different conformations of alkane chain fragment and their nomenclature according to IUPAC.

conditions, molecules tend to adopt the state with lowest energy. Representation with Newman projections or perspective formulas is possible only for small molecules. When the length of molecular backbone grows, probability of the same conformation in individual segments becomes smaller. Even more conformational differences occur in branched alkenes. In these compounds, optical activity can occur when four different substituents are linked to a single carbon atom. Determining optical activity of "pure" alkanes using polarimetric methods can be difficult since none of the substituents caused appearance of absorption bands in the UV region. However, if a substituent at the asymmetric carbon atom is a chromophore, optical activity for this compound can be determined. In the event, atoms other than hydrogen (or carbon) are attached to carbon atoms of the chain, thus determining stability of particular conformers becomes complicated. In such cases, besides steric interactions, also other interactions of this atom with the rest of the molecule must be taken into consideration. For example, in case of halogen derivatives one needs to bear in mind that not only a large steric hindrance is caused by halogen atom but also its participating in weak hydrogen bonds (weak donor–weak acceptor type bonds) [49]. Occurrence of such interactions was confirmed, among others, by conformational studies of 1-chloropropane molecule. From among the three most favorable conformations depicted in Figure 2.11, the most stable are those of synclinal isomers (they do not differ in energy), in which interactions of C—H··· Cl—C type occur [50,51].

If a compound contains a chiral center (or centers), then in order to assign the configuration of the asymmetric carbon atom the Cahn–Ingold–Prelog system (abbreviated as CIP) is used [52].

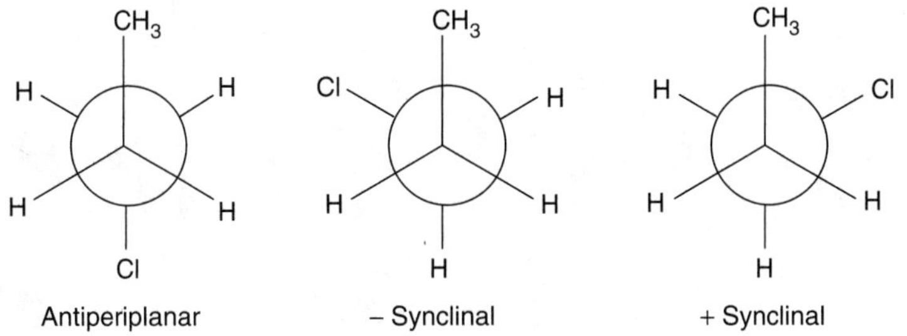

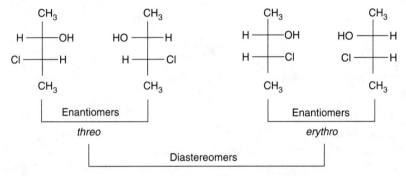

FIGURE 2.11   The most favorable conformations of 1-chloropropane.

FIGURE 2.12   3-Chloro-butane-2-ol as an example of a molecule with two different chiral centers.

Compounds containing several chiral centers can exist in different stereoisomeric forms. Various stereoisomers usually possess different physicochemical properties, except for isomers that are enantiomers (the remaining isomers with different properties are diastereoisomers). If a molecule contains two identical chiral centers (as in, e.g., 2,3-butanediol, 2,3-dichlorobutane, or tartaric acid), it can occur in the *meso* form that is not optically active, or in the form of two optically active enantiomers. If a molecule contains two different chiral centers, then it exists as two enantiomeric pairs that are diastereoisomeric with respect to each other (Figure 2.12).

Individual pairs of enantiomers are called *threo* or *erythro*. The *threo* form has substituents on opposite sides of the main backbone chain represented as Fischer projection, whereas the *erythro* form has them on the same side, respectively. These terms function correctly in case of linear sugar molecules, but in other cases there are often problems with determining what is actually the "main chain" [53]. In such cases, individual chiral carbon atoms can always be characterized using configurational notation (*R* and *S*).

Stereochemistry of cyclic compounds should be considered from the point of view of both configuration and conformation. The smallest cyclic compound having a carbon backbone is cyclopropane. Because it is a flat molecule, its

(a)   (b)   (c)

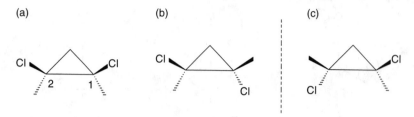

FIGURE 2.13   Stereoisomers of 1,2-dichlorocyclopropane.

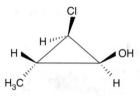

FIGURE 2.14   Trisubstituted   cyclopropane: *c*-2-chloro-*t*-3-methyl-*r*-hydroxycyclo-propane.

conformational aspects are not considered. Monosubstituted cyclopropane is an achiral molecule, whereas disubstituted one (1,2-dichlorocyclopropane, as shown in Figure 2.13) can exist in the *meso* form (a) or as two enantiomers (b and c) (Figure 2.13).

The derivative (a) commonly called *cis* has carbon atoms in 1*R*,2*S* configuration, while the *trans* enantiomer (b) has configuration 1*S*,2*S*, and isomer (c) — 1*R*,2*R*, respectively, according to the CIP system. Compound (a) is achiral, while isomers (b) and (c) are chiral. The attachment of the next, third substituent requires indicating a "leading" or "fiducial" substituent in the molecule, which is denoted with the -*r* prefix. The remaining substituents are described with -*c* (*cis*) or -*t* (*trans*) prefixes. Figure 2.14 shows, as an example, *c*-2-chloro-*t*-3-methyl-*r*-hydroxycyclopropane.

When applying the CIP system, the discussed compound would be described as 1(*S*)-hydroxy-2(*R*)-chloro-3(*S*)-methylcyclopropane. However, reconstructing a proper structural formula based on this notation is far more complicated; therefore, for cyclic compounds, it is the former notation that is used.

Molecules of cyclobutane, cyclopentane, and cyclohexane are nonplanar. Besides determining configuration at each carbon atom in these compounds, conformational aspects of the whole molecule should also be considered. Figure 2.15 shows the most stable conformations of rings in these molecules. Of course, such molecules can also exist in any "intermediary" conformations between those shown. In monosubstituted rings of cyclobutane, cyclopentane, and cyclohexane there is no asymmetric carbon atom. In disubstituted isomers of these cyclic compounds, optical activity can appear even if the substituents are identical. The *cis* forms will not be optically active in any case since they exist in *meso* form. On the other hand, the *trans* forms will be optically active (they occur as two

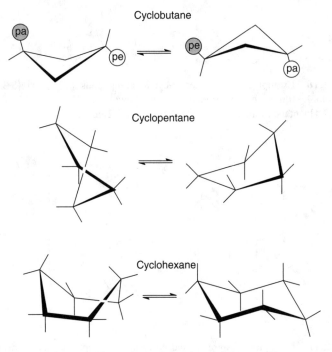

FIGURE 2.15 The most stable conformations of cyclobutane, cyclopentane, and cyclo-hexane rings (pa and pe denote pseudoaxial and pseudoequatorial positions, respectively).

separable enantiomers), as in the case of 1,2-disubstituted cyclobutane and 1,2- and 1,3-disubstituted cyclopentane or cyclohexane. For cyclohexane, 1,4-isomer is also possible, but it has a symmetry plane. In the event a molecule has two different substituents, its *cis* isomers will also occur as two enantiomers (except for 1,4-cyclohexane).

With an increase in the number of substituents in a molecule, the number of possible substitutions also rises, but it will depend on the kinds of substituents (i.e., whether they are identical or not) and their mutual location. A similar situation exists in saturated polycyclic systems. Let us consider some condensed bicyclic systems, such as those shown in Figure 2.16. In these compounds, introduction of any substituent in any location, except for carbon atoms common for both rings, will generate three chiral centers: at the substituted carbon atom and at carbon atoms common for both rings. The appearance of a second substituent will cause more chiral centers to appear, except for situations where *meso* structures form. Stereoisomeric considerations of fused, bridged, and caged ring systems are discussed in *Basic Organic Stereochemistry* [53].

Another interesting group of compounds where chirality often occurs is spir-anes (Figure 2.17). It is sufficient to substitute any two hydrogen atoms in both spirane rings with any substituents (they can be identical) to make this molecule optically active.

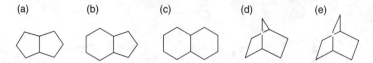

FIGURE 2.16 Examples of saturated bicyclic compounds (a) bicyclo[3.3.0]octane, (b) bicyclo[4.3.0]nonane (hydrindane), (c) bicyclo[4.4.0]decane (decalin), (d) bicyclo[2.2.1]heptane (norbornene), and (e) bicyclo[2.2.2]octane).

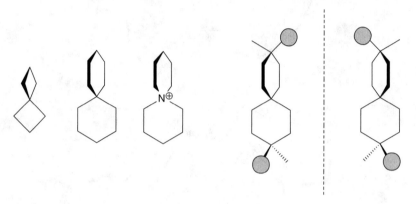

FIGURE 2.17 Examples of spiranes and an example of optically active disubstituted spirane.

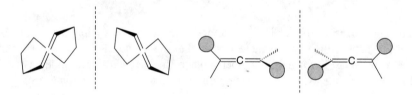

FIGURE 2.18 Examples of optically active compounds with double bonds: *trans*-cyclooctene and allene, a simplest cumulene.

2.5.2 ALKENES

Alkenes or polyenes with isolated or coupled double bonds that are devoid of chiral atoms are not optically active. Such activity occurs in cyclic alkenes where double bonds occur in the *trans* form, such as in *trans*-cyclooctene (Figure 2.18). Another group of alkenes that includes representatives having optical activity is that of cumulenes. The name refers to cumulation of double bonds in such molecules. The best-characterized group of these compounds is that of allenes in which two double bonds occur next to each other [54]. Compounds of this type have a so-called chirality axis determined by cumulated double bonds. Besides allenes, higher optically active cumulenes are also known. An example of optically active cyclic allene is provided by 1,2-cyclononadiene, which was synthesized in 1972 [55].

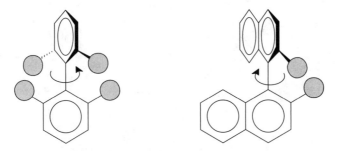

FIGURE 2.19 Schematic representation of biphenyl and binaphtyl compounds with bulky substituents inhibiting rotation around the single bond.

Besides compounds with double bonds, which were described in the preceding paragraph one should also mention compounds that contain such bonds and, in addition, possess chiral quaternary carbon atoms. In such cases, the molecular fragment containing a double bond "acts" as a substituent at the chiral carbon atom or another heteroatom.

2.5.3 AROMATIC COMPOUNDS

Optical activity can exist in several classes of aromatic compounds. The activity can result from:

- Inhibited rotation of two aromatic fragments linked by a single bond or bonds (biphenylenes, binaphtylenes, 9-arylfluorenes); we then speak of atropoisomers
- Helical structure of condensed benzene rings (aromatic)
- Planar chirality occurring in cyclophanes and *ansa*-type compounds
- Presence of chiral side chain

Atropoisomerism denotes stereoisomerism resulting from limited rotation around a single bond. For biphenyls and binaphtyls, this is the single bond between the two aromatic fragments (Figure 2.19).

Optical activity in this type of compounds exists when aromatic rings are not coplanar (i.e., they are not positioned in the same plane). Substitution at *ortho* position of biphenyl or binaphtyl derivatives (one *ortho* position is substituted with cumulated benzene ring) causes steric hindrance and makes free rotation of rings difficult relative to each other. Inhibition of rotation around the single bond joining both aromatic fragments leads to the appearance of two different conformers (Figure 2.20).

When several aromatic rings are linked in such a way that three adjacent rings are condensed in *ortho* position, the corresponding compounds are called *helicenes*. Since outer rings are not condensed with each other during the formation

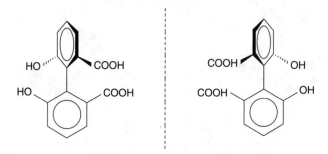

FIGURE 2.20   Two different conformers of a biphenyl derivative that are mirror images.

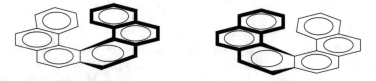

FIGURE 2.21   The two forms of [6]helicene.

of these compounds, the existing steric hindrance can cause them adopting a helical form that is devoid of symmetry. They are thus aromatic compounds that are not planar [56]. Right- and left-rotating forms of helicenes can form. As an example, [6]helicene, the first optically active helicene obtained [57], is shown in Figure 2.21.

Helicenes are rather easily racemized [58,59]. [5]Helicene racemizes at room temperature [60], whereas total racemization of [9]helicene at 380°C takes place within 10 min [61]. Substitution of helicenes often results in higher racemization energy, leading to their greater stability during thermal racemization [62]. Heterocyclic (tiophenone-benzene) [63,64] and pyrrole [65] analogues of helicenes were obtained as optically active isomers.

Planar chirality occurs in cyclophanes and *ansa* type compounds. [2.2]Paracyclophane is optically inactive. Its aromatic monosubstituted ring analogues can exist as optically active enantiomers despite the fact that there is no chiral carbon atom in these compounds [66] (Figure 2.22).

In cyclophanes having few carbon atoms in their bridges, free rotation of benzene rings is inhibited and results in the optical activity.

A similar situation is encountered in *ansa*-type compounds (e.g., benzene derivatives in which *para* or *meta* positions are linked via heteroatoms containing bridge). In these compounds, as in cyclophanes, when a bridge is sufficiently short, the benzene ring rotation can be hindered, making feasible separation of individual optically active isomers. Figure 2.23 shows schematic representation of isomers of an *ansa*-type compound.

The last group of optically active aromatic compounds to be mentioned includes those in which chiral groups are linked to the aromatic ring. Compounds of this type are frequently encountered in nature (e.g., alanine, mandelic acid,

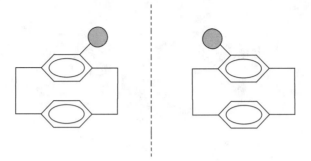

FIGURE 2.22 Schematic representation of two enantiomers of monosubstituted [2.2]paracyclophane.

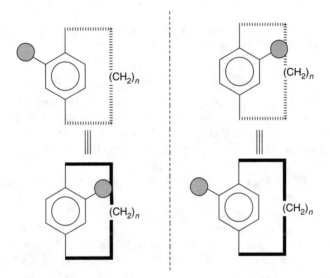

FIGURE 2.23 Schematic representation of enantiomers of *ansa*-type compounds.

chinine, and ephedrine). In addition, numerous synthetic drugs belong to this category (e.g., the previously mentioned ibuprofen, naproxen, and ketoprofen). The aromatic fragment occurring in these compounds, being a chromophore, enables spectropolarimetric investigations (CD, ORD).

2.5.4 HETEROCYCLIC COMPOUNDS

Heterocyclic compounds, as their carbon analogs, may exist in various conformations, and this has been the most frequently considered aspect of their structure [67]. Chirality of these compounds is most frequently due to the presence of various substituents in the heterocyclic ring. Many compounds of this type occur in nature (Figure 2.24).

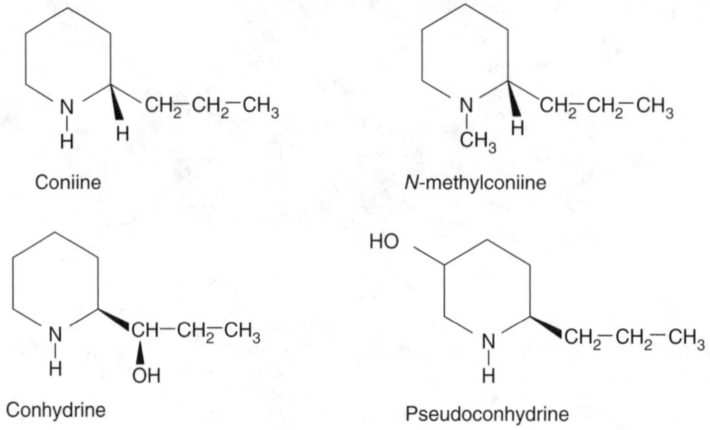

FIGURE 2.24 Examples of piperidine derivatives (alkaloids occurring in *Conium maculatum*) containing a chiral carbon atom in the heterocyclic ring.

In case of compounds with a heterocyclic ring, more substituents at the sp^3 carbon atom, change in the location of one of them leads to altered rotation of polarized light. A typical example of such situation includes spontaneous change of rotation of polarized light that takes place in aqueous solutions of optically pure monosaccharides; it is called *mutarotation* and individual interconverting isomers are called *anomers*.

Heterocycles of high enantiomeric purity are used to prepare other optically active compounds; for example, to obtain chiral amino acids some optically active derivatives of hydantoin or aziridine have been employed [68].

Recently, macrocyclic chiral compounds of crown ether or cyclamen type have been attracting wide interest. These compounds contain numerous heteroatoms in their molecules (mainly oxygen, sulfur, and nitrogen) and can find practical applications, for example, as chiral selectors [69,70] and chiral NMR discriminating agents [71]. Asymmetric substitution of two carbon atoms in the ring of crown ether or cyclamen can lead to many different optically active compounds useful in various branches of supramolecular chemistry. Such substitution can be accomplished with appropriate starting compounds that are optically active, for example, amino acids and polyhydroxy alcohols.

2.5.5 STEREOCHEMISTRY (CHIRALITY) OF ORGANIC COMPOUNDS WITH HETEROATOMS OTHER THAN OXYGEN (E.G., NITROGEN, PHOSPHORUS, SULFUR, AND SILICA)

Substances containing chiral heteroatoms will be discussed using examples from individual classes of compounds with chiral nitrogen, silica, phosphorus, and sulfur. Compounds containing these elements occur with higher frequency

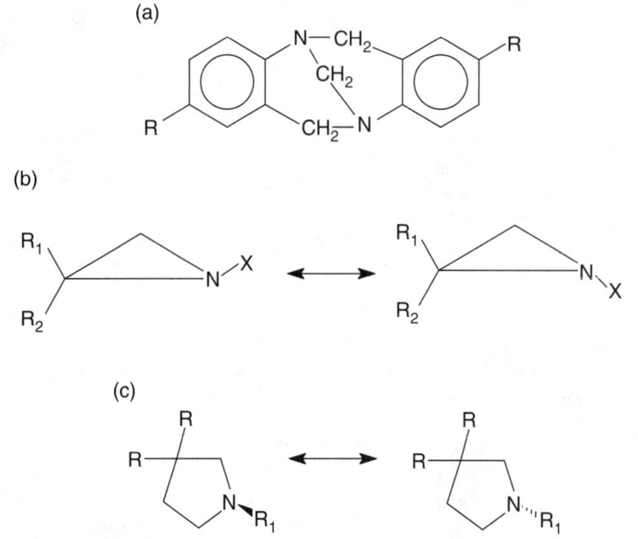

FIGURE 2.25 Optically active nitrogen compounds: a — Tröger's base, b — aziridine derivatives, and c — pyrrolidine derivatives.

compared with those that contain other chiral elements (e.g., antimony and selenium).

Nitrogen: Tertiary amines that do not possess elements of symmetry should occur in the form of optically active compounds. However, it turns out that their enantiomers interconvert rapidly and cannot be separated in pure form. Nonetheless, if tertiary amine nitrogen is rigidly locked in the ring, then its configuration is stable (it can exist as an asymmetric atom) and individual enantiomers can be separated. Tröger's bases, *N*-substituted aziridine derivatives, and *N*-substituted piperidine (Figure 2.25) are typical examples. A Tröger's base was successfully separated into individual enantiomers [72]. In case of *N*-aziridine derivatives, in which R_1 and R_2 substituents are identical (Figure 2.25), the only reason for occurrence of optical activity is the presence of asymmetric nitrogen atom. A similar situation is encountered with piperidine derivatives. Optically active are also *N,N*-disubstituted derivatives of aniline in which both nitrogen substituents are different. In such compounds inhibition of rotation around the single C_{arom}—N bond takes place (especially when substituents increasing steric hindrance in the molecule are in *ortho* positions with respect to nitrogen), similar as in substituted biphenyl derivatives.

Compounds containing nitrogen atom that are capable of forming four bonds and having different substituents may be optically active. In such compounds, there is a certain analogy to a chiral tetrahedral carbon atom. This type of compounds includes ammonium salts, amine oxides, and spiranes where nitrogen atom is a central one (Figure 2.26).

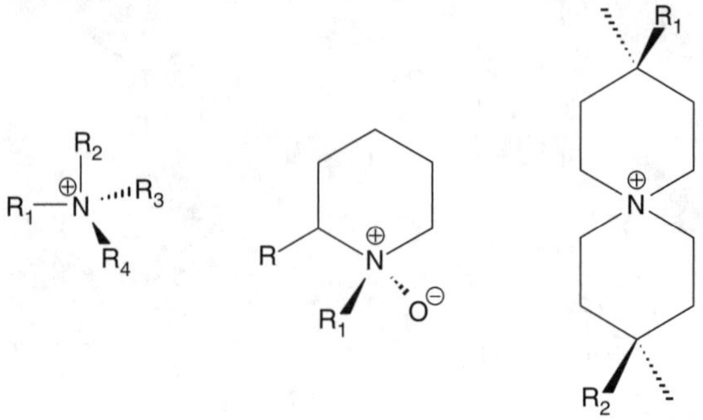

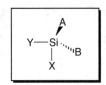

FIGURE 2.26 Examples of compounds containing tetravalent nitrogen, which may exhibit optical activity.

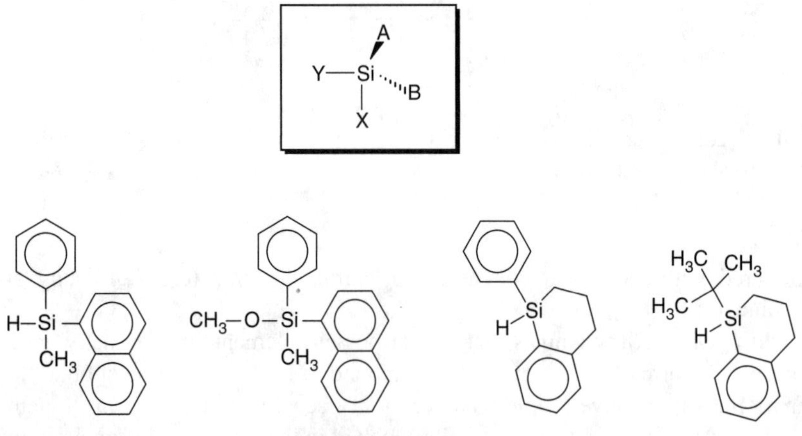

FIGURE 2.27 Structure of an optically active tetravalent silicon-containing compound having different substituents and examples of such optically active compounds.

Silicon: This element, similar to carbon, forms compounds having tetrahedral configuration. At the beginning of the twentieth century, it was found that silicon compounds of suitable structure might be optically active.

One of the first optically active silicon compounds obtained was phenylmethoxymethyl-α-naphtylsilane (Figure 2.27). It was prepared from phenyldimethoxymethylsilane, a precursor (starting compound) of many other optically active silicon compounds.

Silanes having hydrogen atoms are used for hydrosilylation of alkenes in the presence of palladium catalysts. An interesting example is provided by compound used for hydrosilylation of norbornene. In this reaction, a chiral silicone reagent induces formation, with large enantiomeric excess, of chirality on the carbon atom [73,74] (Figure 2.28). Similar compounds were also obtained in the case of menthol [75]. To determine absolute configuration of chiral oligosilicone compounds, CD was used, as for carbon compounds [76].

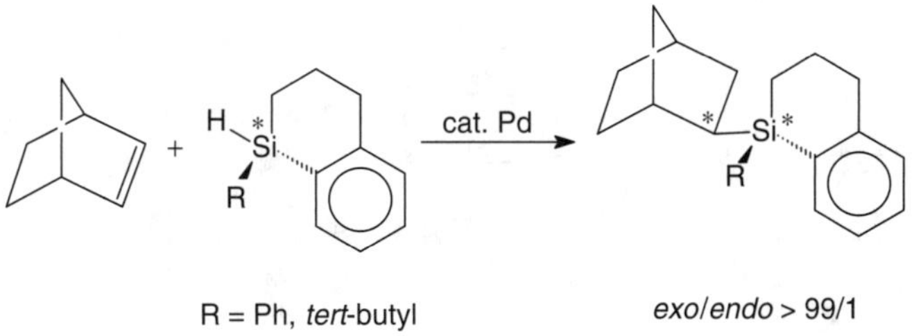

R = Ph, *tert*-butyl   *exo*/*endo* > 99/1

FIGURE 2.28   Hydrosilylation of norbornene.

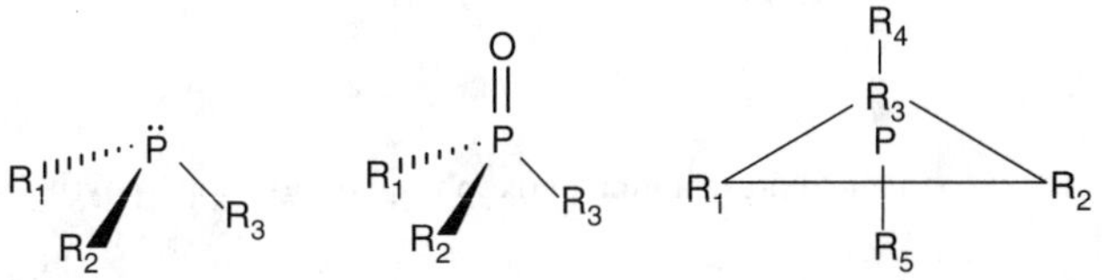

FIGURE 2.29   Examples of configuration of optically active phosphorus compounds.

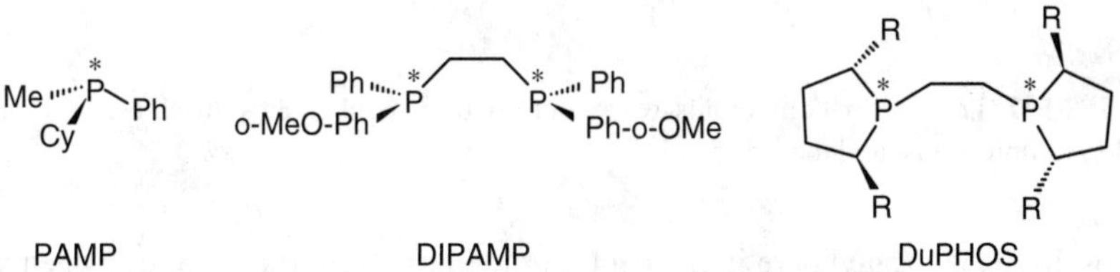

PAMP   DIPAMP   DuPHOS

FIGURE 2.30   Examples of commercially available phosphine ligands.

Phosphorus: Phosphorus may form optically active compounds in all of its oxidation states. In the third oxidation state, phosphorus-containing compounds occur in pyramidal configuration (similar to that of nitrogen), whereas in the fifth oxidation state they occur in tetrahedral form or as trigonal bipyramid (Figure 2.29).

Organic phosphorus compounds play an important role in biological systems. In chemical investigations, chiral phosphorus compounds have found extensive applications as chiral ligands occurring in many catalysts used in asymmetric syntheses [77,78].

Configurations of organic phosphorus compounds are more stable compared with their nitrogen equivalents [79]. Numerous optically active compounds of phosphorus with different coordination numbers were obtained. Optically active phosphines and diphosphines are also available commercially (e.g., methyl-cyclohexylphenylphosphine — PAMP; *bis*[(2-methoxyphenyl)phenylphosphino] ethane — DIPAMP; 1,2-*bis*(2,5-di-*R*-phospholano)ethane — DuPHOS) (Figure 2.30).

Optically active phosphonium salts were also reported [80]. Methods of obtaining them are described in literature [81]. An interesting example of optically active phosphorus compound is its complex with ethioporphyrin [82] (Figure 2.31).

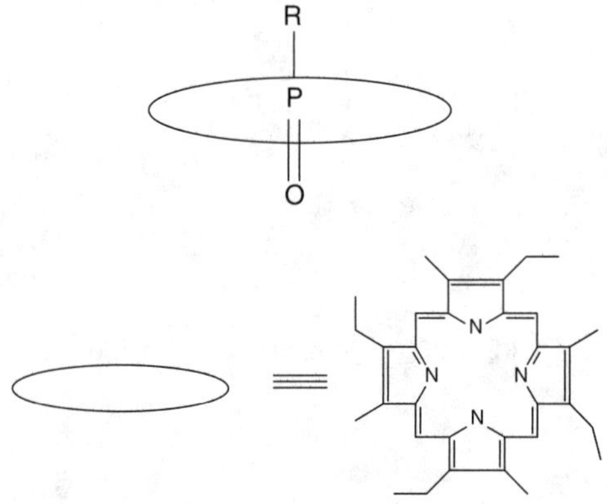

FIGURE 2.31 Schematic representation of oxophosphorous etioporphyrin (R = methyl, phenyl).

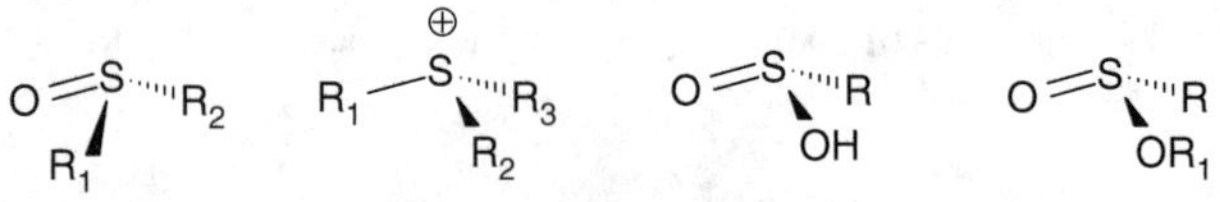

FIGURE 2.32 Sulfur compounds with coordination number 3: sulfoxides, sulfonium salts, sulfinic acids, and their derivatives.

Phosphorus is located here in the center of the porphyrin ring, while oxygen and ethyl or phenyl groups occupy axial positions. CD and NMR spectra confirmed the existence of this compound as an enantiomeric mixture and its separation was accomplished using chiral HPLC.

Sulfur: Structure of organic sulfur compounds with coordination number 3 is similar to that of previously described analogous compounds of nitrogen and phosphorus. Sulfur compounds are, however, much more stable and do not exhibit propensity toward easy inversion. This class of compounds includes, among others, sulfoxides, sulfonium salts, sulfinic acids, and esters (Figure 2.32).

Conformation of compounds belonging here has been investigated since 1950s using various chemical and spectroscopic methods [83,84]. Numerous reactions of optically active sulfur compounds can occur with both inversions as well as with retention of configuration. The latter can also happen if free electron pair from asymmetric sulfur atom participates in the reaction, without the involvement of existing intramolecular bonds [85].

2.6 CHIRALITY OF COORDINATION COMPOUNDS AND CHIRAL POLYMERS

Chirality of coordination compounds is associated with the presence in their molecules of various organic or inorganic ligands. The magnitude of Cotton effect,

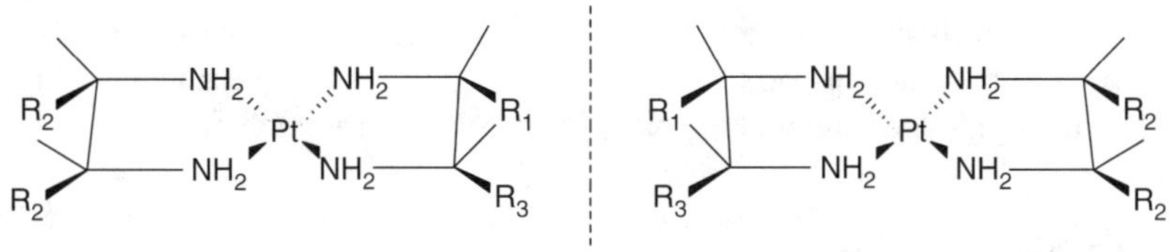

FIGURE 2.33 Schematic representation of two enantiomers of a platinum chelate planar molecule.

which determines the character of optical activity of coordination compounds, depends on the sum of different effects arising from ligands surrounding the central atom. The sign of this effect depends on configuration around such atom. If the ligands possess their own chiral centers then they will have strong influence upon d–d transitions in metal complexes, which furnish extensive information on optical activity of complexes. Particularly, interesting are chelates in which the metal atom forms cyclic systems with ligands. By appropriate choice of ligands, one can obtain optically active complexes (chelates) as shown in Figure 2.33. This example shows that even flat complexes can occur in optically active form.

$$R_1 \neq R_2 \neq R_3$$

Nonetheless, the most common and the best-studied complex compounds are octahedral ones. Numerous optically active octahedral coordination compounds with bidentate ligands were obtained [86,87]. Likewise, many optically active metalocenes species were procured. Ferrocene is an example: if one of its rings has two different substituents, the molecule can exist as two enantiomers [88].

Besides naturally occurring chiral polymers, many synthetic ones are known. Such compounds are under active investigation since they have found many practical uses, among other as organic synthesis catalysts, in chirality-based separations, in nonlinear optical applications, etc. [89–92]. They are usually obtainable from other synthetic compounds although in some instances naturally occurring sugars or amino acids have also been used. Recently, synthetic methods became available that are based on biotechnological processes [93].

2.7 CONCLUSION

This short chapter is by no means an exhaustive or even advanced study of subjects concerning optically active compounds but rather a brief recapitulation of methods that have been employed in their study. It is also a concise source of information about stereochemistry of such compounds. The references provided therein are selective and should be regarded as guidelines only in broadening the reader's knowledge about the subject matter. Information concerning stereochemistry of various compounds and their optical activity can be found in numerous sources, not even wholly devoted to stereochemistry. Whereas the textbook by March as well as many other excellent general textbooks discuss subjects related to stereochemistry at various depths, the answers to more detailed questions concerning chirality of particular compounds should be looked for in relevant research papers.

REFERENCES

1. Hargittai, I. and Hargittai, M., Eds., *Symmetry Through the Eyes of a Chemist*, VCH Publishers, Inc., New York, 1987.
2. Gibb, B.C., *J. Supramol. Chem.*, 2, 123, 2002.
3. Siegel, J.S., Ed., *Supramolecular Stereochemistry*, Kluwer, Dordrecht, 1995.
4. Matsuura, T. and Koshima, H., *J. Photochem. Photobiol. C Photochem. Rev.*, 6, 7, 2005.
5. Eliel, E.L., Wilen, S.H. and Doyle, M.P., *Basic Organic Stereochemistry*, Wiley-Interscience, New York, 2001.
6. Roth, H.J., Muller, C.E. and Folkers, G., *Stereochemie & Arzneistoffe: Grundlagen-Betrachtungen-Auswirkungen*, Wissenschaftliche Verlagsgesellschaft, Stuttgart, 1998.
7. Popa, R., *J. Mol. Evol.*, 44, 121, 1997.
8. Ladik, J.J. and Szekeres, Z., *J. Mol. Model*, 12, 462, 2006.
9. Maier, N.M., Franco, P. and Lindner, W., *J. Chromatogr. A*, 906, 3, 2001.
10. Ong, A.I., Kamaruddin, A.H. and Bhatia, S., *Process Biochem.*, 40, 3526, 2005.
11. Collins, A.N., Sheldrake, G.N. and Crosby, J., Eds., *Chirality in Industry*, John Wiley & Sons, Chichester, 1992.
12. Legrand, M. and Rougier, M.J., Application of the optical activity to stereochemical determinations, in *Stereochemistry: Fundamentals and Methods*, Vol. 2, Kagan, H.B., Ed., Georg Thieme Publishers, Stuttgart, 1977.
13. Tarczay, G., Magyarfalvi, G. and Vass, E., *Angew. Chem. Int. Ed.*, 45, 1775, 2006.
14. Kagan, H.B., Ed., *Stereochemistry: Fundamentals and Methods*, Vol. 1, Georg Thieme Publishers, Stuttgart, 1977.
15. Parker, D., *Chem. Rev.*, 91, 1441, 1991.
16. Morill, T.C., Ed., *Lanthanide Shift Reagents in Stereochemical Analysis (Methods in Stereochemical Analysis)*, John Wiley & Sons, Canada, 1987.
17. Lafontaine, E., Bayle, J.P. and Courtieu, J., *J. Am. Chem. Soc.*, 111, 8294, 1989.
18. Canet, I., et al., *J. Am. Chem. Soc.*, 117, 6520, 1995.
19. Fukui, H., Fukushi, Y. and Tahara, S., *Tetrahedron Lett.*, 46, 5089, 2005.
20. Fukui, H., Fukushi, Y. and Tahara, S., *Tetrahedron Lett.*, 44, 4063, 2003.
21. Mendelbaum, A., Application of mass spectrometry to stereochemical problems, in *Stereochemistry: Fundamentals and Methods*, Vol. 1, Kagan, H.B., Ed., Georg Thieme Publishers, Stuttgart, 1977, pp. 137–180.
22. Schalley, C.A., *Int. J. Mass Spectrom.*, 194, 11, 2000.
23. Speranza, M., *Int. J. Mass Spectrom.*, 232, 277, 2004.
24. Filippi, A., et al., *Int. J. Mass Spectrom.*, 198, 137, 2000.
25. Mancel, V., et al., *Int. J. Mass Spectrom.*, 237, 185, 2004.
26. Vékey, K. and Czira, G., *Anal. Chem.*, 69, 1700, 1997.
27. Sawada, M., et al., *Int. J. Mass Spectrom.*, 193, 123, 1999.
28. Shizuma, M., et al., *Int. J. Mass Spectrom.*, 210/211, 585, 2001.
29. Yao, Z.-P., et al., *Anal. Chem.*, 72, 5383, 2000.
30. Gal, J.F., Stone, M. and Lebrilla, C.B., *Int. J. Mass Spectrom.*, 222, 259, 2003.

31. Grigorean, G., Cong, X. and Lebrilla, C.B., *Int. J. Mass Spectrom.*, 234, 71, 2004.
32. Augusti, D.V. and Augusti, R., *Tetrahedron: Asymmetry*, 16, 1881, 2005.
33. Lu, H.-J. and Guo, Y.-L., *J. Am. Soc. Mass Spectrom.*, 14, 571, 2003.
34. Schug, K.A. and Lindner, W., *J. Am. Soc. Mass Spectrom.*, 16, 825, 2005.
35. Tao, W.A., et al., *Anal. Chem.*, 71, 4427, 1999.
36. Arakawa, R., Kobayashi, M. and Ama, T., *J. Am. Soc. Mass Spectrom.*, 11, 804, 2000.
37. Tao, W.A. and Cooks, R.G., *Angew. Chem. Int. Ed.*, 40, 757, 2001.
38. Guo, J., et al., *Angew. Chem. Int. Ed.*, 38, 1755, 1999.
39. Wolucka, B.A., et al., *Anal. Biochem.*, 255, 244, 1998.
40. Parthasarathy, R., The determination of relative and absolute configurations of organic molecules by x-ray diffraction methods, in *Stereochemistry: Fundamentals and Methods*, Kagan, H.B., Ed., Georg Thieme Publishers, Stuttgart, Vol. 1, 1977.
41. March, J., *Advanced Organic Chemistry, Reactions, Mechanisms, and Structure*, 4th ed., John Wiley & Sons, Inc., 1992.
42. Eliel, E.L., Wilen, S.H. and Doyle, M.P., Eds., *Separation of Stereoisomers, Resolution, and Racemisation in Basic Organic Stereochemistry*, Wiley-Interscience, New York, 2001, 197 pp.
43. Kamphuis, J., et al., The production and uses of optically pure natural and unnatural amino acids in Chirality in Industry, in *The Commercial Manufacture and Applications of Optically Active Compounds*, Collins, A.N., Sheldrake, G.N. and Crosby, J., Eds., John Wiley & Sons, Chichester, 1992, 187 pp.
44. Gübitz, G. and Schmid, M.G., *Biopharm. Drug Dispos.*, 22, 291, 2001.
45. Eliel, E.L., Wilen, S.H. and Doyle, M.P., Eds., Determination of enantiomer and diastereomer composition, in *Basic Organic Stereochemistry*, Wiley-Interscience, New York, 2001, 142 pp.
46. Akabori, S., et al., *Nature*, 178, 323, 1956.
47. Izumi, Y., *Angew. Chem.*, 83, 956, 1971.
48. Ager, D., Ed., *Handbook of Chiral Chemicals*, 2nd ed., Taylor & Francis, Boca Raton, 2005.
49. Desiraju, G.R. and Steiner, T., *The Weak Hydrogen Bond in Structural Chemistry and Biology*, Oxford University Press, New York, 2001.
50. Sarachman, T.N., *J. Chem. Phys.*, 39, 469, 1963.
51. Yamanouchi, K., *J. Phys. Chem.*, 88, 2315, 1984.
52. Cahn, R.S., Sir Ingold, C. and Prelog, V., *Angew. Chem.*, 78, 413, 1966.
53. Eliel, E.L., Wilen, S.H. and Doyle, M.P., Eds., *Basic Organic Stereochemistry*, Wiley-Interscience, New York, 2001.
54. Landor, S.R., Ed., *The Chemistry of the Allenes*, Academic Press, New York, 1982.
55. More, W.R. and Bach, R.D., *J. Am. Chem. Soc.*, 94, 3148, 1972.
56. Hopf, H., *Classics in Hydrocarbon Chemistry, Syntheses, Concepts, Perspectives*, Wiley-VCH Verlag GmbH, Weinheim, 2000.
57. Newman, M.S. and Lednicer, D., *J. Am. Chem. Soc.*, 78, 4765, 1956.
58. Grimme, S. and Peyerimhoff, S.D., *Chem. Phys.*, 204, 411, 1996.
59. Katzenelson, O., Edelstein, J. and Avnir, D., *Tetrahedron: Asymmetry*, 11, 2695, 2000.
60. Goedicke, Ch. and Stegemeyer, H., *Tetrahedron Lett.*, 937, 1970.
61. Martin, R.H. and Marchant, M.J., *Tetrahedron Lett.*, 3707, 1972.

62. Ogawa, Y., et al., *Tetrahedron Lett.*, 44, 2167, 2003.
63. Tanaka, K., et al., *Tetrahedron Lett.*, 36, 915, 1995.
64. Maiorana, S., et al., *Tetrahedron*, 59, 6481, 2003.
65. Pischel, I., et al., *Tetrahedron: Asymmetry*, 7, 109, 1996.
66. Grimme, S. and Bahlmann, A., Electronic circular dichroism of cyclophene, in *Modern Cyclophane Chemistry*, Gleiter, R. and Hopf, H., Eds., Wiley-VCH Verlag GmbH, Weinheim, 2004, pp. 311.
67. Eliel, E.L., Wilen, S.H. and Doyle, M.P., Eds., Configuration and conformation of cyclic molecules, in *Basic Organic Stereochemistry*, Wiley-Interscience, New York, 2001, 421 pp.
68. Zwanenburg, B. and Thijs, L., *Pure Appl. Chem.*, 68, 735, 1996.
69. Lee, W., Jin, J.Y. and Baek, C.-S., *Microchem. J.*, 80, 213, 2005.
70. Yuan, Q., et al., *Tetrahedron Lett.*, 43, 3935, 2002.
71. Nakatsuji, Y., et al., *Tetrahedron Lett.*, 46, 4331, 2005.
72. Prelog, V. and Wieland, P., *Helv. Chim. Acta*, 27, 1127, 1944.
73. Oestreich, M. and Rendler, S., *Angew. Chem. Int. Ed.*, 44, 1661, 2005.
74. Oestreich, M., *Chem. Eur. J.*, 12, 30, 2006.
75. Oestreich, M., et al., *Synthesis*, 2725, 2003.
76. Oh, H.-S., Imae, I. and Kawakami, Y., *Chirality*, 15, 231, 2003.
77. Imamoto, T., *Pure Appl. Chem.*, 72, 373, 2001.
78. Genet, J.P., *Pure Appl. Chem.*, 74, 77, 2002.
79. Horner, L., *Pure Appl. Chem.*, 9, 225, 1964.
80. Wittig, G. and Braun, H., *Liebigs Ann. Chem.*, 751, 27, 1971.
81. Hudson, R.F. and Green, M., *Angew. Chem.*, 75, 47, 1963.
82. Konishi, K., Suezaki, M. and Aida, T., *Tetrahedron Lett.*, 40, 6951, 1999.
83. Mikołajczyk, M. and Drabowicz, J., *Top. Stereochem.*, 13, 333, 1982.
84. Stirling, C.J.M., *The Chemistry of the Sulphonium Group*, Wiley, New York, 1988, 55 pp.
85. Cram, D.J., et al., *J. Am. Chem. Soc.*, 92, 7369, 1970.
86. Belser, P., et al., *Coord. Chem. Rev.*, 159, 1, 1997.
87. Yamada, S., *Coord. Chem. Rev.*, 190–192, 537, 1999.
88. Schlögl, K., *Pure Appl. Chem.*, 23, 413, 1970.
89. Aglietto, M., et al., *Pure Appl. Chem.*, 60, 415, 1988.
90. Reggelin, M., et al., *Proc. Natl Acad. Sci. USA*, 101, 5461, 2004.
91. Jiang, Z., Adams, S.E. and Sen, A., *Macromolecules*, 27, 2694, 1994.
92. Wilson, A.J., et al., *Angew. Chem. Int. Ed.*, 44, 2275, 2005.
93. Bui, V.P. and Hudlicky, T., *Tetrahedron*, 60, 641, 2004.

3 Commercial Precoated Layers for Enantiomer Separations and Analysis

Joseph Sherma

CONTENTS

3.1 INTRODUCTION

As stated in Chapter 1, the separation of optical isomers has become increasingly important in the pharmaceutical industry, because one enantiomer of a drug may be therapeutically active, whereas the other may be nonactive, have different activity, or be toxic. Pesticide potency can also depend on the enantiomer present, making chiral analysis important also in the agrochemical field. There is a variety of approaches to the separation of enantiomers using different kinds of thin layer chromatography (TLC) systems. These include the use of chiral stationary phases with achiral mobile phases, achiral layers with impregnated chiral selectors (sometimes termed *chiral-coated phases*) and an achiral mobile phase, or achiral layers developed with a chiral mobile phase. Enantiomers have been separated on conventional layers, such as silica gel and reversed phase (RP) chemically bonded silica gel, by using diastereomeric derivatives produced by reaction with an optically pure reagent. The reaction may increase separability of the compounds as well as sensitivity of detection and ease of identification.

Enantiomers are separated without prior synthetic formation of diastereomeric adducts by development in chiral systems having adequate enantioselectivity. The chiral stationary phases that have been most widely used are the polysaccharide cellulose and acetylated cellulose. Impregnated chiral layers can be commercial (Chiralplate) or silica gel or amino-bonded silica gel manually impregnated in the laboratory, for example, with selectors such as D-galacturonic acid, (+)-tartaric acid, (−)-brucine, L-aspartic acid, erythromycin, and vancomycin. Chiral mobile phases have been used in conjunction with normal phase (NP) layers (silica gel and diol) and RP layers (chemically bonded C-2, C-18, C-18W, and diphenyl).

In situ quantification of enantiomers can be carried out by densitometry in visible absorption, ultraviolet (UV) absorption (fluorescence quenching), or fluorescence modes. Chiralplates and octylsiloxane-bonded silica (C-8) layers have been employed for these analyses. Densitometry can be used to determine the quantitative proportions of each enantiomer based on calibration curves (scan area vs. percent composition of the two enantiomers) prepared from mixtures of the individual compound standards, as well as their detection limit. It can also provide critical instrumental confirmation of the successful TLC separation of two antipodes by scanning chromatogram lanes and comparing their *in situ* absorption or fluorescence spectra acquired in the spectral mode of the densitometer [1].

Homemade layers manually coated in the laboratory from synthesized or commercial bulk sorbents usually are not of sufficient quality for highest-quality TLC separations and analysis. Standardized, commercially prepared plates are available from a number of commercial sources, the most notable being EMD Chemicals (Gibbstown, NJ, USA; an affiliate of Merck KGaA, Darmstadt, Germany); Macherey-Nagel (Dueren, Germany); Whatman (Florham Park, NJ, USA); J.T. Baker (Phillipsburg, NJ, USA); and Analtech (Newark, DE, USA). To remove extraneous materials that may be present in commercial precoated plates due to manufacture, shipping, or storage conditions, it is generally advisable to preclean plates before applying the samples. EMD Chemicals [2] recommends total immersion of the plate in the mobile phase or allowing the mobile phase to migrate up the plate in a development chamber overnight. The following two-step high-performance TLC (HPTLC) plate cleaning method has been proposed [3] for surface residue removal in critical applications when optimum sensitivity is required for detection and quantification: develop the plate to the top with methanol, air dry for 5 min, totally immerse the plate in a tank filled with methanol, air dry for 5 min, oven dry for 15 min at 80°C, and cool in a desiccator before use. The routine activation of plates, for example, at 120°C for 30 min in a clean drying oven followed by equilibration with the laboratory atmosphere (temperature and relative humidity) in a suitable container providing protection from dust and fumes, is often proposed in the literature [4], but this treatment is not usually necessary for commercial plates unless they have been exposed to high humidity. RP plates do not usually require activation before use. Chiralplates are activated for 15 min at 110°C before use [5]. Suggestions for initial treatment, prewashing, activation, and conditioning of different types of glass- and foil-backed layers have been published [6].

Literature coverage in this chapter will not be comprehensive, but selected, important examples of enantiomer separations and analyses utilizing all of the types of systems involving commercial, precoated plates that are mentioned above will be given with corresponding references. Mechanisms of separations with chiral layers and mobile phases are described in Chapter 7 and will not be covered in detail in this chapter. Applications of homemade plates from commercial batch sorbents will not be covered in this chapter.

3.2   CHIRAL STATIONARY PHASES

The following layers have been used for direct separations of enantiomers.

3.2.1   Unmodified Native and Microcrystalline Cellulose

Cellulose for chromatography is a linear polymer of optically active D-glucopyranose units connected through 1,4-β linkages in which the pyranose residues assume the energetically favored chair conformation; its chains are arranged on a partially crystalline fiber structure with helical cavities. The dominant mechanism for chiral resolution is believed to be inclusion, with contributions from hydrogen bonding and dipolar interactions [7].

Thin Layer Chromatography plates are prepared from native cellulose with polymerization of 400–500 glucose units, fiber length 2–20 μm, specific surface area about 2 m^2/g, and molecular weight 2.5×10^5 to 1×10^6 or higher, and from Avicel (microcrystalline cellulose manufactured by American Viscose Division of FMC Corp., Philadelphia, PA, USA) with an average degree of polymerization between 40 and 200. Fibrous cellulose is purified by washing with acid and organic solvents and crystalline cellulose is formed by hydrolysis of high-purity cellulose with hydrochloric acid (HCl) to remove amorphous material. Both types have low-specific surface area and are used mainly in NP partition chromatography (adsorption effects cannot be excluded) for the separation of relatively polar compounds, such as amino acids, other carboxylic acids, and carbohydrates, and they are natural chiral sorbents for direct enantiomer separations. Cellulose layers usually do not require a binder to adhere to the support. As examples of available products, native fibrous cellulose plates from Macherey-Nagel (MN 300) have layer thicknesses of 0.10, 0.25, and 0.50 mm on glass and 0.10 mm on polyester or aluminum, and microcrystalline cellulose (Avicel MN 400) is sold as 0.10 mm layers on glass or polyester. Analtech offers Avicel Uniplates with a 0.25 mm layer on glass, and fibrous and microcrystalline 0.1 mm layers on plastic. Cellulose layers with fluorescent indicator (designated F-layer or UV-layer) allow viewing of substances in shortwave (254 nm) UV light; the greenish fluorescence is quenched by substances with absorption above about 230 nm, and these appear as dark zones against a fluorescent background. Compounds separated on microcrystalline cellulose tend to form more compact zones than on fibrous cellulose layers.

The first cellulose enantioseparation was of a racemic amino acid by paper chromatography [8]. Early cellulose TLC studies were aimed at repeating paper chromatography enantiomer separations more quickly and with better resolution. Cellulose TLC has mostly been used for amino acids, their derivatives, and peptides and the following are examples of successful enantioseparations. Aromatic amino acids were separated on polyester-backed 20 × 20 cm plates by elution with methanolic aqueous 0.1 M HCl (A) or ethanol/pyridine/water (1:1:1) (B). Typical respective R_f values were 0.75/0.81 for L- and D-tyrosine in mobile phase A and 0.37/0.46 for L- and D-5-methyl tryptophan in B [9,10].

Long development times (e.g., 5–10 h) are often required for cellulose layers, but separation of racemic 3,4-dihydroxyphenylalanine (dopa), tryptophan, and 5-hydroxytryptophan was achieved in only 2 h on a Merck 0.1 mm cellulose HPTLC plate developed for 17 cm in the ascending direction in a saturated chamber with methanol/water (3:2) [5]. Ninhydrin was used as the detection reagent, and respective R_f values of the D- and L-isomers of the three compounds were 0.58/0.53, 0.51/0.44, and 0.40/0.32. Figure 3.1 shows densitograms of the separations.

n-Butanol/acetic acid/water (1:1:1) was the mobile phase used for separation of D,L-phenylalanine-4-sulfonic acid on DC-Alufolein cellulose-F plates (Merck). Separated zones at R_f 0.71 (L)/0.74 (D) were detected by using ninhydrin reagent [11].

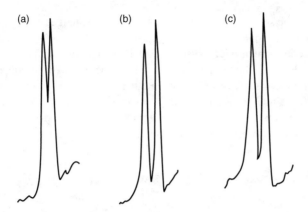

FIGURE 3.1 Separations on HPTLC cellulose of (a) D,L-dopa, (b) D,L-tryptophan, and (c) D,L-hydroxytryptophan. Densitograms were obtained using a CD 60 chromatogram spectrophotometer (Desaga) in the reflectance mode at 565 nm with a 6 × 0.2 mm slit. (Reprinted from Guenther, K. and Moeller, K., in *Handbook of Thin Layer Chromatography*, 3rd edn., Sherma, J. and Fried, B., Eds., Marcel Dekker, Inc., New York, NY, 2003, pp. 471–533. With permission.)

3.2.2 Acetylated Cellulose

Acetylated cellulose (AC) is used for RP-TLC. It is prepared by esterification of cellulose with acetic acid, and up to three hydroxy groups/cellulose units can be acetylated. The acetyl content can range from a few percent up to the maximum of 44.8%, which corresponds to cellulose triacetate or triacetyl cellulose. (Many literature sources use the term triacetyl cellulose or cellulose triacetate, but the acetyl content of the layer material is below 44.8%, so acetylated cellulose would be correct.) As the acetyl content increases, the hydrophobic character of the layer increases. Commercial plates are available from Macherey-Nagel with 0.1 mm fibrous cellulose layers having 10 or 20% acetyl content on glass or with 10% on polyester sheets. It has been found that chiral discrimination by AC increases with increasing crystallinity [12]. Enantiomer separations have been reported in the literature on MCTA (microcrystalline cellulose triacetate) layers, but these have been mostly on noncommercial layers. For example, the separation of racemic Tröger's base was carried out on homemade MCTA plates [13], and analysis of a variety of chiral compounds, including flavanone derivatives and pyrethroids, was carried out on layers prepared by mixing MCTA for column HPLC with silica gel 60GF, development with organic mixtures containing methanol, ethanol, or 2-propanol, and detection by scanning densitometry [14]. The following R_f values were reported for enantiomer separations on Macherey-Nagel CEL 300-10/AC-20% TLC glass plates using ethanol/water (80:20) mobile phase at 25°C and detection under UV light: (+/−)-1,1′-binaphthyl-2.2′-diamine, 0.33/0.40; (+/−)-7,8,9,10-tetrahydrobenzo[α]pyren-7-ol, 0.29/0.40; (+/−)-γ-(2-naphthyl)-γ-butyrolactone, 0.46/0.50; (+/−)-1-(2-naphthyl)ethanol, 0.72/0.74 [15].

Enantiomeric separations of the benzyl ester of the analgesic agent oxindanac, 2-phenylcyclohexanone, and (R,S)-2,2,2-trifluoro-1-(9-anthryl)-ethanol were reported on commercial microcrystalline MCTA OPTI-T.A.C. TLC plates L.254 from Antec (Bennwil, Switzerland) [5,16]. One-dimensional development without chamber saturation was carried out using ethanol/water (80:20 or 85:15) mobile phase for 10 cm (1.3 h). Zones were detected under 254 and 366 nm UV light and absorbance was scanned with a densitometer set at 254 nm or fluorescence at 366 nm (excitation) and 420 nm (emission).

3.2.3 HOME-IMPREGNATED COMMERCIAL LAYERS

3.2.3.1 Silica Gel

Silica gel is by far the most frequently used layer material for TLC, and its usual mode of separation without modification is NP adsorption. Some characteristic properties, including porosity, flow resistance, particle size, optimum velocity, and plate height, have been tabulated for three popular brands of silica gel TLC and HPTLC plates [17]. Typical properties of TLC silica gel are (a) a silanol group (Si—O—H) level of ca. 8 μmol/m^2, and pore diameter 4, 6, 8, and 10 nm; (b) specific pore volumes 0.5–2.0 ml/g; and (c) specific surface area 200–800 m^2/g. The most widely used TLC silicas are those with 6 nm (60 Å) pores. Siloxane groups (Si—O—Si) are also contained on the silica surface but do not represent adsorption-active centers; silanol active centers can differ slightly depending on whether they occur as isolated, vicinal, or geminal silanols. HPTLC plates (10 × 10 or 10 × 20 cm) are produced from sorbents having a narrow pore and particle size distribution and an apparent particle size of 2–10 μm instead of 5–17 μm for 20 × 20 cm TLC plates (values for Macherey-Nagel precoated plates). Layer thickness is usually 100–200 μm for HPTLC plates compared with 250 μm for TLC. HP layers are more efficient, leading to tighter zones, better resolution, and more sensitive detection. Flow resistance is higher (migration time/cm is slower), but overall development time is shorter because smaller migration distances are used for HPTLC compared with TLC (typically 3–8 cm vs. 10–16 cm). Binders used by manufacturers for commercial precoated plates are usually proprietary organic polymeric compounds related to polyvinyl alcohol, polyvinyl pyrollidone, or similar compounds. Silica gel G plates contain inorganic gypsum binder (calcium sulfate hemihydrate).

Impregnation of layers with a chiral selector for use in enantiomer separations has usually been carried out by mixing the compound with silica gel to prepare a slurry for in-house preparation of the layer. For example, Bhushan and Martens [18] reported use of this method for preparing layers containing selectors such as (+)-tartaric acid or (+)-ascorbic acid, erythromycin, vancomycin, L-lysine and L-arginine, and (−)-brucine.

The use of impregnated commercial precoated silica gel plates has been much less common. Enhanced chiral separations were demonstrated for (+/−)-ibuprofen and propanolol on commercial glass silica gel 60F TLC plates

(0.25 mm layer) impregnated with a 3×10^{-2} M solution of L-arginine by dipping compared with direct addition of L-arginine to the silica gel slurry before self-coating glass plates [1]. The difference in R_f values was 0.03 for one-dimensional development and 0.07 for two-dimensional development using the impregnated commercial plates.

Enantiomers of $(+/-)$-metoprolol tartrate were separated on silica gel 60F plates using the chiral selector D-$(-)$-tartaric acid (11.6 mM solution) both impregnated into the layer and as mobile phase additive. The mobile phase was ethanol/water (70:30) and development temperature was 25°C. The layer was impregnated at ambient temperature for 90 min before sample application. Separated zones were detected by viewing at 254 nm, scanning at 230 nm, and exposure to iodine vapor. R_f values were 0.65 and 0.50 for the S-$(-)$ and R-$(+)$ enantiomers, respectively [19].

3.2.3.2 Amino-bonded Silica Gel

Hydrophilic NH_2-modified layers contain amino groups bonded to silica gel via a propyl group [$-(CH_2)_3-$] as a spacer. They are compatible with aqueous mobile phases and exhibit multimodal mechanisms, that is, NP and weak anion exchange. In NP-TLC, compounds are retained on amino layers by hydrogen bonding as with silica gel, but the selectivity is different. Charged substances such as nucleotides or sulfonic acids can be separated by ion exchange using acidic mobile phases. There is limited retention in RP-TLC.

A two-phase "Pirkle" modified amino-bonded phase for enantiomer separation was prepared by partially immersing a commercial HPTLC NH_2 F-plate (Merck) in a solution of the chiral selector N-(3,5-dinitrobenzoyl)-L-leucine. The portion of the plate modified with chiral selector was used to separate the enantiomers of the model compounds 2,2,2-trifluoro-(9-anthryl)ethanol and 1,1'-binaphthol with the mobile phase hexane/isopropanol (80:20). The separated enantiomers were then eluted using continuous development onto the unmodified portion of the plate. The absence of the selector on this segment of the layer allowed the detection of the separated compounds by fluorescence quenching [20].

3.2.4 COMMERCIALLY IMPREGNATED CHIRALPLATES

Commercial layers are available for separation of enantiomers by the mechanism of chiral ligand exchange under the name Chiralplate (Macherey-Nagel). Chiralplates consist of a glass plate coated with a 0.25 mm layer of RP-modified (C-18) [12] silica gel coated (not chemically bonded) with a chiral selector, $(2S, 4R, 2'RS)$-N-(2'-hydroxydodecyl)-4-hydroxyproline, and Cu(II) ions. The chiral selector on the plate and the enantiomers to be separated form diastereomeric mixed chelate complexes with the cupric ion; complexes of different antipodes have different stabilities, thus achieving TLC separation. The chiral phase is bonded to the stationary phase strongly through hydrophobic interactions so that mobile phases with a high degree of organic composition can be used without bleeding of the

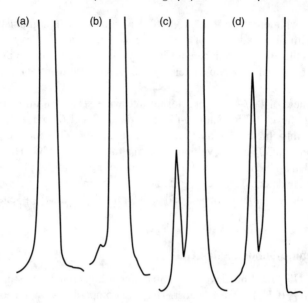

FIGURE 3.2   Densitograms obtained in the quantitative determination of TLC-separated enantiomers of *tert*-leucine: (a) L-*tert*-leucine, (b) L-*tert*-leucine+0.1% D-*tert*-leucine, (c) L-*tert*-leucine + 1% D-*tert*-leucine, and (d) external reference standard. Layer, Chiralplate; mobile phase, methanol/water (10:80); detection, dip in 0.3% ninhydrin solution in acetone; quantification, scanning at 520 nm. (Reprinted from Guenther, K. and Moeller, K., in *Handbook of Thin Layer Chromatography*, 3rd edn., Sherma, J. and Fried, B., Eds., Marcel Dekker, Inc., New York, NY, 2003, pp. 471–533. With permission.)

copper complex from the support [7]. A similar layer called HPTLC-CHIR was available from Merck but is no longer produced; it was a high performance layer rather than TLC and contained a concentrating zone [21].

Amino acid derivatives have been separated on Chiralplates using the mobile phases methanol/water/acetonitrile (50:50:200, developing time about 30 min) or (50:50:30, 60 min). Enantiomers separated in these mobile phases are amino acids, dipeptides, α-methyl amino acids, N-alkyl and N-formyl amino acids, halogenated amino acids, and other classes of compounds. Additional mobile phases that have been successful include the following: methanol/water (10:80, 90 min) for leucine (Figure 3.2); acetone/methanol/water (10:2:2, 50 min) for alanine, serine, and pipecolic acid; 1 mM cupric acetate solution containing 5% methanol (pH 5.8) for N-carbamyl-trypthophan; dichloromethane/methanol (45:5, 20 min) for enantiomeric α-hydroxycarboxylic acids; acetonitrile/methanol/water (4:1:1 or 2) and acetonitrile/methanol/water/diisopropylethylamine (4:1:2:0.1) for phenylalanine, tyrosine, histidine, and tryptophan analogs and analogs containing tetralin or 1,2,3,4-tetrahydroisoquinoline skeletons [22]; and n-hexane/ethyl acetate (1:9) and water/methanol/acetonitrile (1:1:3) for E-Z isomers of pyrazole, pyrimidine, and purine derivatives with potential cytokinin activity [23]. pH was shown to be an important variable in separations on Chiralplate; at pH 3–4 or 6–7, the highest enantioselectivity was obtained and drift and disturbance of chromatogram baselines were smallest for the separation of amino acids using a horizontal sandwich DS chamber, acetonitrile/methanol/aqueous pyridine buffer mobile phases, ninhydrin detection, and scanning at 518 nm with a diode array detector densitometer [24]. R_f values have been tabulated for many compounds in these mobile phases [22,23,25].

For optimal separations and reproducible R_f values on Chiralplates, the manufacturer recommends activation of plates before spotting (15 min at 100°C) and developing in a saturated chamber. In some cases, good separations can be achieved in 5 min without activation and chamber saturation. Chiralplates have been used with forced flow development [26] as well as the usual ascending capillary-controlled flow. Chiralplates are compatible with ninhydrin reagent (0.3% in acetone) applied by dipping or spraying and followed by drying at 110°C for 5 min for the detection of proteinogenic and nonproteinogenic amino acids as red zones on a white background, and with vanadium pentoxide for the detection of α-hydroxycarboxylic acids (blue zones on a yellow background). Use of these reagents before quantitative evaluation of antipodes by densitometric scanning can increase specificity and sensitivity.

The following selected applications were taken, with permission, from a Macherey-Nagel brochure [25] and the Applications Database on the Macherey-Nagel homepage (www.mn-net.com). The database search was carried out by choosing "applications by phase," then "TLC," and then "Chiralplate." The application number from the database or reference to the brochure is cited in each case.

3.2.4.1 Control of Optical Purity of L-Dopa

Mobile phase: Methanol/water/acetonitrile (50:50:30).
Detection: 0.1% Ninhydrin spray reagent.
Results: Figure 3.3.
Application No. 400600.

3.2.4.2 Enantiomer Separation of D,L-Penicillamine

Mobile phase: Methanol/water/acetonitrile (15:15:60), 12 cm, 30 min.
Detection: 0.1% Ninhydrin spray reagent, densitometry at 565 or 595 nm.
Results: R_f values, 0.52 for D-5,5-dimethyl-4-thiazolidinecarboxylic acid and 0.72 for the L-isomer; Figure 3.4.
Application No. 400610.

3.2.4.3 Enantiomer Separation of α-Hydroxycarboxylic Acids

Mobile phase: Dichloromethane/methanol (45:5), 13 cm.
Detection: Dip in vanadium pentoxide solution.

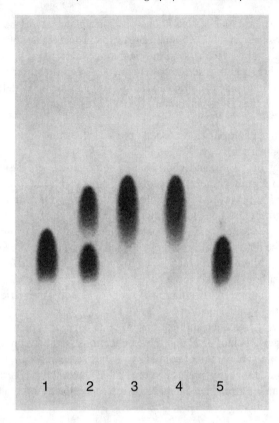

FIGURE 3.3 Separation of D,L-dopa on Chiralplate. Zones: (1) L-dopa, (2) D,L-dopa, (3) D-dopa, (4) 3% L-dopa in D-dopa, and (5) 3% D-dopa in L-dopa. (Reprinted from the Machery-Nagel Application Database. With permission.)

Results: R_f value (configuration), mandelic acid 0.46/0.53 (L); hydroxyisoleucine 0.56/0.63 (L); hydroxyleucine 0.56/0.60 (L); hydroxymethionine 0.52/0.58 (L); hydroxyphenylalanine 0.56/0.62 (L); hydroxyvaline 0.52/0.60 (L). Application No. 400630.

3.2.4.4   Enantiomer Separation of Proteinogenic Amino Acids

Mobile phase: Methanol/water/acetonitrile (50:50:200), 13 cm, saturated chamber. Detection: 0.1% Ninhydrin spray reagent.
Results: R_f value (configuration), glutamine 0.41 (L)/0.55 (D); isoleucine 0.47 (D)/0.58 (L); methionine 0.54 (D)/0.59 (L); valine 0.54 (D)/0.62 (L); phenylalanine 0.49 (D)/0.59 (L); tyrosine 0.58 (D)/0.66 (L); tryptophan 0.51 (D)/0.61 (L); proline 0.41 (D)/0.47 (L). Application No. 400530.

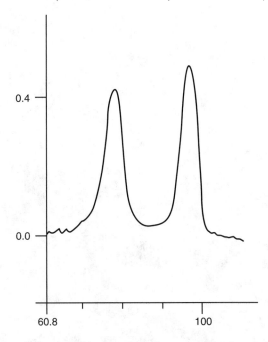

FIGURE 3.4 Densitogram of the separation of D,L-penicillamine on Chiralplate measured at 565 nm. Left peak, D-isomer; right peak, L-isomer. *y*-axis, absorbance units; *x*-axis, mm distance from origin. (Reprinted from the Macherey-Nagel Application Database. With permission.)

3.2.4.5 Optical Purity Control of D-Penicillamine

Mobile phase: Methanol/water/acetonitrile (50:50:200), 30 min.
Detection: 0.1% Ninhydrin spray reagent.
Results: Figure 3.5.
Application No. 400590.

3.2.4.6 Separation of Aspartame and Its Precursor Stereoisomers

Mobile phase: Methanol/water/acetonitrile (50:50:200).
Detection: Ninhydrin.
Results: Figure 3.6.
Application No. 400570.

3.2.4.7 Enantiomeric Separation of Dipeptides

Mobile phase: Methanol/water/acetonitrile (50:50:30), 13 cm, chamber saturation.
Detection: Ninhydrin.

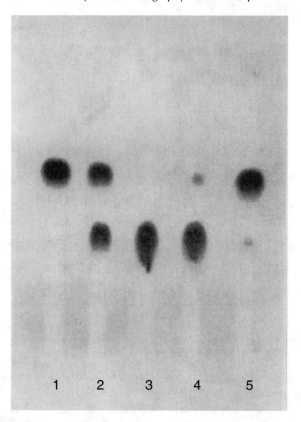

FIGURE 3.5 Chromatograms on Chiralplate containing the following zones: (1) L-5,5-dimethyl-4-thiazolidine carboxylic acid (L-3), (2) D,L-5,5-dimethyl-4-thiazolidine carboxylic acid (D,L-3), (3) D-5,5-dimethyl-4-thiazolidine carboxylic acid (D-3), (4) 3% L-3 in D-3, and (5) 3% D-3 in L-3. (Reprinted from the Macherey-Nagel Application Database. With permission.)

Results: R_f values, Glycine (Gly)-D,L-phenylalanine 0.57 (L)/0.63 (D); Gly-D,L-leucine 0.53 (L)/0.60 D; Gly-D,L-isoleucine 0.54 (L)/0.61 (D); Gly-D,L-valine 0.58 (L)/0.62 (D); Gly-D,L-tryptophan 0.48 (L)/0.55 (D).
Reference: [25].

3.2.4.8   Separation of Enantiomeric α-Methylamino Acids

Mobile phase: Methanol/water/acetonitrile (50:50:200) (A) or 50:50:30 (B), chamber saturation.
Detection: Ninhydrin.
Results: R_f values (in mobile phase A), α-methylmethionine 0.56 (D)/0.64 (L); α-methyltyrosine 0.63 (D)/0.70 (L); α-methylphenylalanine 0.53 (L)/0.66 (D). R_f values (in mobile phase B), α-methylserine 0.56 (L)/0.67 (D); α-methyldopa 0.46 (L)/0.66 (D) [25].

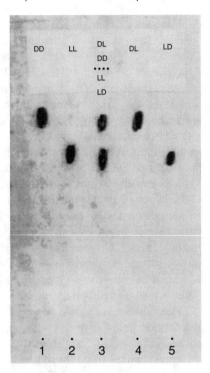

FIGURE 3.6 Separation of aspartame and its precursor stereoisomers on Chiralplate. D,D-, D,L-aspartame, $R_f = 0.62$; L,L-, L,D-aspartame, $R_f = 0.50$. (Reprinted from the Macherey-Nagel Application Database. With permission.)

3.2.4.9 Separation of Enantiomeric *N*-Alkyl and *N*-Formyl Amino acids

Mobile phase: Methanol/water/acetonitrile (50:50:200), chamber saturation.
Detection: Ninhydrin.
Results: R_f values, *N*-methylleucine 0.49 (L)/0.57 (D); *N*-methylphenylalanine 0.50 (D)/0.61 (L); *N*-formal-*tert*-leucine 0.48 (+)/0.61 (−) [25].

3.2.4.10 Separation of Halogenated Amino Acid Enantiomers

Mobile phase: Methanol/water/acetonitrile (50:50:200), chamber saturation.
Detection: Ninhydrin.
Results: R_f values, 4-iodophenylalanine 0.45 (D)/0.61 (L); thyroxine 0.38 (D)/0.49 (L) [25].

3.3 NONCHIRAL PLATES USED WITH CHIRAL MOBILE PHASES

Direct separations of enantiomers have been made using a variety of commercial achiral precoated TLC and HPTLC plates developed with a chiral mobile phase.

The layers used include unmodified silica gel (described in Section 3.2.3) and C-2 (dimethyl), C-18 (octadecyl), C-18W, diol, and diphenyl chemically bonded silica gel.

Chemically bonded layers are prepared by reacting silanol groups on the silica gel surface with various organosilane derivatizing reagents (monofunctional, bifunctional, or trifunctional), leading to layers from different manufacturers with varying average particle size, pore size, carbon loading, and percent silanol groups reacted. Poole [27] has tabulated characteristic properties of chemically bonded layers from different manufacturers. As a specific example, the Macherey-Nagel catalog lists two C-18 plates, Nano-Sil C-18-50 and C-18-100, both with 0.20 μm thick layers made from silica gel with 60 Å pore size and 2–10 μm mean particle diameter; the former is 50% silanized and the latter 100%.

C-18 or RP-18 layers are the main layers used for RP-TLC. Problems of wettability and lack of migration of mobile phases with high proportions of water have been solved by using "water-wettable" C-18 layers (C-18W) with a slightly larger particle size, less exhaustive surface bonding, and a modified binder. The latter layers with a low degree of surface coverage and more residual silanol groups exhibit partially hydrophilic as well as hydrophobic character and can be used for RP-TLC and NP-TLC. Chemically bonded phenyl layers are also classified as RP. C-18 layers are more hydrophobic than C-8 or C-2.

Diol plates are prepared by reaction of the silica matrix with a silane derived from glycerol. The resulting alcoholic hydroxyl residues bonded to the silica by a propyl spacer can operate with NP- or RP-TLC mechanisms, depending on the mobile phase and solutes. Polar compounds show reasonable retention by hydrogen bond and dipole-type interactions in the former mode, and in the RP mode retention is low but higher than with amino layers.

3.3.1 SILICA GEL

Cyclodextrins (CDs) have been used for enantiomer separations in the form of homemade-bonded plates (commercial CD plates do not exist) and mobile phase additives. As an example of the latter, arginine, glutamic acid, histamine, lysine, and valine enantiomers were resolved with differences of R_f values in the range 0.04–0.11 on Merck silica gel 60F plates developed with 6.5×10^{-3} M hydroxypropyl-β-CD in water/acetonitrile (2:1.5 and 2.5:1). Zone detection was performed by spraying with salicylaldehyde reagent (1.5 g in 100 ml of toluene) and heating for 10 min at 50°C [28].

Enantiomers of phenylpropanolamine, octopamine, pindolol, norphenylephedrine, propanolol, and isoproterenol were separated with R_f differences of 0.05–0.24 by development of Whatman HP-KF HPTLC plates with dichloromethane/methanol (75:25) containing different amounts of N-benzoxycarbonyl (BOC)-alanyl-L-proline (ZAP), BOC-isoleucyl-L-proline (ZIP), BOC-L-proline (ZP), BOC-glycyl-L-proline (ZGP), (1R)-(−)-ammonium-10-camphorsulfonate (CSA), and triethylamine (TEA). Visualization was under 254 nm UV light [29].

3.3.2 C-18 AND C-18W BONDED SILICA GEL

β-CD (0.133–0.231 M) in saturated aqueous solutions of urea containing 3.5% NaCl–methanol or acetonitrile was used as mobile phases with Whatman KC18F layers to separate dansyl derivatives of leucine, valine, methionine, glutamic acid, aspartic acid, phenylalanine, serine, threonine, and tryptophan with differences in R_f values of 0.02–0.09 between D- and L-isomers. Zones were detected by viewing under 254 nm light and scanning at 230, 254, and 280 nm [30]. Similar mobile phases containing aqueous urea, NaCl, and acetonitrile were employed with Sil C-18-50/UV plates (Macherey-Nagel) to separate enantiomers of 11 dansyl amino acids with R_f differences of 0.05–0.08 [31]. Urea increases the solubility of β-CD in water, and NaCl stabilizes the binder of RP plates. Sil C-18-50/UV layers were also applied to separate enantiomers of flavanone, 2′-hydroxyflavanone, 4′-hydroxyflavanone, and 4′-methoxyflavanone by developing with 0.15 M aqueous β-CD + 32% urea + 2% NaCl/acetonitrile (80:20) (0.04–0.06 R_f differences) [32], and 4-benzyl-3-propionyl-2-oxazolidinone and N-α-(2,4-dinitro-5-fluorophenyl)valinamide with 0.2 M hydroxypropyl-γ-CD in water/acetonitrile (90:10) (0.01–0.03 R_f differences) [33].

Racemic dinitropyridyl, dinitrophenyl, and dinitrobenzoyl amino acids were resolved on C-18W/UV TLC and HPTLC plates (Macherey-Nagel) developed with 2% aqueous isopropanol containing 2–5% BSA. Development times were 1–2 h, and visualization was under 254 nm UV light. R_f differences ranged from 0.06 to 0.49 [34]. The same plates with mobile phases composed of isopropanol + BSA + sodium tetraborate, acetic acid, or sodium carbonate served to separate enantiomeric D,L-methylthiohydantoin and phenylthiohydantoin derivatives of amino acids, kynureyne, 3-(1-naphthyl)alanine, lactic acid derivatives, alanine and leucine p-nitroanilides, and 2,2,2-trifluoro-1-(9-anthryl)ethanol [35], and with mobile phases composed of water with 5–7% BSA + 2% isopropanol to separate dansyl amino acid derivatives [36].

Sil C18-50/UV plates (Macherey-Nagel) developed for 7 cm (1–1.5 h) with 0.1 M acetate buffer (pH 4.86) containing isopropanol (12–36%) and BSA (5–6%) were used to separate the enantiomers of 12 N-α-(9-fluorenylmethoxycarbonyl) (Fmoc) amino acids. Detection was by irradiation with 366 nm UV light. The α values (selectivity) ranged from 1.35 to 2.79 and resolution (R_s) factors from 1.1 to 2.7 for the L- and D-enantiomers [37]. These same layers with mobile phases containing sodium tetraborate, isopropanol, and BSA were used for the TLC of enantiomeric tryptophans [38].

3.3.3 C-2 BONDED SILICA GEL

Whatman KC2F bonded silica gel developed with 0.4 M maltosyl-β-CD and 0.6 M NaCl in acetonitrile/water (30:70) separated enantiomers of D,L-alanine-β-naphthylamide, dansyl D,L-leucine, dansyl D,L-valine, D,L-methionine-β-naphthylamide, and N′-(2-naphthylmethyl)nornicotine, and the diastereomers cinchonidine/cinchonine and quinidine/quinine. Zone visualization was performed by using a 254 nm UV lamp. R_f values differed by 0.05–0.25 units [39].

3.3.4 DIPHENYL-BONDED SILICA GEL

Whatman RP diphenylsiloxane-bonded silica gel plates have 8.5% carbon loading and 10–14 μm average particle size; they are endcapped but still contain a relatively high concentration of silanol groups [27]. These layers were used with the macrocyclic antibiotic, vancomycin, as a mobile phase additive to resolve 6-aminoquinolyl-N-hydroxysuccinimidyl carbamate (AQC) derivatized amino acids, racemic drugs, and dansyl amino acids. The mobile phase was acetonitrile–0.6 M NaCl–1% triethylammonium acetate buffer (pH 4.1) with 0.025–0.08 M vancomycin in various ratios. The development distance was 10 cm, development time 1–3 h, and detection by fluorescence at 245 and 365 nm. Typical R_f values were bendroflumethiazide (0.02/0.06), AQC-$allo$-isoleucine [0.14 (L)/0.21 (D)], and dansyl-norleucine [0.04 (L)/0.16 (D)] [40].

3.3.5 DIOL-BONDED SILICA GEL

The enantiomeric separation of the amino alcohol β-blocking drugs alprenolol and propranolol was carried out on Merck LiChrosorb diol-F plates developed in a short bed/continuous development (SB/CD) chamber (Regis Technologies Corp., Morton Grove, IL, USA) with dichloromethane containing 0.4 mM ethanolamine and 5 mM N-carbobenzoxyglycyl-L-proline. Detection was by viewing and scanning under 280 or 300 nm UV light. With continuous development, regular R_f values cannot be calculated, but propranolol retained more strongly than alprenolol and the S-isomer moved more slowly than the R-isomer for each compound [41]. In later work, the same authors used the Merck diol-F plate with a mobile phase consisting of dichloromethane containing 5 mM ZGP to separate (R,S)-metoprolol, -propranolol, and -alprenolol [42].

Diol HPTLC plates (Alltech) with mobile phases consisting of dichloromethane/isopropanol (95:5) plus different amounts of ZAP, ZIP, ZP, ZGP, CSA, or TEA were used to separate timolol with R_f values of 0.26/0.51. Visualization was under 254 nm UV light [29].

3.4 ENANTIOMER SEPARATIONS USING DIASTEREOMERIC DERIVATIVES

Indirect enantiomer separations on nonchiral layers via the formation of diastereomers have been carried out mostly using silica gel plates, and also on C-8 and C-18 bonded silica gel. The silica gel TLC and HPTLC plates were usually Merck silica gel 60 with fluorescent indicator so that compounds could be detected by fluorescence quenching under 254 nm UV light.

C-8 layers are prepared by reacting silanol groups with dichlorooctylmethylsilane; they are more hydrophobic than C-2 but less than C-18. Most of the published work focuses on reactions of racemic compounds with $NH_2(NH)$, OH, and COOH functional groups with a variety of reagents, including ready-to-use reagents employed earlier in HPLC.

3.4.1 SILICA GEL

The following chiral reagents were employed for diastereomer formation before sample application and chromatography on silica gel or silica gel G TLC plates: (L)-leucine N-carboxyanhydride for D,L-dopa-carboxyl-^{14}C separated with ethyl acetate/formic acid/water (60:5:35) mobile phase and detected by ninhydrin [R_f 0.38 (D)/0.56 (L)] [43]; N-trifluoroacetyl-L-prolyl chloride for D,L-amphetamine separated with chloroform/methanol (197:3) and detected by sulfuric acid/formaldehyde (10:1) (R_f 0.49 (D)/0.55 (L)) [44]; N-benzyloxycarbonyl-L-prolyl chloride for D,L-methamphetamine separated with n-hexane/ethyl acetate/acetonitrile/diisopropyl ether (2:2:2:1) and detected by sulfuric acid/formaldehyde (10:1) [R_f 0.57 (L)/0.61 (D)] [44]; (1R,2R)-(−)-1-(4-nitrophenyl)-2-amino-1,3-propanediol (levobase) and its enantiomer dextrobase for chiral carboxylic acids separated with chloroform/ethanol/acetic acid (9:1:0.5) and detected under UV (254 nm) light (R_f values 0.63 and 0.53 for S- and R-naproxen, respectively) [45]; (S)-(+)-α-methoxyphenylacetic acid for R,S-ethyl-4-(dimethylamino)-3-hydroxybutanoate (carnitine precursor) with diethyl ether mobile phase [R_f 0.55 (R)/0.79 (S)] [46]; and (S)-(+)-benoxaprofen chloride with toluene/acetone (100:10, ammonia atmosphere) mobile phase and fluorescence visualization (Zeiss KM 3 densitometer; 313 nm excitation, 365 nm emission) (respective R_f values of R- and S-isomers of metoprolol, oxprenolol, and propranolol were 0.24/0.28, 0.32/0.38, and 0.32/0.39) [47].

Silica gel HPTLC plates were used with the following chiral reagents and analytes: R-(−)-1-(1-naphthyl)ethyl isocyanate for six R,S-β-blocking agents with benzene/ethyl ether/acetone (88:10:5) mobile phase and UV absorption and fluorescence detection (R_f differences 0.05–0.09) [48]; (R)-(+)-1-phenylalanine hydrochloride for ketoprofen, suprofen, and indoprofen with benzene/methanol (93:7) and chloroform/ethyl acetate (15:1) mobile phases and UV (254 nm) detection (R_f differences were 0.04–0.10, with the R-isomer migrating more slowly for each compound) [49]; and (S)-(+)-naproxen chloride for the methyl esters of 10 amino acids with toluene/dichloromethane/tetrahydrofuran (5:1:2) mobile phase and UV detection (R_f differences 0.04–0.10, with the R-isomer always having a lower value than the S-isomer) [50].

3.4.2 REVERSED PHASE CHEMICALLY BONDED SILICA GEL

Marfey's reagent (1-fluoro-2,4-dinitrophenyl-5-L-alanine amide; FDAA) was used to prepare diastereomeric derivatives of 22 amino acids, which were then separated on Whatman C-18F layers developed for 83 mm in an SB/CD chamber with mobile phase consisting of methanol and 0.3 M aqueous sodium acetate adjusted to pH 4 by the addition of 2 M HCl. Zone detection was made under 254 nm UV light.

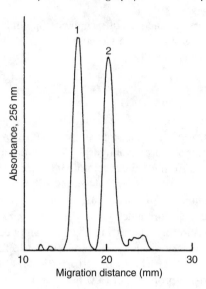

FIGURE 3.7   Densitogram of (1) R-(+)-pindolol and (2) S-(−)-pindolol. (Reprinted from Spell, J.C. and Stewart, J.T., *J. Planar Chromatogr.-Mod. TLC*, 10, 222–224, 1997. With permission.)

Maximum R_f differences between D- and L-isomers ranged from 0.06 to 0.22 at the optimum mole fraction of methanol/buffer for each compound, which were determined to be 0.07–0.53 [51].

C-8F HPTLC layers (Merck) developed with 24 ml of water/isopropanol (70:30 + 0.3 g NaCl) in a saturated chamber served to separate pindolol isomers that were derivatized using 2,3,4,6-tetra-O-acetyl-β-D-glucopyranosylisothiocyanate (GITC). Detection was carried out by densitometric scanning of UV absorbance (Scanner II; Camag, Wilmington, NC, USA). The separation of the enantiomers is shown in Figure 3.7. Quantification of the enantiomers is described in Section 3.5.2 [52].

3.5   QUANTITATIVE ANALYSIS OF TLC-SEPARATED ENANTIOMERS BY DENSITOMETRY

Densitometry is invaluable for determining detection limits and relative amounts of separated isomers based on *in situ* scanning of the visible absorption, UV absorption, or fluorescence of zones. Examples of procedures that can be used are described in the following two sections.

3.5.1   CHIRALPLATES

Figure 3.2 shows the separation of D- and L-*tert*-leucine enantiomers on a Chiralplate and Figure 3.8 shows the calibration plot prepared for quantification. Low

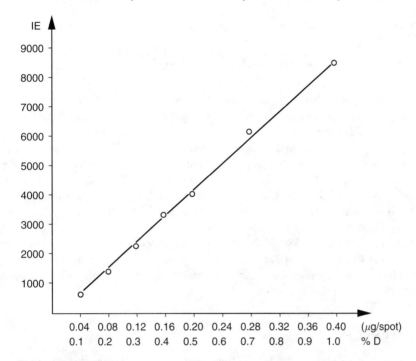

FIGURE 3.8 Calibration plot for D-*tert*-leucine. IE = integration units (peak area); regression equation: $y = -375 + 21.984x$; correlation coefficient (r) = .9980; slope = .0060 μg/spot. (Reprinted from Guenther, K. and Moeller, K., in *Handbook of Thin Layer Chromatography*, 3rd edn., Sherma, J. and Fried, B., Eds., Marcel Dekker, Inc., New York, NY, 2003, pp. 471–533. With permission.)

microliter volumes of test sample and standard solutions were applied manually with Microcaps (Drummond, Broomall, PA, USA), or with a Nanomat II or Linomat IV instrument (Camag). TLC was carried out as described in Section 3.2.4, and zones were scanned using a Shimadzu (Columbia, MD, USA) CS 930 or Desaga (Wiesloch, Germany) CD 60 densitometer at 520 nm [5]. It can be seen that the analysis of the D-enantiomer in the L-amino acid is highly sensitive.

3.5.2 C-8 CHEMICALLY BONDED SILICA GEL

Pindolol enantiomers separated on C-8 silica gel (Figure 3.7) were quantified by densitometry [52]. Samples and standards (9 μl) were applied by means of a Linomat IV instrument (Camag) in 5 mm bands, developed as described in Section 3.4.2, and scanned at 256 nm in the absorbance mode with a slit 4 mm long and 3 mm wide (aliquot scanning method). To determine the amount of each enantiomer in each band of a drug substance or dosage form, the assay utilized external standard methodology:

$$\% \text{ Enantiomer} = \frac{[(PH_{sam}/PH_{std})(SM)]}{2},$$

where PH_{sam} is the height of the sample scan peak, PH_{std} the height of the standard peak, and SM the standard mass of the enantiomer in the standard solution (ca. 1 mg). The linear regression graphs for R- and S-pindolol were prepared in the 60–120 ng/band range, and respective r-values were .9967 and .9995. Assay of samples spiked at 80 and 110 ng/band gave recoveries of $103.2\pm0.93\%$ and $104.8\pm 0.22\%$ ($n = 3$) for R-pindolol and $104.9 \pm 0.50\%$ and $104.0 \pm 0.32\%$ ($n = 3$) for S-pindolol, respectively. For actual assay of R- and S-pindolol in pharmaceuticals, the GITC-derivatized sample (9 μl) was applied in 5 mm bands. HPTLC analysis ($n = 4$) showed S-($-$)-pindolol and R-($+$)-pindolol present in drug substance at levels of 51.51 ± 0.58 and $49.04 \pm 0.39\%$, respectively, and in dosage form 51.32 and 49.24%, respectively. The same approach used for this indirect analysis involving prior chemical derivatization can be used for quantification in direct enantiomer analyses.

REFERENCES

1. Sajewicz, M., Pietka, R., and Kowalska, T., *J. Liq. Chromatogr. Relat. Technol.*, 28, 2499–2513, 2005.
2. *TLC Catalog*, EMD Chemicals Inc. (A Division of Merck KGaA, Darmstadt, Germany), 480 S. Democrat Road, Gibbstown, NJ, October 2005 revision.
3. Maxwell, R.J. and Lightfield, A.R., *J. Planar Chromatogr.-Mod. TLC*, 12, 109–113, 1999.
4. Reich, E. and Schibli, A., *J. Planar Chromatogr.-Mod. TLC*, 17, 438–443, 2004.
5. Guenther, K. and Moeller, K., Enantiomer separations, in *Handbook of Thin Layer Chromatography*, 3rd edn., Sherma, J. and Fried, B., Eds., Marcel Dekker, Inc., New York, NY, 2003, pp. 471–533.
6. Hahn-Dienstrop, E., *J. Planar Chromatogr.-Mod. TLC*, 6, 313–318, 1993.
7. Bereznitski, Y., Thompson, R., O'Neill, E., and Grinberg, N., *J. AOAC Int.*, 84, 1242–1251, 2001.
8. Dent, C.E., *Biochem. J.*, 43, 169–180, 1948.
9. Lederer, M., *J. Chromatogr.*, 510, 367–371, 1990.
10. Yuasa, S., Shimada, A., Kameyama, K., Yasui, M., and Adzuma, K., *J. Chromatogr. Sci.*, 18, 311–314, 1980.
11. Bach, K. and Haas, H.J., *J. Chromatogr.*, 136, 186–188, 1977.
12. Lepri, L., Del Bubba, M., and Cincinelli, A., Chiral separations by TLC, in *Planar Chromatography, A Retrospective for the Third Millennium*, Nyiredy, Sz., Ed., Springer Scientific Publisher, Budapest, Hungary, 2001, pp. 518–549.
13. Hesse, G. and Hagel, R., *Chromatographia*, 6, 277–280, 1973.
14. Lepri, L., Del Bubba, M., and Masi, F., *J. Planar Chromatogr.-Mod. TLC*, 10, 108–113, 1997.
15. Macherey-Nagel Application Database (http://www.mn-net.com), TLC Application No. 400560.

16. Faupel, M., *Proc. Fourth Int. Symp. Instrumental HPTLC*, Selvino, Italy, 147, 1987.
17. Poole, C.F. and Poole, S.K., *Anal. Chem.*, 66, 27A–37A, 1994.
18. Bhushan, R. and Martens, J., Amino acids and their derivatives, in *Handbook of Thin Layer Chromatography*, 3rd edn., Sherma, J. and Fried, B., Eds., Marcel Dekker, Inc., New York, NY, 2003, pp. 373–415.
19. Lucic, B., Radulovic, D., Vujic, Z., and Agbaba, D., *J. Planar Chromatogr.-Mod. TLC*, 18, 294–299, 2005.
20. Witherow, L., Spurway, T.D., Ruane, R.J., Wilson, I.D., and Longdon, K., *J. Chromatogr.*, 553, 497–501, 1991.
21. Mack, M., Hauck, H.E., and Herbert, H., *J. Planar Chromatogr.-Mod. TLC*, 1, 304–308, 1988.
22. Darula, Z., Torok, G., Wittmann, G., Mannekens, E., Iterbeke, K., Toth, G., Tourwe, D., and Peter, A., *J. Planar Chromatogr.-Mod. TLC*, 11, 346–349, 1998.
23. Kowalska, T., Sajewicz, M., Nishikawa, S., Kus, P., Kashimura, N., Kolodziejczyk, M., and Inoue, T., *J. Planar Chromatogr.-Mod. TLC*, 11, 205–210, 1998.
24. Polak, B., Golkiewicz, W., and Tuzimski, T., *Chromatographia*, 63, 197–201, 2006.
25. *Solutions for Chiral Chromatography*, brochure available from Macherey-Nagel, Dueren, Germany, pp. 22–25.
26. Nyiredy, Sz., Dallenbach-Toelke, K., and Sticher, O., *J. Chromatogr.*, 450, 241–252, 1988.
27. Poole, C.F., *The Essence of Chromatography*, Elsevier, Amsterdam, Netherlands, 2003, pp. 499–567.
28. Hao, A.-Y., Tong, L.-H., Zhang, F.-S., Gao, X.-M., and Inou, Y., *Anal. Lett.*, 28, 2041–2048, 1995.
29. Duncan, J.D., Armstrong, D.W., and Stalcup, A.M., *J. Liq. Chromatogr.*, 13, 1091–1103, 1990.
30. Armstrong, D.W., He, F.-Y., and Han, S.M., *J. Chromatogr.*, 448, 345–354, 1988.
31. Lepri, L., Coas, V., Desideri, P.G., and Checchini, L., *J. Planar Chromatogr.-Mod. TLC*, 3, 311–316, 1990.
32. Lepri, L., Del Bubba, M., Coas, V., and Cincinelli, A., *J. Liq. Chromatogr. Relat. Technol.*, 22, 105–118, 1999.
33. Lepri, L., Boddi, M., Del Bubba, M., and Cincinelli, A., *Biomed. Chromatogr.*, 15, 196–201, 2001.
34. Lepri, L., Coas, V., and Desideri, P.G., *J. Planar Chromatogr.-Mod. TLC*, 5, 175–178, 1992.
35. Lepri, L., Coas, V., Desideri, P.G., and Pettini, L., *J. Planar Chromatogr.-Mod. TLC*, 5, 364–367, 1992.
36. Lepri, L., Coas, V., Desideri, P.G., and Santianni, D., *Chromatographia*, 36, 297–301, 1993.
37. Lepri, L., Coas, V., and Desideri, P.G., *J. Planar Chromatogr.-Mod. TLC*, 5, 294–296, 1992.
38. Lepri, L., Coas, V., Desideri, P.G., and Zocchi, A., *J. Planar Chromatogr.-Mod. TLC*, 5, 234–238, 1992.
39. Duncan, J.D. and Armstrong, D.W., *J. Planar Chromatogr.-Mod. TLC*, 3, 65–67, 1990.
40. Armstrong, D.W. and Zhou, Y.W., *J. Liq. Chromatogr.*, 17, 1695–1707, 1994.

41. Tivert, A.-M. and Backman, A., *J. Planar Chromatogr.-Mod. TLC*, 2, 472–473, 1989.
42. Tivert, A.-M. and Backman, A., *J. Planar Chromatogr.-Mod. TLC*, 6, 215–219, 1993.
43. Barooshian, A.V., Lautenschleger, M.J., and Harris, W.G., *Anal. Biochem.*, 49, 569–571, 1972.
44. Eskes, D., *J. Chromatogr.*, 117, 442–444, 1976.
45. Slegel, P., Vereczkey-Donath, G., Ladanyi, L., and Toth-Lauritz, M., *J. Pharm. Biomed. Anal.*, 5, 665–673, 1987.
46. Comber, R.N. and Brouillette, W.J., *J. Org. Chem.*, 52, 2311–2314, 1987.
47. Pflugmann, G., Spahn, H., and Mutschler, E., *J. Chromatogr.*, 416, 331–339, 1987.
48. Guebitz, G. and Mihellyes, S., *J. Chromatogr.*, 314, 462–466, 1984.
49. Rossetti, V., Lombard, A., and Buffa, M., *J. Pharm. Biomed. Anal.*, 4, 673–676, 1986.
50. Buyuktimkin, N. and Buschauer, A., *J. Chromatogr.*, 450, 281–283, 1988.
51. Ruterbories, K.J. and Nurok, D., *Anal. Chem.*, 59, 2735–2736, 1987.
52. Spell, J.C. and Stewart, J.T., *J. Planar Chromatogr.-Mod. TLC*, 10, 222–224, 1997.

4 Planar Chromatographic Enantioseparations on Noncommercial CSPs

Luciano Lepri, Alessandra Cincinelli, and Massimo Del Bubba

CONTENTS

4.1 INTRODUCTION

Till now, numerous chiral stationary phases (CSPs) were used for the direct separation of optical isomers in planar chromatography, as reported in various reviews published in the past 10 years [1–3].

However, only few stationary phases (i.e., cellulose and cellulose triacetate) have been commercialized as precoated plates because of the high cost of most chiral adsorbents and the difficulty to detect analytes on stationary phases showing a strong adsorbance during exposition to UV lights.

The commercial plates should be preferred to the homemade plates, because they offer more reproducibility and do not need further treatments.

However, these plates could have low versatility depending on kind or composition of the eluent or could be constituted by chiral adsorbents with low enantioselectivity. Taking into account all these reasons, noncommercial plates are useful to resolve racemic compounds, which otherwise would not be resolved by this technique, to better understand enantioresolution mechanisms of chiral adsorbents in different experimental conditions and, therefore, to stimulate industrial production of cheap and performant chiral precoated plates.

4.2 CELLULOSE

The first chiral plates were those on cellulose [4–6], which were used to obtain the same paper chromatographic separations [1,7–9]. In particular, in 1965, Contractor and Wragg [4] prepared cellulose homemade plates (20×20 cm; Whatman, USA) to resolve racemic amino acids. Later, other researchers [5] used noncommercial microcrystalline cellulose plates (20×20 cm) to separate optical isomers of tryptophan and kynurenine, and their derivatives (see Table 4.1). All the investigated racemates were well resolved showing α-values in the range 1.13 to 1.60. The enantiomeric amino acid sequence was reversed by changing the kind of the eluent.

With respect to kynurenine and its derivatives, the D-isomer resulted more retained respect to the L-isomer even on commercial cellulose plates (Merck, Germany) eluted with methanol/water (3:2, v/v) [10]. On homemade microcrystalline cellulose plates (5×20 cm) the enantiomers of D,L-trisethylene–diamine–cobalt (III) complex were resolved with a separation factor $\alpha = 2.56$ [6]. Further studies were performed on commercially available native and microcrystalline cellulose plates, which substituted the noncommercial layers keeping the same performances and high resolution of the homemade plates [1].

TABLE 4.1

Retention (hR_{f1}, hR_{f2}) and Resolution (α) Data for Some Enantiomers on Noncommercial Cellulose Plates

Racemate	hR_{f1}*	hR_{f2}*	α**	Eluents and remarks	Ref.
D,L-Tryptophan (Trp)	52 (L)	62 (D)	1.50	Plate: cellulose powder (Whatman)	[4]
D,L-5-HydroxyTrp	47 (L)	54 (D)	1.32	Butanol/pyridine/water (1:1:1, v/v/v)	
D,L-6-HydroxyTrp	36 (L)	41 (D)	1.23	Development time: 1 h Detection: Ehrlich's reagent	
D,L-Trp	50 (D)	53 (L)	1.13	Plate: microcrystalline cellulose (Avicel SF, Japan)	[5]
D,L-5-HydroxyTrp	25 (D)	31 (L)	1.35	Methanol/butanol/benzene/water	
D,L-Kynurenine	54 (D)	61 (L)	1.33	(2:1:1:1, v/v/v/v) and other mixtures	
D,L-3-Hydroxy-kynurenine	47 (D)	53 (L)	1.27	Development time: 2 h 30 min	
D,L-5-Hydroxy-kynurenine	20 (D)	26 (L)	1.40	Detection: UV (365 nm)	
D,L-3-Methoxy-kynurenine	55 (D)	62 (L)	1.34		
D,L-N-a-Acetyl-kynurenine	74 (D)	82 (L)	1.60		

*$hR_f = R_f \times 100$.

**$\alpha = [(1/R_{f1}) - 1]/[(1/R_{f2}) - 1]$.

4.3 CELLULOSE TRIACETATE

The first paper on noncommercial plates of cellulose triacetate for the Tröger's base enantiomeric separation was published in 1973 [11]. Later, applications on commercial cellulose triacetate plates containing a fluorescent indicator (Antec, Switzerland) and on cellulose triacetate at 20, 30, and 40% acetylation degree (w/w) were carried out. However, the use of these plates was limited by their little success using different solvent systems.

It was noted that cellulose triacetate tends to swell with a large number of solvents and even if this characteristic plays an important role in the enantiomer resolution, it could determine the stop of the eluent during the chromatographic development.

A comparison between retention and resolution factors of commercial and noncommercial plates of acetylated cellulose using ethanol/water (80:20, v/v) as mobile phase was performed by Lepri et al. [12] in 1994. Results of this study are reported in Table 4.2. Ready-to-use cellulose triacetate plates with 20% acetylation degree (AC-20) and 40% acetylation degree (AC-40) showed a different behavior during the development process. In fact, AC-40 plates swelled strongly and stationary phase broke away from the glass support and could not be used. On the contrary, AC-20 plates did not show this inconvenience, but their use was limited to the resolution of four racemates, such as 1,1'-binaphthyl-2,2'-diamine,

TABLE 4.2

Retention (hR_{f1}, hR_{f2}) and Resolution (α, R_S) Data of Racemic Solutes on Commercial and Noncommercial Acetylated Cellulose Plates Eluted with Ethanol/Water (80:20, v/v) at 25°C [12]

Racemate	AC-20 precoated plates[a]				AC-20 + CMC[b]				MCTA + CMC[c]				MCTA + Silica gel 60GF$_{254}$[d]			
	hR_{f1}*	hR_{f2}*	α**	R_S***	hR_{f1}*	hR_{f2}*	α**	R_S***	hR_{f1}*	hR_{f2}*	α**	R_S***	hR_{f1}*	hR_{f2}*	α**	R_S***
1,1'-Binaphthyl-2,2'-diamine	33 (R)	40 (S)	1.35	1.0	41 (R)	50 (S)	1.43	1.6	10 (R)	20 (S)	2.25	2.0	13 (R)	23 (S)	1.99	3.3
1,1'-Bi-2-naphthol	48	48	1.00	—	64	64	1.00	—	33	33	1.00	—	—	—	—	—
7,8,9,10-Tetrahydrobenzo[α]pyren-7-ol	29	40	1.63	1.6	30	44	1.83	1.6	6	12	2.13	1.4	14	25	2.04	2.8[e]
γ-(2-Naphthyl)-γ-butyrolactone	46	56	1.49	2.4	52	64	1.64	2.6	14	34	3.16	4.5	22	36	1.99	4.6
1-(2-Naphthyl)ethanol	72 (S)	74 (R)	1.10	0.5	79 (S)	82 (R)	1.21	1.0	52 (S)	62 (R)	1.51	2.2	39 (S)	47 (R)	1.39	3.3[f]
1-(1-Naphthyl)ethanol	70	70	1.00	—	75	75	1.00	—	44	44	1.00	—	33	33	1.00	—[f]
Aminoglutethimide	68	68	1.00	—	75	75	1.00	—	44	48	1.17	1.0	40	44	1.18	1.7[f]
Benzoin	63	63	1.00	—	68 (S)	71 (R)	1.15	0.7	36 (S)	39 (R)	1.13	0.7	26 (S)	29 (R)	1.16	1.2[f]
Benzoin methyl ether	68	68	1.00	—	77	77	1.00	—	43	46	1.13	0.7	30	34	1.20	1.2[f]
Flurbiprofen	70	70	1.00	—	82	82	1.00	—	68	68	1.00	—	45	49	1.17	1.2[f]
3-(4-Nitrophenyl)glycidol	68	68	1.00	—	74	74	1.00	—	49 (−)	52 (+)	1.12	0.6	42 (−)	45 (+)	1.13	1.2[f]

*$hR_f = R_f \times 100$.

**$\alpha = [(1/R_{f1}) - 1]/[(1/R_{f2}) - 1]$.

***$R_S = 2 \times$ (distance between the centers of two adjacent spots)/(sum of the widths of the two spots in the direction of development).

a 300-10/AC-20 plates (20 × 20 cm), Macherey-Nagel, Germany; development time: 150 min; migration distance: 12.5 cm.

b Noncommercial cellulose triacetate (300AC-20; Macherey-Nagel, Germany) plates with sodium carboxymethylcellulose (1.18 mg/g; Carl Schleicher and Schuell, Germany) as binder. Development time: 90 min; migration distance: 13.5 cm.

c Noncommercial microcrystalline cellulose triacetate plates with sodium carboxymethylcellulose as binder (development time: 150 min; migration distance: 10.7 cm).

d Noncommercial microcrystalline triacetate plates with silica gel 60GF254 (Merck) as binder (development time: 150 min; migration distance: 10 cm).

e Ethanol/water (90:10, v/v).

f Ethanol/water (70:30, v/v).

7,8,9,10-tetrahydrobenzo[*a*]pyren-7-ol, γ-(2-naphthyl)γ-butyrolactone, and 1-(2-naphthyl)ethanol, out of the eleven tested. The use of homemade AC-20 plates with carboxymethylcellulose as binder (AC-20 + CMC) gave better resolutions as shown by the higher α-values observed for the aforementioned racemates and racemic benzoin.

Best results were obtained with noncommercial AC-40 + CMC plates and, in particular, on layers of microcrystalline cellulose triacetate (MCTA; Fluka, Switzerland) + CMC for all the investigated racemates with the exception of 1,1'-bi-2-naphthol, 1-(1-naphthyl)ethanol, and flurbiprofen. The addition of silica gel GF$_{254}$ to plates prepared with MCTA and the use of aqueous organic eluents, containing high percentages of water, also yielded the resolution of ($\pm$)-flurbiprofen (see Table 4.2).

4.3.1 PREPARATION OF NONCOMMERCIAL LAYERS OF MCTA + SILICA GEL 60GF$_{254}$

Homemade MCTA plates were made by mixing 3 g silica gel 60GF$_{254}$ (particle size 15 μm; Merck, Germany) with 15 ml distilled water under magnetic stirring for 5 min. Then, 9 g of MCTA for HPLC (particle size <10 μm; Fluka, Switzerland) and 35 ml ethanol were added to the previous suspension, stirred again for 5 min, before spreading the plates (10 × 20 cm or 20 × 20 cm, 0.25 mm thickness) by using a Chemetron automatic apparatus (Camag, Switzerland). Plates were dried at room temperature (20–22°C) and used within few hours (2–5 h). If plates were dried for too long, the chiral recognition of the stationary phase could be reduced and mobile phase could change its migration rate. Ascending development was used in a 22 × 22 × 6 cm (Desaga, Heidelberg, Germany) thermostatic chamber. Moreover, if room temperature was higher than 22°C, a water–propan-2-ol mixture would be suggested during the slurry preparation to avoid a too fast drying of layers, which could involve the formation of breaks in the surface and affect chiral discrimination of the stationary phase.

The spots were detected by UV illumination (λ = 254 or 366 nm) or spraying specific detection reagents on the plates.

For this reason, noncommercial MCTA plates were preferably used for enantiomeric separations with respect to the commercial ones.

4.3.2 ANALYTICAL APPLICATIONS

The enantioselectivity of MCTA is well known, because it was largely used in column chromatography (HPLC) to separate various optical antipodes [13].

On MCTA plates, hydrophobic and polar interactions and hydrogen bonding can be involved and the simultaneous participation of several chiral sites or several polymeric chains is possible.

Regarding results obtained on column chromatography [14,15], the use of *n*-hexane + 2-propanol mixtures as eluents on MCTA plates was not suggested, because elongated spots for all the racemates were observed.

Mixtures of aqueous organic eluents, such as water/methanol, water/ethanol, and water/2-propanol in different ratios, were used as mobile phases, because they showed round and compact spots.

Many different racemates were resolved on these layers as reported in Table 4.3.

The α-values ranged between 1.08 for ($\pm$)-γ-(trityloxymethyl)-γ-butyrolactone and 7.92 for ($\pm$)-2-phenylbutyrophenone with a mean separation factor of 1.58.

Other racemates showing high α-values (≥ 2.00) are Tröger's base (2.64), 1-(9-fluorenyl)ethanol (2.24), phenylthiohydantoin (PTH-Pro) (2.23), 2-phenylcycloheptanone (2.20), 7,8,9,10-tetrahydrobenzo[a]pyren-7-ol (2.04), and trans-4-chlorostilbeneoxide (2.00). It should be noted that α-values of 7.92 are exceptional in chiral planar chromatography, whereas α-values exceeding 12 (ranging between 12.7 and 37) were previously obtained for six specific solutes on MCTA, BSA–silica, and (S)-N-(2-naphthyl)alanine–silica columns [16].

It is worthy to note that α-values for Tröger's base obtained on commercial and noncommercial MCTA plates, using ethanol or ethanol/water (80:20, v/v) as mobile phase, were similar (2.64, 2.67, and 2.96; see Table 4.3). Taking into account the role of chemical properties of solutes on the chiral recognition abilities of MCTA, it should be noted that chromatographic studies on benzoin, benzoin methyl ether, benzoin oxime, and hydrobenzoin evidenced that only the first two, containing a carbonyl group, were resolved and showed that this group played an important role in enantioseparation. Moreover, α-values for methylthiohydantoin-DL-proline (MTH-DL-Pro) and phenylthiohydantoin-DL-proline (PTH-DL-Pro), having two rings fused together, show that a chiral center on a rigid structure plays a key role to improve enantioresolution (see Table 4.3). The separation factor increased as the size of the hydrophobic group at position 3 (methyl group to phenyl group) increased. Methylthiohydantoin-DL-phenylalanine (MTH-DL-Phe) and methylthiohydantoin-DL-tyrosine (MTH-DL-Tyr), in fact, were baseline resolved owing to the presence of an aromatic ring in the side chain of amino acids close to the asymmetric carbon atom. The substitution of a hydrogen atom by a —OH group in the aromatic ring made the optical resolution worse.

Bearing in mind the considerations outlined above, the lack of resolution of methylthiohydantoin-DL-leucine (MTH-DL-Leu) could be attributed to the absence of an aromatic ring in the side chain of the amino acid. This behavior is common to all the methylthiohydantoin derivatives of aliphatic amino acids.

The separation of the 1,1,2-triphenyl-1,2-ethandiol enantiomers was interesting, because this compound contains two —OH groups, as the hydrobenzoin, but no carbonyl groups. The substitution of a hydrogen atom by a phenyl group increased the hydrophobic characteristic of the compound and made the chiral separation possible.

With regard to this aspect, data reported in Table 4.4 through 4.6 related to homogeneous groups of racemates, as flavanones, oxiranes, and oxazolidinones, could give useful indications. Among the investigated flavanones [23], two different behaviors on MCTA plates could be distinguished. On the contrary to the results for polysubstituted compounds, no chiral discrimination was observed for monosubstituted flavanones, such as 2'-hydroxy, 4'-hydroxy, and 4'-methoxyflavanone,

TABLE 4.3

Retention (hR_{f1}, hR_{f2}) and Resolution (α, R_S) Data for Various Racemic Solutes Partially or Baseline Resolved on Noncommercial Microcrystalline Cellulose Triacetate Plates with Silica Gel 60GF$_{254}$ as Binder

Racemate	hR_{f1}*	hR_{f2}*	α**	R_S***	Eluent	Ref.
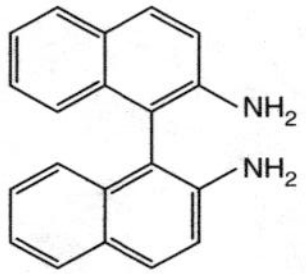 1,1'-Binaphthyl-2,2'-diamine	13 (R)	23 (S)	1.99	3.3	Ethanol/water (80:20, v/v)	[12]
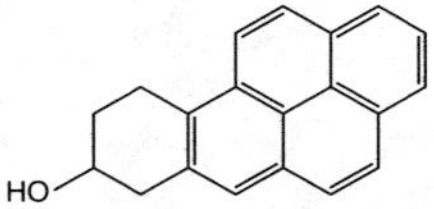 7,8,9,10-Tetrahydrobenzo[a]pyren-7-ol	14	25	2.04	2.8	Ethanol/water (80:20, v/v)	[12]
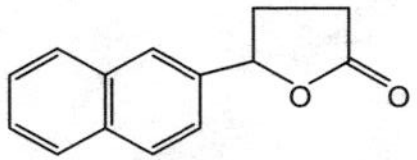 γ-(2-Naphthyl)-γ-butyrolactone	22	36	1.99	4.6	Ethanol/water (80:20, v/v)	[12]
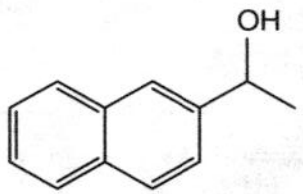 1-(2-Naphthyl)ethanol	15 (S)	20 (R)	1.41	3.2	Ethanol/water (50:50, v/v)	[17]
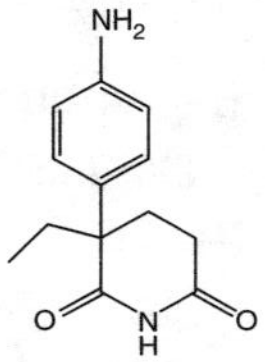 Aminoglutethimide	23	29	1.36	3.0	Ethanol/water (50:50, v/v)	[12]

(Continued)

TABLE 4.3
(Continued)

Racemate	hR_{f1}*	hR_{f2}*	α**	R_S***	Eluent	Ref.
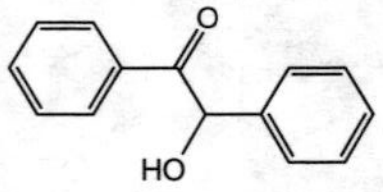 Benzoin	39 (S)	45 (R)	1.27	1.8	2-Propanol/water (80:20, v/v)	[12]
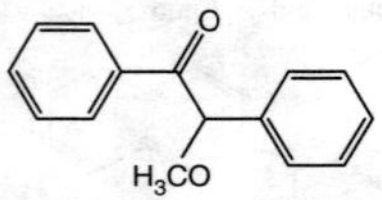 Benzoin methyl ether	47	54	1.32	2.0	2-Propanol/water (80:20, v/v)	[12]
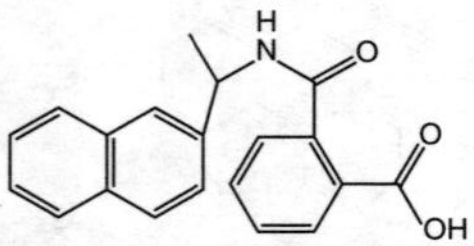 N-[1-(1-Naphthyl)ethyl]phthalamic acid	54 (R)	58 (S)	1.17	1.6	Ethanol/water (70:30, v/v)	[12]
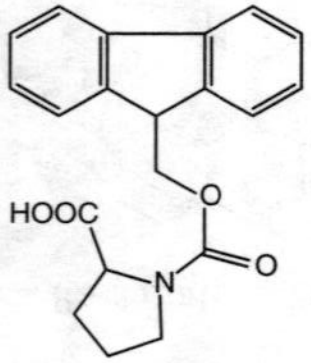 Fmoc-Pro	34 (D)	40 (L)	1.29	2.0	2-Propanol/water (60:40, v/v)	[12]

Compound					Mobile phase	Ref.
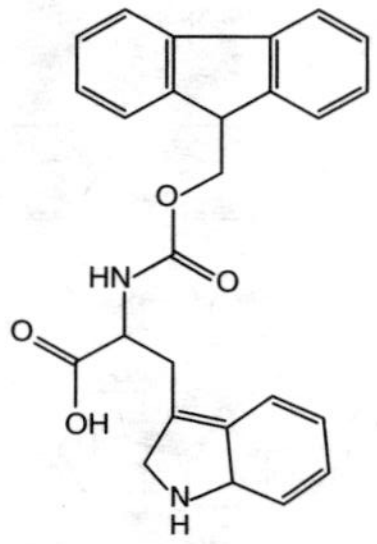 Fmoc-Trp	18 (L)	21 (D)	1.21	0.8	2-Propanol/water (60:40, v/v)	[12]
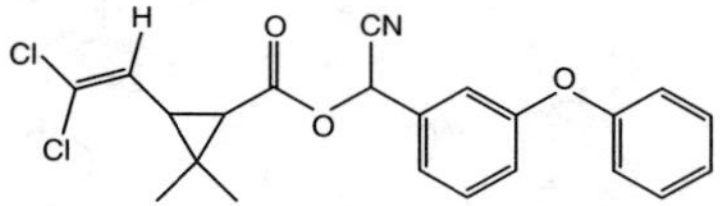 Alphametrin[a]	23	29	1.37	1.7	Ethanol/water (80:20, v/v)	[17]
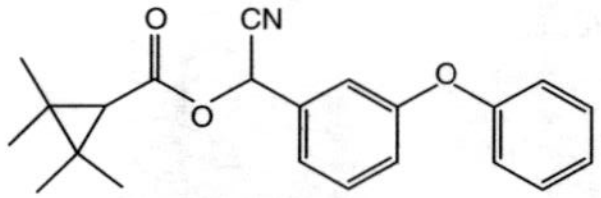 Fenpropathrin	30	34	1.20	1.2	Ethanol/water (80:20, v/v)	[17]
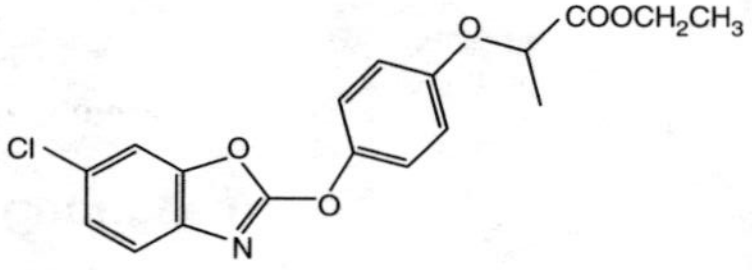 Fenoxaprop-ethyl	36 (R)	46 (S)	1.52	2.2	2-Propanol/water (80:20, v/v)	[17]

(Continued)

TABLE 4.3
(Continued)

Racemate	hR_{f1}^*	hR_{f2}^*	α^{**}	R_S^{***}	Eluent	Ref.
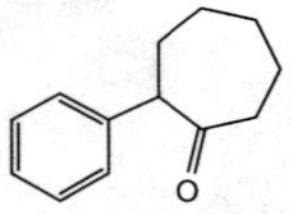 2-Phenylcycloheptanone	17	31	2.20	3.5	2-Propanol/water (60:40, v/v)	[17]
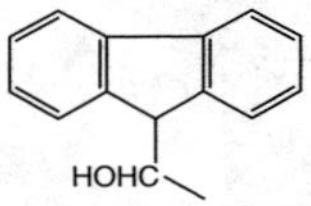 1-(9-Fluorenyl)ethanol	26	44	2.24	3.0	2-Propanol/water (80:20, v/v)	[17]
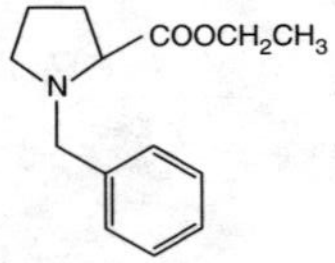 N-Benzylproline ethyl ester	19 (D)	22 (L)	1.20	1.0	2-Propanol/water (40:60, v/v)	[17]
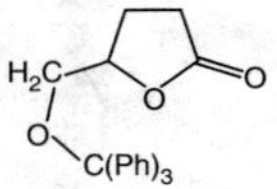 γ-(Trityloximethyl)-γ-butyrolactone	48 (−)	50 (+)	1.08	0.4	2-Propanol/water (70:30, v/v)	[17]
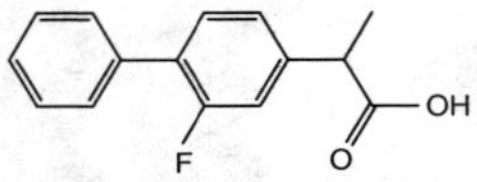 Flurbiprofen	18	24	1.44	2.0	Ethanol/water (40:60, v/v)	[12]

Structure					Mobile phase	Ref.
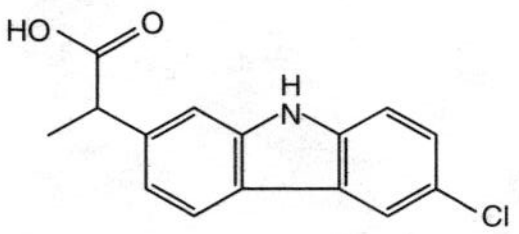 Carprofen	36	41	1.23	1.6	2-Propanol/water (60:40, v/v)	[18]
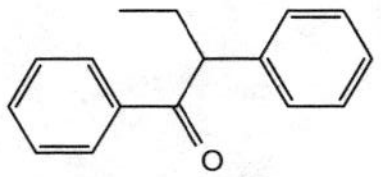 2-Phenylbutyrophenone	19	65	7.92	7.4	2-Propanol/water (80:20, v/v)	[18]
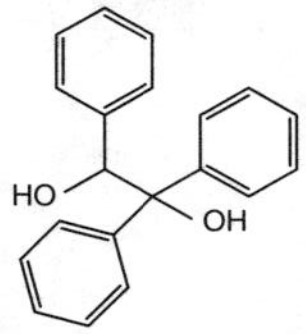 1,1,2-Triphenyl-1,2-ethandiol	34 (R)	42 (S)	1.40	1.8	Ethanol/water (80:20, v/v)	[18]
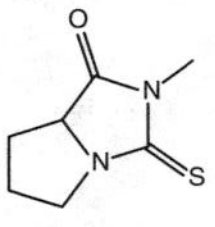 MTH-Proline	33	37	1.19	1.0	2-Propanol/water (60:40, v/v)	[18]
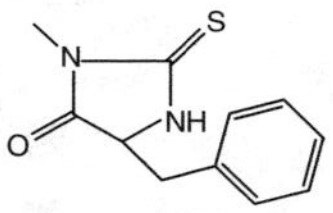 MTH-Phenylalanine	43	49	1.27	1.7	2-Propanol/water (80:20, v/v)	[18]

(Continued)

TABLE 4.3 (Continued)

Racemate	hR_{f1}*	hR_{f2}*	α**	R_S***	Eluent	Ref.
MTH-Tyrosine	42	45	1.13	1.0	2-Propanol/water (60:40, v/v)	[18]
PTH-Proline	13	25	2.23	2.5	2-Propanol/water (60:40, v/v)	[18]
PTH-Phenylalanine	31	36	1.25	1.6	Ethanol/water (80:20, v/v)	[19]
PTH-Tyrosine	53 (L)	64 (D)	1.58	1.6	2-Propanol/water (80:20, v/v)	[19]

Structure	Name					Mobile phase	Ref.
	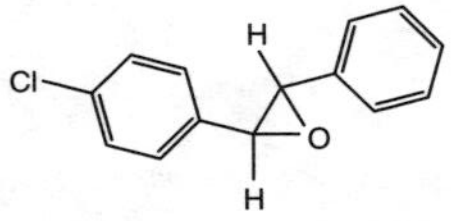 trans-4-Chlorostilbeneoxide	25	40	2.00	4.3	2-Propanol/water (80:20, v/v)	[17]
	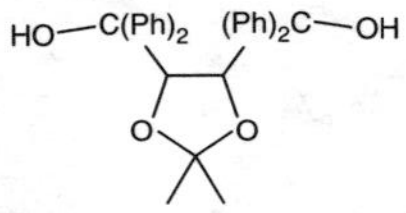 2,3-O-Isopropylidene-1,1,4,4-tetraphenylthreitol	18 (−)	20 (+)	1.14	0.7	2-Propanol/water (50:50, v/v)	[17]
	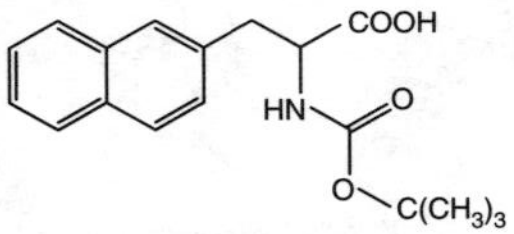 N-tBOC-3-(2-naphthyl)-alanine	69 (D)	72 (L)	1.16	0.8	Ethanol/water (80:20, v/v)	[17]
	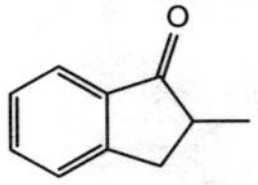 2-Methyl-1-indanone	50	57	1.33	1.8	Methanol/water (80:20, v/v)	[17]
	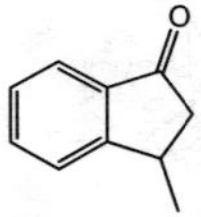 3-Methyl-1-indanone	52	58	1.28	1.6	Methanol/water (80:20, v/v)	[17]

(Continued)

**TABLE 4.3
(Continued)**

Racemate	hR_{f1}*	hR_{f2}*	α**	R_S***	Eluent	Ref.
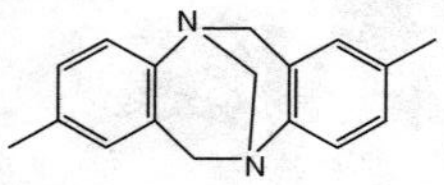 Tröger's base[b]	23 (+) 19 (+)	44 (−) 41 (−)	2.64 2.96	3.6 —	Ethanol/water (80:20, v/v)	[11,17]
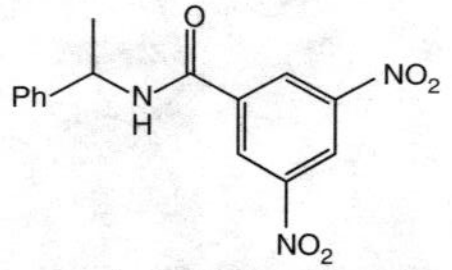 3,5-Dinitro-N-(1-phenylethyl)benzamide	22	24	1.12	0.6	Methanol/water (80:20, v/v)	[17]
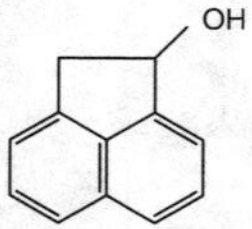 1-Acenaphthenol	57	63	1.28	1.4	2-Propanol/water (80:20, v/v)	[20]
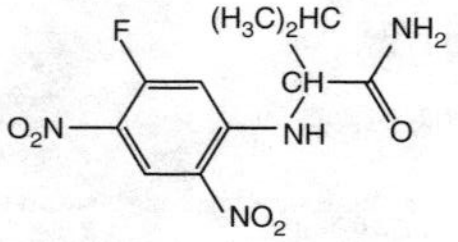 Nα-(2,4-dinitro-5-fluorophenyl)-valinamide	33 (D)	36 (L)	1.14	0.8	Ethanol/water (80:20, v/v)	[21]
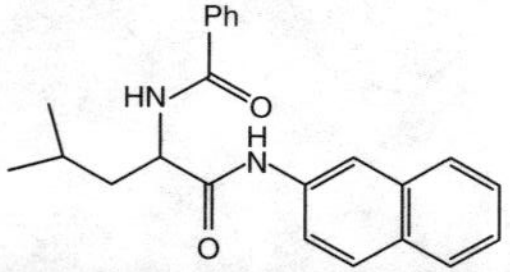	61	66	1.24	1.1	2-Propanol/water (80:20, v/v)	[19]

Compound	hRf[*]	hRf[*]	α[**]	Rs[***]	Solvent	Ref.
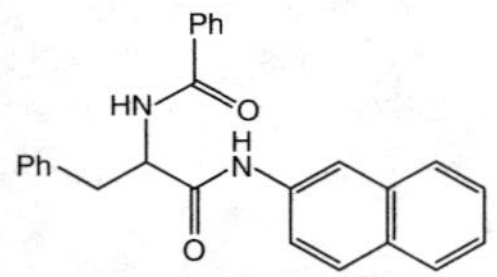 N-Benzoyl-Phe-βNA	38	46	1.39	1.1	2-Propanol/water (80:20, v/v)	[19]
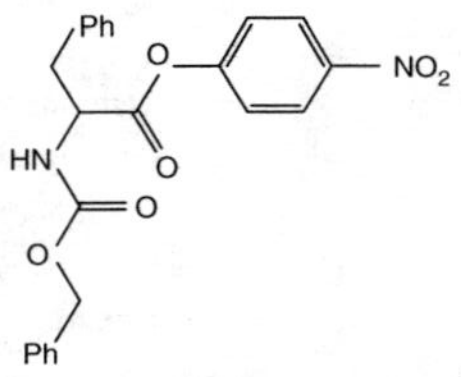 N-CBZ-Phe-ONp	24 (D)	27 (L)	1.17	0.8	2-Propanol/water (80:20, v/v)	[19]
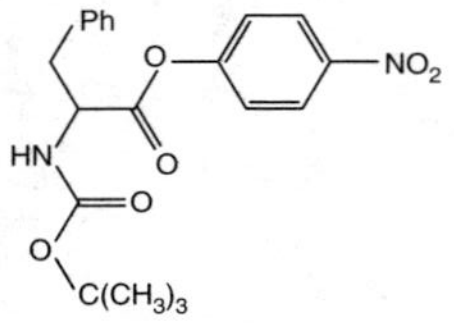 N-tBOC-Phe-ONp	32 (L)	34 (D)	1.09	0.7	Methanol/water (80:20, v/v)	[19]

*hRf = Rf × 100.

**α = [(1/Rf1) − 1]/[(1/Rf2) − 1].

***Rs = 2 × (distance between the centers of two adjacent spots)/(sum of the widths of the two spots in the direction of development).

a Common name alpha-cypermethrin, a racemate comprising (S)-α-cyano-3-phenoxybenzyl (1R,3R)-3-(2,2-dichlorovinyl)-2,2-dimethylcyclopropane-carboxylate and (R)-α-cyano-3-phenoxybenzyl (1S,3S)-3-(2,2-dichlorovinyl)-2,2-dimethylcyclopropanecarboxylate in the direction of development.

b On commercially available 20 × 20 cm cellulose triacetate plates (Antec, Switzerland) the following data were obtained by eluting with ethanol/water (80:20, v/v): hRf(+) = 40; hRf(−) = 64; α = 2.67 [22].

TABLE 4.4

Retention (hR_{f1}, hR_{f2}) and Resolution (α, R_S) Data for Racemic Flavanones Partially or Baseline Resolved on Microcrystalline Cellulose Triacetate Plates with Silica Gel 60GF$_{254}$ as Binder

Structures of racemic flavanones

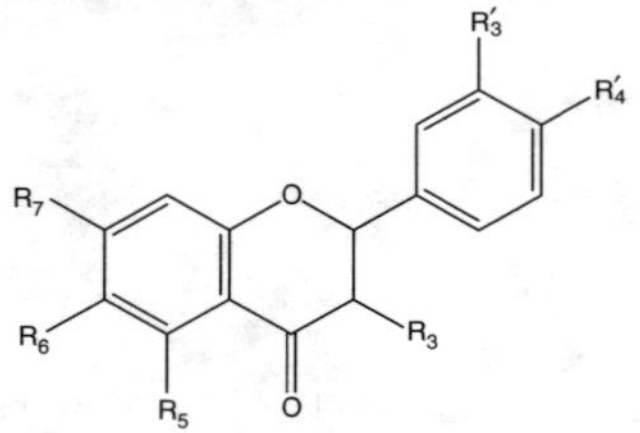

Racemate	R_3	R_5	R_6	R_7	R'_3	R'_4	hR_{f1}*	hR_{f2}*	α**	R_S***	Eluent	Ref.
Flavanone	H	H	H	H	H	H	22	24	1.12	0.4	Ethanol/water (80:20, v/v)[a]	[17]
6-Hydroxyflavanone	H	H	OH	H	H	H	36	39	1.14	0.8	Ethanol/water (80:20, v/v)[a]	[17]
6-Methoxyflavanone	H	H	OCH$_3$	H	H	H	24	27	1.17	0.8	Ethanol/water (80:20, v/v)[a]	[17]
Pinocembrin	H	OH	H	OH	H	H	54	60	1.27	1.8	Ethanol/water (70:30, v/v)[b]	[23]
Naringenin	H	OH	H	OH	H	OH	23	28	1.30	1.6	Methanol/water (80:20, v/v)[c]	[17]
Sakuranetin	H	OH	H	OCH$_3$	H	OH	43	48	1.22	1.2	Ethanol/water (70:30, v/v)[b]	[23]
Isosakuranetin	H	OH	H	OH	H	OCH$_3$	18	21	1.21	1.3	Methanol/water (80:20, v/v)[c]	[23]
Eriodictyol	H	OH	H	OH	OH	OH	26	30	1.21	1.5	Methanol/water (80:20, v/v)[c]	[23]
Homoeriodictyol	H	OH	H	OH	OCH$_3$	OH	23	26	1.17	0.8	Methanol/water (80:20, v/v)[c]	[23]
Hesperetin	H	OH	H	OH	OH	OCH$_3$	23	27	1.24	1.5	Methanol/water (80:20, v/v)[c]	[17]
Taxifolin	OH	OH	H	OH	OH	OH	44	48	1.17	1.3	Methanol/water (80:20, v/v)[c]	[17]

*$hR_f = R_f \times 100$.

**$\alpha = [(1/R_{f1}) - 1]/[(1/R_{f2}) - 1]$.

***$R_S = 2 \times$ (distance between the centers of two adjacent spots)/ (sum of the widths of the two spots in the direction of development).

[a] Migration distance: 12 cm.

[b] Migration distance: 14 cm.

[c] Migration distance: 16 cm.

TABLE 4.5
Retention (hR_{f1}, hR_{f2}) and Resolution (α, R_S) Data for Racemic Oxiranes Partially or Baseline Resolved on Noncommercial Microcrystalline Cellulose Triacetate Plates with Silica Gel 60GF$_{254}$ as Binder

Racemate	hR_{f1}*	hR_{f2}*	α**	R_S***	Ethanol/ water ratio	Ref.
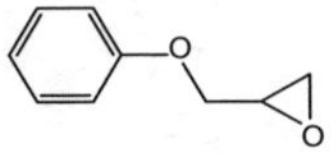 Glycidyl phenyl ether	38	47	1:45	1.9	80:20	[20]
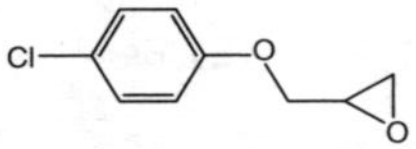 4-Chlorophenyl glycidyl ether	46	50	1:17	1.1	80:20	[20]
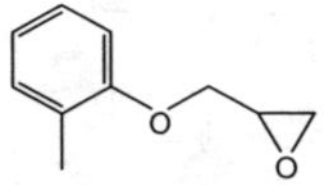 Glycidyl 2-methylphenyl ether	48	51	1:13	1.0	80:20	[20]

(Continued)

TABLE 4.5
(Continued)

Racemate	hR_{f1}^*	hR_{f2}^*	α^{**}	R_S^{***}	Ethanol/water ratio	Ref.
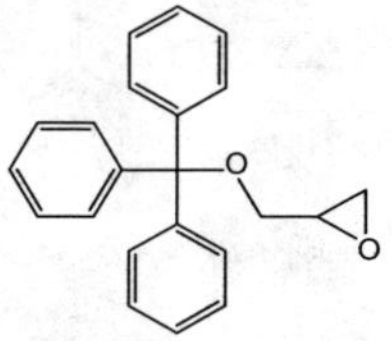 Glycidyl trityl ether	63 (R)	66 (S)	1:14	0.6	80:20	[20]
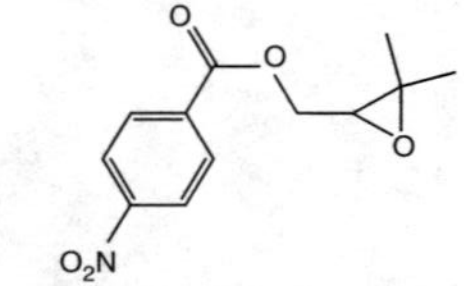 (3,3-Dimethylglycidyl)-4-nitrobenzoate	23 (2R)	29 (2S)	1:37	1.8	80:20	[18]
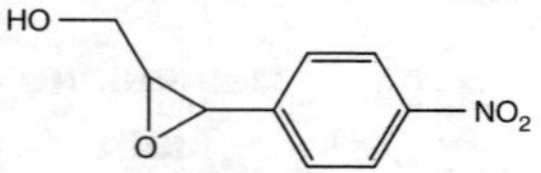 3-(4-Nitrophenyl)glycidol	24 (2S, 3S)	27 (2R, 3R)	1:17	0.8	50:50	[12]

*$hR_f = R_f \times 100$.

**$\alpha = [(1/R_{f1}) - 1]/[(1/R_{f2}) - 1]$.

***$R_S = 2 \times$ (distance between the centers of two adjacent spots)/(sum of the widths of the two spots in the direction of development).

TABLE 4.6
Retention (hR_{f1}, hR_{f2}) and Resolution (α, R_S) Data for Racemic Oxazolidinones and Imidazolidinones Partially or Baseline Resolved on Noncommercial MCTA Plates with Silica Gel GF$_{254}$ as Binder

Racemate	hR_{f1}*	hR_{f2}*	α**	R_S***	Eluent and remarks	Ref.
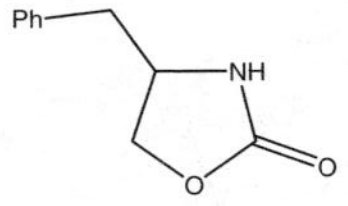 4-Benzyloxazolidin-2-one	35 (S)	37 (R)	1.09	0.6	2-propanol/water (40:60, v/v) Development time: 6 h Temperature: 25°C	[18]
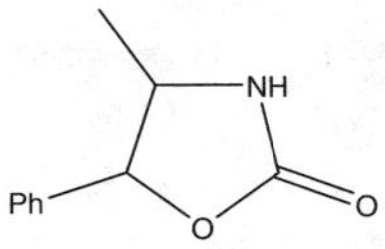 4-Methyl-5-phenyloxazolidin-2-one	62 (4S,5R) (−)	72 (4R, 5S) (+)	1.58	2.6	Methanol/water (80:20, v/v) Migration distance: 16 cm Separation time: 2.5 h Temperature: 25°C	[17]
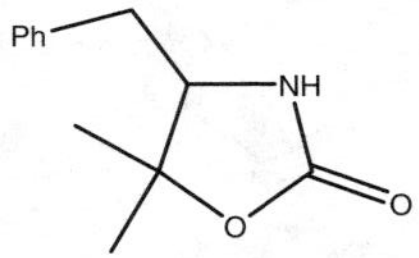 4-Benzyl-5,5-dimethyloxazolidin-2-one	60 (S)	62 (R)	1.08	0.5	Methanol/water (80:20, v/v) Migration distance: 16 cm Separation time: 2.5 h Temperature: 22°C	[21]
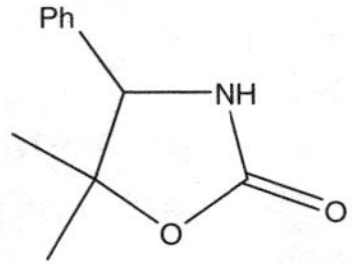 5,5-Dimethyl-4-phenyloxazolidin-2-one	70 (S)	73 (R)	1.16	1.0	Ethanol/water (80:20, v/v) Migration distance: 15 cm Separation time: 3 h Temperature: 22°C	[21]

(Continued)

TABLE 4.6
(Continued)

Racemate	hR^*_{f1}	hR^*_{f2}	α^{**}	R^{***}_S	Eluent and remarks	Ref.
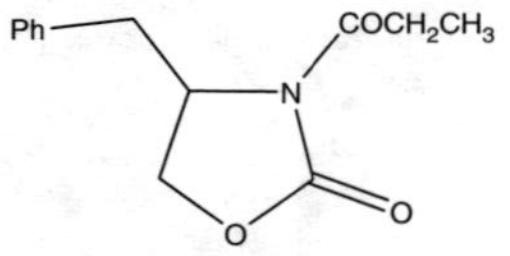 4-Benzyl-3-propionyloxazolidin-2-one	37 (S)	50 (R)	1.70	3.9	Ethanol/water (80:20, v/v) Migration distance: 15 cm Separation time: 3 h Temperature: 22°C	[21]
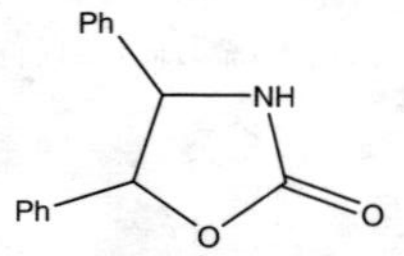 cis-4,5-Diphenyloxazolidin-2-one	54 (4S, 5R) (−)	58 (4R, 5S) (+)	1.18	1.2	Ethanol/water (95:5, v/v) Migration distance: 15 cm Separation time: 3 h Temperature: 22°C	[21]
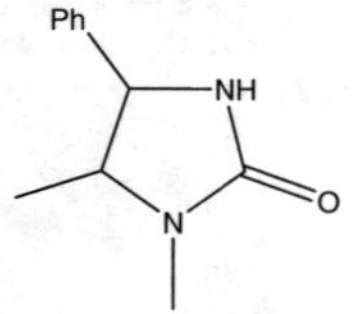 1,5-Dimethyl-4-phenylimidazolidin-2-one	75 (4R, 5S) (−)	86 (4S, 5R) (+)	2.06	2.5	Methanol/water (80:20, v/v) Migration distance: 16 cm Separation time: 2.5 h Temperature: 25°C	[17]

$^*hR_f = R_f \times 100$.

$^{**}\alpha = [((1/R_{f1}) - 1]/[(1/R_{f2}) - 1]$.

$^{***}R_S = 2 \times$ (distance between the centers of two adjacent spots)/(sum of the widths of the two spots in the direction of development).

evidencing that the presence of —OH and —OCH$_3$ groups in these positions were not sufficient for resolution of relative racemates.

A partial resolution was obtained for flavanone, 6-methoxyflavanone, and 6-hydroxyflavanone using ethanol/water (80:20, v/v) as mobile phase but other monosubstituted derivatives in the benzene ring fused with the hetero ring, such as 5-methoxyflavanone and 7-hydroxyflavanone, were not resolved at all.

These results showed that MCTA plates present a chiral discrimination for the flavanone molecule and this particular property does not change even with —OH or —OCH$_3$ groups in position 6.

The resolved polysubstituted flavanones have two hydroxyl groups in positions 5 and 7 with the exception of the 7-methoxy substituted sakuranetin. However, this last compound has a —OH group in position 4′, which could justify the different behavior on MCTA with respect to the not resolved pinocembrin-7-methylether. A comparable behavior was observed on MCTA columns for polysubstituted flavanones [24]. In particular, the comparison of separation factors (α) obtained on MCTA plates with those on MCTA columns, using similar water–alcohol mixtures as mobile phases, evidenced lower α-values on TLC (ranging from 0.4 to 0.8) with respect to those observed on HPLC, except for sakuranetin for which similar α-values (1.22) were obtained by both techniques [3].

An interesting difference was the lack of resolution for flavanone, 6-hydroxyflavanone, and 6-methoxyflavanone on MCTA columns, which were baseline resolved on MCTA plates with two successive developments using ethanol/water (80:20, v/v) [17].

The lower α-values observed on MCTA plates with respect to HPLC could be due to the use of a stationary phase constituted of a mixture 3:1 (w/w) with silica gel instead of pure MCTA.

In fact, when racemates were chromatographated on columns of silica coated with MCTA by HPLC, α-values usually lower than those observed with pure MCTA were achieved. For example, hesperitin was not resolved ($\alpha = 1.01$) on silica coated with MCTA by column chromatography [25], whereas it was well resolved with pure MCTA [24].

Data of closely related oxiranes reported in Table 4.5 evidenced that these compounds could be resolved on MCTA plated eluting with ethanol/water in different ratios.

Under the investigated experimental conditions, racemic benzyl glycidyl ether and 4-nonylphenyl glycidyl ether were not resolved. From the comparison of the hR_f values for the first four oxiranes reported in Table 4.5, it was apparent that less hydrophobic compounds were most retained. The hR_f sequence for oxiranes was trityl > 2-methylphenyl $\cong$ 4-chlorophenyl > phenyl. This trend was exactly the opposite of that predicted for reversed phase chromatography and could be attributed to steric hindrance, which means that some oxiranes were more excluded from the interior of swelled MCTA, the greater was their size. With regard to the effect of the chemical structure on enantiomeric separation, the highest α-value (1.45) was obtained for glycidyl phenyl ether (phenoxymethyloxirane) in accordance with the results obtained on MCTA columns ($\alpha = 1.80$) eluted with 95:5 ethanol/water [13].

The introduction of a substituent group in the aromatic ring reduced the MCTA resolution and such reduction was more pronounced when the group was in *ortho* rather than in *para* position. However, the nonresolution of benzyl glycidyl ether was not explained.

Further information on the role played by the chemical structure of the solute on the chiral resolution of MCTA could be obtained from the data reported in Table 4.6 for a group of differently substituted 2-oxazolidinones and 2-imidazolidinones. This class of compounds was suitable for studies on MCTA plates, because all the enantiomeric pairs were baseline or partially resolved.

Experimental data showed that two successive developments with the same eluent [i.e., ethanol/water (80:20, v/v)] improved the resolution of *cis*-4,5-diphenyl-2-oxazolidinone [$R_f(-) = 0.77$; $R_f(+) = 0.83$, $\alpha = 1.46$, $R_S = 1.6$]. The results reported in Table 4.6 confirmed that compounds having a benzyl or phenyl group in position 4 had a poor chance of enantiomeric discrimination. In fact, 4-benzyl- ($\alpha = 1.09$), 4-benzyl-5,5-dimethyl- ($\alpha = 1.08$), and 5,5-dimethyl-4-phenyl-2-oxazolidinone ($\alpha = 1.16$) showed worse resolution than when the same aromatic group was present at position 5. Moreover, 4-methyl-5-phenyloxazolidin-2-one ($\alpha = 1.58$) and, on MCTA column [26], 5-phenyl-2-oxazolidinone ($\alpha = 1.62$) were baseline resolved. The behavior of *cis*-4,5-diphenyl-2-oxazolidinone was intermediate between the two aforementioned situations.

The best results were obtained when the substituent group was in position 3, such as 4-benzyl-3-propionyl-2-oxazolidinone ($\alpha = 1.70$). An analogous behavior was observed on MCTA columns for 3-isopropyl-5-*p*-(2-methoxyethyl)phenyloxymethyl- and 3-isopropyl-5-(α-naphthyl)oxymethyl-2-oxazolidinone with α-values of 1.84 and 1.39, respectively [27]. In such cases, the substitution of the hydrogen in position 3 was essential for chiral recognition. It should be noted that oxazolidinones obtained by derivatization with phosgene of β-amino-alcohols such as ephedrine, pseudoephedrine, and norephedrine, which were too polar compounds to be directly resolved on MCTA plates, had an aromatic group in position 5, such as those formed by cyclization of β-blockers, which also contained an alkyl group in position 3 (see Figure 4.1).

Therefore, all these oxazolidinones had a very good chance to be optically resolved on MCTA plates. It should be mentioned that oxazolidinones were easily cleaved by dilute alkali to regenerate β-amino-alcohols without racemization.

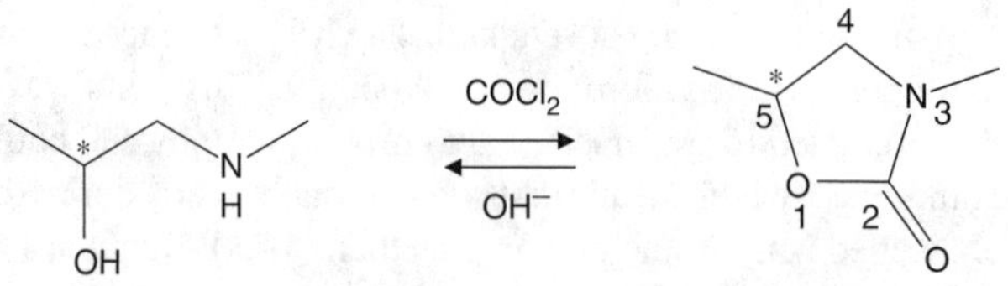

FIGURE 4.1 Structures of β-amino-alcohols and the reactions used to produce oxazolidinone derivatives and to regenerate the original solute.

The results achieved showed that dipole–dipole interactions between the carbonyl group of the solute and the ester carbonyl group of MCTA were essential for enantiomeric resolution even if small variation in the chemical structure of oxazolidinone could strongly reduce the enantioselectivity of this stationary phase.

A review by Siouffi et al. [3] underlined the importance of data obtained on homemade MCTA plates for 64 baseline or partially resolved different racemates because of the low cost and commercial availability of this adsorbent.

Detection and quantitative analysis of separated enantiomers on these plates were carried out by using densitometric methods.

Studies on MCTA plates seem to be interesting not only for the strong correlation between these data and those obtained on MCTA columns, but also because it is possible to obtain optical separations (i.e., flavanone, 6-hydroxyflavanone, and 6-methoxyflavanone) that are not still achieved in column. In addition, on MCTA plates, it was possible to separate some nonsteroidal anti-inflammatory drugs (NSAIDs), such as flurbiprofen and carprofen, without derivatization of the carboxylic group, an essential condition for their resolution on MCTA columns. Such separations are very important if we take into account that these compounds are clinically administered as racemates, but their therapeutic activity is due to only an enantiomer, while the other one is responsible for side effects, such as gastrointestinal irritation [28].

Finally, numerous racemates were resolved on MCTA plates, but, to our better knowledge, no studies on columns filled with the same stationary phase were reported in literature.

4.4   CELLULOSE TRIBENZOATE

Cellulose tribenzoate (CTB) is an excellent stationary phase for chiral resolution of a large number of racemic alcohols and amides by column chromatography [29,30]. However, its use in planar chromatography was strongly limited because of high-UV adsorption background and the difficulty to find inexpensive materials.

The plates coated with this stationary phase were recently prepared in laboratory and proposed by Lepri et al. [31] by using a CTB suspension for column chromatography. Until now, researchers have studied particularly applications about racemic alcohols containing one or more aromatic rings; in such cases, CTB showed high resolution. Precoated plates are not still available to allow a larger use of such chiral adsorbent.

4.4.1   PREPARATION OF NONCOMMERCIAL LAYERS OF CTB + SILICA GEL 60GF$_{254}$

A slurry of 40% tribenzoylcellulose (#02388; Fluka, Switzerland) in methanol/water (95:5, v/v) was filtered off on a fritted glass dish, washed with methanol, and dried at room temperature to constant weight.

TABLE 4.7
The Different Mixtures of CTB/Silica Gel Studied [31]

CTB/silica gel ratio	Amount of silica gel and CTB used
3:1	Silica gel (3 g in 15 ml water) + CTB (9 g in 35 ml 2-propanol)
2:1	Silica gel (4 g in 20 ml water) + CTB (8 g in 25 ml 2-propanol)
1:1	Silica gel (4 g in 20 ml water) + CTB (4 g in 15 ml 2-propanol)
1:2	Silica gel (6 g in 30 ml water) + CTB (3 g in 10 ml 2-propanol)

Plates were prepared by mixing CTB with silica gel 60 GF_{254} (particle size 15 μm; Merck, Germany) in different proportions (see Table 4.7) using the following procedure:

Silica gel was suspended in distilled water followed by 5 min of magnetic stirring; afterwards, the right amount of CTB was added to the suspension together with 2-propanol. After agitation for 5 min, the slurry was transferred, as a 250 μm layer, to 10 × 20 cm plates by using a Chemetron automatic apparatus (Camag, Switzerland).

The plates were dried at room temperature (about 22°C) and used within 2–5 h. Chromatograms were obtained in a 22 × 22 × 6 cm thermostatic chamber (Desaga, Germany), previously saturated with mobile phase vapor for 1 h at 22°C.

Amount of solute applied to the layer: 2–5 μg.

Visualization of solutes was performed by exposing the plates to iodine vapor for 24 h. Some compounds (1-acenaphthenol, 1-indanol, and α-tetralol) appeared as dark brown spots on a paler brown background while the remaining alcohols were detected as white spots on a yellow background after partial removal of the iodine by exposure of the plates to the air. Tröger's base was visualized by UV light ($\lambda = 254$ nm).

Both water–alcohol and n-hexane–2-propanol mixtures were used as eluents on those layers, other than that observed on MCTA plates, where the use of n-hexane–2-propanol mixtures gave rise to more or less elongated spots. For this reason, we can study the behavior of the tested solutes on CTB plates in different experimental conditions, that is, with aqueous and nonaqueous eluents.

4.4.2 ANALYTICAL APPLICATIONS

The CTB/silica gel ratio was the most important parameter for chiral discrimination and solute detection, taking into account that, for CTB/silica gel ratio $\geq$2:1, the difficulty to determine analytes with iodine vapor increased, especially for those compounds that produced white spots on a yellow background. In addition, chiral recognition decreased with decreasing CTB/silica gel ratio and for numerous racemates, no resolution was observed when the ratio is 1:2. Thus, it was important to use plates when the CTB/silica gel ratio was $\geq$1:1. Retention and resolution data of racemic alcohols obtained with n-hexane–2-propanol and 2-propanol–water as

TABLE 4.8
Retention (hR_{f1}, hR_{f2}) and Resolution (α, R_S) Data for Racemic Solutes Partially or Baseline Resolved on Noncommercial CTB/Silica Gel Plates

Racemate	CTB/silica gel ratio	hR_{f1}*	hR_{f2}*	α**	R_S***
Tröger's base	3:1	27 (+)	40 (−)	1.80	3.1
α-Tetralol	2:1	58	68	1.54	3.4
1-Indanol	2:1	55	65	1.52	3.4
1-Acenaphthenol	2:1	41	47	1.27	1.7
1-Phenyl-1,2-ethanediol	2:1	48 (+)	55 (−)	1.32	1.0
trans-2-Phenyl-1-cyclohexanol	2:1	74 (+)	76 (−)	1.11	0.4
Benzoin	1:1	28	36	1.44	1.9
Thiochroman-4-ol	1:1	45	52	1.32	2.2
1-Phenylethanol[a]	1:1	28	30	1.10	0.7
2-Phenyl-3-butyn-2-ol	1:1	39	43	1.18	0.8

The mobile phase was 2-propanol–water (80:20, v/v) for benzoin and thiochroman-4-ol and n-hexane–2-propanol (80:20, v/v) for the others [31,32].

Development distance: 17 cm; temperature: 22°C.

*$hR_f = R_f \times 100$.

**$\alpha = [(1/R_{f1}) - 1]/[(1/R_{f2}) - 1]$.

***$R_S = 2 \times$ (distance between the centers of two adjacent spots)/(sum of the widths of the two spots in the direction of development).

[a] n-hexane–2-propanol (90:10, v/v).

eluents were reported in Table 4.8. The structures of racemic compounds tested were shown in Figure 4.2.

The baseline resolution of the first three alcohols reported in Table 4.8 was also obtained by eluting with different mixtures of n-hexane–2-propanol and isopropanol–water (80:20, v/v). It should be noted that enantiomers of Tröger's base and 1-acenaphthenol were also resolved on MCTA plates with α- and R_S-values comparable to those found on CTB plates for alcohol and higher for the base.

The comparison between the separation factors α obtained on CTB plates and on column was affected by the lower content of chiral selector on the plates.

When the CTB/silica gel ratio is high (i.e., 3:1), we obtained similar results using both techniques (see Tröger's base resolution). When the CTB/silica gel ratio is lower than 3:1, α-values obtained by planar chromatography were lower than those achieved on column and were staggered in a constant way [3].

An improvement in the use of CTB layers was obtained from the measure of UV adsorption spectra of CTB [32] that showed a strong absorbance between 200 and 280 nm but no absorbance at higher wavelengths. It gave us the possibility to study and quantitatively determine solutes, such as benzoin and its derivatives, which strongly absorb at $\lambda = 295$ nm and similar to thiochroman-4-ol that showed a significant absorption at 300 nm (see Figure 4.3).

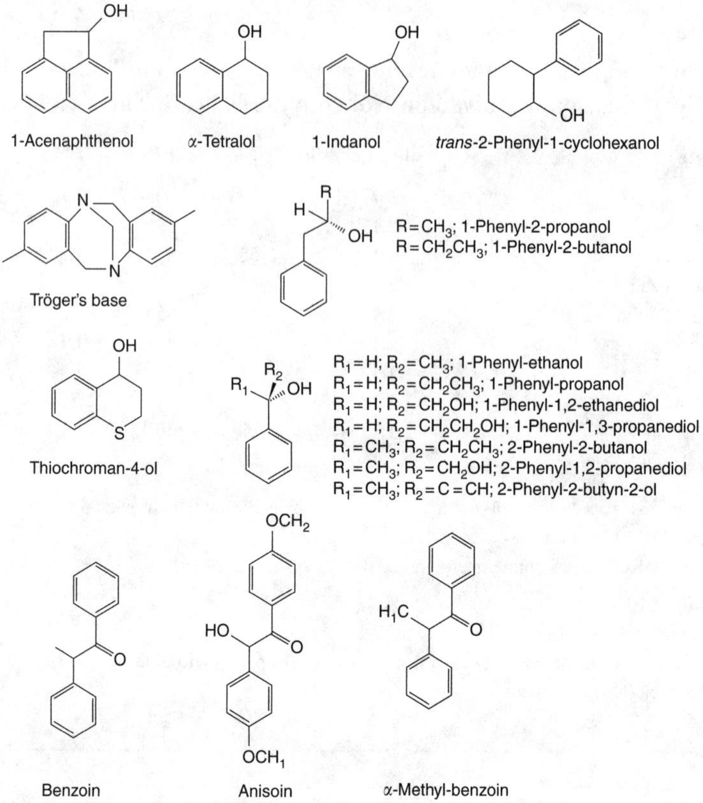

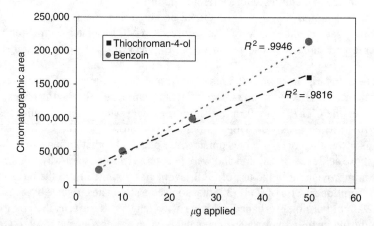

FIGURE 4.2 Chemical structures of racemic alcohols (and Tröger's base) tested on CTB/silica gel plates.

FIGURE 4.3 Sum of the chromatographic peak area of the enantiomers of thiochroman-4-ol and benzoin vs. the amounts applied to 1:2 (w/w) CTB/silica gel layers.

Densitometric measures enabled the determination of 2 μg of each enantiomer. A linear regression was observed plotting the sum of chromatographic area of the two enantiomers as a function of the amount of analyte applied to the layer.

Several studies demonstrated that the chiral recognition mechanism of CTB was based on the formation of inclusion complexes and, simultaneously, of hydrogen bonds between alcoholic hydrogen of the analyte and the carbonyl group of the CTB [33]. The type and shape of substituents attached to the stereogenic center strongly affected the permeability of the enantiomers into chiral cavities and, therefore, resolution.

However, the aforementioned information can give us only preliminary indications about enantioselectivity of CTB because small variations in the chemical structure of solutes might change the CTB resolving abilities without any apparent reason similar to those observed with the good separation of benzoin and the lack of separation of anisoin and α-methyl-benzoin.

Despite the variation in the eluent composition and in the CTB/silica gel ratio, the two last alcohols were not resolved, probably because the researchers were not able to determine the optimum swelling of the sorbent to have their stereospecific inclusion between the laminae of CTB.

4.5   CELLULOSE TRICARBAMATE

Cellulose triphenylcarbamates were widely used to separate a large number of enantiomers in column chromatography [34,35], and only recently these CSPs were applied in planar chromatography.

In 1997, Suedee and Heard [36] proposed the use of various cellulose triphenylcarbamate derivatives to separate some enantiomeric β-blockers on noncommercial plates. Cellulose derivatives tested are as follows: (a) trisphenylcarbamate, (b) *tris*(2,3-dichlorophenyl)carbamate, (c) *tris*(2,4-dichlorophenyl)carbamate, (d) *tris*(2,6-dichlorophenyl)carbamate, (e) *tris*(2,3-dimethylphenyl)carbamate, (f) *tris*(3,4-dichlorophenyl)carbamate, (g) tris(3,5-dichlorophenyl)carbamate, and (h) *tris*(3,5-dimethylphenyl)carbamate.

All these stationary phases show a strong adsorption when exposed to UV lights.

Recently, Kubota et al. [37] prepared cyclohexylcarbamates of cellulose and amylose to evaluate their chiral recognition abilities as stationary phases for HPLC and TLC.

4.5.1   PREPARATION OF CELLULOSE TRICARBAMATE DERIVATIVES AND NONCOMMERCIAL PLATES

Preparation of the previously cited triphenylcarbamates was carried out by reaction of microcrystalline cellulose with the specific phenyl isocyanate in pyridine at 120°C for 6 h, according to the procedure reported by Okamoto et al. [34]. Each product was washed with methanol to remove the unreacted isocyanate.

The cellulose derivatives contained all three —OH groups of each glucose unit replaced by carbamate bonds. For the preparation of plates, 300 mg of each derivative (particle size 10–100 μm) and 300 mg of cellulose microcrystalline were mixed with 3 ml water and a small amount of ethanol, as wetting agent, to obtain a suspension that was stratified on 2.6 × 7.6 cm glass microslides. Plates were dried in an oven at 105°C for 5 min to obtain layers having a thickness of about 0.25 mm [36].

Ascending development was carried out at room temperature in a saturated 250 ml glass beaker containing 10 ml of mobile phase. Migration distance was 6 cm. Spots were detected by drying the plates at 110°C, spraying with anisaldehyde reagent and then heating for further 10–15 min. Solutes appeared as yellow-brown spots on a paler yellow-brown background. The ternary phase ethyl acetate/propan-2-ol/water (65:23:12, v/v/v) and three n-hexane–propan-2-ol binary mixtures, in the ratios of 90:10, 80:20, and 70:30 (v/v), were used as eluents.

Synthesis of cellulose and amylose cyclohexyl carbamates was performed with a similar process: amylose derivative was prepared by dissolving amylose (1 g) in N,N-dimethylacetamide (10 ml), containing LiCl (1 g), at 80°C for 24 h; then, an excess of cyclohexyl isocyanate (3.3 g, 26 mmol) and 20 ml pyridine were added to the amylose solution. The reaction was continued at 80°C for 24 h.

Preparation of the stationary phase was performed by dissolving 0.75 g of amylose *tris*(cyclohexyl carbamate) in 10 ml of tetrahydrofuran. Macroporous silanized silica gel (3 g) [made from silica gel, Daiso Gel SP-1000, pore size 100 nm, particle size 7 μm, silanized using (3-aminopropyl)triethoxysilane in benzene at 80°C] was added to the solution and the resulting suspension was dried under vacuum. A portion of this material (1.5 g) and fluorescent indicator (Merck) (0.1 g) were mixed with methanol (3 ml). The slurry was applied to microslides (2.6 × 7.6 cm, thickness 0.3 mm) and the plates were dried in an oven at 110°C for 30 min [37].

4.5.2 ANALYTICAL APPLICATIONS

Some chromatograms showing the best resolutions obtained on layers of cellulose and amylose tricarbamates (a, b, and c) at different eluent compositions were reported in Figure 4.4.

Cellulose phenylcarbamate derivatives (Figures 4.4a,b) showed high α-values but low R_S-values because of the elongated spots. In particular, R,S-propranolol (Figure 4.4a) yielded $\alpha = 5.5$ and R_S near the unity, whereas bupranolol (Figure 4.4b) showed $\alpha = 2.53$ and R_S-value approaching the unity.

In the case of bupranolol, α-values higher than 4.13 were obtained on the same stationary phase eluting with n-hexane–propan-2-ol (80:20) ($R_{fR} = 0.21$ and $R_{fS} = 0.09$) but resolution was worse because of the formation of very elongated spots.

The use of microcrystalline cellulose and cellulose phenylcarbamate mixtures in the ratio 1:1 (w/w) involved a better adherence of the stationary phase to the glass microslide, and a better visualization of solutes.

(a)   (b)   (c)

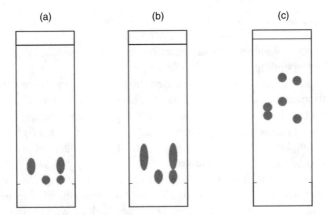

FIGURE 4.4 Chromatograms showing chiral resolution of racemic compounds on (a) *Microcrystalline cellulose*: *tris*(3,5-dimethylphenyl)carbamate of cellulose in the 1:1 (w/w) ratio as stationary phase. *Eluent*: *n*-hexane/propan-2-ol (80:20, v/v). *Visualization*: anisaldehyde reagent. Migration distance = 6 cm at room temperature. Enantiomers and racemic propranolol [36]. (b) *Microcrystalline cellulose*: *tris*(2,3-dimethylphenyl)carbamate of cellulose in the 1:1 (w/w) ratio as stationary phase. *Eluent*: *n*-hexane/propan-2-ol (90:10, w/w). Enantiomers and racemic bupranolol [36]. (c) Aminopropyl silanized silica gel (3 g) coated with *tris*(cyclohexyl)carbamates of amylose (0.75 g) as stationary phase. *Eluent*: *n*-hexane/propan-2-ol (90:10, v/v). *Racemates*: (a) 2,2,2-trifluoro-(9-antryl)ethanol, (b) Tröger's base, and (c) benzoin ethyl ether. *Visualization*: UV (254 nm). Migration distance = 6.2 cm at room temperature [37].

Layers of aminopropyl silanized gel coated with *tris*(cyclohexyl)carbamate of amylose (Figure 4.4c) involved a large number of advantages with respect to the previously described materials, such as a better solute detection owing to the absence of a UV background and the formation of round and compact spots.

Thus, it was also possible to resolve racemic compounds such as 2,2,2-trifluoro-(9-anthryl)ethanol with low separation factor ($\alpha = 1.22$). In the case of Tröger's base ($\alpha = 2.00$) and benzoin ethyl ether ($\alpha = 2.72$), very good resolutions were obtained. It should be noted that α-values are lower than those obtained on column chromatography with the same eluent but have the same sequence [1.32 for 2,2,2-trifluoro-(9-anthryl)ethanol, 2.54 for Tröger's base, and 3.89 for benzoin ethyl ether] [37]. This difference could be related to the larger volume of stationary phase obtained in HPLC with respect to the mobile phase [37].

ΔR_f for enantiomeric couples resolved on *tris*(cyclohexyl)carbamate of amylose were 0.05 for 2,2,2-trifluoro(9-anthryl)ethanol, 0.16 for Tröger's base, and 0.14 for benzoin ethyl ether, respectively.

4.6 CHITIN AND CHITOSAN

These biopolymers, poly(*N*-acetyl-1,4-β-D-glucopyranosamine) (chitin) and poly(1,4-β-D-glucopyranosamine) (chitosan), are widely abundant in nature. Their use as chiral phases was initially studied on HPLC as phenylcarbamates. In particular, a column filled with chitosan phenylcarbamate resulted idoneous to separate 2,2,2-trifluoro-1-(9-anthryl)ethanol enantiomers giving rise to a better resolution than the corresponding cellulose derivative [38].

Rozylo and Malinowska [39] proposed the use of these polymers to separate optical antipodes in TLC in view of the complexing property of Cu(II) retained on the chitin surface, in analogy with chiral ligand exchange chromatography (CLEC). These plates were not considered as CCPs (see Chapter 5) because they were constituted by a chiral sorbent impregnated with an achiral selector, in the opposite way to the CCPs formation.

Results were amazing with α-values for enantiomers of aliphatic amino acids ranging from 4.91 for D,L-leucine to 17.3 for D,L-threonine eluting with methanol/water (1:1, v/v). Excellent separations were obtained with ternary mixtures, such as methanol/water/acetonitrile [(1:1:x, v/v/v), with x between 0.1 and 0.7].

However, such results were uncertain if we take into account that nonenantiomeric glycine gave rise to two well-separated spots of different colors in the same way as racemic amino acids.

Later, the same authors [40] investigated the performances of chitin and chitosan beds modified by different metal ions solutions [Cu(II), Co(II), Ni(II), Hg(II), Ag(I)] for the resolution of five racemic amino acids (threonine, alanine, valine, leucine, and phenylglycine). Results showed that only stationary phases of chitin modified by Cu(II) ions may be used for separation of the aforementioned racemates using both binary and ternary mobile phases. In particular, the best results were obtained with methanol/water/acetonitrile mixtures (1:1:0.1, v/v/v) for which D,L-valine showed the highest α-value ($\alpha = 12.9$) and D,L-phenylglycine the lowest ($\alpha = 4.56$).

Preparation of chitin surface modified by Cu(II) was obtained by mixing 100 g of 20% deacetylated chitin with 500 ml of an aqueous solution of 0.5 M Cu(II).

The suspension was shaken at room temperature for 12 h, filtered, washed with water, and dried at 80°C for 3 h. Plates were made from aqueous suspension of the aforementioned material and dried in the air for 3 days. Chromatographic experiments were carried out at 21°C. Migration distance was 10 cm. Visualization: ninhydrin solution.

4.7 MOLECULAR IMPRINTING TECHNIQUES

The synthesis and characterization of polymers with specific cavities that can recognize only the template molecule miming the enzyme's binding site were first proposed by Wulff and Sarhan [41] in 1972.

The preparation of molecularly imprinted polymers (MIPs) showing chiral activities follows three steps (see Figure 4.5):

1. The chiral template was dissolved in a suitable solvent together with monomers giving rise to a covalent or self-assembly complex.

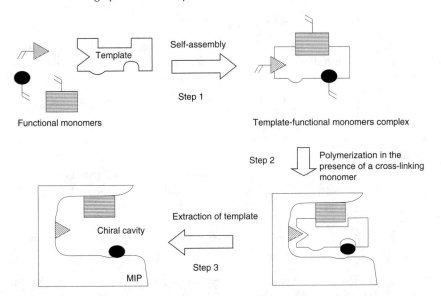

FIGURE 4.5 Schematic representation of the preparation of molecularly imprinted polymers.

2. The assembled complex was incorporated into the polymer matrix by polymerization of the monomers together with a cross-linker.
3. The template was removed from the polymer forming chiral cavities.

There were two different methods for the preparation of site-specific polymers: covalent and self-assembly (noncovalent) approaches. The first method involved the formation of more homogeneous chiral cavities with respect to the second approach, because monomer and template were held together by covalent bonds during polymerization. Disadvantages of this procedure depended on the chance to obtain very elongated or tailing spots in particular for the more-retained optical isomer.

For this reason, the most popular strategy for synthesizing MIPs as stationary phases for molecular recognition was the approach based on noncovalent imprinting in which monomers and templates were held via weak intermolecular interactions such as hydrogen bonding and $\pi-\pi$, dipole, hydrophobic and ionic interactions. After polymerization, the removal of print molecules was carried out by washing with suitable solvents.

The disadvantage of such a method was the possibility to obtain heterogeneous binding sites. In such cases, a lower selectivity and tailing spots were expected.

The first paper on the employment of such technique for the separation of optical isomers by column chromatography was carried out by Wulff et al. [42] in 1977 while more recently (1994) Kriz et al. [43] used MIPs for resolution of racemates by planar chromatography.

4.7.1 PREPARATION OF MIPs STATIONARY PHASES

Scientists used similar approaches to prepare MIPs for planar chromatography. Self-assembly imprinting involved the use of methacrylic acid (MAA) or itaconic acid (ITA) and 4-vinylpyridine (4-VIP), as functional monomers, for basic and acidic templates, respectively.

Taking into account that the MIPs preparation approach was the same both in planar and column chromatography, the optimal experimental conditions to obtain the best chiral discrimination of the polymers in TLC were those used on column chromatography [44,45].

In particular, well-defined ratios between monomer and template concentrations and between monomer/template/cross-linker were proposed by various scientists [44,46] and used to prepare MIPs for planar chromatography.

4.7.1.1 Preparation of Synthetic Polymers Imprinted with L- or D-Phenylalanine Anilide (L- or D-Phe-An) [43]

The composition of the polymerization mixture was as follows: L- or D-Phe-An [43], as print molecule (1.956 mmol); MAA, as functional monomer (7.86 mmol); ethylene glycol dimethacrylate (EDMA), as cross-linker (39.3 mmol); 2,2'-azobis(2-methylpropionitrile) (AIBN), as initiator (0.57 mmol); acetonitrile, as solvent (12 ml). The MAA/print molecule ratio was 4:1.

In this connection, it should be underlined that the best MAA/L-Phe-An ratio found in column chromatography was 4:1, using a mixture of 10% acetic acid in acetonitrile as eluent, while no separation of optical antipodes was observed for 2:1 ratio [44]. In the latter case, a monomer/template complex (1:1) was obtained, while in the former case a 2:1 complex was observed, which was able to give selective binding sites by ionic bonding formation with the free primary amine and hydrogen bonds with amide (see Figure 4.6) [45].

The template was dissolved in an ultrasonic bath and the mixture was then cooled on ice, degassed under vacuum, and then purged with nitrogen for 5 min.

Polymerization was effected by photolytic initiation with UV lights (366 nm) at 4°C for 18 h.

Polymers were ground for 20 min in a mortar; sifted particles were washed with acetonitrile and then floated in the same solvent. The suspension was allowed to settle for 15 min; sediment was discharged and procedure was repeated twice.

Then, the suspension was let to settle for 24 h to obtain the desired fraction.

This fraction was first washed with 20% acetic acid in methanol and then with acetone.

After drying, white particles were obtained (size 1–20 μm).

4.7.1.2 Preparation of Synthetic Polymers Imprinted with Quinine [47]

Methacrylic acid (MAA) and itaconic acid (ITA) were used as functional monomers to obtain two different complexes with quinine since they contained one and two carboxyl groups, respectively.

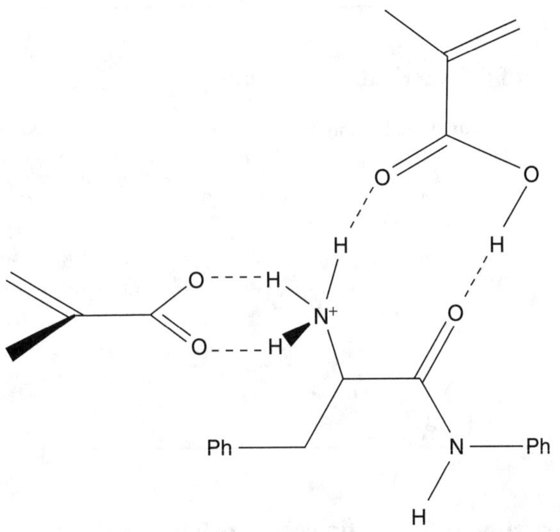

FIGURE 4.6 Chemical structure of 2:1 MAA/L-Phe-An complex, as proposed by Sellergren et al. [45].

The composition of the polymerization mixture was as follows: quinine (0.5 mmol); MAA or ITA (18 mmol); EDMA (540 mmol); AIBN (0.25 mmol); and tetrahydrofuran, as solvent. The MAA or ITA/quinine ratio was 36:1. After degasification and purging under nitrogen for 5 min, polymerization was carried out by photolytic initiation with UV light (λ = 366 nm) at 4°C for 24 h.

Polymer was ground to a fine powder and sieved through a 100 μm mesh. The removal of quinine was performed by immersing the polymer in a solution of 10% acetic acid in acetonitrile for 24 h and then washing with acetonitrile. Finally, the material was dried under vacuum overnight.

4.7.1.3 Preparation of Synthetic Polymers Imprinted with (+)-Ephedrine, (+)-Pseudoephedrine, and (+)-Norephedrine [48]

The composition of the polymerization mixture was as follows: template (3 mmol); MAA (12 mmol); EDMA (310 mmol); AIBN (0.1 mmol); dichloromethane, as solvent (40 ml). Polymers were obtained by thermal polymerization at 40°C for 16 h.

The resulting material was ground and sieved through a 100 μm mesh (size distribution 30–40 μm). The removal of template was performed as described previously [47].

Imprinted polymers were also prepared from (−)-pseudoephedrine and (−)-norephedrine as templates and MAA and ITA as functional monomers [49].

TABLE 4.9
Composition of Polymerization Mixtures

Template (mmol)	Functional monomer (mmol)	EDMA (mmol)	AIBN (mmol)	Solvent (mmol)	Ref.
R-(+)-propranolol[a] (0.56)	MAA (3)	20	0.14	CHCl$_3$ (15)	[50]
S-(+)-naproxen[a] (2.17)	VDP (13)	65	0.76	CHCl$_3$ (20)	[50]
S-(−)-timolol[b] (0.5)	MAA (18)	540	0.25	THF	[51]

[a] Thermal polymerization at 60°C for 6 h under nitrogen.
[b] Photoinitiation at 4°C for 24 h.

4.7.1.4 Preparation of Synthetic Polymers Imprinted with R-(+)-Propranolol, S-(+)-Naproxen, and S-(−)-Timolol [50,51]

The composition of the polymerization mixtures was reported in Table 4.9. All other conditions were analogous to those described previously [47,49].

4.7.2 PREPARATION OF NONCOMMERCIAL LAYERS OF MIPS

Plates were prepared from slurry obtained by mixing the polymer (100 mg) and CaSO$_4$ · 1/2 H$_2$O (100 mg) (gypsum), as binder, with 1.4 ml water and 10 μl ethanol, as wetting agent [43]. After sonication (3 min), the slurry was poured on glass microslides (7.6 × 2.6 cm) and dried at room temperature overnight (thickness about 0.5 mm).

Some plates were made by adding a fluorescent indicator to the previous mixture (Merck).

Suedee et al. [47,49] used similar procedures by mixing polymer (1 g) and gypsum (1 g) with 3 ml water and a drop of ethanol. The 7.6 × 2.6 cm microslides (thickness 0.25 mm) were dried for 24 h at room temperature.

4.7.3 ANALYTICAL APPLICATIONS

Table 4.10 shows the retention and resolution data for L- and D-Phe-An both on MIPs and not on imprinted plates eluting with a mobile phase consisting of 5% acetic acid in acetonitrile, which represented the eluent that gave rise to the highest α-values.

The mobile phase composition was very important in the enantioseparation, because the polymer could swell, resulting in a different accessibility of the cavities with consequent variation in chiral discrimination.

TABLE 4.10
Retention (hR_f) and Resolution (α, R_S) Data for Phenylalanine Anilide Enantiomers, on Plates Containing Imprinted and Nonimprinted Polymers at Room Temperature (20°C)

Template	$hR^*_{f(L)}$	$hR^*_{f(D)}$	α^{**}	R^{***}_S	Visualization
L-phenylalanine anilide	10 (L)	35 (D)	4.86	0.9	Plates were dried, sprayed with
D-phenylalanine anilide	35 (L)	10 (D)	4.86	0.9	0.05% fluorescamine in acetone
None	65 (L)	65 (D)	1.00	—	and observed at UV light (366 nm)

The mobile phase was 5% acetic acid in acetonitrile [43].

*$hR_f = R_f \times 100$.

**$\alpha = [(1/R_{f1}) - 1]/[(1/R_{f2}) - 1]$.

***$R_S = 2 \times$ (distance between the centers of two adjacent spots)/(sum of the widths of the two spots in the direction of development). Calculated from original chromatograms.

An opposite behavior was shown from L- and D-enantiomers of phenylalanine anilide on L- and D-Phe-An MIP plates. Elution time was 7 min. Although very high α-values were obtained, the resolution between the two enantiomers on both MIPs was low, because very elongated spots were obtained, independently on their high or less retention.

Elongated spots were also observed for both enantiomers on nonimprinted layers, probably owing to the high migration rate of the mobile phase (about 1 cm/min).

This behavior could be due to slow mass-transfer kinetics between the stationary and mobile phases according to the presence of remarkable interactions between solutes and the polymer matrix.

Further studies on L-Phe-An MIP plates with the enantiomers of various compounds, such as anilide, dansyl, methyl and ethyl esters, and amide derivatives of amino acids, showed that polymer interacts more weakly with enantiomers of these compounds and, in the case of dansyl derivatives, no chiral resolution was observed, probably due to the dansyl group that blocks the ammine group of phenylalanine.

Thus, the polymer was able to recognize and, in some cases, separate enantiomers of molecules having structures similar to that of original template.

MIPs prepared with quinine as template [47] were employed to separate quinine/quinidine and cinchonine/cinchonidine diastereomers and also ($\pm$)-ephedrine and ($\pm$)-norephedrine enantiomers. Eluents, such as acetic acid/acetonitrile and acetic acid/methanol, were used in various proportions.

UV light ($\lambda = 366$ nm) was used for spot detection, because the residual print molecules that were irreversibly adsorbed on the polymer matrix did not interfere in the detection of analytes.

The best separation of quinine/quinidine was obtained on MIP prepared with MAA as functional monomer eluting with 10% acetic acid in acetonitrile

($\alpha = 2.26$). Elongated spots were observed for both diastereomers with R_f 0.36 (quinine) and 0.56 (quinidine) and R_S 1.2.

Baseline resolution of ($\pm$)-ephedrine was observed in acetonitrile without acetic acid, using MIP made with MAA [R_f values 0.05 and 0.12 for ($-$)-ephedrine and ($+$)-ephedrine, respectively].

These results evidenced that this stationary phase was also able to separate other enantiomers, structurally related to quinine, which contained vicinal amine and hydroxyl functionalities. Taking into account that MAA/quinine ratio was 36:1, the high excess of monomer could be responsible for the formation of strong binding sites in the MIP showing selectivity for the optical antipodes of ephedrine.

Successively, Suedee et al. [48] used MIP plates prepared by thermal polymerization methods and imprinted with ($+$)-ephedrine, ($+$)-pseudoephedrine, and ($+$)-norephedrine for the separation of α-agonists, β-agonists, and β-antagonists using mixtures of 5 or 10% acetic acid in methanol or dichloromethane as mobile phases in planar chromatography. Propanolol, pindolol, and oxprenolol were baseline separated by all the polymers imprinted with the aforementioned α-agonists with α-values ranging from 1.43 to 2.16.

Ninhydrin, acidified potassium permanganate, and anisaldehyde were chosen as detection reagents according to the different classes of investigated compounds. Moreover, the use of acetonitrile as eluent was more appropriate for α- and β-agonists, whereas a less polar solvent (such as dichloromethane) was used for the separation of β-blockers.

Data in Tables 4.11 and 4.12 show that MIPs distinguish between closely related compounds evidencing the key role of specific substituent groups in the enantioseparation of imprinted polymers. In particular, separation of the optical isomers was observed for all α- and β-adrenergic agonists, independently of the number of stereogenic centers and of modification in their chemical structure. No resolution was observed for nadolol, probably excluded from the stereospecific cavity of MIPs owing to its large size. It should be noted that MIP, prepared from ($+$)-norephedrine, failed in the resolution of racemic norephedrine. This result might be correlated with the thermal polymerization method used, which, evidently, produced a polymer with less chiral characteristic similar to that obtained by the photo-initiation method.

MIPs imprinted with ($-$)-norephedrine were able to separate enantiomers of such molecule rather than other α-agonists [49].

For example, best results were obtained for norephedrine enantiomers on plates imprinted with ($-$)-norephedrine, using ITA as functional monomer and eluting with acetic acid (1 M) in methanol. Later, Suedee et al. [50] extended their research to MIPs prepared with (R)-($+$)-propranolol, which separated (R,S)-propranolol giving rise to very tailing spots ($\alpha = 7.11$ and $R_S = 1.0$) (see Table 4.12) and to MIPs imprinted with (S)-($+$)-naproxen that resolved racemic (R,S)-ketoprofen.

To our best knowledge, the last paper on this subject was published by Aboul-Enein et al. [51], who prepared MIPs imprinted with (S)-timolol for the resolution of (R,S)-timolol and other cardiovascular drugs (i.e., propanolol, atenolol, timolol,

TABLE 4.11

Retention (hR_{f1}, hR_{f2}) and Resolution (α, R_S) Data for α-Adrenergic Agonists on MIPs Prepared from (+)-Ephedrine (A), (+)-Pseudoephedrine (B), and (+)-Norephedrine (C) with Thermal Polymerization [48], and (−)-Pseudoephedrine (D) and (−)-Norephedrine (E) with Photo-Initiation [49]

Racemate	MIP	hR_{f1}*	hR_{f2}*	α**	R_S***	Eluents and remarks
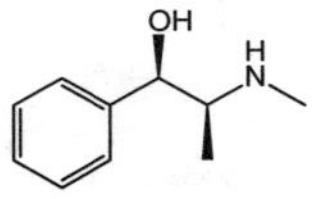 OH / N̄H (Ephedrine)	A	22 (+)	34 (−)	1.82	—	A, B, C: 5 or 10% acetic acid in acetonitrile or dichloromethane
	B	34 (+)	47 (−)	1.73	—	Methacrylic acid as functional monomer [48]
	C	22 (+)	22 (−)	1.00	—	
	D	40 (−)	54 (+)	1.76	—	
	E	32 (−)	44 (+)	1.67	—	
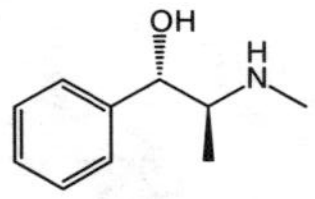 OH / N̄H (Pseudo-ephedrine)	A	31 (−)	51 (−)	2.31	1.2	D, E: 5 or 10% acetic acid in acetonitrile
	B	44 (+)	58 (−)	1.75	—	Itaconic acid as functional monomer [49]
	C	40 (+)	60 (−)	2.24	—	
	D	20 (−)	29 (+)	1.63	—	
	E	74 (−)	88 (+)	2.57	—	

(Continued)

TABLE 4.11
(Continued)

Racemate	MIP	hR_{f1}*	hR_{f2}*	α**	R_S***	Eluents and remarks
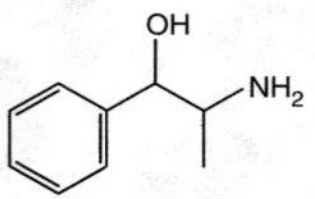 Norephedrine	A	16 (+)	25 (−)	1.75	—	
	B	24 (+)	42 (−)	2.29	1.1	
	C	18 (+)	18 (−)	1.00	—	
	D	22 (−)	24 (+)	1.12	—	
	E	38 (−)	50 (+)	1.63	1.5	
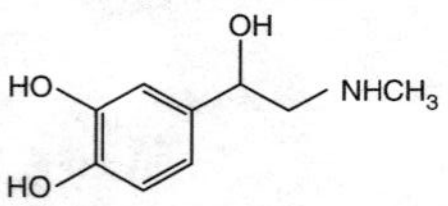 Epinephrine	A	67 (+)	67 (−)	1.00	—	
	B	34 (+)	64 (−)	3.46	—	
	C	31 (+)	31 (−)	1.00	—	
	D[a]	14 (−)	24 (+)	1.94	—	
	E[a]	8 (−)	20 (+)	2.87	—	

* $hR_f = R_f \times 100$.

** $\alpha = [(1/R_{f1}) - 1]/[(1/R_{f2}) - 1]$.

*** $R_S = 2 \times$ (distance between the centers of two adjacent spots)/(sum of the widths of the two spots in the direction of development). Calculated from original chromatograms.

[a] Eluent: 1% acetic acid in acetonitrile.

TABLE 4.12

Retention (hR_{f1}, hR_{f2}) and Resolution (α, R_S) Data for Racemic β-Blockers on MIPs Prepared from (+)-Ephedrine (A), (+)-Pseudoephedrine (B), (+)-Norephedrine [48] (C), (+)-Propranolol [50] (D), and (−)-S-Timolol [51] (E)

Racemate	MIP	hR_{f1}^{*}	hR_{f2}^{*}	α^{**}	R_S^{***}	Eluents and remarks
Propanolol	A	56 (−)	73 (+)	2.16	1.0	A, B, C: 5 or 7% acetic acid in dichloromethane
	B	45 (−)	54 (+)	1.43	—	Detection: anisaldehyde reagent (blue or purple-blue spots)
	C	57 (−)	71 (+)	1.84	—	Thermal polymerization at 40°C for 16 h [48]
	D	32 (+)	77 (−)	7.11	1.0	
	E	29 (−)	44 (+)	1.92	—	
Pindolol	A	14	22	1.73	—	D: 5% acetic acid in acetonitrile
	B	24	36	1.48	—	Detection: anisaldehyde reagent
	C	31	40	2.16	—	Thermal polymerization at 60°C for 6 h [50]
Oxprenol	A	40	53	1.69	1.2	E: 5% acetic acid in acetonitrile
	B	44	58	1.76	—	Detection: UV light (254 nm)
	C	42	60	2.07	—	Photopolymerization at 4°C [50]
Timolol	E	28 (−)	45 (+)	2.10	—	
Atenolol	E	29 (−)	46 (+)	2.08	—	

Solute: 1 μg of 3 mg/ml solution in methanol or 70% ethanol.

$^{*}hR_f = R_f \times 100$.

$^{**}\alpha = [(1/R_{f1}) − 1]/[(1/R_{f2}) − 1]$.

$^{***}R_S = 2 \times$ (distance between the centers of two adjacent spots)/(sum of the widths of the two spots in the direction of development). Calculated from original chromatogram.

nadolol, nifepidine, and verapamil). Some of these separations are reported in Table 4.12.

4.8 β-CYCLODEXTRIN (β-CD)-BONDED STATIONARY PHASES

As other stationary phases used in chiral planar chromatography, β-CD-bonded TLC plates were also prepared [52] after the success obtained by this chiral phase in column chromatography to resolve a large number of optical isomers [53,54].

In addition to the paper of Alak and Armstrong [52] in 1986, only a second work was published on this subject [55], probably because analogous results could be obtained in planar chromatography using precoated silanized silica gel plates and chiral mobile phase containing β-CD or its methyl or acetyl derivates [2].

The separation technique was based on the formation of reversible inclusion complexes of different streight between enantiomers and β-CD, a cyclic oligosaccharide (cycloheptamylose). This stationary phase can present chiral discrimination for optical solutes tightly complexed by β-CD and containing substituent groups idoneous to interact with the 2-hydroxyl groups at the mouth of β-CD cavity.

4.8.1 PREPARATION OF NONCOMMERCIAL LAYERS OF β-CD-BONDED SILICA GEL

The stationary phase was prepared by DeMond and Armstrong [53] according to the procedure used for column chromatography.

Plates (5 × 20 cm) were obtained by mixing 1.5 g of β-CD-bonded silica gel in 15 ml of 1:1 (v/v) water–methanol mixture containing 0.002 g of a polymeric binder (ASTEC "all solvent binder"). Plates (thickness <1 mm) were dried at room temperature and heated at 75°C for 15 min, before use.

Different kinds of silica gel and binder were tested and best results (i.e., maximum coverage of β-CD and high efficiency) were obtained with Mackery-Nagel silica gel.

Visualization of tested racemates was performed by UV light (254 nm) or by ninhydrin reagent. Zhu et al. [55] prepared β-CD-bonded stationary phases (dried at 110°C) with 10 g β-CD and 5 g silica gel H, dissolving β-CD in 120 ml anhydrous dimethylformamide (DMF) containing a little amount of sodium and stirring the resulting solution at 90°C for 2–3 h. After filtration and separation of precipitate, dried silica gel H and 2 g of 3-glycidoxypropylmethoxysilane were added to the solution that was stirred at 90°C for 12–18 h. The β-CD-bonded phase was filtered, washed with several solvents (DMF, toluene, methanol, water, and methanol again) and dried at room temperature.

Glass microslides (7.5 × 2.5 cm, thickness 0.2–0.4 mm) were prepared using a slurry obtained by mixing 2 g β-CD-bonded silica gel with 6 ml aqueous solution of 0.3% sodium carboxymethylcellulose as binder.

TABLE 4.13
Retention (hR_{f1}, hR_{f2}) and Resolution (α) Data for Several Racemates on β-CD-Bonded Silica Gel Plates [52,55]

Racemate	hR_{f1}*	hR_{f2}*	α**	Remarks	Eluents
					Methanol/1% TEAA (pH 4.1) ratio (v/v)
Dns-ala	25 (D)	33 (L)	1.48	Plates: β-CD-bonded silica gel with 0.002 g of ASTEC "all solvent binder" [52]	25:75
Dns-met	28 (D)	48 (L)	2.37		25:75
Dns-val	31 (D)	42 (L)	1.54		25:75
Dns-leu	49 (D)	66 (L)	2.02		40:60
Ala-βNA	16 (D)	25 (L)	1.75		30:70
Met-βNA	16 (D)	24 (L)	1.66		30:70
($\pm$)-1-Ferrocenyl-1-methoxyethane	31 ($-$)	42 (+)	1.54		90:10
($\pm$)-1-Ferrocenyl-2-methylpropanol	33 ($-$)	39 (+)	1.30		90:10
($\pm$)-S-(1-ferrocenylethyl)thioglycolic acid	37 ($-$)	44 (+)	1.34		90:10
Olloxacin	46	54	1.38	Plates (2.5 × 7.5 cm): β-CD-bonded silica gel with sodium carboxymethyl-1-cellulose as binder [55]	50:50
Isoprenaline	26	39	1.82		60:30
Carvediol	34	54	2.28		50:50
					Methanol/acetonitrile/TEAA ratio (v/v/v)
Promethazine	80	89	2.02		2:0.2:0.1
Nimodipine	93	97	2.43		1.5:0.2:0:1
					Petroleum ether/ethyl acetate/methanol ratio (v/v/v)
R,S-10,10'-Dihydroxy-9,9'-biphenanthrene	22	31	1.59		2:1:2
R,S-1,1'-Bi(2-naphthol)	21	33	1.85		3:2:1
R,S-2,2'-Dimethoxy-1,1'-binaphthalene	60	73	1.80		1:2:1
R,S-7,7'-Diethoxy-1,1'-bi(2-naphthol)	28	41	1.79		1:1:1

*$hR_f = R_f \times 100$.

**$\alpha = [(1/R_{f1}) - 1]/[(1/R_{f2}) - 1]$.

Before use, plates were heated at 80°C for 1 h. Development was effected in a 4 × 4 × 8 cm glass chamber for a distance of 5 cm. Visualization was performed by UV light (365 nm) or iodine vapor.

4.8.2 ANALYTICAL APPLICATIONS

Table 4.13 shows the results obtained on β-CD-bonded stationary phases for the resolution of a large number of racemates having very different chemical

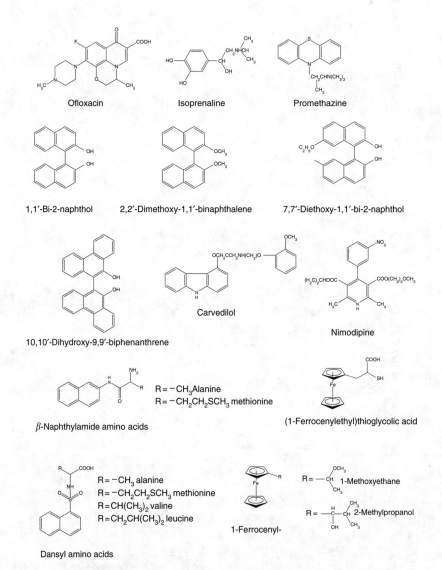

FIGURE 4.7 Structures of racemates on β-CD-bonded silica gel plates.

characteristics, including four atrapisomer couples. The chemical structures of tested compounds were shown in Figure 4.7.

In particular, derivatives of amino acids and ferrocene were also separated on column chromatography and α-values were higher in planar than in column chromatography [56].

All the investigated enantiomeric pairs were baseline resolved on such stationary phase.

The resolution factors (R_S) of Dns-amino acids, calculated from original chromatograms, were ≥ 1.2.

All racemates had a hydrophobic moiety, generally constituted by the presence of one or more aromatic groups, which was able to give an inclusion complex with β-CD, and substituents groups, which gave rise to hydrogen bonds with $-OH$ groups at the entrance of the β-CD cavity (see Figure 4.6).

Compact spots were obtained, because tailing could be minimized by the use of specific eluents and, particularly, of buffer, such as 1% aqueous triethylammonium acetate (TEAA), pH $=$ 4.1, which increased resolution and efficiency. The disadvantages of this technique were the time spent to prepare homemade plates and stationary phases, which are not commercially available.

REFERENCES

1. Günther, K., Enantiomer separations, in *Handbook of Thin Layer Chromatography*, Sherma, J. and Fried, B., Eds., Marcel Dekker Inc., New York, 1991, p. 541.
2. Lepri, L., Del Bubba, M. and Cincinelli, A., Chiral separations by TLC, in *Planar Chromatography, A Retrospective View for the Third Millenium*, Nyiredy, Sz., Ed., Springer Scientific Publishers, Budapest, Hungary, 2001, p. 517.
3. Siouffi, A.M., Piras, P. and Roussel, C., Some aspects of chiral separations in planar chromatography compared with HPLC, *J. Planar Chromatogr.*, 18, 5–13, 2005.
4. Contractor, S.F. and Wragg, J., Resolution of the optical isomers of DL-tryptophan, 5-hydroxy-DL-tryptophan and 6-hydroxy-DL-tryptophan by paper and thin-layer chromatography, *Nature*, 208, 71–72, 1965.
5. Kido, R., Noguchi, T., Tsuji, T., Kawamoto, M. and Matsumura, Y., Resolution of optical isomers of kynurenine and its derivatives by thin-layer chromatography, *Wakayama Med. Rep.*, 11, 129–131, 1967.
6. Haworth, D.T. and Hung, Y.-W., Thin-layer chromatography of an optically active complex, *J. Chromatogr.*, 75, 314–315, 1973.
7. Kotake, M., Sakan, T., Nakamura, N. and Senoh, S., Resolution into optical isomers of some amino acids by paper chromatography, *J. Am. Chem. Soc.*, 73, 2973–2974, 1951.
8. Bonino, G.B. and Carassiti, V., Separation of the optical antipodes of racemic β-naphthol benzylamine by paper chromatography, *Nature*, 167, 569–570, 1951.
9. Dalgliesh, C.E., The optical resolution of aromatic amino acids on paper chromatography, *J. Chem. Soc.*, 137, 3940–3942, 1952.

10. Bach, K. and Haas, H.J., Thin layer chromatographic separation of racemates of some amino acids, *J. Chromatogr.*, 136, 186–188, 1977.

11. Hesse, G. and Hagel, R., Complete separation of a racemic mixture by elution chromatography on cellulose triacetate, *Chromatographia*, 6, 277–280, 1973.

12. Lepri, L., Coas, V., Desideri, P.G. and Zocchi, A., Reversed phase planar chromatography of enantiomeric compounds on triacetylcellulose, *J. Planar Chromatogr.*, 7, 376–381, 1994.

13. Francotte, E. and Wolfe, N., Preparation of chiral building blocks and auxiliaries by chromatography on cellulose triacetate (CTA I): indications for the presence of multiple interaction sites in CTA I, *Chirality*, 2, 16–31, 1990.

14. Blaschke, G., Chromatographic resolution of chiral drugs on polyamides and cellulose triacetate, in *Optical Resolution by Liquid Chromatography*, Hara, S. and Cazes, J., Eds., Marcel Dekker Inc., New York, Vol. 9, 1986, pp. 341–368.

15. Miller, L. and Weyker, C., Effects of compound structure and temperature on the resolution of enantiomers of cyclopentenones by liquid chromatography on derivatized cellulose chiral stationary phases, *J. Chromatogr.*, 653, 219–228, 1993.

16. Allenmark, S.T., Chromatographic enantioseparation, in *Methods and Applications*, Eds., John Wiley & Sons, New York, 1988, p. 205.

17. Lepri, L., Del Bubba, M. and Masi, F., Reversed phase planar chromatography of enantiomeric compounds on microcrystalline cellulose triacetate (MCTA), *J. Planar Chromatogr.*, 10, 108–113, 1997.

18. Lepri, L., Reversed phase planar chromatography of enantiomeric compounds on microcrystalline triacetylcellulose, *J. Planar Chromatogr.*, 8, 467–469, 1995.

19. Lepri, L., Cincinelli, A. and Del Bubba, M., Reversed phase planar chromatography of optical isomers on microcrystalline cellulose triacetate, *J. Planar Chromatogr.*, 12, 298–301, 1999.

20. Lepri, L., Del Bubba, M., Cincinelli, A. and Boddi, L., Inclusion planar chromatography of enantiomeric and racemic compounds, *J. Planar Chromatogr.*, 13, 384–387, 2000.

21. Lepri, L., Boddi, L., Del Bubba, M. and Cincinelli, A., Reversed phase planar chromatography of some enantiomeric aminoacids and oxazolidinones, *Biomed. Chromatogr.*, 15, 196–201, 2001.

22. Gunther, K., Zeller, M., Degussa, A.G. and Hanau, Germany.

23. Lepri, L., Del Bubba, M., Coas, V. and Cincinelli, A., Reversed-phase planar chromatography of racemic flavanones, *J. Liq. Chrom. Rel. Technol.*, 22, 105–118, 1999.

24. Krause, M. and Galensa, R., Direct enantiomeric separation of racemic flavanones by high-performance liquid chromatography using cellulose triacetate as a chiral stationary phase, *J. Chromatogr.*, 441, 417–422, 1988.

25. Ficarra, P., Ficarra, R., Bertucci, C., Tommasini, S., Cobro, M.L., Costantino, D. and Carulli, M., Direct enantiomeric separation of flavanones by high performance liquid chromatography using various chiral stationary phases, *Planta Med.*, 61, 171–176, 1995.

26. Rizzi, A.M., Band broadening in high-performance liquid chromatographic separations of enantiomers with swollen microcrystalline cellulose triacetate packings: I. Influence of capacity factor, analyte structure, flow velocity and column loading, *J. Chromatogr.*, 478, 71–86, 1989.

27. Isaksson, R. and Lamm, B., Semipreparative separation of cyclic carbamates of β-blocking agents by liquid chromatography on swollen microcrystalline triacetylcellulose, *J. Chromatogr.*, 362, 436–438, 1986.
28. Jamali, F., Mehvar, R. and Pasutto, F.M., Eantioselective aspect of drug action and disposition: therapeutic pitfalls, *J. Pharm. Sci.*, 78, 695–715, 1989.
29. Wainer, I.W. and Alembik, M.C., Resolution of enantiomeric amides on a cellulose-based chiral stationary phase: steric and electronic effects, *J. Chromatogr.*, 358, 85–89, 1986.
30. Francotte, E. and Wolf, R.M., Benzoyl cellulose beads in the pure polymeric form as a new powerful sorbent for the chromatographic resolution of racemates, *Chirality*, 3, 43–55, 1991.
31. Lepri, L., Del Bubba, M., Cincinelli, A. and Boddi, L., Direct resolution of aromatic alcohols by planar chromatography on tribenzoylcellulose as chiral stationary phase, *J. Planar Chromatogr.*, 14, 134–136, 2001.
32. Lepri, L., Del Bubba, M., Cincinelli, A. and Bracciali, M., Quantitative determination of enantiomeric alcohols by planar chromatography on tribenzoylcellulose, *J. Planar Chromatogr.*, 15, 220–222, 2002.
33. Wainer, I.W., Stittin, R.M. and Shibata, T., Resolution of enantiomeric aromatic alcohols on a cellulose tribenzoate high-performance liquid chromatography chiral stationary phase: a proposed chiral recognition mechanism, *J. Chromatogr.*, 411, 139–151, 1987.
34. Okamoto, Y., Kawashima, M. and Hatada, K., Useful chiral packing materials for high-performance liquid chromatographic resolution of enantiomers: phenyl carbamates of polysaccharides coated on silica gel, *J. Am. Chem. Soc.*, 106, 5357–5359, 1984.
35. Okamoto, Y., Aburatani, R., Fukumoto, T. and Hatada, K., Useful chiral stationary phases for HPLC. Amylose *tris*(3,5-dimethylphenylcarbamate) and *tris*(3,5-dichlorophenylcarbamate) supported on silica gel, *Chem. Lett.*, 8, 1857–1860, 1987.
36. Suedee, R. and Heard, C.M., Direct resolution of propranolol and bupranolol by thin-layer chromatography using cellulose derivatives as stationary phase, *Chirality*, 9, 139–144, 1997.
37. Kubota, T., Yamamoto, C. and Okamoto, Y., *Tris*(cyclohexylcarbamate)s of cellulose and amylose as potential chiral stationary phase for high-performance liquid chromatography and thin layer chromatography, *J. Am. Chem. Soc.*, 122, 4056–4059, 2000.
38. Shibata, T., Okamoto, Y. and Ishii, K., Chromatographic optical resolution on polysaccharides and their derivatives, in *Optical Resolution by Liquid Chromatography*, Hara, S. and Cazes, J., Eds., Marcel Dekker Inc., New York, Vol. 9, 1986, pp. 313–340.
39. Rozylo, J.K. and Malinowska, I., The possibilities of optical isomer separation by TLC on layers of chitin and its derivatives, *J. Planar Chromatogr.*, 6, 34–37, 1984.
40. Malinowska, I. and Rozylo, J.K., Separation of optical isomers of amino acids on modified chitin and chitosan layers, *Biomed. Chromatogr.*, 11, 272–275, 1997.
41. Wulff, G. and Sarhan, S., Use of polymers with enzyme-analogous structures for the resolution of racemates, *Angew. Chem. Int. Ed.*, 11, 341, 1972.
42. Wulff, G., Wesper, R., Grobe-Einsler, R. and Sarchan, A., Enzyme-analogue built polymers: (4) on the synthesis of polymers containing chiral cavities and

their use for the resolution of racemates, *Macromol. Chem.*, 178, 2799–2816, 1977.

43. Kriz, D., Kriz, C.B., Andersson, L.I. and Mosbach, K., Thin-layer chromatography based on the molecular imprinting technique, *Anal. Chem.*, 66, 2636–2639, 1994.

44. O'Shannessy, D.Y., Ekberg, B., Andersson, L.I. and Mosbach, K., Recent advances in the preparation and use of molecularly imprinted polymers for enantiomeric resolution of amino acid derivatives, *J. Chromatogr.*, 470, 391–399, 1989.

45. Sellergren, B., Legisto, M. and Mosbach, K., Highly enantioselective and substrate-selective polymers obtained by molecular imprinting utilizing non covalent interactions. NMR and chromatographic studies on the nature of recognition, *J. Am. Chem. Soc.*, 110, 5853–5860, 1988.

46. Ansell, R.J. and Kuah, K.L., Imprinted polymers for chiral resolution of (±)-ephedrine: understanding the pre-polymerization equilibrium and the action of different mobile phase modifiers, *Analyst*, 130, 179–187, 2005.

47. Suedee, R., Songkram, C., Petmoreekul, A., Sangkunakup, S., Sankasa, S. and Kongyarit, N., Thin-layer chromatography using synthetic polymers imprinted with quinine as chiral stationary phase, *J. Planar Chromatogr.*, 11, 272–276, 1998.

48. Suedee, R., Srichana, T., Saelim, J. and Thavornpibulbut, T., Chiral determination of various adrenergic drugs by thin-layer chromatography using molecularly imprinted chiral stationary phases prepared with α-agonists, *Analyst*, 124, 1003–1009, 1999.

49. Suedee, R., Songkran, C., Petmoreekul, A., Sangkunakup, S., Sankara, S. and Kongyarit, N., Direct enantioseparation of adrenergic drugs via thin-layer chromatography using molecularly imprinted polymers, *J. Pharm. Biomed. Anal.*, 19, 519–527, 1999.

50. Suedee, R., Srichana, T., Saelim, J. and Thavornpibulbut, T., Thin-layer chromatographic separation of chiral drugs on molecularly imprinted chiral stationary phases, *J. Planar Chromatogr.*, 14, 194–198, 2001.

51. Aboul-Enein, H.Y., El-Awady, M.I. and Heard, C.M., Direct enantiomeric resolution of some cardiovascular agents using synthetic polymers imprinted with (−)-*S*-timolol as chiral stationary phase by thin layer chromatography, *Pharmazie*, 57, 169–171, 2002.

52. Alak, A. and Armstrong, D.W., Thin-layer chromatographic separation of optical geometrical and structural isomers, *Anal. Chem.*, 58, 582–584, 1986.

53. Armstrong, D.W. and DeMond, W., Cyclodextrin bonded phases for the liquid chromatographic separation of optical, geometrical and structural isomers, *J. Chromatogr.*, 22, 411–415, 1984.

54. Armstrong, D.W., DeMond, W., Alak, A., Hinze, W.L., Riehl, T.E. and Bui, K.H., Liquid chromatographic separation of diastereoisomers and structural isomers on cyclodextrin-bonded phases, *Anal. Chem.*, 57, 234–237, 1985.

55. Zhu, Q., Yu, P., Deng, Q. and Zeng, L., β-Cyclodextrin-bonded chiral stationary phase for thin-layer chromatographic separation of enantiomers, *J. Planar Chromatogr.*, 14, 137–139, 2001.

56. Ward, T.J. and Armstrong, D.W., Improved cyclodextrin chiral phases: a comparison and review, in *Optical Resolution by Liquid Chromatography*, Hara, S. and Cazes, D.J., Eds., Marcel Dekker, Inc., New York, Vol. 9, 1986, pp. 407–423.

5 Planar Chromatography Enantioseparations on Noncommercial CCSPs

Luciano Lepri, Alessandra Cincinelli, and Massimo Del Bubba

CONTENTS

5.1 INTRODUCTION

Chiral-coated stationary phases (CCSPs) consist of specific chiral selectors permanently adsorbed onto the surface of achiral thin layer chromatography (TLC) materials without covalently modifying their chemical characteristics.

The preparation of CCSP plates was effected in the laboratory with the exception of commercially available Chiralplate (Macherey-Nagel) and HPTLC-Chir (Merck) constituted by silanized silica gel impregnated with the copper (II) complex of $(2S,4R,2'RS)$-N-$(2'$-hydroxydodecyl)-4-hydroxyproline. These plates

111

were able to resolve many structural-related racemic compounds, such as amino acids and their derivatives, dipeptides, and heterocyclic compounds [1].

A large number of CCSPs were prepared in laboratory by several researchers owing to the ease of the impregnation procedures and the variety of chiral selectors used as impregnating agents (i.e., acids, bases, amino acids, and complexes with metal ions). In addition, as only little amounts of chiral selector were necessary for the impregnation of TLC sorbents, this technique was very cheap in comparison with chiral stationary phases (CSPs).

Among the different methods used for CCSPs preparation (i.e., immersion of the plates in a suitable solution, development with an eluent containing the impregnating reagent prior to application of solutes, spraying with the appropriate reagent, and mixing of the specific selector with the adsorbent), immersion of the achiral precoated plate in a solution of the chiral impregnating agent and mixing of the chiral selector with the inert support during plate making were the most frequently employed.

In chiral chromatography, both in CSPs and CCSPs, two labile diastereomers, which differ in their thermodynamic stability, were formed during elution.

The substantial difference between these two techniques was the chiral interaction capacity. In CSPs, chiral sites were incorporated into the stationary phases and several of them were able to simultaneously interact with solutes. On the contrary, in CCSPs, the chiral sites were distributed on the surface of the achiral sorbent relatively far from each other; therefore, only bimolecular interactions with the optical antipodes should be expected.

5.2 SILANIZED SILICA GEL PLATES IMPREGNATED WITH METAL COMPLEXES OF A CHIRAL SELECTOR

This separation technique, usually known as chiral ligand exchange chromatography (CLEC), was based on the "three-point interaction rule," postulated by Dalgliesh [2] in 1952.

According to this theory, ternary diastereomeric complexes of different stability were formed among the ligands of opposite configuration, and the metal ion and the chiral selector were absorbed on the layer. The formation of five term rings allows us to obtain stable ternary copper complexes; in this regard, α-amino acids and α-hydroxy acids were the most suitable while β-amino acids, which gave rise to less stable six term rings, were difficult to resolve by using CLEC. Chiral selectors used as copper (II) complexes between 1984 and 1991 were reported in Figure 5.1 and arranged on the basis of publication year; after 1991, to our knowledge, no paper concerning this topic was published.

5.2.1 PREPARATION OF NONCOMMERCIAL PLATES AND THEIR APPLICATIONS

The first resolutions of racemic analytes in planar chromatography by using CLEC and laboratory-prepared plates were reported as early as 1984 by Gunther

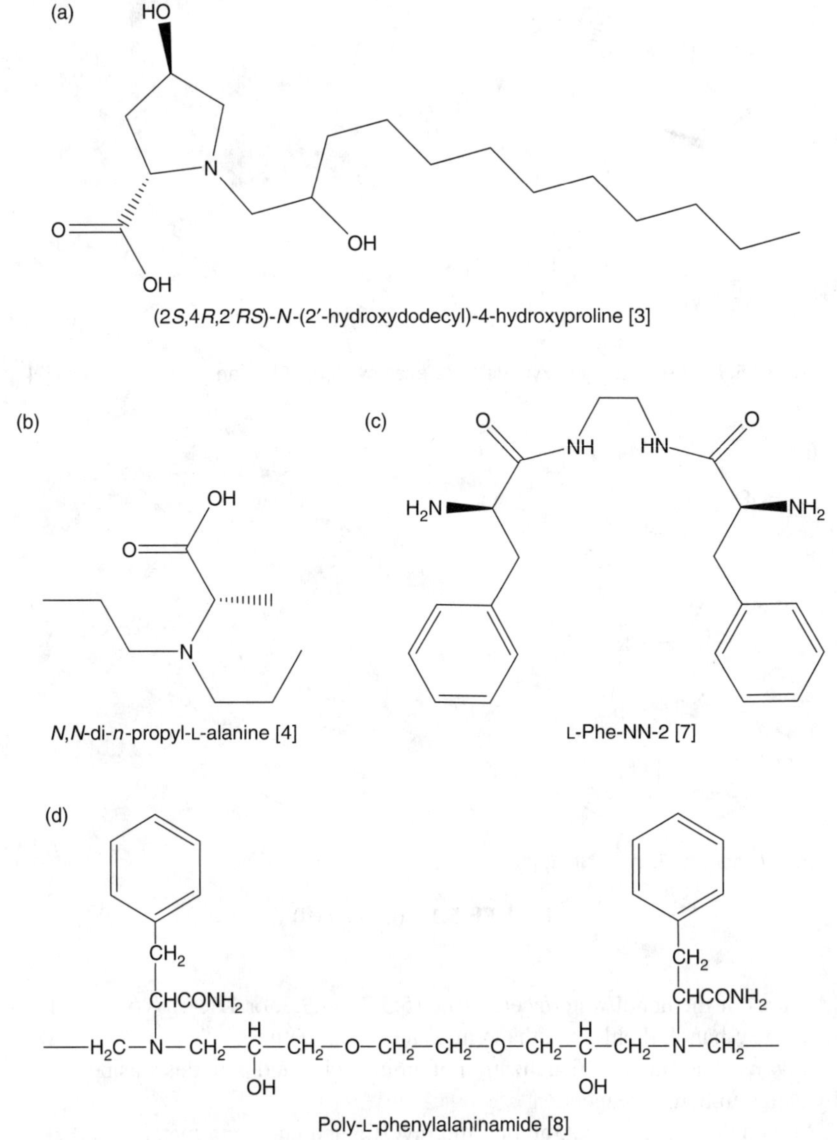

(a)

(2S,4R,2′RS)-N-(2′-hydroxydodecyl)-4-hydroxyproline [3]

(b)

N,N-di-n-propyl-L-alanine [4]

(c)

L-Phe-NN-2 [7]

(d)

Poly-L-phenylalaninamide [8]

FIGURE 5.1 Chiral selectors used as copper (II) complexes for impregnation of silanized silica plates in CLEC.

et al. [3] and Weinstein [4]. Gunther et al. [3] used the following procedure to immobilize (2S,4R,2′RS)-N-(2′-hydroxydodecyl)-4-hydroxyproline (see Figure 5.1a) on commercially available RP-18 plates (Merck, Germany). The achiral plates were dipped in a 0.25% copper (II) acetate solution in methanol/water (1:9, v/v) and dried. Afterwards, the plates were immersed in 0.8% methanol solution of the chiral reagent for 1 min, and dried in the air before use.

(e)

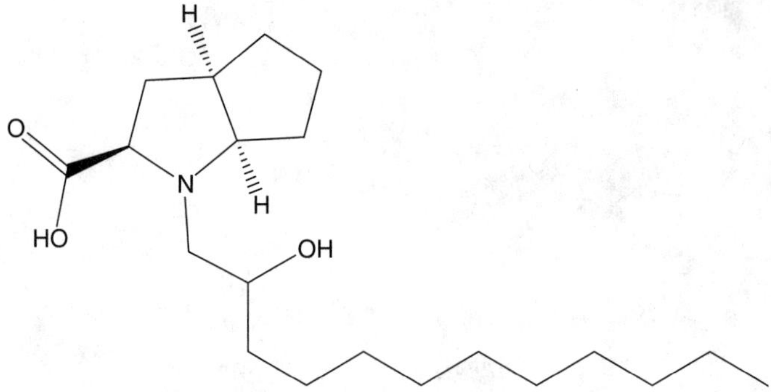

(1*R*,3*R*,5*R*,2'*RS*)-*N*-(2'-hydroxydodecyl)-2-azabicyclo[3.3.0] octan-3-carboxylic acid [9]

(f)

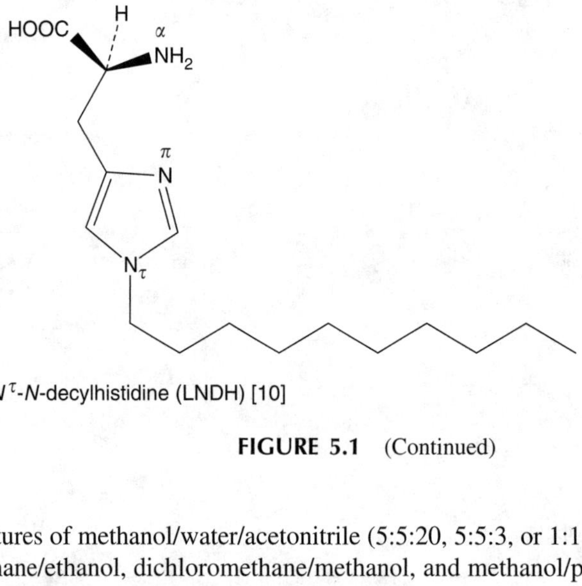

L-*N*$^\tau$-*N*-decylhistidine (LNDH) [10]

FIGURE 5.1 (Continued)

Mixtures of methanol/water/acetonitrile (5:5:20, 5:5:3, or 1:1:4, v/v/v), dichloro-methane/ethanol, dichloromethane/methanol, and methanol/propanol/water were mainly used as eluents. Visualization of amino acids and peptides was performed by using ninhydrin reagent.

It should be noted that, among the investigated chelating metal ions [Cu(II), Ni(II), Zn(II), Hg(II), Co(III), and Fe(III)], copper (II) gave rise to most stable complexes and, therefore, appeared to be most suitable for this chromatographic technique.

Applications of CLEC in planar chromatography are concerned with the enantiomeric separation of several polar organic compounds, particularly α-amino acids and their derivatives, α-hydroxy acids, and peptides. These applications were almost the same tested on the commercially available Chiralplates, whose preparation was identical to that adopted for homemade plates. The D-enantiomer was retained more than L-enantiomer for all bidentate compounds, such as neutral amino acids lacking other polar groups. On the contrary, the elution order of enantiomeric α-methyl amino acids cannot be predicted owing to specific steric effects.

Note that the amounts of solute usually applied on the plates (10–20 μg) were about one magnitude order higher than that generally used in HPTLC. Poor resolution was observed for the enantiomers of serine, whereas no separation was found for threonine and basic amino acids with all the eluents tested.

Weinstein [4] dipped RP-18 TLC precoated plates (Merck, Darmstadt, Germany) in a solution of copper (II) complex of N,N-di-n-propyl-L-alanine (DPA) (see Figure 5.1b) [8 mM chiral selector and 4 mM cupric acetate in acetonitrile/water (97.5:2.5, v/v)] for 1 h and up to overnight. RP-18 plates were predeveloped with 0.3 M sodium acetate solution in acetonitrile/water (40:60, v/v; buffer A), adjusted to an apparent pH of 7 with glacial acetic acid. Before impregnation the plates were fan dried. The chiral selector proposed by Weinstein was commercially available from Aldrich.

All proteinogenic amino acids, with the only exception of proline, were resolved into their enantiomers as dansyl (DNS) derivatives by eluting with buffer A, with or without 4 mM DPA and 1 mM cupric acetate. The use of buffer A with the copper (II) complex of the chiral selector as eluent was necessary to maintain a concentration of the chiral complex on the surface of silanized silica gel sufficiently high to resolve some specific racemates. This suggested that the copper (II)–DPA complex was less adsorbed by silanized silica gel than copper (II)–hydroxydodecylproline complex proposed by Gunther et al. [3]. Detection was effected with UV light (360 nm) and dansyl amino acids appeared as fluorescent yellow-green spots.

As shown in Table 5.1, ΔR_f values of the optical antipodes were included between 0.02 (dansyl-DL-threonine, Thr, and dansyl-DL-valine, Val) and 0.06 (dansyl-DL-aspartic acid, Asp); however, a good resolution of the enantiomers was observed owing to the compactness of the spots. The L-enantiomer was always more retained than the D-enantiomer. This technique allowed the determination of little amounts of Dns-D-enantiomer in the L-form (i.e., it was possible to quantify impurities of Dns-D-glutamic acid in the L-enantiomer as low as 1%).

Further studies [5,6] concerned the use of two-dimensional reversed-phase chromatography for the simultaneous resolution of multicomponent mixtures of Dns-amino acids. In the first direction the mixture was resolved into its components on RP-18 plates in a nonchiral mode. After drying, the layer strips containing the Dns-amino acids were covered with a glass plate and the remaining plate was sprayed with the copper (II) complex of DPA. Development in the second direction was performed by eluting with buffer A containing the copper (II) complex of DPA. A further improvement of the separation was achieved by using a temperature gradient of 6.2°C/cm [6]. Under these experimental conditions, five pairs of Dns-amino acids (aspartic acid, serine, methionine, alanine, and phenylalanine) were resolved with no overlap of the different optical isomers.

Chiral diaminodiamide copper (II) complexes were also used as chiral selectors to impregnate commercial layers and to separate racemic dansyl amino

TABLE 5.1

Retention (hR_{f1}, hR_{f2}) and Resolution (α, R_S) Data of DL-Dansyl-Amino Acids (Dns-AA) on Silanized Silica Gel Precoated Plates (RP-18; Merck) Impregnated with the Copper (II) Complex of Various Chiral Selectors

DL-Dns-AA	hR_{f1}*,a	hR_{f2}*,a	α*	R_S***,a	Eluent and remarks	Ref.
Dns-Glu	73 (L)	77 (D)	1.24	1.6	• RP-18 plates (5 × 20 cm Whatman and 10 × 20 cm Merck) impregnated with the copper (II) complex of DPA (see text)	[4]
Dns-Asp	60 (L)	66 (D)	1.29	4.0	• Buffer A (see text) with or without 4 mM DPA and 1 mM cupric acetate. 25% Acetonitrile is preferred for Glu, Asp, Ser, and Thr derivatives	
Dns-Ser	52 (L)	55 (D)	1.13	1.5	• Sample volume, 1 μl of 5 × 10⁻⁴ M solution	
Dns-Thr	51 (L)	53 (D)	1.08	1.3		
Dns-Ala	48 (L)	53 (D)	1.22	1.5		
Dns-Met	43 (L)	47 (D)	1.18	2.5		
Dns-Val	44 (L)	46 (D)	1.08	1.3		
Dns-Leu	39 (L)	43 (D)	1.18	2.5		
Dns-Phe	36 (L)	40 (D)	1.18	2.5		
Dns-Phe	35 (L)	41 (D)	1.29	—	• RP-18 F$_{254}$s HPTLC plates (10 × 10 cm; Merck) impregnated with 4 mM copper (II) complex of Phe-NN-2 (see text)	[7]
Dns-Glu	24 (L)	40 (D)	2.11	1.2	• Eluents: water/acetonitrile mixtures at different pH; 33% acetonitrile and pH 6.8 is preferred for Dns-Glu and Dns-Asp, whereas 50% acetonitrile and pH 7.5 is used for others. The enantiomer separation of the first six amino acids needs the use of an eluent containing 2 mM chiral complex	
Dns-Asp	9 (L)	21 (D)	2.69	—		
Dns-Ser	41 (L)	51 (D)	1.50	1.0		
Dns-Thr	47 (L)	52 (D)	1.22	—		
Dns-Met	34 (L)	40 (D)	1.29	—		
Dns-Nva	36 (D)	42 (L)	1.29	—		
Dns-Leu	38 (L)	47 (L)	1.45	1.0		
Dns-Nle	33 (D)	42 (L)	1.47	—		
Dns-Glu	55 (L)	63 (D)	1.39	1.50b	• RP-18 F$_{254}$s TLC plates (5 × 20 or 10 × 10 cm; Merck) impregnated with the copper (II) complex of poly-L-phenylalaninamide (see text)	[8]
Dns-Asp	35 (L)	40 (D)	1.24	—	• Eluents: water/acetonitrile in the ratio 55:45 (v/v). 30% Acetonitrile is preferred for the separation of Dns-Glu, Dns-Asp, and Dns-Thr	
Dns-Ser	40 (L)	40 (D)	1.00	—	• Sample volume, 1 μl of 1 × 10⁻³ M solution	
Dns-Thr	43 (L)	46 (D)	1.13	—		
Dns-Met	18 (D)	25 (D)	1.52	2.00b		
Dns-Val	33 (L)	37 (D)	1.19	1.14b		
Dns-Nva	36 (L)	41 (D)	1.23	1.66b		
Dns-Leu	19 (L)	24 (D)	1.35	1.42b		
Dns-Phe	17 (L)	20 (D)	1.22	—		

Visualization: UV light (360 and 365 nm).

a Calculated from original chromatograms.

b Calculated from original chromatograms after eluting with water/acetonitrile (50:50, v/v).

*hR_f = R_f × 100.

** $\alpha = [(1/R_{f1}) - 1]/[(1/R_{f2}) - 1]$.

*** R_S = 2 × (distance between the centers of two adjacent spots)/(sum of the widths of the two spots in the direction of development).

acids [7]. In this type of ligands, two L-amino acids were linked via an amide bond by dimethylene or trimethylene bridges. The best results were obtained by using Phe-NN-2 (see Figure 5.1c), two molecules of L-phenylalanine linked through two methylene groups. Among the other investigated amino acids, valine showed fairly good results, whereas for alanine no enantiomeric separation was achieved.

Chiral ligands were synthesized as dihydrochlorides by reaction of the (Z)-N-amino acid hydroxysuccinimide esters with 1,2-diaminoethane or 1,3-diaminopropane in a triethylamine solution, hydrogenolysis and successive treatment with hydrochloric acid in methanol.

The 10×20 cm RP-18 $F_{254}s$ HPTLC plates (Merck) were coated by dipping them for 1 h in buffer A, adjusted to an apparent pH of 7.5 with glacial acetic acid; afterwards the layers were dried at $100°C$ for 1 h, cooled at room temperature and immersed for 2 h in a 4 mM Phe-NN-2 and 4 mM copper (II) acetate solution in water/acetonitrile (10:90, v/v). Before use, the plates were dried at $60°C$ for 1 h. Water/acetonitrile mixtures with and without the Cu (II)–Phe-NN-2 complex and apparent pH included between 6 and 8 were employed as eluents. Dns-amino acids were detected by using a fluorescent lamp (365 nm).

As shown in Table 5.1, nine racemic Dns-amino acids were resolved, that is, a number of successful separations similar to that obtained by Weinstein [4]. Separation factors (α) were higher than those reported by Weinstein [4], but the actual resolution was worse owing to the lower compactness of the spots. Again Dns-aspartic acid showed the highest α-value, followed by Dns-glutamic acid. Note that the use of chiral diaminodiamide copper (II) complexes as chiral selectors gave rise to an unexpected, inverted elution order for the enantiomers of leucine (Leu), norleucine (Nle), and norvaline (Nva), with a lower R_f value for the D-enantiomer. Under these experimental conditions, AA-NN-2 ligands formed two complexes with copper acetate, such as $(Cu_2L_2H_{-2})^{2+}$ and $CuLH_{-2}$.

The same group of Dns-amino acids was studied on plates of RP-18 $F_{254}s$, 5×20 cm and RP-18 F_{254} HPTLC, 10×10 cm (Merck) coated with the copper (II) complex of poly-L-phenylalaninamide [8], whose structure was reported in Figure 5.1d. The polymer was prepared by adding 2 ml (12.8 mmol) of ethylene glycol diglycidyl ether to 1.5 g (9.1 mmol) of L-phenylalaninamide dissolved in 3 ml of methanol, under magnetic stirring at room temperature. The product was dissolved in methanol and fractionated on a Bio-Sil TSK 250 column (300×75 mm, i.d.) (Bio-Rad) eluted with 0.01 M phosphate buffer (pH 3.5)/methanol mixture in the ratio 9:1 (v/v). The layer impregnation was carried out by dipping the plates in a 0.3 M sodium acetate solution in water/acetonitrile (70:30, v/v), adjusted to an apparent pH of 7 with glacial acetic acid. After gentle drying, plates were coated with water/acetonitrile (10:90 v/v) solution containing 3 mM copper (II) acetate and 6 mM poly-L-phenylalaninamide (expressed as monomeric units) for at least 1 h. Then the plates were dried in the open air. Water/acetonitrile mixtures in different ratios were used as eluents.

The results were reported in Table 5.1 and showed that the adopted chiral complex was suitable to baseline-resolve various DL-Dns-amino acids, although it failed in the separation of Dns-serine enantiomers. The best-resolved racemates were those with marked acidic characteristics, such as Dns-aspartic acid and Dns-glutamic acid. For these reasons the method was considered useful for fossil bone dating, based on racemization of aspartic acid. It should be noted that the amount of solute applied on the layer was one magnitude order higher than that used by Marchelli et al. [7].

With respect to the enantioseparation mechanism, it is worth to note that the same plates impregnated with the copper (II) complex of L-phenylalaninamide in the molar ratio 1:2 did not give rise to optical antipodes separation. Authors suggested that the polymerization of L-phenylalaninamide with an oxirane could allow the formation of an additional coordination bond, involving the hydroxyl group formed after the opening of the epoxy ring. Such hypothesis was supported by the fact that the hydroxyl group was in γ-position with respect to the chiral center, analogously to what was observed for the hydroxyl group of hydroxyproline proposed by Gunther et al. [3].

Using a pharmaceutical industrial waste, such as (1R,3R,5R)-2-azabicyclo-[3.3.0]octan-3-carboxylic acid, as starting material, Martens et al. [9] proceeded to alkylate at the nitrogen atom so as to form (1R,3R,5R,2'RS)-N-(2'-hydroxydodecyl)-2-azabicyclo[3.3.0]octan-3-carboxylic acid (see Figure 5.1e), coated with copper (II) complex was proposed for carrying out CLEC of some racemic imino acids, including the industrial compound (see Table 5.2). High α-value was achieved for both the investigated imino acids.

Remelli et al. [10] proposed a new chiral selector, L-N-decylhistidine (LNDH), prepared through selective alkylation with 1-iododecane of L-histidine at the pyrrolic nitrogen atom of the heterocyclic ring (see Figure 5.1f). As shown in Table 5.3, the copper (II) complex of this chiral selector exhibited a high selectivity toward several enantiomers of aromatic amino acids, while no significant separation was observed for DL-valine, DL-alanine, and DL-leucine enantiomers. With respect to the behavior of aromatic amino acids, the D-enantiomer was the most

TABLE 5.2
Retention (hR_{f1}, hR_{f2}) and Resolution (α) Data of Some Iminoacids on Layers of RP-18 (Merck) Coated with the Copper (II) Complex of (1R,3R,5R,2'RS)-N-(2'-Hydroxydodecyl)-2-Azabicyclo[3.3.0]Octan-3-Carboxylic Acid [9]

Racemate	hR_{f1}*	hR_{f2}*	α**	Eluent
2-Azabicyclo[3.3.0]-octan-3-carboxylic acid	51 (1R,3R,5R)	60 (1S,3S,5S)	1.44	Acetonitrile/methanol/water (5:3:3, v/v/v)
5,5-Dimethyl-3-thiazolin-4-acetic acid	16	32	2.47	Acetonitrile/methanol/water (3:5:5, v/v/v)

*hR_f = R_f × 100.
**α = [(1/R_{f1}) − 1]/[(1/R_{f2}) − 1].

TABLE 5.3
Retention (hR_{f1}, hR_{f2}) and Resolution (α, R_S) Data of Racemic Aromatic Amino Acids on 10 × 10 cm RP-18W F$_{254}$s HPTLC Plates Impregnated with Copper (II) Complex of L-N-Decylhistidine under Optimized Elution Conditions [10]

Racemate	hR_{f1}*,a	hR_{f2}*,a	α**	R_S***,b
Tryptophan (Trp)	18 (D)	24 (L)	1.44	1.0
α-MethylTrp	16 (D)	22 (L)	1.48	—
5-MethylTrp	15 (D)	20 (L)	1.42	—
6-MethylTrp	16 (D)	19 (L)	1.23	—
Phenylalanine (Phe)	36 (D)	46 (L)	1.51	1.0
α-MethylPhe	28 (D)	40 (L)	1.71	—
Tyrosine (Tyr)	43 (D)	52 (L)	1.44	—
α-MethylTyr	37 (D)	47 (L)	1.51	—
Dihydroxyphenylalanine (DOPA)	36 (D)	45 (L)	1.45	—

Eluent: Methanol/acetonitrile/tetrahydrofuran/water (7.3:5.9:33.9:52.9, v/v/v/v).

*$hR_f = R_f \times 100$.

**$\alpha = [(1/R_{f1}) - 1]/[(1/R_{f2}) - 1]$.

***$R_S = 2 \times$ (distance between the centers of two adjacent spots)/(sum of the widths of the two spots in the direction of development).

[a] Calculated from R_M values.

[b] Calculated from an optimized chromatogram *in situ* by the evaluation of red-violet spots using TLC/HPTLC Scanner II (Camag), reflectance mode (550 nm).

retained, as it gave rise to ternary diastereomeric complexes more stable than the corresponding L-form. Under the best experimental conditions, resolution factors (R_S) of DL-phenylalanine (Phe) and DL-tryptophan (Trp) were 1.0 and, therefore, these racemates were almost baseline resolved. Such values were lower than those expected based on high separation factors (1.44 for Trp and 1.51 for Phe), owing to tailed spots and the high amount (>2 μg) of solute necessary to reveal the analytes on these layers, having a strong background. On the other hand, it was impossible to increase the resolution of the different racemates by increasing the chiral selector concentration at the surface, as ninhydrin developed a background red color even with LNDH alone. Visualization was performed by dipping the plates in 0.15% (w/v) ninhydrin in acetone and then warming at 110°C for 3 min.

Note that layer preparation was quite simple (i.e., RP-18 WF$_{254}$s HPTLC plates were immersed for 5 min in 0.125% (w/v) copper (II) acetate in water/methanol (90:10) and, after drying at room temperature, in 0.4% (w/v) LNDH in methanol), but synthesis and characterization of the chiral selector took a very long time.

5.3 SILICA GEL PLATES COATED WITH THE COPPER (II) COMPLEX OF ENANTIOMERIC AMINO ACIDS

To our knowledge, three papers were published on the use of silica gel plates coated with copper (II) complex of an enantiomeric amino acid for the separation of optical antipodes [11–13].

L-proline was the amino acid initially employed as coating agent [11] and the layers (20 × 20 cm, thickness 0.5 mm) were prepared by spreading a slurry of 50 g silica gel G (Merck) into a solution made with 50 ml of 5×10^{-3} M L-proline and 50 ml of 2.5×10^{-3} M copper sulfate, adjusted to a pH of 7 with ammonium acetate. Plates were dried overnight at 60°C. Standard solutions of amino acids (10^{-3} M) were prepared in 70% aqueous ethanol. The amount of each analyte applied on the layer was about 0.5 μg. Detection was performed by spraying the plates, dried at 40°C, with 0.2% freshly prepared ninhydrin solution in acetone and heating at 100°C for 10 min.

Three eluents gave the best results for the separation of racemic amino acids: (a) n-butanol/acetonitrile/water (6:2:3, v/v/v), (b) chloroform/methanol/propionic acid (15:6:4, v/v/v), and (c) acetonitrile/methanol/water (2:2:1, v/v/v).

These layers enabled the enantiomer separation of four amino acids out of the eleven tested, three of which contained aromatic moieties (see Table 5.4). Again the D-enantiomer was the most retained and the mechanism involved in the chiral recognition could still be based on CLEC, even if a reversed-phase chromatography was not operating, in contrast to the previous investigations. Very high α- and R_S-values were obtained, particularly for the three aromatic amino acids owing to the special compactness of the spots; the best resolution was found for DL-tryptophan.

R_S-value for DL-isoleucine (Ile), calculated from the original chromatogram, appeared too high (2.3) when it was compared with the corresponding α (1.21) and the R_f of the two enantiomers (0.28 and 0.32, respectively).

On the same plates, obtained from a slurry consisting of 50 g silica gel G in 10^{-3} M copper sulfate solution and 2×10^{-3} M (1R,3R,5R)-2-azabicyclo[3.3.0]octan-3-carboxylic acid [12] mixed in 1:1 ratio, three racemic amino acids were resolved (see Table 5.4). This nonproteinogenic amino acid, used as chiral selector, was available in the enantiomeric form as secondary product in the industrial manufacture of Ramipril [14]. Excellent separations were obtained, as demonstrated by the high-R_S values, calculated on the basis of original chromatograms.

A disadvantage common to all layers coated with copper (II) complexes of amino acids was the progressive increase in the color depth of the layer sprayed with ninhydrin. In particular, after about 30 min the plate becomes dark pink and the spots of amino acids have to be marked as soon as they are visible.

Silica gel G layers (10 × 20 cm, 0.5 mm thickness) coated with copper (II) complex of L-arginine were used for the separation of β-adrenergic blocking agents [13]. For their preparation, a slurry consisting of 50 g of adsorbent and 100 ml of water/methanol (90:10, v/v) containing 1 mM copper (II) acetate, 2 mM L-arginine

TABLE 5.4
Retention (hR_{f1}, hR_{f2}) and Resolution (α, R_S) Data of DL-Amino Acids on 20 × 20 cm Plates of Silica Gel G (Merck) Coated with Copper (II) Complex of L-Proline [11] and (1R,3R,5R)-2-Azabicyclo[3.3.0]Octan-3-Carboxylic Acid [12] as Chiral Selectors

Racemate	hR_{f1}*	hR_{f2}*	α**	R_S***,a	Eluent	Ref.
Phe	32 (D)	39 (L)	1.36	2.0	A[b]	[11]
Tyr	36 (D)	48 (L)	1.64	2.3	A[b]	
Ile	28 (D)	32 (L)	1.21	2.3	B[b]	
Trp	40 (D)	72 (L)	3.85	3.0	C[b]	
His	11 (D)	36 (L)	4.55	4.4	Acetonitrile/methanol/water (3:1:1, v/v/v)	[12]
Ser	6 (D)	18 (L)	3.44	3.5	Acetonitrile/methanol/water (7:1:1, v/v/v)	
Trp	38 (D)	63 (L)	2.78	4.5	Acetonitrile/methanol/water (8:1:1, v/v/v)	

*$hR_f = R_f \times 100$.
**$\alpha = [(1/R_{f1}) - 1]/[(1/R_{f2}) - 1]$.
***$R_S = 2 \times$ (distance between the centers of two adjacent spots)/(sum of the widths of the two spots in the direction of development).

[a] Calculated from the original chromatograms.
[b] See text.

and ammonia up to a pH value of 7, was used. Before use, the plates were activated overnight at 60°C.

Aliquots of 10 μl of 10^{-2} M racemic atenolol, metoprolol, and propranolol were applied to the layers and chromatograms were developed at 17°C in chambers equilibrated with the eluents for 25 min. Development time was 20 min and the migration distance was 15 cm. The spots were located with iodine vapors. Eluents were mixtures of acetonitrile/methanol/water in different ratios. Excellent separations were obtained for metoprolol and, above all, for propranolol, whereas the resolution of atenolol was much more difficult (see Table 5.5).

5.4 SILICA GEL COATED WITH ACID OR BASIC CHIRAL SELECTORS

In this section, we discuss the use of silica gel layers coated with acid or basic chiral selectors, whereas the amino acids employed as chiral impregnating agents will be discussed in the following section.

Paris et al. [15] in 1967 used plates constituted by a mixture of silica gel and D-galacturonic acid as impregnating agent. They obtained an excellent separation of ephedrine enantiomers, with an α-value of 2.37, employing an aqueous alcoholic eluent containing D-galacturonic acid (see Table 5.6). A successful separation of

TABLE 5.5
Retention (hR_f) and Resolution (α, R_S) Data for Racemic Propranolol, Metoprolol, and Atenolol on Plates of Silica Gel G (Merck) Impregnated with Cu (II)-L-Arginine Complex [14]

Racemate	hR_{f1}*	hR_{f2}*	α**	R_S***,a	Acetonitrile/methanol/water ratio (v/v/v)
Propranolol	33 (−)	51 (+)	2.11	2.8	15:2:1
Metoprolol	20 (−)	28 (+)	1.56	1.5	16:2[b]
Atenolol	16 (−)	20 (+)	1.31	1.0	18:4:2

Eluents: acetonitrile/methanol/water in different ratios. Temperature 17°C, development time 20 min, migration distance 15 cm.

*hR_f = R_f × 100.

**α = [(1/R_{f1}) − 1]/[(1/R_{f2}) − 1].

***R_S = 2 × (distance between the centers of two adjacent spots)/(sum of the widths of the two spots in the direction of development).

[a] Calculated from original chromatograms.

[b] Acetonitrile/methanol.

TABLE 5.6
Retention (hR_{f1}, hR_{f2}) and Resolution (α, R_S) Data for (±)-Ephedrine on Plates of Silica Gel Coated with Acid Chiral Selectors

hR_{f1}*	hR_{f2}*	α**	R_S***,a	Eluent and remarks	Ref.
34 (−)	55 (+)	2.37	—	• 10 × 20 cm plates of silica gel mixed with D-galacturonic acid • *Eluent*: isopropanol-1 M D-galacturonic acid (94:3, v/v)	[15]
10	18	1.97	1.8[a]	• *Plate*: silica gel G (Merck) coated with 0.5% L-(+)-tartaric acid • *Eluent*: acetonitrile/methanol/water (12:5:0.5, v/v/v)	[16]

*hR_f = R_f × 100.

**α = [(1/R_{f1}) − 1]/[(1/R_{f2}) − 1].

***R_S = 2 × (distance between the centers of two adjacent spots)/(sum of the widths of the two spots in the direction of development).

[a] Calculated from the original chromatogram.

the optical antipodes of ephedrine was also obtained with the same eluent on nonimpregnated layers, even if a lower selectivity was observed. Therefore, the impregnation of silica appeared to be less important than the use of a mobile phase containing the chiral selector. In 2001, the ephedrine racemate was resolved by Bhushan et al. [16] by coating the plates with L-(+)-tartaric acid; however, this chiral agent was not useful for the resolution of racemic atropine. The impregnated layers were obtained by spreading slurry of 50 g silica gel G (Merck) in 100 ml of 0.5% aqueous solution of recrystallized L-(+)-tartaric acid with a Stahl apparatus to obtain 20 × 20 cm plates (thickness 0.5 mm). The slurry had a pH included between 7 and 8. This kind of neutral silica gel (containing 13% gypsum, iron, and chloride 0.03% each) was adopted by Bhushan and coworkers in all the researches concerning coated stationary phases. The plates were dried at 60°C overnight and 10 μl of racemate or optical isomer solutions (10^{-3} M in 70% aqueous ethanol, corresponding to about 2 μg of analyte) were applied on the layers. Migration time was about 20 min and development distance was 13 cm. After development, the plates were dried at 60°C and visualization was performed by exposing the plates to iodine vapors. The results are reported in Table 5.6. High values of α (1.97) and R_S (1.80) were obtained. With respect to the temperature effect, the best resolution was achieved at 18°C, whereas no resolution was observed at 10°C, and elongated spots were obtained at 25°C. According to the authors [16] and as postulated by Dalgliesh [2], chiral discrimination occurred as a result of simultaneous interactions between three active positions of the chiral selector and as many active sites of the enantiomer. In particular, ionic interactions, hydrogen bonding, and van der Waals forces were certainly effective and allowed the *in situ* formation of diastereomeric salts.

Bhushan and Ali [17] resolved eight racemic phenylthiohydantoin (PTH) amino acids, out of the nine tested, on CCSPs similar to those aforementioned and prepared by impregnating 50 g of silica gel G with a 0.3% L-(+)-tartaric acid solution. The plates were dried at 60°C for 6–8 h. A solute amount equal to 0.5 μg was applied on the layer using a 10^{-4} M solution in ethyl acetate for each racemate. Before the chromatographic development in chloroform/ethyl acetate/water (28:1:1, v/v/v), the plates were dried at 60°C for 10 min. Visualization was performed by exposing the layers to iodine vapors. The results, reported in Table 5.7, showed that the only one unresolved racemate was PTH-Trp. The separation factors were extremely high; most of them were even one magnitude order higher than the highest α-values ever obtained in chiral liquid chromatography [18]. In this regard, however, a strong incongruence emerged by examining the photo of the original chromatogram that showed the resolution of the different racemates. Each racemate gave rise to two spots in which dimension and color intensity were very different. In particular, all the spots concerning D-enantiomers (the less retained) were much greater than those relating to the L-forms. This experimental evidence was not discussed by the authors [17] who, on the other hand, mentioned the light sensitivity of PTH-amino acids and the racemization ease of their optical antipodes.

More recently, the optical isomers of tartaric acid were used for impregnation of silica gel layers to separate optical antipodes of metoprolol tartrate in Presolol

TABLE 5.7

Retention (hR_{f1}, hR_{f2}) and Resolution (α, R_S) Data for Racemic PTH-Amino Acids on Plates of Silica Gel G (Merck) Coated with L-(+)-Tartaric Acid

Racemate	hR_{f1}*	hR_{f2}*	α**	R_S***,a	Ref.
PTH-Alanine (Ala)	12	55	8.96	6.3	[17]
PTH-Serine (Ser)	10	88	47.4	10.6	
PTH-Threonine (Thr)	30	84	13.2	7.0	
PTH-Valine (Val)	21	80	15.0	7.2	
PTH-Methionine (Met)	16	83	26.2	9.7	
PTH-Isoleucine (Ile)	15	92	70.8	11.1	
PTH-Phenylalanine (Phe)	15	85	32.2	8.8	
PTH-Tyrosine (Tyr)	16	95	101	11.1	
PTH-Tryptophan (Trp)	95	95	1.00	—	

Eluent: chloroform/ethyl acetate/water (28:1:1, v/v/v). Migration distance 10 cm.

*$hR_f = R_f \times 100$.

**$\alpha = [(1/R_{f1}) - 1]/[(1/R_{f2}) - 1]$.

***$R_S = 2 \times$ (distance between the centers of two adjacent spots)/(sum of the widths of the two spots in the direction of development).

[a] Calculated from the original chromatogram.

tablets, each containing 100 mg of such β-blocker [19]. No separation was obtained using L-(+)-tartaric acid, whereas the D-enantiomer gave rise to satisfactory results. Chromatography was effected by ascending technique, at 25°C, on 5 × 10 cm precoated plates of silica gel 60 $F_{254}s$ with concentrating zone (Merck), impregnated with 11.6 mM D-(−)-tartaric acid in 70% aqueous ethanol by predeveloping the plates for 90 min in Camag twin-trough TLC chambers, saturated with the mobile phase, were used. Migration distance was 6 cm and development time about 50 min. Visualization was performed both with UV light (254 nm) and iodine vapors. The eluent composition was similar to that of the impregnating solution. Under these experimental conditions, the retention values are as follows: $hR_f(+)(S) = 50$ and $hR_f(−)(R) = 65$. Values of α and R_S, calculated from the original chromatogram, were 1.86 and 2.0, respectively. With respect to the chiral discrimination mechanism, the authors suggested that monoethyl tartrate was the actual chiral selector adsorbed on silica gel surface that produced the separation of optical antipodes of metoprolol tartrate. The best esterification conditions included warming of impregnating solution at 70°C for 15 min, successive cooling and chromatographic elution as described above. According to this hypothesis, after solute application, the diastereomeric couples (−)-metoprolol/(−)-ethyl tartrate and (+)-metoprolol/(−)-ethyl tartrate, could be formed *in situ*.

An interesting application of this technique concerned the use of silica gel precoated plates impregnated with unusual chiral selectors, such as L-lactic acid

TABLE 5.8
Retention ((hR_{f1}, hR_{f2}) and Resolution (α) Data for Neutral Racemic Amino Acids on Silica Gel Plates Coated with (−)-Brucine

Racemate	pI	hR_{f1}*	hR_{f2}*	α**
Alanine	6.1	18	53	5.13
Serine	5.7	12	50	7.33
Threonine	5.7	16	29	2.14
Valine	6.0	25	25	1.00
Methionine	6.0	16	35	2.92
Isoleucine	5.7	18	29	1.86
Phenylalanine	5.9	27	40	1.80
Tyrosine	5.6	22	29	1.45
Tryptophan	5.9	17	31	2.19

Eluent: n-butanol/acetic acid/chloroform (3:1:4, v/v/v) [21].

*hR_f = R_f × 100.

**$\alpha = [(1/R_{f1}) - 1]/[(1/R_{f2}) - 1]$.

and L-phenylcarbamoyllactic acid, for the separation of enantiomeric *cis* and *trans* receptors found on tetrahydrobenzoxanthene skeleton [20]. The results showed that the layers coated with 1% of L-phenylcarbamoyllactic acid produced a good separation between the enantiomers of both receptors by eluting with chloroform/diethyl ether (9:1, v/v).

Bhushan and Ali [21] proposed the use of a basic chiral selector, (−)-brucine, for the resolution of racemic amino acids on silica gel layers coated with this reagent. In particular, the authors prepared 20 × 20 cm homemade plates (thickness 0.5 mm) by spreading a slurry consisting of 50 g silica gel G in 50 ml of 70% ethanol containing 0.1 g of (−)-brucine; the pH of the suspension was adjusted to 7.1–7.2 by adding 0.1 M sodium hydroxide. Plates were dried at 60°C for 12 h.

Visualization was performed by using ninhydrin (0.2% in acetone) after the layers were sprayed with 0.1% hydrochloric acid and kept at 60°C for 10 min. An aliquot of 0.5 μg of each amino acid was applied on the layers.

As shown in Table 5.8, nine neutral amino acids were tested with isoelectric point (pI) lower than pH of the suspension (pI range 5.7–6.1) so as to be present as anions and give rise to *in situ* formation of diastereomeric salts with the protonated form of the chiral selector, such as (+)AA/(−)brucine and (−)AA/(−)brucine. According to the authors [21], the two diastereomers had different solubility in the mobile phase and gave rise to different R_f values. Under such experimental conditions, very high α-values were obtained for eight enantiomeric couples, above all, for alanine (Ala) (5.13) and serine (Ser) (7.33). However, valine enantiomers were not separated and no explanation concerning this behavior was suggested. The D-enantiomer was always more retained than the L-form.

Later, the same research team used a suspension similar to the one aforementioned [30 g neutral silica gel G, 60 ml of aqueous ethanol, and 0.1 g (−)-brucine] to prepare CCSPs [22] suitable to resolve racemic profens. By using the two-dimensional technique, the optical isomers of two nonsteroidal anti-inflammatory drugs (NSAIDs), such as ibuprofen and flurbiprofen, were separated at 0.1 μg levels by eluting with acetonitrile/methanol (16:3, v/v) in the first direction and acetonitrile/methanol/water (16:3:0.4, v/v/v) in the second direction. This separation was of great importance, because such drugs were only administered as racemates, even if their optical isomers had different pharmacologic activity [23]. The resolution of the two racemates was observed only on layers prepared at pH ranging between 6 and 7. Poor or no resolution was obtained for more acidic pH (included between 4 and 5) or more basic pH (included between 9 and 10). This finding suggested that, as mentioned previously, the chiral recognition mechanism was based on the *in situ* formation of diastereomers with (−)-brucine. For these reasons, it was fundamental the presence of electrostatic interactions between the carboxylic group of the acids and the protonated basic group of the chiral selector.

An innovative procedure [24] was based on the deposition of 10 μl of a solution containing 0.24 mmol of racemic ibuprofen and 0.12 mmol of (−)-brucine on an achiral layer of silica gel G, not impregnated with (−)-brucine. Development was effected at 28°C for 20 min with acetonitrile/methanol (5:1, v/v), as mobile phase, in a saturated chamber. Spots were detected by iodine vapor. Two round and compact spots were obtained with R_f values of 0.71 and 0.85 for (−) and (+) isomer, respectively. Calculated separation factor was 1.32. The enantiomeric separation of ibuprofen was much better on achiral layers than on impregnated plates. Note the absence of other spots regarding the excess of racemate applied on the layer. Molar ratios with higher percentages of brucine were tested [i.e., 0.15 and 0.20 mmol of (−)-brucine] without satisfactory results. According to authors [24], these data confirmed that chiral recognition on silica gel G layers coated with (−)-brucine was due to the *in situ* formation of noncovalent bonded diastereomers, which were not broken during the chromatographic development and were separated owing to their different partition coefficients between mobile and stationary phases.

5.5 SILICA GEL COATED WITH AMINO ACIDS AS CHIRAL SELECTORS

The use of amino acids as chiral agents involved further possibilities of enantiomeric resolution owing to the simultaneous presence of acidic and basic groups. Such compounds can be classified in relation to the number of acidic and basic groups present in the molecule.

Amino acids used as chiral selectors were reported in Figure 5.2, together with their pIs and structures of zwitterions. The pI of proline (6.3) was reported instead of that of 2-azabicyclo[3.3.0]octane-3-carboxylic acid, as the latter value was not known and the structures of the two amino acids were similar.

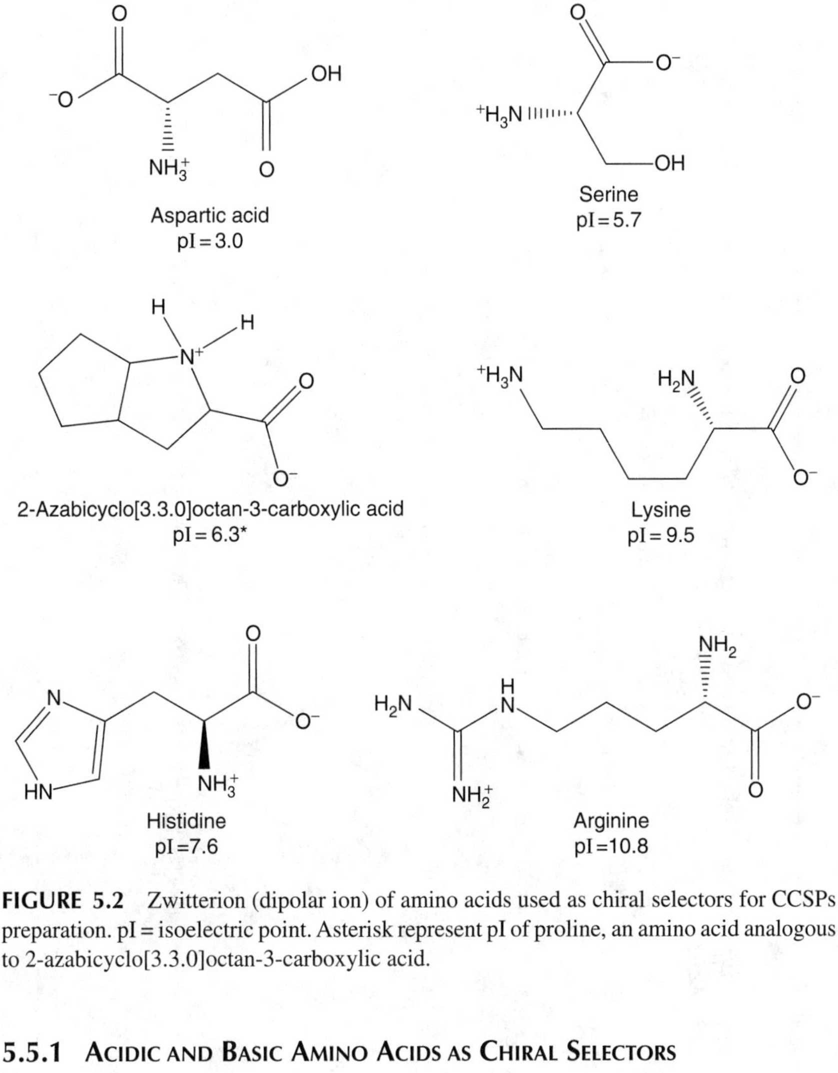

Aspartic acid
pI = 3.0

Serine
pI = 5.7

2-Azabicyclo[3.3.0]octan-3-carboxylic acid
pI = 6.3*

Lysine
pI = 9.5

Histidine
pI = 7.6

Arginine
pI = 10.8

FIGURE 5.2 Zwitterion (dipolar ion) of amino acids used as chiral selectors for CCSPs preparation. pI = isoelectric point. Asterisk represent pI of proline, an amino acid analogous to 2-azabicyclo[3.3.0]octan-3-carboxylic acid.

5.5.1 ACIDIC AND BASIC AMINO ACIDS AS CHIRAL SELECTORS

Bhushan and Ali [25] were the first to propose the use of silica gel G layers coated with acidic L-amino acids for the resolution of racemic alkaloids such as hyoscyamine (or atropine) and colchicine. Preparation of the plates, chromatographic development, and solute detection were performed as reported previously. The slurry of silica gel G was prepared in 100 ml of water containing 0.3 g of L-aspartic acid and chromatograms were developed at 0°C for 3.5 h. Migration distance was 10 cm eluting with n-butanol/chloroform/acetic acid/water (3:6:4:1, v/v/v/v). The extremely low temperature necessary for the enantiomeric separation should be noted as no resolution was observed at 5°C. Based on the results obtained (see Table 5.9), layers impregnated with this chiral selector were suitable for the resolution of the two racemates and particularly of atropine ($\alpha = 1.86$).

TABLE 5.9
Retention (hR_{f1}, hR_{f2}) and Resolution (α, R_S) Data for Racemic Alkaloids and β-Blockers on Layers of Silica Gel (Merck) Coated with L-Aspartic Acid

Racemate	hR_{f1}*	hR_{f2}*	α**	R_S***,a	Eluent and remarks	Ref.
Hyoscyamine	35	50	1.8	—	• *Slurry:* 50 g silica gel G, 100 ml water, and 0.3 g L-aspartic acid	[25]
	(+)	(−)	6			
Colchicine	65	70	1.2	—	• *Eluent: n*-butanol/chloroform/acetic acid/water (3:6:4:1, v/v/v/v). Temperature: 0°C	
	(+)	(−)	6			
Atenolol	25	40	2.0	3.0[b]	• *Slurry:* 50 g silica gel G, 100 ml water, and 0.5 g L-aspartic acid, some drops of ammonia. Temperature: 17°C	[26]
	(+)	(−)	0			
Metoprolol	32	42	1.5	1.3[c]		
	(+)	(−)	4			
Propranolol	25	41	2.0	2.1[d]		
	(+)	(−)	8			

*$hR_f = R_f \times 100$.

**$\alpha = [(1/R_{f1}) - 1]/[(1/R_{f2}) - 1]$.

***$R_S = 2 \times$ (distance between the centers of two adjacent spots)/(sum of the widths of the two spots in the direction of development).

[a] Calculated from the original chromatograms.

[b] Acetonitrile/methanol/water (18:4:2) as eluent.

[c] Acetonitrile/methanol/water (10:4:1) as eluent.

[d] Acetonitrile/methanol/water (16:5:0.5) as eluent.

Aspartic acid has a very acid pI (3.0) and was present in the anionic form at the eluent pH. Under such experimental conditions diastereomeric ion pairs can be formed between L-aspartic acid and the enantiomers of alkaloids, which were largely in the cationic form. However, electrostatic interactions, even if necessary, did not seem to be sufficient to explain the formation of diastereomers leading to enantiomeric separation. In fact, the use of L-glutamic acid as chiral selector did not produce any separation. Therefore, it was deduced that for the resolution of alkaloids, further interactions such as hydrogen bonding and steric interactions are required.

Bhushan and Arora [26] prepared layers similar to those described previously by using a slurry containing higher amounts of L-aspartic acid (0.5 g) and pH adjusted to 6–7 by ammonia, such that the chiral selector was anionic. The authors [26] used these layers for the resolution of racemic β-blockers, developing the chromatograms at 17°C for 20 min. As shown in Table 5.9, all the investigated β-blockers were baseline resolved exhibiting very high R_S-values (calculated from the original chromatograms), in agreement with the good compactness of the spots, especially for atenolol. Using this simple and reliable technique the detection of only 0.25 μg of each compound was achieved.

Basic amino acids were the most investigated as chiral selectors and many interesting separations were obtained for acid and basic enantiomers.

In 1996, Bhushan and Parshad [27] resolved the optical antipodes of an acid compound (ibuprofen) on silica gel G layers impregnated with a 0.5% aqueous solution of basic L-arginine. As the amino acid pI was very high (10.8), trace amounts of acetic acid were needed to obtain arginine in the cationic form. Aliquots (10 μl) of 10^{-3} M racemic ibuprofen in 70% ethanol were applied to the layer. Visualization was performed by using iodine vapors. In addition, after iodine evaporation, detection of the two separated diastereomers was effected by spraying with hydrochloric acid, heating for 15 min and again spraying with 0.2% ninhydrin in acetone. Although the whole arginine impregnated plate reacted with ninhydrin, the diastereomers appeared like characteristic colored spots on a pink background. An acetonitrile/methanol/water (5:1:1, v/v/v) mixture was used as an eluent, both for one- and two-dimensional chromatography. After development at 32°C for 15 min in the first direction and for 20 min in the second direction, calculated hR_f values were 77 for (−) and 80 for (+) isomer, with a ΔR_f equal to 0.03.

This separation was deeply investigated in 2004 by Kowalska and coworkers [28] with an aim to improve the method accuracy and precision by using standardized and commercially available plates, such as 20 × 20 cm silica gel 60 F$_{254}$ (#1.05715; Merck), thickness 0.25 mm, and adopting experimental conditions reported previously [27]. The plates were predeveloped in methanol/water (9:1, v/v), dried at room temperature, and coated by dipping for 2 s in a 3 × 10^{-2} M solution of L-arginine in methanol. Visualization of ibuprofen was performed by UV light (λ = 210 nm) by using the Desaga model CD 60 densitometer (Heidelberg, Germany) with Windows compatible ProQuant software. Elution was effected using the mixture reported by Bhushan and Parshad [27], modified by adding acetic acid up to pH 4.8 to have arginine in the cationic form. Both one- and two-dimensional modes were employed with a migration distance of 15 cm. Using two-dimensional technique, a significant enhancement of the resolution was observed for the two enantiomers [$hR_f(-) = 76$; $hR_f(+) = 83$]. In fact, ΔR_f was 0.07 vs. the value of 0.03 obtained by Bhushan and Parshad [27]. However, the experimental conditions adopted within the two methods cannot be compared, since the paper by Kowalska and coworkers [28] in 2004 did not report the working temperature, which was a fundamental parameter. On the other hand, in the paper by Bhushan and Parshad [27], the pH of the slurry was not reported making impossible to know if the same acidity conditions were adopted. Such information was crucial since arginine contained a large number of basic groups (see Figure 5.2).

The same technique was used by Kowalska and coworkers [29] to investigate the oscillatory transenantiomerization of some profens [i.e., S-(+)-ibuprofen, S-(+)-naproxen, and S,R-(±)-2-phenylpropionic acid] when stored in 70% ethanol at two different temperatures (6°C and 22°C). In this work, the instability of various optical antipodes was pointed out and attributed to change their steric configurations from the S-form to the R-form, and vice versa. Such phenomenon, which took place by a keto-enol tautomerism, was much more remarkable at low temperatures.

According to the same authors [30], tautomeric isomerization proceeded through the migration of a proton from the moiety of a profen molecule to another in aqueous alcoholic medium. A suppression of the oscillatory transenantiomerization of profens was observed storing them in dichloromethane, owing to their much less pronounced electrolytic dissociation.

Successively, Kowalska and coworkers [31] investigated the ability of profens to change their chiral configuration suggesting that it was due to their gelating properties and to a consequent increase in solution viscosity. Adequate evidences supporting the hypothesis that keto-enol tautomerism was the cause of the observed oscillations were given. The reaction rates of keto-enol tautomerism increased in a basic environment and were hampered in an acid medium. The investigation was focused on S-(+)-naproxen stored in basic and acid solutions and pointed out that, in an acid medium, the S-(+) form was stabilized and the transenantiomeric oscillations were hampered. On the contrary, base-catalyzed tautomerism facilitated the partial transformation of S-(+)-naproxen to the R-(−)-form. Finally, two-dimensional chiral separation of S-(+)-naproxen from the R-(−)-form, produced during the storage in basic and acidic media, was reported.

Sajewicz et al. [32] studied the marked and systematic deviations from the vertical of the migration tracks of enantiomers of 2-arylpropionic acids during experiments performed by TLC in the ascending mode. This behavior was evidenced for all three investigated racemates (ibuprofen, naproxen, and 2-phenylpropionic acid) by using precoated silica gel 60 F_{254} plates impregnated with L-arginine. The maximum deviation of the migration tracks increased with increasing resolution of the enantiomer pairs. In fact, deviation from the vertical was 4(±2) mm for ibuprofen ($\Delta R_f = 0.03$), 6(±2) mm for naproxen ($\Delta R_f = 0.04$), and 7(±2) mm for 2-phenylpropionic acid ($\Delta R_f = 0.10$).

It should be noted that, while deviations (left or right) of ibuprofen and naproxen enantiomers were equal, different values were obtained for optical antipodes of 2-phenylpropionic acid [S-(+): 5 ± 1 mm right; R-(−): 2 ± 1 mm left].

Basic amino acids were also tested as chiral selectors for the separation of enantiomeric bases such as β-blockers [33,34] and atropine [16]. In particular, Bhushan and Thiong'o [33] employed silica gel G layers coated with 0.5% L-arginine or L-lysine solutions to separate the optical isomers of β-adrenergic blocking agents as propranolol, metoprolol, and atenolol. Different binary mixtures of acetonitrile/methanol were used as eluents. The first two racemates were resolved at 22°C, while atenolol at 15°C. Note that the resolution of propranolol was possible in a wide range of temperatures, up to 6°C. As shown by the results reported in Table 5.10, high-resolution factors were obtained for the three racemates, particularly for propranolol, on layers impregnated with two basic amino acids. However, taking into account spots, compactness and dimension, the best resolutions for the three drugs were achieved employing silica gel layers coated with acidic amino acids [26], as shown by the data of Table 5.9. Even in this paper [33] the *in situ* formation of diastereomeric salts among racemates (in the cationic form) and the specific chiral selector (in the anionic form) was hypothesized. In fact, the pH of the slurry was adjusted to values greater than pI of lysine (9.5) and arginine (10.8), by adding some ammonia drops, to maintain the chiral selectors in the anionic form. However, at pH values above the pI of arginine (10.8), very low percentages of β-blockers were in the cationic form, owing to their high-pK_a values, included

---

TABLE 5.10

Retention (hR_{f1}, hR_{f2}) and Resolution (α, R_S) Data for Some Racemic β-Blockers on Silica Gel G (Merck) Layers Impregnated with L-Lysine (Lys) or L-Arginine (Arg)

Racemate	Chiral selector	hR_{f1}*	hR_{f2}*	α**	R_S***,a	Eluent: acetonitrile/ methanol (v/v)
Atenolol	L-Lys	3 (−)	10 (+)	3.59	1.2	16:4
	L-Arg	22 (−)	30 (+)	1.52	—	14:6
Propranolol	L-Lys	4 (−)	15 (+)	4.23	—	16:2
	L-Arg	15 (−)	39 (+)	3.62	—	15:4
Metoprolol	L-Lys	15 (−)	25 (+)	1.89	—	15:5
	L-Arg	6 (−)	17 (+)	3.21	—	15:3

Migration distance 13 cm; development time about 40 min; visualization: iodine vapor [33].

*$hR_f = R_f \times 100$.

**$\alpha = [(1/R_{f1}) - 1]/[(1/R_{f2}) - 1]$.

***$R_S = 2 \times$ (distance between the centers of two adjacent spots)/(sum of the widths of the two spots) in the direction of development).

a Calculated from the original chromatograms.

between 9.14 for propranolol and 9.17 for metoprolol. Thus, the formation of diastereomeric ion pairs must be excluded.

An investigation aimed to assess the best experimental conditions for the resolution of racemic propranolol was carried out by Kowalska and coworkers [34] on precoated silica gel 60 F_{254} plates (#1.05715; Merck, Darmstadt, Germany) impregnated with L-arginine, maintaining as much as possible the working conditions reported by Bhushan and Thiong'o [33]. In particular, the effect of the eluent pH on chiral resolution was investigated by using an acetonitrile/methanol mixture (15:4, v/v) containing different amounts of ammonia (pH included between 7.75 and 11.0). S-(−)-propranolol gave rise to R_f values included between 0.01 and 0.04 (the highest at pH 10.9), while the R-(+) form showed increasing R_f from 0.11 to 0.18 with the increase of pH. Visualization was performed at 210 nm with densitometric scanning using a Desaga CD 60 densitometer (Heidelberg, Germany).

Again, the purpose of this research was to obtain a chiral resolution of this important β-blocker employing standardized commercially available plates and densitometric detection of analytes so as to improve the precision and robustness of the procedure.

Resolution of racemic atropine ($\alpha = 1.88$ and $R_S = 1.5$) was observed on layers of silica gel G coated with 0.5% L-histidine solution, where pI is much lower (7.6) than those of the basic amino acids mentioned previously. This allowed to adopting pH of the slurry included between 7 and 8, so as to ensure the presence of anionic histidine and cationic atropine. Acetonitrile/methanol/water mixture (11.3:5.4:0.6, v/v/v) was used as an eluent. The best resolution of atropine enantiomers was observed at 18°C [16].

5.5.2 NEUTRAL AMINO ACIDS AS CHIRAL SELECTORS

The nonproteinogenic (1R,3R,5R)-2-azabicyclo[3.3.0]octan-3-carboxylic acid was used as a chiral selector for the resolution of both derivatized and nonderivatized racemic amino acids [12,35]. In particular, Bhushan et al. [12] used 0.5% aqueous solution of this selector for the impregnation of silica gel G layers, using the same procedure described previously. Employing binary or ternary mixtures, consisting of 0.5 M sodium chloride/acetonitrile or 0.5 M sodium chloride/acetonitrile/methanol, the separation of the optical antipodes for six Dns-amino acids was achieved (see Table 5.11). High α-values were associated with lower R_S factors, owing to tailing spots, particularly for Dns-Phe.

Much more interesting results were achieved on plates impregnated with the same chiral selector under different experimental conditions with respect to those reported previously (see Table 5.11), which led to the separation of enantiomers of some basic and neutral amino acids [35].

In such a case, α-values as high as R_S were observed, as round and compact spots were obtained.

Besides amino acids reported in Table 5.11, racemic asparagine, isoleucine, methionine, norleucine, phenylalanine, and alanine were also partially resolved, showing a typical eight shaped spot, by elution with acetonitrile/methanol/water

TABLE 5.11
Retention (hR_{f1}, hR_{f2}) and Resolution (α, R_S) Data for Racemic Compounds on Silica Gel G Plates Impregnated with (1R,3R,5R)-2-Azabicyclo[3.3.0]Octan-3-Carboxylic Acid [12,35]

Racemate	hR_{f1}*	hR_{f2}*	α**	R_S***,a	Eluent	Ref.
					0.5 M NaCl/acetonitrile/ MeOH (v/v/v)	
Dns-Val[b]	38 (L)	49 (D)	1.57	—	15:1	[12]
Dns-Nva[b]	56 (L)	61 (D)	1.23	—	17:2:0.4	
Dns-Leu[b]	64 (L)	68 (D)	1.19	1.0	10:4:1	
Dns-Phe[b]	50 (L)	65 (D)	1.86	1.2	15:2	
Dns-Trp[b]	23 (L)	34 (D)	1.72	—	20:0.5	
Dns-Asp[b]	30 (L)	52 (D)	2.53	—	20:0.5	
					Acetonitrile/methanol/ water (v/v/v)	
Val[c]	38 (D)	50 (L)	1.63	—	10:5:2	[35]
Leu[c]	11 (D)	21 (L)	2.15	3.1	10:4:3	
His[c]	18 (D)	25 (L)	1.52	—	7:6:2	
Lys[c]	15 (D)	31 (L)	2.55	2.8	10:52	
Arg[c]	6 (D)	13 (L)	2.34	—	7:6:3	

*hR_f = R_f × 100.

**$\alpha = [(1/R_{f1}) - 1]/[(1/R_{f2}) - 1]$.

***R_S = 2 × (distance between the centers of two adjacent spots)/(sum of the widths of the two spots in the direction of development).

[a] Calculated from original chromatograms.

[b] Slurry of silica gel G (50 g) in 100 ml water containing 0.5 g of the chiral selector. Detection: UV light (254 nm); migration distance 10 cm; development time about 30 min. Temperature: not reported.

[c] Slurry of silica gel G (30 g) in 60 ml methanol/water (8.3:91.7, v/v) containing 0.011 M chiral selector. Detection: 0.2% ninhydrin in acetone; migration distance 8.5 cm; development time about 35 min. Temperature: 20°C for valine (Val) and 26°C for leucine (Leu), histidine (His), lysine (Lys), and arginine (Arg).

(10:4:2, v/v/v). As pH of the slurry was adjusted to 8 by ammonia (at pH 4 and 6 no separation of neutral and basic amino acids was observed), the authors [35] proposed not only the usual mechanism of diastereomeric ion pairs formation for more basic analytes such as lysine and arginine (i.e., ionic interactions between carboxylate group of the chiral selector and analyte in the cationic form), but also suggested that ionic interactions between the protonated amino group of the chiral selector (?) and the carboxylate group of hystidine, leucine, and valine occurred, because at pH 8 such analytes exist as anions.

Successively, in 2003, Heard and coworkers [36] investigated the resolution of some profens (see Figure 5.3 for their structures) on plates prepared from a

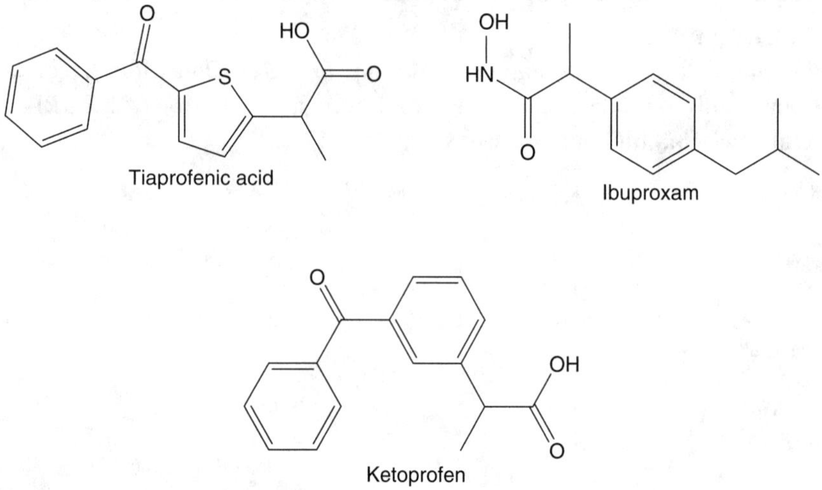

Tiaprofenic acid

Ibuproxam

Ketoprofen

FIGURE 5.3　Chemical structures of 2-arylpropionic acids tested on silica gel G layers impregnated with L-serine.

slurry of 30 g silica gel G layers (Fluka, Switzerland) in 70 ml of water/ethanol solution (97:3, v/v) containing 0.1 g of L-serine. The results were reported in Table 5.12. Both α- and R_S-values were very high, evidencing the excellent separations obtained under these experimental conditions. With respect to tiaprofenic acid and ibuproxam, α factors were found to be exceptionally high owing to the very high R_f values of $S(-)$ forms that moved near to the solvent front. The best resolution was observed at pH included between 6 and 7, at 25°C. The change in temperature showed poorer or no enantioresolution.

With respect to the effect of the impregnating reagent concentration, no resolution was obtained for tested racemates when L-serine was increased or decreased to 0.2 or 0.05 g, respectively.

The authors [36] explained this behavior by invoking the formation of diastereomeric ion pairs within the confines of the pores of silica gel to achieve enantioresolution. The excess of chiral agent on the plate permitted to form diastereomers also outside of the pores, without simultaneous steric interactions; this involved a significant worsening of chiral discrimination capacity of the impregnated layers.

5.5.3　Mechanism of Chiral Recognition on Silica Gel Layers Coated with Acids, Bases, or Amino Acids as Chiral Selectors

The enantioresolution mechanism proposed by Bhushan and Martens [37] for silica gel layers coated with acid, basic, or amphionic chiral agents involved the *in situ* formation of noncovalent-bonded diastereomers between the A(+) and

TABLE 5.12
Retention (hR_{f1}, hR_{f2}) and Resolution (α, R_S) Data for Some 2-Arylpropionic Acids on Silica Gel G Plates (20 × 20 cm × 0.5 mm) Impregnated with L-Serine

Racemate[a]	hR_{f1}*	hR_{f2}*	α**	R_S***,[b]	Eluent: acetonitrile/ methanol/ water (v/v/v)
Tiaprofenic acid	53 (R)	93 (S)	11.8	4.5	16:3:0.5
Ibuproxam	28 (R)	95 (S)	49.5	7.2	16:4:0.5
Ketoprofen	57 (R)	83 (S)	3.68	3.0	16:4:0.5

Migration distance 15 cm; development time 45 min; visualization: iodine vapor. Temperature: 25°C [36].

*$hR_f = R_f \times 100$.

**$\alpha = [(1/R_{f1}) - 1]/[(1/R_{f2}) - 1]$.

***$R_S = 2 \times$ (distance between the centers of two adjacent spots)/(sum of the widths of the two spots in the direction of development).

[a] 10 μl volume of 10^{-3} M solution in 70% ethanol.

[b] Calculated from original chromatograms.

A(−) enantiomers and the chiral selector (CS) adsorbed onto the support. The two diastereomers have different interphase distribution, leading to different mobilities.

The diastereomeric ion pairs [(A(+)/CS) and (A(−)/CS)] were held together through ionic interactions, such as carboxylate group/ammonium ion interactions, hydrogen bonding, steric interactions, and van der Waals forces. The fundamental parameters determining the success of chiral resolution on the aforementioned CCSPs were the following:

1. Ratio between the amount of support and that of the chiral selector used for impregnation
2. pH of the slurry employed for the preparation of the coated layer
3. Temperature of chromatographic development

Type and composition of the mobile phase (usually constituted by ternary organic or aqueous organic mixtures such as n-butanol/acetic acid/chloroform and acetonitrile/methanol/water) play a role as well. Even binary and quaternary mixtures, consisting of acetonitrile/methanol and n-butanol/acetic acid/chloroform/water, were used with various proportions.

The aforementioned parameters were experimentally determined each time since the best resolution was observed for specific w/w ratios between the support and the chiral agent, as well as within a very narrow range of the slurry pH and temperature. Among these parameters, accurate information was only available about pH of the slurry, which has to ensure the simultaneous presence of analyte and chiral selector oppositely charged.

With acidic (or basic) impregnating agents the separation of basic (or acidic) analytes can be hypothesized at a pH near to neutrality, in order to have large percentages of analyte and chiral selector in opposite ionic forms. In the case of chiral agents consisting of amino acids, attention should be paid to their pI. With respect to the resolution of racemic bases containing an aliphatic amino group, the acidic amino acids employed, such as L-aspartic acid, did not involve substantial differences with the use of acids such as (−)-tartaric acid, because at a pH value close to neutrality, aspartic acid (pI = 3.0) will be surely in the anionic form.

The use of basic amino acids for the resolution of acidic optical antipodes needed weakly acid pH so as to have the chiral selector in the cationic form (i.e., pH 4.8 [28] for the resolution of ibuprofen with L-arginine). With respect to the resolution of β-blockers, which contained an aliphatic amino group, the choice of experimental conditions appeared to be more complex, as basic amino acids such as L-lysine (pI = 9.5) and, above all, L-arginine (pI = 10.8) showed pI values higher than pK_a of β-blockers (about 9.15). pH values equal to or little greater than 10.8 ensured the presence of small amounts of the anionic form of L-arginine (being predominant the dipolar ion) and, overall, of the cationic form of β-blockers. Under such conditions ionic interactions similar to those previously hypothesized were not possible.

This aspect of the problem was not investigated by Bhushan and Thiong'o [33], even though they performed the separations of β-blockers with L-arginine at a pH value higher than pI of the amino acid; furthermore, without pH value was also reported. The problem was studied in-depth by Kowalska and coworkers [34], who hypothesized the formation of two pseudo-diastereomers between the enantiomers of unionized propranolol and anionic L-arginine, based on hydrogen bonding as shown in the following scheme:

$$L\text{-}Arg^{(anion)} + S\text{-propranolol} \overset{K_1}{\rightleftharpoons} \text{H-bonded pseudo-diastereomer}$$

$$L\text{-}Arg^{(anion)} + R\text{-propranolol} \overset{K_2}{\rightleftharpoons} \text{H-bonded pseudo-diastereomer}$$

The two reactions were characterized by different thermodynamic equilibrium constants, K_1 and K_2, which determined different R_f values for the two optical antipodes of propranolol.

This behavior was different from what was occurring between ibuprofen enantiomers and L-arginine, which gave rise to diastereomeric ion pairs (or salts).

In effects, enantiomeric separations of neutral and basic amino acids, obtained in 2000 on layers impregnated with a neutral amino acid such as (1R,3R,5R)-2-azabicyclo[3.3.0]octan-3-carboxylic acid [35], pointed out that the species involved in the formation of diastereomers could be more different, in contrast to what was hypothesized by Bhushan et al. [35]. In particular, working at pH 8, a value higher than pI of the nonproteinogenic amino acid, and the anionic form together with that of the dipolar ion were predominant for the chiral selector. Under such experimental conditions, ionic interactions between the anionic impregnating agent and the more basic amino acids lysine and arginine, which were in cationic form, were expected but not those involving less basic amino acids such as hystidine or even more so neutral amino acids, with pI $\leq$ 6. In fact, at pH 8, the latter compounds were present in the anionic form together with that of the dipolar ion. An anion/cation interaction between analyte and chiral selector to form diastereomeric ion pairs cannot therefore occur, while it is much more probable that anion/dipolar ion or even zwitterion/zwitterion interactions took place.

Amounts of chiral selector equal to 0.5 or 0.3 g were generally used for the impregnation of 50 or 30 g of silica gel G, respectively; however, ratios between the support and the chiral selector higher than 100 were also adopted for layer impregnation [17,21,22,25].

The influence of molar ratio racemate/chiral selector on the enantioseparation was scarcely investigated even though, in the case of silica gel achiral layers, it appeared crucial for enantiomeric resolution of ibuprofen [24]. In fact, only diastereomeric ion pairs formed with a molar ratio ibuprofen/($-$)-brucine (2:1) gave rise to well-resolved spots.

Also, Heard et al. [36] highlighted the importance of the chiral agent amount present on the layer when they observed that the increase in L-serine concentration over a certain level gave rise to a worsening of the chiral resolution. The authors [36] hypothesized that to obtain a chiral discrimination, the formation of diastereomers should occur within the pores of silica gel, as steric interactions are also in operation. When the concentration of chiral selector exceeded the pore saturation level, interactions between racemate and impregnating agent occurred outside the pores without the separation of optical isomers, as the steric component was absent. Unfortunately, no data concerning pore volume and average pore diameter of the adopted support was reported in literature, although amorphous silica prepared via spontaneous polymerization and dehydration of aqueous silicic acid was a porous material with a pore diameter ranging from 10 to 15,000 Å [38]. Silica gel of medium (60–100 Å) pores was generally used.

As experiments performed on L-serine impregnated layers [36] demonstrated that the chiral selector was uniformly immobilized on silica gel G even after elution, it was hypothesized a resolution mechanism based on the formation of diastereomers with different thermodynamic stability within the silica pores, where the impregnating agent was immobilized. Therefore, the behavior of CCSPs and CSPs could be considered similar.

Chiral interactions between analyte and chiral selector appeared to be influenced by temperature [39] and this parameter placed a fundamental role in the chiral separations on impregnated layers. The best conditions cannot be absolutely foreseen, as experimental studies demonstrated that temperatures can also differ in about 30°C, according to the considered separations, such as 0°C for racemic hyoscyamine with L-aspartic acid [25] and 32°C for racemic ibuprofen with L-arginine [27]. These results appeared to be in agreement with a chiral discrimination mechanism based on the diverse thermodynamic stability of the diastereomers formed within the pores of silica gel, rather than on their different interphase distributions. In fact, as suggested by Heard and coworkers [36], the change in temperature involved a higher or a less swelling of silica gel and a corresponding variation of pore volume and diameter. Only pores with certain dimensions show optimum steric effects and a tendency for complete separation of the optical antipodes.

5.6 SILICA GEL IMPREGNATED WITH MACROCYCLIC ANTIBIOTIC AS CHIRAL SELECTORS

Bhushan and Parshad [40] used silica gel G layers impregnated with the macrocyclic antibiotic (−)-erythromycin for the resolution of enantiomeric dansyl amino acids. The structure proposed for this compound was reported in Figure 5.4a. Among erythromycins, in fact, only erythromycin A was certified with $pK_a = 8.6$ and specific optical rotation $[\alpha]_{25,D} = -78°C$ ($c = 1.9$ in methanol). Erythromycin employed by the authors [40], purchased from Ranbaxy (India), was not well characterized but purified before use by dissolving 1 g of material in the minimum chloroform volume at 40°C and successively recrystallizing it at −15°C; thus, after vacuum drying, colorless crystals were obtained.

Layer impregnation was performed as usual by mixing 50 g of silica gel G in 100 ml of water containing 0.05 g of (−)-erythromycin without specifications concerning pH of the slurry [probably, pH was near to neutrality, to have (−)-erythromycin in the cationic form].

Results are reported in Table 5.13. Ten racemic amino acids were resolved, with the exception of threonine and the group of basic amino acids. High α-values (included in the range 1.31–1.76) and satisfactory spot compactness were obtained.

Table 5.13 also showed the separations obtained for the same compounds on layers coated with a slurry at pH 6, consisting of 30 g silica gel G in 60 ml water containing 0.34 mM (−)-vancomycin (Abbot Laboratories, Chicago, IL, USA). Vancomycin is an amphoteric glycopeptide with a molecular weight of 1449 (see Figure 5.4b). There are three macrocyclic portions and five aromatic rings in the molecule. The resolution of threonine enantiomers was achieved by slurry, where pH value was adjusted to 4 by adding some drops of acetic acid. The eluents were acetonitrile/0.5 M aqueous NaCl (10:4 and 13:4, v/v) and acetonitrile/0.5 M aqueous NaCl/2-propanol in different proportions for serine and threonine. Visualization was performed with UV light at 254 nm, working temperature at 18°C, and the migration distance was 8.5 cm [41]. Under these experimental conditions nine racemic dansyl amino acids with the exception of Dns-Asp and Dns-Glu were resolved using separation factors higher than those found with (−)-erythromycin, even if the obtained spots were scarcely compact and, in some cases, strongly tailed.

As (−)-vancomycin has a pI = 7 and is stable only within the pH range from 4 to 9, separations were obtained at pH values lower than pI to have more positive than negative charges.

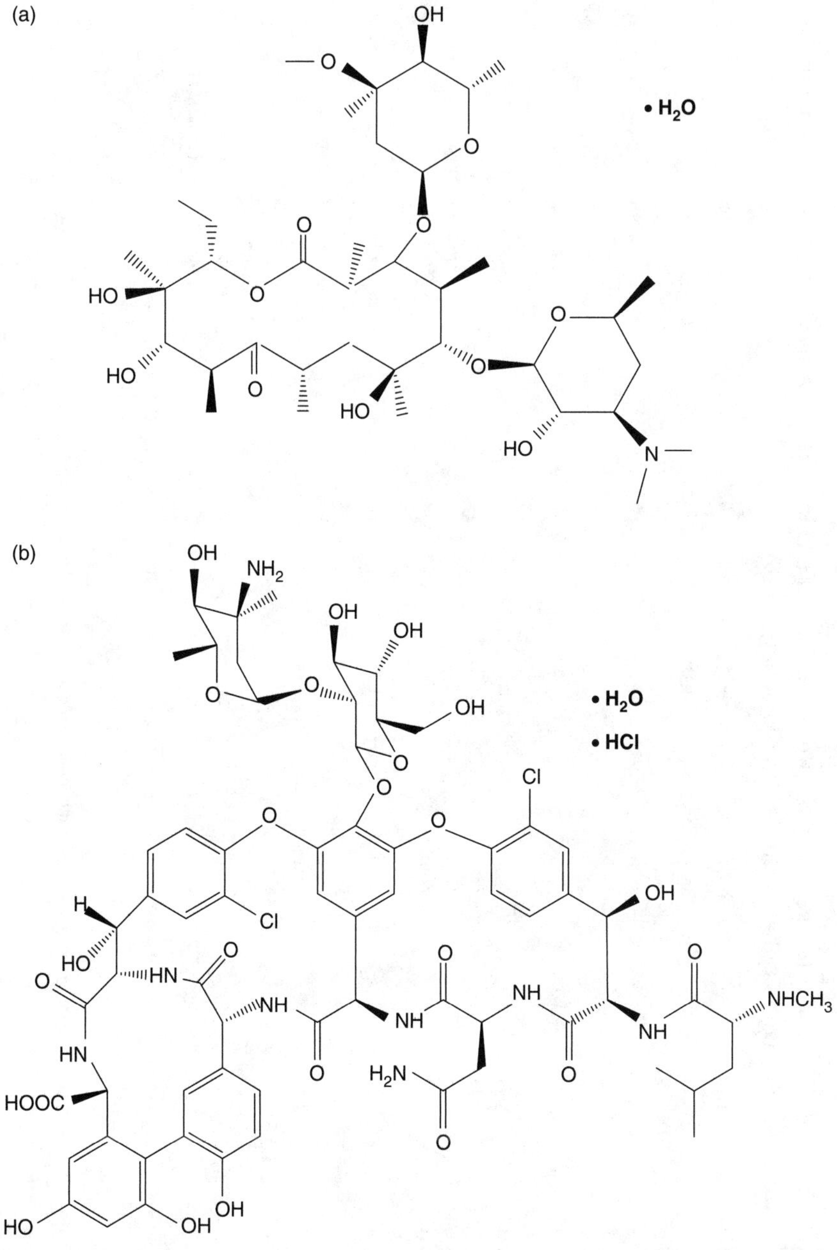

(a)

(b)

FIGURE 5.4 Chemical structures of (a) (−)-erythromycin and (b) (−)-vancomycin hydrochloride.

TABLE 5.13

Retention (hR_{f1}, hR_{f2}) and Resolution (α, R_S) Data for Resolved Racemic Dansyl (Dns)-Amino Acids and Verapamil on Silica Gel Plates Coated (CCSP) with Erythromycin (A) and Vancomycin (B)

Racemate	CCSP	hR_{f1}*	hR_{f2}*	α**	R_S***,a	Eluent and remarks
Dns-Ser	A	30 (L)	36 (D)	1.31	—	15:1:1; $T = 25°C$
	B	82 (L)	94 (D)	3.44	—	ACN + 0.5 M NaCl + 2-propanol (10:4:1)
Dns-Thr	B	75 (L)	88 (D)	2.45	—	ACN + 0.5 M NaCl + 2-propanol (15:3:1)
Dns-Val	A	22 (L)	30 (D)	1.52	—	15:1:0; $T = 25°C$
	B	73 (L)	87 (D)	2.48	—	ACN + 0.5 M NaCl (10:4)
Dns-Nva	B	73 (L)	87 (D)	2.48	—	ACN + 0.5 M NaCl (10:4)
Dns-n-But[b]	A	42 (L)	51 (D)	1.44	—	12:1:0; $T = 25°C$
	B	73 (L)	87 (D)	2.48	—	ACN + 0.5 M NaCl (10 + 4)
Dns-Leu	A	24 (L)	32 (D)	1.49	—	15:1:0; $T = 25°C$
	B	73 (L)	87 (D)	2.48	2.0	ACN + 0.5 M NaCl (10:4)
Dns-Nle	A	63 (L)	71 (D)	1.44	—	16:1:0:0.4; AcOH $T = 25°C$
	B	73 (L)	87 (D)	2.48	1.8	ACN + 0.5 M NaCl (10:4)
Dns-Met	A	56 (L)	63 (D)	1.34	—	25:2:0.5; $T = 34°C$
Dns-Phe	A	50 (L)	65 (D)	1.76	—	15:2:0; $T = 25°C$
	B	68 (L)	79 (D)	1.77	—	ACN + 0.5 M NaCl (10:4)
Dns-Trp	A	38 (L)	47 (D)	1.45	—	18:1:0.25; $T = 25°C$
	B	73 (L)	87 (D)	2.48	—	ACN + 0.5 M NaCl (10:4)
Dns-Asp	A	50 (L)	63 (D)	1.70	—	28:1.5:0.5; $T = 25°C$
Dns-Glu	A	45 (L)	56 (D)	1.55	—	22:1:0.5; $T = 25°C$
Verapamil[c]	B	39 (−)	49 (+)	1.50	1.2	ACN + MeOH + H$_2$O (15:2.5:2.5)

Eluent for (A) 0.5 M NaCl + acetonitrile (ACN) + methanol (MeOH) in different proportions. Temperature for (B): 18°C. Detection: UV light (254 nm) [41,42].

*$hR_f = R_f \times 100$.

**$\alpha = [(1/R_{f1}) - 1]/[(1/R_{f2}) - 1]$.

***$R_S = 2\times$ (distance between the centers of two adjacent spots)/(sum of the widths of the two spots in the direction of development).

a Calculated from original chromatograms.

b α-Amino-n-butyric acid.

c Detection: iodine vapor.

The layers coated with (−)-vancomycin were also successfully used for the separation of enantiomers of verapamil, a calcium channel blocker usually administered as racemic compound, even though the S-(−)-isomer was 20–30 times pharmacologically more active than the R-(+)-form [42]. The results are reported in Table 5.13. Amounts as low as 0.075 μg of each enantiomer were detected by iodine vapor at working temperature 18°C.

The enantioseparation mechanism of silica gel G layers coated with macrocyclic antibiotics could be a consequence of various kinds of interactions such as hydrogen bonding, coulombic interactions, steric interactions, and inclusion in a hydrophobic pocket, which led to the formation of two diastereomers having different thermodynamic stability.

5.7 SILICA GEL IMPREGNATED WITH OPTICALLY ACTIVE *N*-(3,5-DINITROBENZOYL)AMINO ACIDS (PIRKLE SELECTORS)

Pirkle selectors were characterized by π–π interactions (aromatic π–π bonding interactions) as crucial aspect of the retention process. Such interactions were well known and occurred between two types of molecules: the so-called π-donors and π-acceptors. A π-donor will tend to loose an electron since the resulting positive charge will be stabilized through the π-system. An opposite behavior will occur for the π-acceptor.

In this way, a π-donor/π-acceptor pair will form a complex in which a charge can be transferred from the donor to the acceptor, according to the so-called charge-transfer (CT) complexation technique.

A milestone in the use of this technique was the introduction of the N-(3,5-dinitrobenzoyl)amino acids as chiral immobilized CT–acceptor ligands. In particular, Pirkle et al. [43,44] prepared CSPs by covalently binding the chiral selector to a silica gel support and used them for the separation of several optical isomers by column chromatography.

Successively, Wainer et al. [45] employed the π-acceptor N-(3,5-dinitrobenzoyl)-R-(−)-α-phenylglycine (CS1), bound to an aminopropyl silanized silica gel through ionic interactions, for the resolution of racemic 2,2,2-trifluoro-1-(9-anthryl)ethanol (5 μg) by planar chromatography eluting with n-hexane/isopropanol (9.5:1, v/v). CS1 was bound to Zorbax BP-NH$_2$ plates (DuPont, USA) by continuous development with a tetrahydrofuran solution (20 ml) containing 1 g of the chiral selector.

Two fluorescent spots were observed with R_f values of 0.49 and 0.59 and α-value of 1.50, which was higher than the 1.33 reported by Pirkle et al. [44] on column chromatography.

Further studies described the use of silica gel 60 HPTLC NH$_2$ F$_{254}$s plates (10 × 10 cm; Merck), modified both ionically and covalently with the same chiral selector, to separate the aforementioned aromatic alcohol by eluting with n-hexane/isopropanol mixtures in different proportions [46]. This study highlighted the main reasons concerning the limited use in planar chromatography of some CSPs that, on the contrary, were largely employed in HPLC. Although the precoated plates contained a fluorescent indicator, the treatment with CS1 destroyed the fluorescence due to such indicator as well as that due to fluorescent solutes, when they were not present in adequate amounts.

The major disadvantage related to the use of covalent-bonded phases was the background "noise" due to the darkening of the layer, which strongly reduced the use of densitometry for the *in situ* determination of analytes.

Ionically bonded plates, on the contrary, were slightly pink colored and gave rise to minor problems for solute visualization. The darkening of the plates can be strongly reduced by storing them in the dark. In particular, when concentrations of racemic 2,2,2-trifluoro-1-(9-anthryl)ethanol higher than 1% were used, well-visible yellow spots were obtained, also allowing the quantitative determination of the two optical antipodes.

The best resolution of this racemate on both types of layer by changing the relative percentages of isopropanol and *n*-hexane is reported in Table 5.14. The α-values were in agreement with that (1.50) obtained by using planar chromatography on ionically bonded plates [45]. A worsening of the resolution was observed with the increase of 2-propanol percentage, while identical R_f values were obtained for the two enantiomers for alcohol percentages higher than 60%.

In substance, ionically bonded plates had less background color than covalently bonded stationary phases and gave better resolution of racemic test material. In addition, preparation of impregnated plates was very simple and fast: plates were cut into 5×10 cm sizes and placed in a large crystallizing dish, submerged with 30 ml tetrahydrofuran (THF) containing 1 g CS1. After 2 min, the plates were removed and washed with pure THF.

TABLE 5.14

Retention (hR_{f1}, hR_{f2}) and Resolution (α) Data for Racemic 2,2,2-Trifluoro-1-(9-Anthryl)Ethanol on Aminopropyl Silanized Silica Plates Modified with N-(3,5-Dinitrobenzoyl)-R-(−)-α-Phenylglycine (CCSP1, Ionically Bonded and CSP1, Covalently Bonded)

Chiral phase	hR_{f1}*	hR_{f2}*		Eluent and remarks
CCSP1	30 (+)	37 (−)	1.37	• *Eluent*: *n*-hexane/propan-2-ol (90:10, v/v)
CSP1	30 (+)	35 (−)	1.26	• *Visualization*: plates were dipped in 1% solution of vanillin in 50% sulfuric acid and heated at 120°C for 2–3 min (pale yellow spots, on a brown-grey background)

Solute volume $= 0.2$ μl of 0.3% solution in 2-propanol. Migration distance: 7 cm.

*$hR_f = R_f \times 100$.

**$\alpha = [(1/R_{f1}) - 1]/[(1/R_{f2}) - 1]$.

Wall [47] compared the chiral resolution ability of the aforementioned layers with that of plates modified with N-(3,5-dinitrobenzoyl)-L-leucine (CS2) toward two test compounds, such as racemic 2,2,2-trifluoro-1-(9-anthryl)ethanol and 1,1'-bi-2-naphthol, whose optical activity derived from dissymmetry within the molecule (atropisomerism).

Ionically bonded plates gave better resolution of racemates than covalently bonded stationary phases with CS1 while the reverse was observed for CS2. A separation factor of 1.56 was obtained for 1,1'-bi-2-naphthol enantiomers on plates impregnated with CS1 eluting with n-hexane/isopropanol in the ratio 75:25 (v/v).

Other optically active substances, such as hexobarbital, oxazepam, and lorazepam, were also resolved on CS1 or CS2 CSPs without derivatization. Such compounds had to be applied to the layers at high concentrations (i.e., 0.2 μl of 1% solution) as their detection levels were weak. On the contrary, β-blocking agents, such as propranolol, atenolol, and metoprolol, had to be derivatized with 1-isocyanonaphthalene before chromatographic separation and were easily detected even with 0.2% solutions.

An improvement in the use of phases impregnated with Pirkle selectors in planar chromatography was proposed by Witherow et al. [48], which prepared a two-phase modified amino-bonded stationary phase by immersing the lower portion of the precoated silica gel 60 HPTLC NH_2 $F_{254}s$ plate (Merck) in a solution of CS2. The layer portion ionically modified with the chiral selector was employed for the enantioseparation of the two aforementioned test materials.

After separation, the enantiomers were eluted on the above-layer portion by using the continuous development method and were determined by fluorescence quenching at 254 nm, allowing to perform *in situ* quantitative measurements by using a Shimadzu CS9000 scanning densitometer.

The proposed method was precise and rapid, as the modified layer portion was prepared by dipping the plate for a few seconds in 0.05 M THF solution of the chiral selector. Afterwards, the plate was dipped in pure THF to eliminate the CS2 excess and was let to dry at ambient temperature. Mixtures of n-hexane/isopropanol in the 80:20 and 20:80 ratios were used as eluents for the separation of 2,2,2-trifluoro-1-(9-anthryl)ethanol and of 1,1'-bi-2-naphthol, respectively.

REFERENCES

1. Gunther, K., Thin-layer chromatographic enantiomeric resolution via ligand exchange, *J. Chromatogr.*, 448, 11–30, 1988.
2. Dalgliesh, C.E., The optical resolution of aromatic amino acids on paper chromatograms, *J. Chem. Soc.*, 137, 3940–3942, 1952.
3. Gunther, K., Martens, J. and Schickedanz, M., Thin layer chromatographic enantiomer separation by using ligand exchange, *Angew. Chem.*, 96, 514–515, 1984.
4. Weinstein, S., Resolution of optical isomers by thin layer chromatography, *Tetrahedron Lett.*, 25, 985–986, 1984.

5. Grinberg, N. and Weinstein, S., Enantiomeric separation of Dns-amino acids by reversed-phase thin-layer chromatography, *J. Chromatogr.*, 303, 251–255, 1984.

6. Grinberg, N., Thin-layer chromatographic separations using a temperature gradient, *J. Chromatogr.*, 333, 69–81, 1985.

7. Marchelli, R., Virgili, R., Armani, E. and Dossena, A., Enantiomeric separation of D,L-Dns-amino acids by one- and two-dimensional thin-layer chromatography, *J. Chromatogr.*, 355, 354–357, 1986.

8. Sinibaldi, M., Messina, A. and Girelli, A.M., Separation of dansylamino acid enantiomers by thin-layer chromatography, *Analyst*, 113, 1245–1247, 1988.

9. Martens, J., Lubben, S. and Bushan, R., Synthese eines neuen chiralen selektors fur die dunn-schichtchromatographische enantiomerentrennung nach dem ligandenaustauschprinzip, *Tetrahedron Lett.*, 30, 7181–7182, 1989.

10. Remelli, M., Piazza, R. and Polidori, F., HPTLC separation of aromatic α-amino acid enantiomers on a new histidine-based stationary phase using ligand exchange, *Chromatographia*, 32, 278–287, 1991.

11. Bhushan, R., Reddy, G.P. and Joshi, S., TLC resolution of DL-amino acids on impregnated silica gel plates, *J. Planar Chromatogr.*, 7, 126–128, 1994.

12. Bhushan, R., Martens, J., Walbaum, S., Joshi, S. and Parshad, V., TLC resolution of enantiomers of amino acids and dansyl derivatives using (1*R*, 3*R*, 5*R*)-2-azabicyclo[3.3.0]octan-3-carboxylic acid as impregnating reagent, *Biomed. Chromatogr.*, 11, 286–288, 1997.

13. Bhushan, R. and Gupta, D., Ligand-exchange TLC resolution of some racemic beta-adrenergic blocking agents, *J. Planar Chromatogr.*, 19, 241–245, 2006.

14. Metzger, H., Maier, R., Sitter, C. and Stern, H.O., 2-[*N*-[(*S*)-1-ethoxycarbonyl-3-phenylpropyl]-L-alanyl]-(1*S*, 3*S*, 5*S*)-2-azabicyclo[3.3.0]octane-3-carboxylic acid (Hoe 498) — a new and highly effective angiotensin I converting enzyme inhibitor, *Arzenium. Forsch. Drug Res.*, 34, 1402–1406, 1984.

15. Paris, R.R., Sarsunova, M. and Semonsky, M., La chromatographie en couche mince comme methode de dedoublement des alcaloides racemiques. Application aux medicaments contenant de l'ephedrine, *Ann. Pharm. Franc.*, 25, 177–180, 1967.

16. Bhushan, R., Martens, J. and Arora, M., Direct resolution of (+)-ephedrine and atropine into their enantiomers by impregnated TLC, *Biomed. Chromatogr.*, 15, 151–154, 2001.

17. Bhushan, R. and Ali, I., Resolution of enantiomeric mixtures of phenyl-thiohydantoin amino acids on (+)-tartaric acid-impregnated silica gel, *J. Chromatogr.*, 392, 460–463, 1987.

18. Allenmark, S.G., *Chromatographic Enantioseparation: Methods and Applications*, John Wiley & Sons, New York, 1988, p. 205.

19. Lucic, B., Radulovic, D., Vujic, Z. and Agbada, D., Direct separation of the enantiomers of (+)-metoprolol tartrate on impregnated TLC plates with D-(−)-tartaric acid as a chiral selector, *J. Planar Chromatogr.*, 18, 294–299, 2005.

20. Oliva, A.I., Simon, L., Muniz, F.M., Sanz, F. and Moran, J.R., Enantioselective lutidine-tetrahydrobenzoxanthene receptors for carboxylic acids, *Eur. J. Org. Chem.*, 1698–1702, 2004.

21. Bhushan, R. and Ali, I., TLC resolution of enantiomeric mixtures of amino acids, *Chromatographia*, 23, 141–142, 1987.

22. Bhushan, R. and Thiong'o, G.T., Direct enantiomeric resolution of some 2-arylpropionic acids using (−)-brucine-impregnated thin-layer chromatography, *Biomed. Chromatogr.*, 13, 276–278, 1999.

23. Jamali, F., Mehvar, R. and Pasutto, F.M., Enantioselective aspects of drug action and dispodition: therapeutic pitfalls, *J. Pharm. Sci.*, 78, 695–715, 1989.

24. Bhushan, R. and Gupta, D., Resolution of (+)-ibuprofen using (−)-brucine as a chiral selector by thin layer chromatography, *Biomed. Chromatogr.*, 18, 838–840, 2004.

25. Bhushan, R. and Ali, I., Resolution of racemic mixtures of hyoscyamine and colchicine on impregnated silica gel layers, *Chromatographia*, 35, 679–680, 1993.

26. Bhushan, R. and Arora, M., Direct enantiomeric resolution of (+)-atenolol, (+)-metoprolol and (+)-propranolol by impregnated TLC using L-aspartic acid as chiral selector, *Biomed. Chromatogr.*, 17, 226–230, 2003.

27. Bhushan, R. and Parshad, V., Resolution of (+)-ibuprofen using L-arginine-impregnated thin-layer chromatography, *J. Chromatogr. A*, 721, 369–372, 1996.

28. Sajewicz, M., Pietka, R. and Kowalska, T., Chiral separation of S(+)- and R(−)-ibuprofen by thin-layer chromatography. An improved analytical procedure, *J. Planar Chromatogr.*, 17, 173–176, 2004.

29. Sajewicz, M., Pietka, R., Pieniak, A. and Kowalska, T., Application of thin-layer chromatography (TLC) to investigating oscillatory instability of the selected profen enantiomers, *Acta Chromatogr.*, 15, 131–149, 2005.

30. Sajewicz, M., Pietka, R., Pienak, A. and Kowalska, T., Application of thin-layer chromatography to investigate oscillatory instability of the selected profen enantiomers in dichloromethane, *J. Chromatogr. Sci.*, 43, 542–548, 2005.

31. Sajewicz, M., Pietka, R., Drabik, G. and Kowalska, T., On the mechanism of oscillatory changes of the retardation factor (RF) and the specific rotation [a]D with selected solutions of S-(+)-naproxen, *J. Liquid Chromatogr. Rel. Technol.*, 29, 2071–2082, 2006.

32. Sajewicz, M., Pietka, R., Dabrik, G., Namyslo, E. and Kowalska, T., On the stereochemically peculiar two-dimensional separation of 2-arylpropionic acids by chiral TLC, *J. Planar Chromatogr.*, 19, 273–277, 2006.

33. Bhushan, R. and Thiong'o, G.T., Direct enantioseparation of some beta-adrenergic blocking agents using impregnated thin-layer chromatography, *J. Chromatogr. B*, 708, 330–334, 1998.

34. Sajewicz, M., Pietka, R. and Kowalska, T., Chiral separations of ibuprofen and propranolol by TLC. A study of the mechanism and thermodynamics of retention, *J. Liquid Chromatogr. Rel. Technol.*, 28, 2499–2513, 2005.

35. Bhushan, R., Martens, J. and Thiong'o, G.T., Direct thin layer chromatography enantioresolution of some basic DL-amino acids using a pharmaceutical industry waste as chiral impregnating reagent, *J. Pharm. Biomed. Anal.*, 21, 1143–1147, 2000.

36. Aboul-Enein, H.Y., El-Awady, M.I. and Heard, C.H., Enantiomeric resolution of some 2-arylpropionic acids using L-(−)-serine-impregnated silica as stationary phase by thin layer chromatography, *J. Pharm. Biomed. Anal.*, 32, 1055–1059, 2003.

37. Bhushan, R. and Martens, J., Separation of amino acids, their derivatives and enantiomers by impregnated TLC, *Biomed. Chromatogr.*, 15, 155–165, 2001.

38. Geiss, F., *Fundamentals of Thin Layer Chromatography*, Huthig Verlag A, Heidelberg, 1987, p. 226.

39. Armstrong, D.W., Tang, Y., Chen, S., Zhu, Y., Bagwill, C. and Chen, J.R., Macrocyclic antibiotics as a new class of chiral selectors for liquid chromatography, *Anal. Chem.*, 66, 1473–1484, 1994.

40. Bhushan, R. and Parshad, V., Thin-layer chromatographic separation of enantiomeric dansylamino acids using a macrocyclic antibiotic as a chiral selector, *J. Chromatogr. A*, 736, 235–238, 1996.

41. Bhushan, R. and Thiong'o, G.T., Separation of the enantiomers of dansyl-DL-amino acids by normal-phase TLC on plates impregnated with a macrocyclic antibiotic, *J. Planar Chromatogr.*, 13, 33–36, 2000.

42. Bhushan, R. and Gupta, D., Thin-layer chromatography separation of enantiomers of verapamil using macrocyclic antibiotic as chiral selector, *Biomed. Chromatogr.*, 19, 474–478, 2005.

43. Pirkle, W.H. and House, D.W., Chiral high-performance liquid chromatographic stationary phases. 1. Separation of the enantiomers of sulfoxides, amines, amino acids, alcohols, hydroxy acids, lactones, and mercaptans, *J. Org. Chem.*, 44, 1957–1960, 1979.

44. Pirkle, W.H., House, D.W. and Finn, J.M., Broad spectrum resolution of optical isomers using chiral high-performance liquid chromatographic bonded phases, *J. Chromatogr.*, 192, 143–158, 1980.

45. Wainer, I., Brunner, C. and Doyle, T., Direct resolution of enantiomers via thin-layer chromatography using a chiral adsorbent, *J. Chromatogr.*, 264, 154, 1983.

46. Wall, P.E., Preparation and Application of TLC Plates for Enantiomer Separation, *Proc. Int. Symp. Instrumental Thin-Layer Chromatography/Planar Chromatography*, Brighton, Sussex, U.K., 1989, pp. 237–243.

47. Wall, P.E., Preparation and application of HPTLC plates for enantiomer separation, *J. Planar Chromatogr.*, 2, 228–232, 1989.

48. Witherow, L., Sprurway, T.D., Ruane, R.J. and Wilson, I.D., Problems and solutions in chiral thin-layer chromatography: a two-phase "Pirkle" modified amino-bonded plate, *J. Chromatogr.*, 553, 497–501, 1991.

6 Chiral Mobile Phase Additives

Danica Agbaba and Branka Ivković

CONTENTS

6.1 PREFACE

Chiral mobile phase additives (CMPAs) are generally used to perform direct chiral separation in thin-layer chromatography (TLC). This mode offers the advantages of flexibility and low cost as compared to the equivalent chiral stationary phase (CSP). Also, the lack of a wide range of CSPs in TLC resulted in CMPAs becoming a commonly employed approach for enantiomeric separations.

In the CMPAs method, enantiomeric separation is accomplished by the formation of a pair of transient diastereomeric complexes between a racemic analyte and the CMPA. Chiral discrimination is due to the differences in the interphase distribution ration, solvatation in the mobile phase, or binding of the complexes to the achiral/chiral stationary phase. Ion pairing, ligand exchange, inclusion complexes, and protein interactions represent the major approaches in the formation of diastereomeric complexes.

So far, a variety of different racemates, such as those of amino acids, amino acid derivatives, amino alcohols, fatty acids, steroids, pharmacologically active agents, and other substances, have been resolved by this method. The enantiomeric separation has been achieved using normal- or reversed-phase TLC (NP-TLC and RP-TLC, respectively). Commercially available silica gel or chemically bound polar diol F_{254} HP-TLC plates, hydrophobic octadecyl-, ethyl-, and diphenyl-RP-TLC plates, as well as polyamide, represent commonly used achiral stationary phases. Also, homemade and industrially produced microcrystalline cellulose and its derivative microcrystalline cellulose triacetate have been employed as CSPs, whereas achiral-precoated commercially available plates have been produced by Merck (Darmstadt, Germany), Macherey-Nagel (Duren, Germany), Whatman Chemical Division (Clifton, NJ), Alltech Associates, Inc. (Deerfield, IL), and Fisher Scientific (Fair Lawn, NJ).

Till now, different compounds containing one or more chiral centers have been used as mobile phase additives, for example, D-galacturonic acid, (±)10-camphor-sulfonic acid (CSA), (1R)-(−)-ammonium-10-camphorsulfonate, D(−)-tartaric acid, L-alanine, N-benzyloxycarbonyl-L-proline (ZP), N-benzyloxycarbonylglycyl-L-proline (ZGP), N-benzyloxycarbonylisoleucyl-L-proline (ZIP), N-benzyloxycarbonylalanyl-L-proline (ZAP), bovine serum albumin (BSA), macrocyclic antibiotic vancomycin; beta-cyclodextrin, alpha-cyclodextrin, gamma-cyclodextrin (β-CD, α-CD, and γ-CD respectively); and derivatized β-CDs such as maltosyl-, hydroxypropyl-, hydroxyethyl-, methyl-, dimethyl-, hydroxy-trimethylpropylammonium-, and (carboxymethyl)-β-CD, as well as hydrosoluble β-CD polymers. β-CDs have been applied as CMPAs for performing two-dimensional microseparation in coupled system capillary electrophoresis (CE)-TLC.

6.2 MACROCYCLICS AS MOBILE PHASE ADDITIVES

6.2.1 CYCLODEXTRINS AND THEIR DERIVATIVES AS MOBILE PHASE ADDITIVES

Cyclodextrins represent cyclic oligomers containing six (α-), seven (β-), or eight (γ-CD) D-glucose monomers in a chair conformation, connected through α-1,4-linkages. The glucose rings are arranged in the shape of a hollow, truncated cone with a relatively hydrophobic cavity and a polar outer surface where the hydroxyl groups are situated. The diameters of α-, β-, and γ-CD cavities, range from 4.7 to 5.3, 6.0 to 6.5, and 7.5 to 8.3 Å, respectively, whereas the height of the cavity is the same (approximately 7 Å) in all these CD classes. The larger opening of the cone is surrounded by the secondary hydroxyl groups, whereas the primary hydroxyl groups occur at the narrower end of the cone. The primary (C-6) hydroxyl groups can freely rotate, thus partially blocking the narrow entrance of the cavity. The restricted conformational freedom and orientation of the secondary hydroxyl groups encircling the opposite end of the cavity are believed to play an important role in the enantioselectivity of the CDs [1]. There are several advantages of CDs such as stability over the wide range of pH, nontoxicity, resistance to light, and UV transparency within the wavelength range commonly used for chromatographic detections. As the CD molecules are chiral, themselves, they can form a diastereomeric pair of inclusion complexes with each enantiomer of a racemate.

Enantiomeric resolution requires the inclusion of an alkyl or aryl moiety in the cavity and formation of additional hydrogen bonds between secondary hydroxyl groups of a CD and substituents attached or occurring in a close proximity of the analyte's chiral center. The effects of derivatization of analyte molecules were investigated by LeFevre [2], who suggested that variations of the enantioselectivity are presumably due to changes in the ability of the analyte molecules to form hydrogen bonds with hydroxyl groups at the larger cone opening of a CD cavity. The interactions of CDs with chiral molecules are quite complex, because they may include many species of the solute, such as neutral, ionic, and species bound to one or more CD molecules. In addition to inclusion–complexation interactions, many other nonspecific forms of interaction can occur in the chromatographic system [3]. Examples of interaction of the analyte molecules include external absorption to the hydroxyl groups of a CD moiety, interaction with the linkage chain, or binding to free silanol groups. Therefore, the interaction of the analyte molecules with CDs depends on numerous factors, for example, their polarity, hydrophobicity, size, and spatial arrangement [4]. The formation of inclusion complexes, as well as the assessment of various aspects of interactions have been studied by applying different physicochemical methods, such as spectroscopy [5,6], high-performance liquid chromatography (HPLC) [7–9], and TLC [10,11]. Formation of the CD inclusion complexes is strongly affected by pH, temperature, and composition of the mobile phase.

6.2.1.1 β-Cyclodextrin

In respect to best sized complex former, β-CD has mostly been used for chiral separations. The general structure of β-CD is presented in Figure 6.1. Armstrong et al. [12] were the first to report the application of β-CD as a chiral additive for separation on RP-TLC plates. The racemates mostly resolved by CDs as CMPAs are listed in Table 6.1. Taking into account the requirement for the critical concentration of β-CD in mobile phase for chiral separation, as well as its limited solubility in water (0.017 M at 25°C), saturated solution of urea has been frequently used to increase the solubility of this CD [4,12–19]. Parallel to the increase in urea concentration, Lepri et al. [4] noticed a decrease in separation factor for some selected dansyl-amino acid derivatives, due to the competition for the same binding sites in the CD cavity. In the case where chiral separation is performed using RP plates, sodium chloride has been added to mobile phase to protect the binding sites from any damage [4,12–18,20–23]. Acetonitrile [2,4,12–17,19,21–25], methanol, [2,12,14,18,19,23,25,26], and pyridine [19] represent the frequently used organic modifiers in RP-TLC. The formation of inclusion complex was found to be strongest in water, becoming weaker upon the addition of organic modifiers, because of competition between organic modifier molecules with the analyte molecules for binding sites within the cavity of CDs. The interaction of organic modifier molecules with the binding sites is much stronger than that of water molecules. However, their interaction with a CD should be weaker than that of the analyte molecules. Nevertheless, as the organic modifier is always present in great excess, it can still displace the analyte molecules from

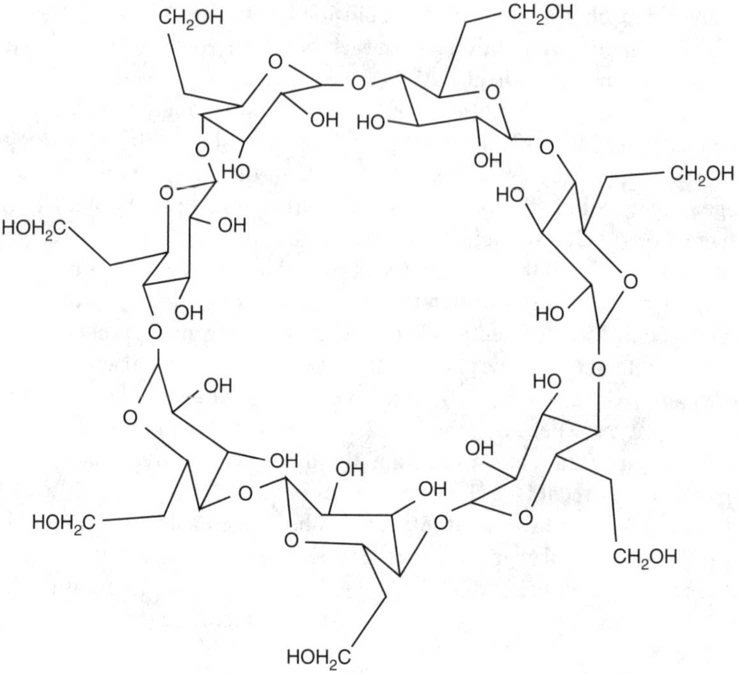

FIGURE 6.1 Structure of β-cyclodextrin.

the binding sites in the CD cavity. In contrast, under normal phase conditions, nonpolar solvents, such as hexane or chloroform, predominantly occupy the CD cavity and cannot be easily displaced by the analyte molecules. The mechanism of chiral recognition under normal phase conditions has not been fully understood [27]. It has been reported that the temperature may influence the kinetics of complexation between an analyte and a CD, and therefore can affect TLC enantioseparation. It was also found that the effect of temperature on the enantiomer separation varies considerably with the substituent and its position, but parallel to increasing temperature, a decrease in enantioseparation was always recorded [20].

The type of structure of chiral compounds, as well as that of the stationary phase, could strongly affect the separation. LeFevre et al. [28] investigated the effect of racemic dansyl-amino acids structure on the resolution using β-CD as a CMPA in the RP-TLC system. The same authors confirmed stereochemistry of the peptide-derived amino acids and detected a partial racemization of several amino acids occurring during hydrolysis of enkephalin, gramicidin, achatin-I, and cyclosporin by using β-CD as a CMPA and RP-TLC [25]. In some cases, β-CD as a CMPA can improve chiral resolution of cellulose and can be used for enantioseparation of some flavanones [16], amino acids, and oxazolidinone [15], as well as of budesonide in pharmaceuticals [29], and some aromatic amino acids, and aromatic amino alcohols [19].

TABLE 6.1
Compounds Separated Using CDs as Chiral Mobile Phase Additives

Compound	Mobile phase	Ref.
Dansyl-DL-alanine	0.20 M β-CD:methanol (65:35)	[2]
Dansyl-DL-*all*-isoleucine	0.20 M β-CD:acetonitrile (68:32)	[2]
Dansyl-DL-asparagine	0.20 M β-CD:acetonitrile (80:20)	[2]
Dansyl-DL-arginine	0.20 M β-CD:acetonitrile (80:20)	[2]
Dansyl-DL-citrulline	0.20 M β-CD:acetonitrile (80:20)	[2]
Dansyl-DL-cystine	0.20 M β-CD:methanol (45:55)	[2]
Dansyl-DL-glutamine	0.20 M β-CD:acetonitrile (80:20)	[2]
Dansyl-DL-histidine	0.20 M β-CD:acetonitrile (80:20)	[2]
Dansyl-DL-isoleucine	0.20 M β-CD:acetonitrile (68:32)	[2]
Dansyl-DL-lysine	Saturated β-CD:methanol (40:60)	[2]
Dansyl-DL-*N*-methyl valine	0.20 M β-CD:methanol (50:50)	[2]
Dansyl-DL-ornithine	Saturated β-CD:methanol (40:60)	[2]
Dansyl-DL-proline	0.20 M β-CD:methanol (50:50)	[2]
Dansyl-DL-tyrosine	Saturated β-CD:methanol (40:60)	[2]
DL-Thyroxine	0.2 M methyl-β-CD in water:acetonitrile (10:90)	[15]
Dansyl-DL-leucine	0.4 M HP-β-CD, acetonitrile:water (30:70)	[24]
	Acetonitrile:0.151 M β-CD in saturated urea solution and 0.6 M NaCl (30:70)	[12]
	0.1 M β-CD in saturated aqueous urea solution containing 3.5% NaCl:acetonitrile (80:20)	[4]
	Acetonitrile:water (30:70) containing 0.4 M HP-β-CD and 0.6 M NaCl	[21]
	Acetonitrile:water (30:70) containing 0.4 M maltosyl-β-CD and 0.6 M NaCl	[22]
	Acetonitrile:0.2 M β-CD in saturated urea solution and 0.6 M NaCl (20:80)	[55]
Dansyl-DL-valine	0.4 M HP-β-CD in acetonitrile:water (30:70)	[24]
	Acetonitrile:0.151 M β-CD (30:70)	[12]
	0.1 M β-CD in saturated aqueous urea solution containing 3.5% NaCl:acetonitrile (80:20)	[4]
	Acetonitrile:water (30:70) containing 0.4 M HP-β-CD and 0.6 M NaCl	[21]

(Continued)

TABLE 6.1
(Continued)

Compound	Mobile phase	Ref.
Dansyl-DL-methionine	Acetonitrile:water (30:70) containing 0.4 M maltosyl-β-CD and 0.6 M NaCl	[22]
	Acetonitrile:0.2 M β-CD in saturated urea solution and 0.6 M NaCl (20:80)	[55]
	0.4 M HP-β-CD, acetonitrile:water (30:70)	[24]
	Acetonitrile:0.151 M β-CD (30:70)	[12]
	0.1 M β-CD in saturated aqueous urea solution containing 3.5% NaCl:acetonitrile (80:20)	[4]
	Acetonitrile:water (30:70) containing 0.4 M HP-β-CD and 0.6 M NaCl	[21]
Dansyl-DL-threonine	0.4 M HP-β-CD, acetonitrile:water (30:70)	[24]
	Methanol:0.151 M β-CD (30:70)	[12]
	0.1 M β-CD in saturated aqueous urea solution containing 3.5% NaCl:acetonitrile (80:20)	[4]
Dansyl-DL-phenylalanine	0.4 M HP-β-CD, acetonitrile:water (30:70)	[24]
	Acetonitrile:0.151 M β-CD (30:70)	[12]
	0.1 M β-CD in saturated aqueous solution of urea containing 3.5% NaCl:acetonitrile (80:20)	[4]
	Acetonitrile:water (30:70) containing 0.4 M HP-β-CD and 0.6 M NaCl	[21]
DL-Methyonine-β-naphthylamide	0.3 M HP-β-CD, acetonitrile:water (35:65)	[24]
	Acetonitrile:water (35:65) containing 0.3 M HP-β-CD and 0.6 M NaCl	[21]
	Acetonitrile:water (30:70) containing 0.4 M maltosyl-β-CD and 0.6 M NaCl	[22]
DL-Alanine-β-naphthylamide	0.3 M HP-β-CD, acetonitrile:water (25:75)	[24]
	0.1 M β-CD in saturated aqueous solution of urea containing 3.5% NaCl:acetonitrile (80:20)	[4]
	Acetonitrile:water (30:70) containing 0.4 M maltosyl-β-CD and 0.6 M NaCl	[22]
Dansyl-DL-glutamic acid	Methanol:0.163 M β-CD (35:65)	[12]
	0.1 M β-CD in saturated aqueous solution of urea containing 3.5% NaCl:acetonitrile (80:20)	[4]
Dansyl-DL-α-amino-n-butyric acid	Acetonitrile:0.151 M β-CD (30:70)	[12]

Compound	Mobile phase	Reference
Dansyl-DL-norvaline	Acetonitrile:0.151 M β-CD (30:70)	[12]
	0.1 M β-CD in saturated aqueous solution of urea containing 3.5% NaCl:acetonitrile (80:20)	[4]
	Acetonitrile:0.2 M β-CD in saturated solution of urea and 0.6 M NaCl (20:80)	[55]
Dansyl-DL-norleucine	Acetonitrile:0.151 M β-CD (30:70)	[12]
	0.1 M β-CD in saturated aqueous urea solution containing 3.5% NaCl:acetonitrile (80:20)	[4]
	Acetonitrile:water (30:70) containing 0.4 M HP-β-CD and 0.6 M NaCl	[21]
	0.4 M HP-β-CD, acetonitrile:water (30:70)	[24]
	Acetonitrile:0.2 M β-CD in saturated urea solution and 0.6 M NaCl (20:80)	[56]
Dansyl-DL-serine	Acetonitrile:0.133 M β-CD (20:80)	[12]
	0.1 M β-CD in saturated aqueous urea solution containing 3.5% NaCl:acetonitrile (80:20)	[4]
Dansyl-DL-aspartic acid	Acetonitrile:0.133 M β-CD (25:75)	[12]
Dansyl-DL-tryptophan	Acetonitrile:0.231 M β-CD (35:65)	[12]
Dansyl-DL-α-aminobutyric acid	0.1 M β-CD in saturated aqueous urea solution containing 3.5% NaCl:acetonitrile (80:20)	[4]
Dansyl-DL-aspartic acid	0.1 M β-CD in saturated aqueous urea solution containing 3.5% NaCl:acetonitrile (80:20)	[4]
DL-Tryptophan	Aqueous 4% α-CD and 1 M NaCl solutions	[20]
	Methanol:β-CD saturated urea solution:33% diethylamine (4:1:0.14)	[19]
	Methanol:β-CD saturated urea solution:33% diethylamine (4:1:0.28)	
	Methanol:β-CD saturated urea solution:33% diethylamine (4:1:0.42)	
	Methanol:formic acid:β-CD saturated urea solution (7:1:2)	
	Methanol:formic acid:0.2 M α-CD urea solution (7:1:2)	
4-Methyl-DL-tryptophan	Aqueous 4% α-CD:1 M NaCl solutions	[20]
5-Methyl-DL-tryptophan	Aqueous 4% α-CD:1 M NaCl solutions	[20]
6-Methyl-DL-tryptophan	Aqueous 4% α-CD:1 M NaCl solutions	[20]
7-Methyl-DL-tryptophan	Aqueous 4% α-CD:1 M NaCl solutions	[20]
4-Fluoro-DL-tryptophan	Aqueous 4% α-CD:1 M NaCl solutions	[20]
5-Fluoro-DL-tryptophan	Aqueous 4% α-CD:1 M NaCl solutions	[20]
6-Fluoro-DL-tryptophan	Aqueous 4% α-CD:1 M NaCl solutions	[20]

(Continued)

TABLE 6.1
(Continued)

Compound	Mobile phase	Ref.
DL-Lysine	2-O-[(R)-2-hydroxypropyl]-β-CD	[30]
DL-Valine	2-O-[(R)-2-hydroxypropyl]-β-CD	[30]
DL-Citrulline	2-O-[(R)-2-hydroxypropyl]-β-CD	[30]
DL-Histidine	2-O-[(R)-2-hydroxypropyl]-β-CD	[30]
DL-Glutamine	2-O-[(R)-2-hydroxypropyl]-β-CD	[30]
DL-Arginine	2-O-[(R)-2-hydroxypropyl]-β-CD	[30]
N-2,4-dinitrophenyl-DL-leucine	HE-β-CD (16.3 g) in water:acetonitrile:acetic acid (45:4:1; 100 ml)	[17]
	Met-β-CD (9.8 g) in water:acetonitrile:acetic acid (35:4:1; 100 ml)	
	HP-β-CD (13.8 g) in water:acetonitrile:acetic acid (45:4:1; 100 ml)	
N-3,5-dinitro-2-pyridyl-DL-leucine	HE-β-CD in water:acetonitrile:acetic acid (45:4:1)	[17]
	Met-β-CD in water:acetonitrile:acetic acid (35:4:1)	
	HP-β-CD in water:acetonitrile:acetic acid (45:4:1)	
N-2,4-dinitrophenyl-DL-phenylalanine	HE-β-CD in water:acetonitrile:acetic acid (45:5:1)	[17]
	Met-β-CD in water:acetonitrile:acetic acid (35:4:1)	
	HP-β-CD in water:acetonitrile:acetic acid (45:5:1)	
Methyl-thiohydantoin-DL-phenylalanine	HE-β-CD in water:acetonitrile:acetic acid (45:5:1)	[17]
	Met-β-CD in water:acetonitrile:acetic acid (35:4:1)	
	HP-β-CD in water:acetonitrile:acetic acid (45:5:1)	
	Acetonitrile:0.2 M β-CD containing urea (32%) and NaCl (3%) (20:80)	[13]
Methyl-thiohydantoin-DL-tyrosine	HE-β-CD in water:acetonitrile:acetic acid (45:5:1)	[17]
	Met-β-CD in water:acetonitrile:acetic acid (35:4:1)	
	HP-β-CD in water:acetonitrile:acetic acid (45:5:1)	
	Acetonitrile:0.2 M β-CD containing urea (32%) and NaCl (3%) (20:80)	[13]

DL-Tyrosine	Methanol:formic acid:β-CD saturated urea solution (7:1:2)	[19]
	Acetonitrile:formic acid:β-CD saturated urea solution (7:1:2)	
	Pyridine:β-CD saturated urea solution (3:2)	
	Methanol:β-CD saturated urea solution:33% diethylamine (4:1:0.14)	
	Methanol:β-CD saturated urea solution:33% diethylamine (4:1:0.28)	
	Methanol:β-CD saturated urea solution:33% diethylamine (4:1:0.42)	
	Methanol:formic acid:0.2 M α-CD urea solution (7:1:2)	
	Methanol:formic acid:0.2 M α-CD urea solution:33% diethylamine (4:1:0.42)	
DL-Phenylalanine	Methanol:β-CD saturated urea solution:33% diethylamine (4:1:0.14)	[19]
	Methanol:β-CD saturated urea solution:33% diethylamine (4:1:0.28)	
	Methanol:β-CD saturated urea solution:33% diethylamine (4:1:0.42)	
	Methanol:formic acid:β-CD saturated urea solution (7:1:2)	
	Methanol:formic acid:0.2 M α-CD urea solution (7:1:2)	
DL-p-Hydroxyphenylglycine	Methanol:formic acid:β-CD saturated urea solution (7:1:2)	[19]
	Methanol:formic acid:0.2 M α-CD urea solution (7:1:2)	
DL-Thyronine	Methanol:formic acid:β-CD saturated urea solution (7:1:2)	[19]
	Methanol:formic acid:0.2 M α-CD urea solution (7:1:2)	
DL-p-amino phenylalanine	Methanol:formic acid:β-CD saturated urea solution (7:1:2)	[19]
	Methanol:formic acid:0.2 M α-CD urea solution (7:1:2)	
L-[1-^{14}C]alanine	Methanol:0.2 M β-CD (40:60)	[23]
L-[U-^{14}C]aspartic acid	Methanol:0.13 M β-CD (30:70)	[23]
L-[U-^{14}C]glutamic acid	Acetonitrile:0.175 M β-CD (20:80)	[23]
L-[3,4,5-^{3}H$_3$]lysine	Acetonitrile:0.175 M β-CD (20:80)	[23]
L-[methyl-^{14}C]methionine	Methanol:0.2 M β-CD (35:65)	[23]
L-[U-^{14}C]phenylalanine	Methanol:0.2 M β-CD (40:60)	[23]
L-[^{3}H(G)]serine	Methanol:0.2 M β-CD (30:70)	[23]
L-[U-^{14}C]threonine	Acetonitrile:0.15 M β-CD (25:75)	[23]
L-[U-^{14}C]tyrosine	Methanol:0.175 M β-CD (40:60)	[23]

(Continued)

**TABLE 6.1
(Continued)**

Compound	Mobile phase	Ref.
L-[ring-3,5-^{3}H$_2$]tyrosine	Methanol:0.175 M β-CD (40:60)	[23]
DL-[1-^{14}C]alanine	Acetonitrile:0.2 M β-CD (25:75)	[23]
DL-Methionine-α-naphthylamide	100 ml aqueous solution containing NaCl (2.5 g), urea (26 g), acetonitrile (20 ml), and 0.15 M β-CD	[14]
3,5-Dinitro-N-(1-phenylethyl) benzamide	HE-β-CD in water:acetonitrile:acetic acid (45:5:1) Met-β-CD in water:acetonitrile:acetic acid (35:4:1) HP-β-CD in water:acetonitrile:acetic acid (45:5:1) 16.3% Hydroxyethyl-β-CD in water:acetonitrile:acetic acid (45:4:1) 13.8% HP-β-CD in water:acetonitrile:acetic acid (45:4:1) 9.8% Methyl-β-CD in water:acetonitrile:acetic acid (35:4:1)	[17]
1,1-Bi-2-naphthol	16.3% Hydroxyethyl-β-CD in water:acetonitrile:acetic acid (45:4:1) 13.8% HP-β-CD in water:acetonitrile:acetic acid (45:4:1) 9.8% Methyl-β-CD in water:acetonitrile:acetic acid (35:4:1)	[17]
(R,S)-5-(4-methylphenyl)-5-phenylhydantoin	Acetonitrile:water (35:65) containing 0.3 M HP-β-CD and 0.6 M NaCl 0.3 M HP-β-CD in water:acetonitrile (35:65)	[21] [24]
DL-3,4-Dihydroxyphenylalanine (DOPA)	Methanol:formic acid:β-CD saturated urea solution (7:1:2) Methanol:formic acid:α-CD saturated urea solution (7:1:2)	[19]
DL-Epinephrine	Methanol:formic acid:β-CD saturated urea solution (7:1:2) Methanol:formic acid:α-CD saturated urea solution (7:1:2)	[19]
DL-Isopropylepinephrine	Methanol:formic acid:β-CD saturated urea solution (7:1:2) Acetonitrile:formic acid:β-CD saturated urea solution (pH 4.5) (7:1:2) Pyridine:β-CD saturated urea solution (3:2)	[19]
Budesonide	1% Aqueous solution of β-CD in methanol	[29]

Compound	Mobile phase	Reference
Aminoglutethimide	0.05 M β-CD:methanol (65:35; 50:50; 40:60; 35:65; 20:80)	[18]
	10% HTMA-β-CD:methanol (50:50)	[18]
	30% HTMA-β-CD:methanol (50:50)	[18]
	40% HTMA-β-CD:methanol (50:50)	
Acetylated aminoglutethimide	30% HTMA-β-CD:methanol (50:50)	[18]
Dansyl aminoglutethimide	0.05 M β-CD:methanol (65:35)	[18]
	15% Carboxymethyl-β-CD:methanol (65:35)	
Fluoxetine	2.5 mM HP-β-CD in methanol:buffer (pH 4.5) (55:45; 50:50)	[26]
Norfluoxetine	2.5 mM HP-β-CD in methanol:buffer (pH 6.0) (55:45; 50:50)	[26]
Promethazine and isopromethazine	2.5 mM HP-β-CD in methanol:buffer (pH 6.0) (60:40)	[26]
Mephenytoin	Methanol:0.308 M β-CD in saturated urea solution and 0.6 M NaCl (35:65)	[12]
	0.4 M HP-β-CD, acetonitrile:water (30:70)	[24]
(±)S-(1-ferrocenyl-2-methylpropyl)–thioethanol	Acetonitrile:0.125 M β-CD in saturated urea solution and 0.6 M NaCl (15:85)	[12]
(±)S-(1-ferrocenylethyl-thiophenol)	Acetonitrile:0.151 M β-CD (30:70) in saturated urea solution and 0.6 M NaCl	[12]
N'-benzylnornicotine	Methanol:0.200 M β-CD (60:40) in saturated urea solution, 0.6 M NaCl and 1% aqueous triethyl ammonium acetate (pH 7.1)	[12]
N'-(2-naphthylmethyl)-nornicotine	Methanol:0.200 M β-CD (60:40) in saturated urea solution, 0.6 M NaCl and 1% aqueous triethyl ammonium acetate (pH 7.1)	[12]
(±)-2-Chloro-2-phenylacetyl chloride	Methanol:0.200 M β-CD (60:40) in saturated urea solution and 0.6 M NaCl	[12]
DL-Alanine-2-naphthylamide hydrochloride	Methanol:0.163 M β-CD (35:65) in saturated urea solution and 0.6 M NaCl	[12]
(1R,2S,5R)-(−)-methyl-(S)-p-toluenesulfinate	Acetonitrile:0.151 M β-CD (30:70) in saturated urea solution and 0.6 M NaCl	[12]
(1S,2R,5S)-(+)-methyl-(R)-p-toluenesulfinate	Acetonitrile:0.151 M β-CD (30:70) in saturated urea solution and 0.6 M NaCl	[12]
(±)-2,2'-Binaphthyldiyl-N-benzylmonoaza-16-crown-5	Methanol:0.265 M β-CD (60:40) in saturated urea solution and 0.6 M NaCl	[12]
(R,S)-4-benzyl-5,5-dimethyl-2-oxazolidinone	0.2 M HP-β-CD in water:acetonitrile (10:90)	[15]
(4S,5R/4R,5S)-cis-4,5-diphenyl-2-oxazolidinone	0.1 M β-CD solution containing 20% urea and 2% NaCl:acetonitrile (20:80)	[15]
	0.1 M HP-β-CD in water:acetonitrile (10:90)	

(Continued)

TABLE 6.1
(Continued)

Compound	Mobile phase	Ref.
(R,S)-4-benzyl-3-propionyl-2-oxazolidinone	0.2 M β-CD solution containing 32% urea and 2% NaCl:acetonitrile (35:65)	[15]
	0.2 M HP-γ-CD in water:acetonitrile (10:90)	
(L,D)-N_α-(2,4-dinitro-5-fluorophenyl)-valinamide	0.15 M β-CD solution containing 32% urea and 2% NaCl:acetonitrile (20:80)	[15]
	0.2 M methyl-β-CD in water:acetonitrile (10:90)	
	0.2 M HP-γ-CD in water:acetonitrile (10:90)	
4'-Methoxyflavanone	0.15 M β-CD aqueous solution containing 32% urea and 2% NaCl:acetonitrile (80:20)	[16]
2'-Hydroxyflavanone	0.15 M β-CD aqueous solution containing 32% urea and 2% NaCl:acetonitrile (80:20)	[16]
4'-Hydroxyflavanone	0.15 M β-CD aqueous solution containing 32% urea and 2% NaCl:acetonitrile (80:20)	[16]
Pinocembrin-7-methylether	0.15 M β-CD aqueous solution containing 32% urea and 2% NaCl:acetonitrile (80:20)	[16]
Flavanone	0.15 M β-CD aqueous solution containing 32% urea and 2% NaCl:acetonitrile (80:20)	[16]
R (+)-Benzyl-2-oxazolidinone	0.3 M HP-β-CD, acetonitrile:water (35:65)	[24]
S (−)-Benzyl-2-oxazolidinone		
N'-(2-naphthylmethyl)-nornicotine	Acetonitrile:water (30:70) containing 0.3 M maltosyl-β-CD and 0.6 M NaCl	[22]

6.2.1.2 Derivatized CDs

Derivatization of the structure of CDs at the hydroxyl groups represents an alternative route to improve their hydrosolubility and enhance enantioselectivity. For that purpose, several CD derivatives have been produced so far and used as CMPAs, for example, hydroxypropyl-β-CD (HP-β-CD) [17,18,24,30], hydroxyethyl-β-CD (HE-β-CD) [17,24], maltosyl-β-CD (Malt-β-CD) [22], dimethyl-β-CD (DIME-β-CD) [22], methyl-β-CD (Met-β-CD) [17], hydroxy-trimethylpropyl-ammonium-β-CD (HTMA-β-CD), carboxymethyl-β-CD (CM-β-CD), and malto-dextrins [18].

The degree of substitution in derivatized CDs may affect the chiral separation mechanism [21]. The changes in enantioselectivity of a derivatized CD were attributed to the loss of hydrogen bonding groups in the α-position to the chiral centers of the CD molecule, or to a gain of additional interaction sites around the base of the derivatized CD molecule [27].

6.2.2 MACROCYCLIC ANTIBIOTICS

Macrocyclic glycopeptides/antibiotics are a relatively new class of chiral selectors with favorable chromatographic properties. So far, macrocyclic antibiotics were frequently employed as chiral bonded phases and CMPAs in HPLC and CE. Among them, the glycopeptides, vancomycin, and teicoplanin were the most commonly used ones. All glycopeptides contain an aglycone portion made of fused macro-cyclic *basket* shape, with carbohydrate groups attached to the aglycone basket [1]. Macrocyclic antibiotics serving as chiral selectors possess several characteristics enabling their interaction with analyte molecules. Similar to other natural products, they have a number of stereogenic centers and functional groups allowing their multiple interactions with chiral molecules such as electrostatic interactions occurring through hydrogen bonding, dipole, hydrophobic and π–π interactions, and steric hindrance. Macrocyclic antibiotics may be acidic, alkaline, or neutral with a negligible or no UV-VIS absorbance [1,31]. There is still a limited body of information available with respect to the application of macrocyclines for TLC enantioseparation. So far, vancomycin (Figure 6.2) has been the most frequently used macrocycline for that purpose. Armstrong et al. [32] were the first to describe the application of vancomycin as a chiral additive for the separation of 6-aminoquinolyl-N-hydroxysuccinimidyl derivatized amino acids and of some dansyl-amino acids. In addition to the separation of some amino acid racemates, the racemates of some drugs (e.g., bendroflumethiazide, warfarin, coumachlor, and indoprofen) were also successfully resolved employing vancomycin as a chiral additive in the RP mode. These authors [32] found that different types of stationary phase and the composition of the mobile phase strongly influenced enantioseparation. The best results were obtained using diphenyl stationary phases and acetonitrile as an organic modifier that produced the most effective separations with the shortest time of development. Bhushan et al. [33] applied erythromycin as a stationary phase modifier and more recently vancomycin [34] for the separation of dansyl-DL-amino acids and ($\pm$)-verapamil, respectively.

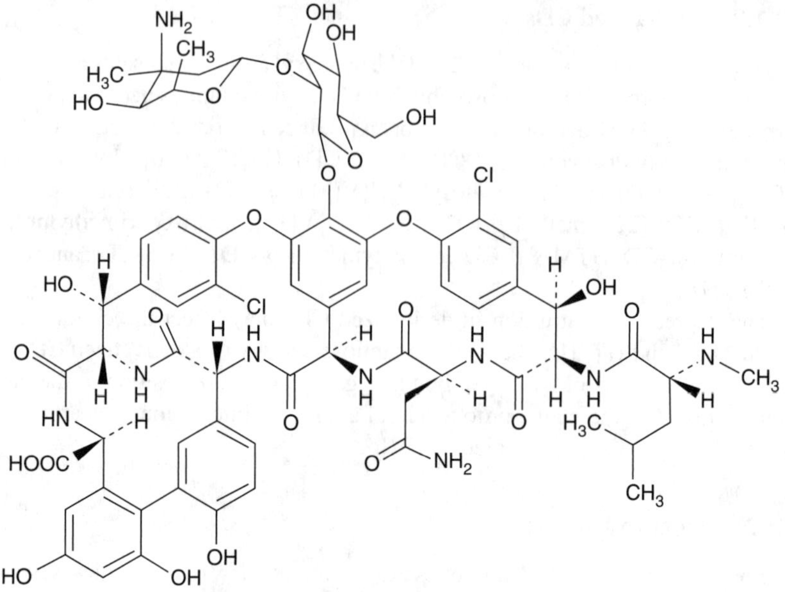

FIGURE 6.2   Structure of vancomycin.

6.3   CHIRAL COUNTERIONS IN MOBILE PHASE

The use of a chiral ion-pairing agent in the mobile phase was introduced to column liquid chromatography by Pettersson et al. [35], who assumed that the diastereomeric complexes are formed between the enantiomers and the chiral counterions and that the separation is achieved on a nonchiral stationary phase. The formed diastereomeric complexes have a different distribution ration between the two phases. The chiral selector should have a structure that makes chromatographic discrimination possible. The interaction between the two species can be based on binding groups, as well as on steric repulsions.

The most commonly used chiral ion-pair additives in TLC are CSA [36,37] and ZGP [38–40]. In order to resolve enantiomers of some aromatic amino alcohols, Gaolan et al. [36] added CSA to the mobile phase consisting of methanol and dichloromethane (the latter in optimum concentration of 60%) and the resolution was performed at a lower temperature (2–4°C). Bazylak and Aboul-Enein [37] investigated the chiral separation of selected β-blockers using CSA as a chiral counterion dissolved in dichloromethane containing different concentrations of methanol or propanol as polar organic modifiers and chemically bonded aminopropylsiloxane silica gel layer as a stationary phase. Duncan et al. [38,39] employed CSA and ZGP as chiral ion interaction agents and HP-TLC diol or HP-TLC silica plates for the chiral resolution of racemates of adrenergic receptor agonists or antagonists (octopamine, norphenylephrine, isoproterenol, propranolol, pindolol, timolol, and metoprolol). These authors have also compared different

N-benzyloxycarbonyl amino acid derivatives, such as ZP, ZIP, ZAP, and ZGP as chiral counterion/mobile phase additives. They established that it was possible to attain some separation with all of the N-CBZ amino acid derivatives, but the ZGP employed as a chiral counterion provided the best results. The same authors also found that the separation could not be achieved on other conventional TLC stationary phases including microcrystal cellulose, alumina, or ordinary silica gel plates and compared TLC results with HPLC ones [39]. Based on the experimental evidence, they emphasized both the low cost and wide versatility of TLC over HPLC. Their results confirmed that TLC can be used as a pilot technique for HPLC as suggested previously [41]. Tivert and Backman [40,42] also used ZGP (mobile phase 5 mM ZGP in dichloromethane) as a chiral ion-pairing mobile phase additive for enantioseparation of alprenolol, propranolol, and metoprolol, and studied the influence of humidity on this procedure. They obtained satisfactory and reproducible results for the migration distance and chromatographic performance after keeping chromatographic LiChrosorb Diol plates for about 20 h or longer at high-relative humidity. It was confirmed that the humidity of chromatographic system is a critical factor for enantioseparation than ethanolamine, previously used as an organic modifier. However, the reasons for beneficial influence of water on enantioseparation are still far from being fully understood. This method was successfully applied for the separation of active substances from controlled release tablets containing metoprolol succinate.

Also, D-(−)-tartaric acid has been used as a CMPA for the separation of (±)-metoprolol tartarate on silica gel plates preimpregnated with the mobile phase (ethanol:water, 70:30, v/v) containing D-(−)-tartaric acid as a chiral selector. The results of experiments performed with different concentrations of D-(−)-tartaric acid (5.8, 11.6, and 23 mmol/l) revealed that the best resolution of the metoprolol tartarate enantiomers was achieved with 11.6 mmol/l D-(−)-tartaric acid in both the mobile phase and the impregnation solution at $25 \pm 2°C$. It has been assumed that tartaric acid (pK_{a1} 2.93) dissolved in excess of ethanol could react with ethanol forming monoethyltartrate, which might play a role of a real chiral selector in this separation system [43]. The structures of CSA, ZGP, and monoethyltartrate as counterions are presented in Figure 6.3.

It should also be mentioned that Szulik and Sowa [44] successfully separated enantiomers of some fatty hydroxyl acids such as DL-α-hydroxypalmitic- and DL-12-hydroxystearic acid using 1% L-alanine and 2% L-alanine as mobile phase additives, respectively, with silica gel plates serving as a stationary phase. Besides, these authors presented enantioseparation of DL-12-hydroxyoleic acid using chiral plates [44].

6.4 PROTEINS AS MOBILE PHASE ADDITIVES

6.4.1 General

Proteins are composed of chiral monomers, organized in different sequences, providing a huge structural diversity. The inherent chirality of the amino acid

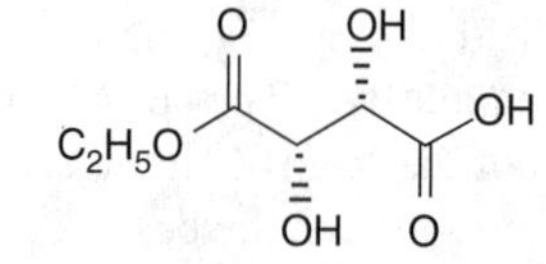

Monoethyltartrate

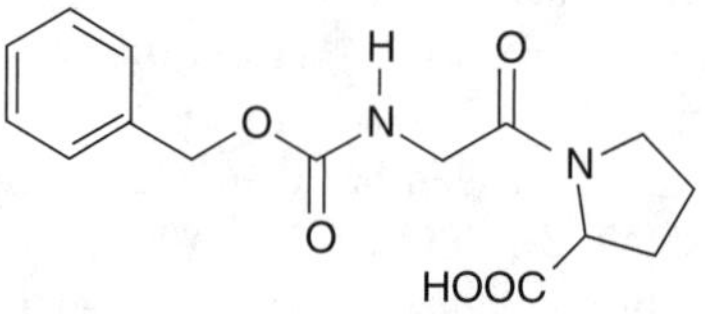

ZGP

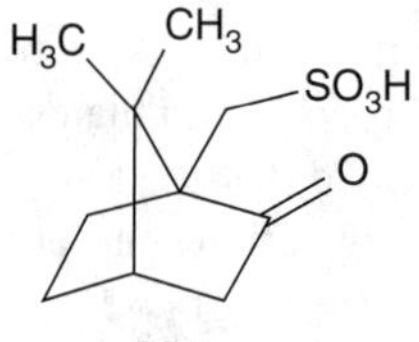

CSA

FIGURE 6.3 Structure of some of the counterions.

residues induces chiral recognition of the protein. Typical proteins used for enantiomeric separations include albumins (bovine and human serum), glycoproteins (α_1-acid glycoprotein, ovoglycoprotein, and ariden), and some enzymes [1]. Proteins immobilized on solid support can be widely used as CSP in HPLC or in solution as CMPAs in TLC. The chiral recognition properties of absorbed or bound proteins are often different from those of proteins in solution because of functional groups blocking or conformational changes.

6.4.2 BOVINE SERUM ALBUMIN AS A MOBILE PHASE ADDITIVE

Bovine serum albumin has been mostly used as CMPA in TLC for chiral separations. It is a globular protein with an isoelectric point of 4.7. This protein consists of 581 amino acids with 17 intrachain disulfide bridges and two cavities significantly differing in selectivity and binding energy [45]. Chiral recognition is affected by the ionic strength of the eluents, nature of organic modifiers, temperature, and pH, similar to the phenomena observed in HPLC. All these parameters can lead to changes in protein conformation, changes in chiral compounds, and the extent of electrostatic interactions. Almost all investigations related to the use of BSA as a chiral selector additive were performed by Lepri and co-workers [15,16,46–54], who have successfully resolved more than 13 different racemates mostly including amino acids and their derivatives, selected drugs, and other chiral compounds listed in Table 6.2. The separations were performed on the reverse-mode stationary

TABLE 6.2
Compounds Separated Using BSA as a Chiral Mobile Phase Additive

Compound	Mobile phase	Ref.
Thyroxine	0.2 M acetate buffer containing 20% 2-propanol and 6% BSA	[15]
N-α-(2,4-dinitro-5-fluorophenyl)-valinamide	0.5 M acetate buffer containing 20% 2-propanol and 7% BSA	[15]
N-α-(9-fluorenylmethoxycarbonyl)-DL-valine	5% BSA in acetate buffer (pH 4.86), 36% 2-propanol	[49]
N-α-(9-fluorenylmethoxycarbonyl)-DL-phenylalanine	5% BSA in acetate buffer (pH 4.86), 36% 2-propanol	[49]
N-α-(9-fluorenylmethoxycarbonyl)-DL-leucine	5% BSA in acetate buffer (pH 4.86), 36% 2-propanol	[49]
N-α-(9-fluorenylmethoxycarbonyl)-DL-norleucine	5% BSA in acetate buffer (pH 4.86), 23% 2-propanol	[49]
N-α-(9-fluorenylmethoxycarbonyl)-DL-proline	5% BSA in acetate buffer (pH 4.86), 23% 2-propanol	[49]
N-α-(9-fluorenylmethoxycarbonyl)-DL-norvaline	5% BSA in acetate buffer (pH 4.86), 23% 2-propanol	[49]
N-α-(9-fluorenylmethoxycarbonyl)-DL-tryptophan	6% BSA in acetate buffer (pH 4.86), 12% 2-propanol	[49]
N-α-(9-fluorenylmethoxycarbonyl)-DL-alanine	6% BSA in acetate buffer (pH 4.86), 23% 2-propanol	[49]
N-α-(9-fluorenylmethoxycarbonyl)- DL-cyclohexylalanine	5% BSA in acetate buffer (pH 4.86), 23% 2-propanol	[49]
N-α-(9-fluorenylmethoxycarbonyl)-DL-methionine	6% BSA in acetate buffer (pH 4.86), 12% 2-propanol	[49]
N-3,5-dinitro-2-pyridyl-DL-norleucine	5% BSA (pH 4.72)	[47]
N-3,5-dinitro-2-pyridyl-DL-leucine	5% BSA (pH 4.72)	[47]
N-2,4-dinitrophenyl-DL-norleucine	4% BSA (pH 6.86)	[47]
N-2,4-dinitrophenyl-DL-leucine	4% BSA (pH 6.86)	[47]
N-2,4-dinitrophenyl-DL-norvaline	4% BSA (pH 4.72)	[47]
N-2,4-dinitrophenyl-DL-methionine	3% BSA (pH 6.86)	[47]
N-2,4-dinitrophenyl-DL-methionine sulfone	5% BSA (pH 4.72)	[47]
N-2,4-dinitrophenyl-DL-methionine sulfoxide	5% BSA (pH 4.72)	[47]
N-3,5-dinitrobenzoyl-DL-leucine	5% BSA (pH 4.72)	[47]
N-3,5-dinitro-2-pyridyl-DL-methionine	2% BSA (pH 4.72)	[47]

(Continued)

TABLE 6.2
(Continued)

Compound	Mobile phase	Ref.
N-3,5-dinitro-2-pyridyl-DL-phenylalanine	2% BSA (pH 4.72)	[47]
N-3,5-dinitrobenzoyl-DL-α-phenylglycine	5% BSA (pH 4.72)	[47]
N-acetyl-5-methyl-DL-tryptophan	3% BSA, 0.5 M phosphate buffer	[52]
N-benzyloxycarbonyl-DL-tryptophan	5% BSA, 0.5 M acetic acid	[52]
N-tert-butyloxycarbonyl-DL-tryptophan	8% BSA, 0.5 M acetic acid	[52]
N-phthalyl-glycine-DL-tryptophan	3% BSA, 0.1 M acetic buffer	[52]
	6% BSA, 0.1 M acetic buffer	[52]
N-tert-butyloxycarbonyl-p-nitro-DL-phenylalanine	6% BSA, 0.05 M $NaHCO_3$/Na_2CO_3	[52]
N-o-nitrophenylsulfenyl-DL-norvaline	6% BSA, 0.1 M acetate buffer	[52]
N-o-nitrophenylsulfenyl-DL-norleucine	6% BSA, 0.1 M acetate buffer	[52]
N-2,4-dinitrophenyl-DL-α-aminobutyric acid	4% BSA, 0.1 M acetate buffer, 10°C	[52]
2,4-dinitrophenyl-DL-pipecolic acid	4% BSA, 0.1 M acetate buffer, 10°C	[52]
2,4-dinitrophenyl-DL-ethionone sulfone	4% BSA, 0.5 M acetic acid, 1% NaCl, 10°C	[52]
N-3,5-dinitro-2-pyridyl-DL-alanine	4% BSA, 0.5 M acetic acid, 1% NaCl, 10°C	[52]
N-3,5-dinitro-2-pyridyl-DL-norvaline	4% BSA, 0.5 M acetic acid, 1% NaCl, 10°C	[52]
4-Fluoro-DL-tryptophan	6% BSA, 0.05 M $NaHCO_3$/Na_2CO_3	[52]
5-Fluoro-DL-tryptophan	6% BSA, 0.05 M $NaHCO_3$/Na_2CO_3	[52]
Binaphthyl-2,2'-diylhydrogen phosphate-1,1'-bi-2-naphthol	5% BSA, 0.5 M acetic acid, 1% NaCl, 2% 2-propanol	[52]
	5% BSA, 0.05 M phosphate buffer, 11% 2-propanol	[52]
	6% BSA, 0.1 M $NaHCO_3$, 20% 2-propanol	[52]
β-Hydrastine	5% BSA, 0.05 M $NaHCO_3$/Na_2CO_3, 10% 2-propanol, 10°C	[52]
p-Nitrophenyl-β-thiofucopyranoside	5% BSA, 0.05 M phosphate buffer, 11% 2-propanol, 10°C	[52]
3,5-Dinitro-N-(1-phenylethyl)-benzamide	5% BSA, 0.1 M $NaHCO_3$/Na_2CO_3, 20% 2-propanol	[52]

Compound	Mobile Phase	Ref.
6-Fluoro-DL-tryptophan	6% BSA, 0.05 M NaHCO$_3$/Na$_2$CO$_3$	[52]
N-2,4-Dinitrophenyl-DL-ethionine	6% BSA, 0.1 M acetate buffer	[53]
N-2,4-Dinitrophenyl-DL-citrulline	6% BSA, 0.1 M acetate buffer	[53]
DL-Amethopterin	8% BSA, 0.5 M acetic acid	[53]
L-(+)-Amethopterin	8% BSA, 0.5 M acetic acid	[53]
(±)-Warfarin	8% BSA, 0.5 M sodium acetate	[53]
(±)-p-Chlorowarfarin	8% BSA, 0.5 M sodium acetate	[53]
DL-Tryptophan	6% BSA in 0.05 M sodium tetraborate, 6% 2-propanol (pH > 9)	[48]
DL-Tryptophanamide	6% BSA in 0.05 M sodium tetraborate, 6% 2-propanol (pH > 9)	[48]
DL-α-Methyltryptophan	6% BSA in 0.05 M sodium tetraborate, 6% 2-propanol (pH > 9)	[48]
DL-1-Methyltryptophan	6% BSA in 0.05 M sodium tetraborate, 6% 2-propanol (pH > 9)	[48]
DL-4-Methyltryptophan	6% BSA in 0.05 M sodium tetraborate, 6% 2-propanol (pH > 9)	[48]
DL-5-Methyltryptophan	6% BSA in 0.05 M sodium tetraborate, 6% 2-propanol (pH > 9)	[48]
DL-6-Methyltryptophan	6% BSA in 0.05 M sodium tetraborate, 6% 2-propanol (pH > 9)	[48]
DL-7-Methyltryptophan	6% BSA in 0.05 M sodium tetraborate, 6% 2-propanol (pH > 9)	[48]
DL-5-Hydroxytryptophan	6% BSA in 0.05 M sodium tetraborate, 6% 2-propanol (pH > 9)	[48]
DL-5-Methoxytryptophan	6% BSA in 0.05 M sodium tetraborate, 6% 2-propanol (pH > 9)	[48]
DL-Glycyltryptophan	6% BSA in 0.05 M sodium tetraborate, 6% 2-propanol (pH > 9)	[48]
DL-Alanine-p-nitroanilide	8% BSA in 0.1 M sodium acetate and 0.1 M acetic acid, 3% 2-propanol	[50]
DL-Leucine-p-nitroanilide	6% BSA in 0.1 M sodium acetate and 0.1 M acetic acid, 8% 2-propanol	[50]
	8% BSA in 0.1 M sodium acetate and 0.1 M acetic acid, 20% 2-propanol	
Methylthiohydantoin-DL-proline	9% BSA in 0.1 M acetate buffer, 2% 2-propanol (pH 4.86)	[50]
Phenylthiohydantoin-DL-tyrosine	9% BSA in 0.1 M acetate buffer, 2% 2-propanol (pH 4.86)	[50]
Phenylthiohydantoin-isoleucine	7% BSA in 0.5 M acetic acid, 2% 2-propanol (pH 3.50)	[50]
Phenylthiohydantoin-DL-methionine	7% BSA in 0.5 M acetic acid, 2% 2-propanol (pH 3.50)	[50]
	8% BSA in 0.1 M sodium acetate, 2% 2-propanol and acetic acid (pH 4.10)	[50]
Phenylthiohydantoin-DL-tryptophan	7% BSA in 0.5 M acetic acid, 2% 2-propanol (pH 3.50)	[50]

(Continued)

TABLE 6.2
(Continued)

Compound	Mobile phase	Ref.
Phenylthiohydantoin-DL-tyrosine	7% BSA in 0.5 M acetic acid, 2% 2-propanol (pH 3.50)	[50]
	8% BSA in 0.1 M sodium acetate, 2% 2-propanol and acetic acid (pH 4.10)	[50]
Phenylthiohydantoin-DL-valine	7% BSA in 0.5 M acetic acid, 2% 2-propanol (pH 3.50)	[50]
Phenylthiohydantoin-DL-phenylalanine	8% BSA in 0.1 M sodium acetate, 2% 2-propanol and acetic acid (pH 4.10)	[50]
DL-Kynurenine	6% BSA in 0.05 M Na$_2$B$_4$O$_7$, 6% 2-propanol (pH 9.30)	[50]
DL-3-(1-Naphthyl)-alanine	6% BSA in 0.05 M NaHCO$_3$ and 0.05 M Na$_2$CO$_3$, 6% 2-propanol (pH 9.80)	[50]
O-Hippuryl-DL-β-phenyl lactic acid	6% BSA in 0.05 M NaHCO$_3$ and 0.05 M Na$_2$CO$_3$, 6% 2-propanol (pH 9.80)	[50]
N-α-benzoyl-DL-arginine-7-amido-4-methylcoumarine	6% BSA in 0.05 M NaHCO$_3$ and 0.05 M Na$_2$CO$_3$, 6% 2-propanol (pH 9.80)	[50]
N-[1-(1-naphthyl)ethyl]phthalamic acid	6% BSA, 0.5 M acetic acid	[54]
	8% BSA, 0.5 M acetic acid	
	8% BSA, 0.5 M acetic acid, 1% NaCl	
	7% BSA, 0.5 M acetic acid	
N-benzoyl-phenylalanine	6% BSA, 4% 2-propanol, 0.5 M acetic acid	[54]
	6% BSA, 20% 2-propanol, 0.1 M acetate buffer	
Hydrobenzoine	7% BSA, 0.5 M acetic acid, 1% NaCl, 4% 2-propanol	[54]
1-(2-Methoxybenzoyl)-2-pyrrolidine methanol	7% BSA, 0.5 M acetic acid, 1% NaCl, 2-propanol	[54]
	6% BSA, 0.1 M acetic buffer, 20% 2-propanol	
α,α-Diphenyl-2-pyrrolidine methanol	7% BSA, 0.5 M acetic acid, 1% NaCl, 2-propanol	[54]
	6% BSA, 0.1 M acetic buffer, 20% 2-propanol	
Benzoin	7% BSA, 0.5 M acetic acid, 1% NaCl, 4% 2-propanol	[54]
2-Hydroxyflavonone	6% BSA, 0.5 M NaHCO$_3$/NaCO$_3$, 12% 2-propanol	[54]
Homoeriodictyol	6% BSA, 0.1 M Na$_2$CO$_3$, 6% 2-propanol	[54]

phase using mobile phases consisting of buffers of different pH and 2-propanol as the organic modifier. Concentrations of 2-propanol varied from 2 to 36% depending on the type of stationary phase used and the properties of chiral compounds. It was established that hydrophobic compounds require a higher concentration of 2-propanol in the eluent [52]. However, the highest concentration of 2-propanol of 36% as an organic modifier used by the aforementioned authors can decrease the stereoselectivity and solubility of BSA. The content of 2-propanol was shown to affect the pH of eluents, that is, higher 2-propanol concentrations led to an increase of the pH value [48]. It was also observed that the BSA concentrations of 4 to 6% were most suitable for enantioseparation and that increasing BSA content resulted in prolonged development time [47,50]. Taking into account both the type of a stationary phase and chemical properties of chiral compounds, the most common pH of buffer solution used range between 3.5 and 9.8. Lepri et al. [48] observed the formation of a pH gradient along the plate during the separation of enantiomers of selected tryptophan derivatives using eluents consisting of 6% BSA and 6% 2-propanol, pH 9.8. The acidity of the layer increased noticeably with increasing distance from the origin. This could be a consequence of the absorption of the hydroxyl ions by the stationary phase and an explanation for the identical behavior of L-forms of some tryptophan derivatives obtained by column chromatography employing acidic BSA solution (pH 6.85) as an eluent. Besides, it has been observed that the ionic strength of the eluent affects enantioseparation by producing more compact zones and increasing the retention of tryptophan derivatives by the stationary phase, probably due to the reduction of the ionic interactions of BSA with the solute molecules [52]. It has also been shown that the addition of 1% sodium chloride to eluents acts to reduce the electrostatic interactions of BSA with optical isomers and increase the hydrophobic interactions of the solute with both the stationary phase and the albumin. This is the reason for a greater retention and the inversion in the order of elution of N-substituted phthalamic acid enantiomers [54]. The stereoselectivity of BSA could be remarkably influenced by temperature, and it was observed that chiral selectivity was significantly increased parallel to the decrease of the temperature [52]. Based on the obtained results, Lepri and associates [53] suggested three possible kinds of sites present in BSA molecule: (a) nonchiral, (b) specific for a given enantiomer, and (c) more selective for one than for the other enantiomer. In favor of BSA use as CMPA vs. BSA bound to support, there is à fact that BSA moves more freely in solution, changes its conformation, influences the ratio of chiral sites number vs. that of nonchiral sites, and finally facilitates the interaction of the sites with the different enantiomers.

ACKNOWLEDGMENTS

The authors wish to express their gratitude to Prof. Teresa Kowalska (Institute of Chemistry, University of Silesia) for her kind invitation to contribute a chapter to this book. We are also very obliged to Prof. Gordana Popović (Faculty of Pharmacy, University of Belgrade) for her help during the preparation of this chapter.

REFERENCES

1. Poole, C.F., *The Essence of Chromatography*, 2d ed., Elsevier, Amsterdam, 2003, 803.
2. LeFevre, J.W., Reversed-phase thin-layer chromatographic separations of enantiomers of dansyl-amino acids using β-cyclodextrin as a mobile phase additive, *J. Chromatogr.*, 653: 293, 1993.
3. Han, S.M. and Armstrong, D.W., Enantiomeric separation by thin-layer chromatography, *Chem. Anal.*, 108: 81, 1990.
4. Lepri, L. et al., Separation of optical and structural isomers by planar chromatography with development by β-cyclodextrin solutions, *J. Planar Chromatogr.*, 3: 311, 1990.
5. Radulović, D.M. et al., A preliminary study of beta-cyclodextrin/metoprolol tartarate inclusion complexes for potential enantiomeric separation, *J. Pharm. Biomed. Anal.*, 24: 871, 2001.
6. Ikeda, Y. et al., NMR spectroscopic characterization of metoprolol/cyclodextrin complex in aqueous solutions: cavity size dependency, *J. Pharm. Sci.*, 93: 1659, 2004.
7. Sadlej-Sosnowska, N., Thermodynamic parameters of the formation of complexes between cyclodextrins and steroid hormones, *J. Chromatogr.*, 728: 89, 1996.
8. Fujimura, K. et al., Reversed-phase retention behavior of aromatic compounds involving β-cyclodextrin inclusion complex formation in the mobile phase, *Anal. Chem.*, 58: 2968, 1986.
9. Fujimura, K. et al., Retention behavior and chiral recognition mechanism of several cyclodextrin-bonded stationary phases for dansyl amino acids, *Anal. Chem.*, 62: 2198, 1990.
10. Cserháti,T. and Forgács, E., Inclusion complex formation of steroidal drugs with hydroxypropyl-β-cyclodextrin studied by charge — transfer chromatography, *J. Pharm. Biomed. Anal.*, 18: 179, 1998.
11. Forgács, E., Interaction of amino acids with hydroxypropyl-β-cyclodextrin, *Proc. Int. Symp. Planar Chromatogr.*, Nyiredy, Sz., Ed., Research Institute of Medicinal Plants, Budakalăsz, Hungary, 2000, p. 195.
12. Armstrong, D.W., He, F.-Y. and Han, S.M., Planar chromatographic separation of enantiomers and diastereoisomers with cyclodextrin mobile phase additives, *J. Chromatogr.*, 448: 345, 1988.
13. Lepri, L., Coas, V. and Desideri, P.G., Planar chromatography of isomers using β-cyclodextrin solutions as mobile phase, *J. Planar Chromatogr.*, 3: 533, 1990.
14. Lepri, L., Coas, V. and Desideri, P.G., Reversed phase planar chromatography of isomers using α- and β-cyclodextrin solutions as eluents, *J. Planar Chromatogr.*, 4: 338, 1991.
15. Lepri, L. et al., Reversed-phase planar chromatography of some enantiomeric amino acids and oxazolidinones, *Biomed. Chromatogr.*, 15: 196, 2001.
16. Lepri, L. et al., Reversed-phase planar chromatography of racemic flavonones, *J. Liq. Chrom. Rel. Technol.*, 22: 105, 1999.
17. Lepri, L., Coas, V. and Desideri, P.G., Planar chromatography of optical and structural isomers with eluents containing modified β-cyclodextrins, *J. Planar Chromatogr.*, 7: 322, 1994.
18. Aboul-Enein, H.Y., El-Awady, M.I. and Heard, C.M., Enantiomeric separation of aminoglutethimide, acetyl aminoglutethimide, and dansyl aminoglutethimide

by TLC with β-cyclodextrin and derivatives as mobile phase additives, *J. Liq. Chrom. Rel. Technol.*, 23: 2715, 2000.

19. Huang, M.-B. et al., Planar chromatographic direct separation of some aromatic amino acids and aromatic amino alcohols into enantiomers using cyclodextrin mobile phase additives, *J. Chromatogr.*, 742: 289, 1996.

20. Xuan, H.T.K. and Lederer, M., Adsorption chromatography on cellulose. XI. Chiral separations with aqueous solutions of cyclodextrins as eluents, *J. Chromatogr.*, 659: 191, 1994.

21. Duncan, J.D. and Armstrong, D.W., A study of the effect of the degree of substitution of hydroxypropyl-β-cyclodextrin used as a chiral mobile phase additive in TLC, *J. Planar Chromatogr.*, 4: 204, 1991.

22. Duncan, J.D. and Armstrong, D.W., Chiral mobile phase additives in reversed-phase TLC, *J. Planar Chromatogr.*, 3: 65, 1990.

23. LeFevre, J.W. et al., Determination of enantiomeric purity of commercial [14]C- and ^{3}H-labeled L-α-amino acids, *J. Labelled Cpd. Radiopharm.*, 41: 477, 1998.

24. Armstrong, D.W., Faulkner, J.R. and Han, S.M., Use of hydroxypropyl-derivatized β-cyclodextrins for the thin-layer chromatographic separation of enantiomers and diastereomers, *J. Chromatogr.*, 452: 323, 1988.

25. LeFevre, J.N. et al., Qualitative reversed-phase thin-layer chromatographic analysis of the stereochemistry of D- and L-α- amino acids in small peptides, *J. Planar Chromatogr.*, 13: 160, 2000.

26. Lambroussi, V., Piperaki, S. and Tsantili-Kakoulidou, A., Formation of inclusion complexes between cyclodextrins as mobile phase additives in RP-TLC, and fluoxetine, norfluoxetine, and promethazine, *J. Planar Chromatogr.*, 12: 124, 1999.

27. Bereznitski, Y. et al., Thin-layer chromatography — A useful technique for the separation of enantiomers, *J. AOAC Int.*, 84: 1242, 2001.

28. LeFevre, J.W. et al., The effect of structures on the resolution of dansyl amino acids using beta-cyclodextrin as a mobile phase additive, *Chromatographia*, 52: 648, 2000.

29. Kizek, J. et al., Determination of budesonide R-(+) and S-(−) isomers in pharmaceuticals by thin-layer chromatography with UV densitometric detection, *Chromatographia*, 56: 759, 2002.

30. Hao, A.Y. et al., Direct thin-layer chromatographic separation of enantiomers of six selected amino acids using 2-O-[(R)-2-hydroxypropyl]-β-cyclodextrin as a mobile phase additive, *Anal. Lett.*, 28: 2041, 1995.

31. Ward, T.J. and Farris III, A.B., Chiral separations using the macrocyclic antibiotics: a review, *J. Chromatogr.*, 906: 73, 2001.

32. Armstrong, D.W. and Zhou, Y., Use of macrocyclic antibiotics as the chiral selectors for enantiomeric separations by TLC, *J. Liq. Chromatogr.*, 17: 1695, 1994.

33. Bhushan, R. and Parshad, V., Thin-layer chromatographic separation of enantiomeric dansyl amino acids using a macrocyclic antibiotic as a chiral selector, *J. Chromatogr.*, 736: 235, 1996.

34. Bhushan, R. and Gupta, D., Thin-layer chromatography separation of enantiomers of verapamil using macrocyclic antibiotic as a chiral selector, *Biomed. Chromatogr.*, 19: 474, 2005.

35. Pettersson, C. and Schill, G., Separation of enantiomeric amines by ion-pair chromatography, *J. Chromatogr.*, 204: 179, 1981.

36. Gaolan, Li. et al., Enantiomeric separation of aromatic alcohol amino drugs by thin-layer chromatography, *Sepu*, 17: 215, 1999.

37. Bazylak, G. and Aboul-Enein, H.Y., Enantioseparation in series of beta-blockers by normal-phase planar chromatography systems with bonded aminopropyl-oxane silica layer and chiral counter ion. *Presented at 22nd Int. Symp. on Chromatography*, Rome, September 13–18, 1998, p. 404.

38. Duncan, J.D., Armstrong, D.W. and Stalcup, A.M., Normal phase TLC separation of enantiomers using chiral ion interaction agents, *J. Liq. Chromatogr.*, 13: 1091, 1990.

39. Duncan, J.D., Chiral separations: a comparison of HPLC and TLC, *J. Liq. Chromatogr.*, 13: 2737, 1990.

40. Tivert, A.M. and Backman, A.E., Enantiomeric separation of amino alcohols by TLC using a chiral counter-ion in the mobile phase, *J. Planar Chromatogr.*, 2: 472, 1989.

41. Schilitt, H. and Geiss, F., Thin-layer chromatography as a pilot technique for rapid column chromatography, *J. Chromatogr.*, 67: 476, 1972.

42. Tivert, A.M. and Backman, A.E., Separation of the enantiomers of β-blocking drugs by TLC with a chiral mobile phase additive, *J. Planar Chromatogr.*, 6: 216, 1993.

43. Lučić, B. et al., Direct separation of the enantiomers of ($\pm$) metoprolol tartarate on impregnated TLC plates with D-($-$)- tartaric acid as a chiral selector, *J. Planar Chromatogr.*, 18: 294, 2005.

44. Szulik, J. and Sowa, A., Separation of selected enantiomers of fatty hydroxyl acids by TLC, *Acta Chromatogr.*, 11: 233, 2001.

45. Jacobson, S. et al., Chromatographic band profiles and band separation of enantiomers at high concentration, *J. Am. Chem. Soc.*, 112: 6492, 1990.

46. Lepri, L., Bubba, M.D. and Cincinelli, A., Chiral separation by TLC, in *Planar Chromatography. A Retrospective View for the Third Millennium*, Nyiredy, Sz., Ed., Springer Scientific Publisher, Budapest, Hungary, 2001, chap. 25.

47. Lepri, L., Coas, V. and Desideri, G.P., Planar chromatography of optical isomers with bovine serum albumin in the mobile phase, *J. Planar Chromatogr.*, 5: 175, 1992.

48. Lepri, L. et al., Reversed phase planar chromatography of enantiomeric trypto-phans with bovine serum albumin in the mobile phase, *J. Planar Chromatogr.*, 5: 234, 1992.

49. Lepri, L., Coas, V. and Desideri, G.P., Reversed phase planar chromatography of optically active fluorenylmethoxycarbonyl amino acids with bovine serum albumin in the mobile phase, *J. Planar Chromatogr.*, 5: 294, 1992.

50. Lepri, L. et al., Reversed phase planar chromatography of enantiomeric com-pounds with bovine serum albumin in the mobile phase, *J. Planar Chromatogr.*, 5: 364, 1992.

51. Lepri, L. et al., Reversed phase planar chromatography of dansyl DL amino acids with bovine serum albumin in the mobile phase, *Chromatographia*, 36: 297, 1993.

52. Lepri, L. et al., Thin layer chromatographic enantioseparation of miscellaneous compounds with bovine serum albumin in the eluent, *J. Planar Chromatogr.*, 6: 100, 1993.

53. Lepri, L. et al., The mechanism of retention of enantiomeric solutes on silanized silica plates eluted with albumin solutions, *J. Planar Chromatogr.*, 7: 103, 1994.

54. Lepri, L. et al., Reversed-phase planar chromatography of optical isomers with bovine serum albumin as mobile phase additive, *J. Planar Chromatogr.*, 12: 221, 1999.
55. De Vault, G.L. and Sepaniak, M.J., Two-dimensional capillary electrophoresis-thin layer chromatography separations of amino acid enantiomers using electro-filament transfer, *J. Micro. Sep.*, 12: 419, 2000.

7 An Overview of the Chiral Separation Mechanisms

Antoine-Michel Siouffi and Patrick Piras

CONTENTS

7.1 INTRODUCTION

Chiral separations have become an important field in separation sciences, since many drugs, agrochemicals, food additives, and fragrances are chiral compounds and their bioactivities are related to their chirality.

A molecule can exist as a pair of enantiomers if it does not exhibit a symmetry plane or a symmetry center. More generally, the nonsuperimposability of the molecule on its mirror image is a necessary and sufficient check to determine whether a molecule is chiral or not chiral (achiral).

Many types of molecular chirality are observed. They can be shortlisted into four categories:

1. *Central chirality*: molecules with quadrivalent atoms (e.g., sp3 hybridized carbon atoms) bearing four different substituents (e.g., nitrogen in quaternary ammonium salts); molecules with trivalent atoms bearing three different substituents at the top of a pyramid.
2. *Axial chirality*: the atoms are placed in two perpendicular planes that are nonsymmetrical and cannot rotate freely (allenes, cumulenes, spiranes, and substituted ortho-biphenyls).
3. *Planar chirality*: with loss of a symmetry plane (paracyclophanes).
4. *Helical chirality*: molecules develop a helicoidal structure (*trans*-cyclooctene and helicenes).

Different ways are used in the literature to name enantiomers: (+)- and (−)-labels are often assigned to each mirror isomers and derive from the ability of the two enantiomers to rotate plane-polarized light in opposite directions. The (+)- and (−)-enantiomers are also, respectively, named D- and L- (for dextrorotatory and levorotatory), and this naming is not related to the D and L convention. D and L naming convention refer historically to the two enantiomers D(+) and L(−) of glyceraldehyde and is mainly used as a standard for the designation of absolute configuration of natural chiral compounds (amino acids and sugars).

Another convention was developed by Cahn et al. [1] who assigned a letter, R or S, to describe the absolute configuration of compounds with a single chiral center.

It is important to outline that all these naming conventions and nomenclatures R/S, +/−, and D/L have been designed from different approaches and accordingly are each other unrelated.

Pasteur was first to separate enantiomers by crystallization. Modern separations are performed by chromatography whereas spectroscopic techniques, such as NMR and x-ray, are useful tools for elucidation of absolute configuration or mechanistic aspects of enantioseparations. NMR allows to assess complexation stoichiometry or association constants.

Enantiomers exhibit different behaviors in an anisotropic environment. Chromatographic enantioseparation requires interaction of the enantiomers with a pure chiral compound, the selector.

The origin of available chiral selectors is natural proteins, oligosaccharides e.g., cyclodextrins (CDs), polysaccharides (e.g., cellulose), antibiotics, and amino acids, semisynthetic (modified oligo- or polysaccharides), and synthetic (Pirkle types, ligand exchange, and synthetic polymers). Development in this field is well reflected by the increasing amount of new selectors introduced in the past few years [2–4].

The earliest reported chromatographic separation is the resolution of a racemic amino acid on paper as stationary phase [5]. Later, a pioneering paper appeared that reported the resolution of enantiomers by gas chromatography (GC) on a chiral stationary phase (CSP) [6].

Soon after, the advent of column liquid chromatography (LC) tremendously expanded the domain of possibilities, and this mode is nowadays the most popular and versatile as exemplified by the 100,000 chiral separations described in Chirbase molecular database [7]. LC allows the use of chiral selector in any phase, mobile or stationary; conversely, the solvents play a significant role since the solvation of either solute or selector (or both) greatly influences the chiral recognition, which is not the case in GC. Anyway, separations, thought in the past to be impossible, are now routine. Despite the popularity of thin layer chromatography (TLC) due to its simplicity, low cost, wide versatility, and ability to screen a large number of samples in a single run, enantiomeric separations are not as numerous as in LC, which is considered more prestigious. Most reviews do not quote planar chromatography (PC) or TLC.

There are two modes to achieve chromatographic enantioseparations:

1. Indirect methods, which involves the synthesis of stable diastereo-isomers by a chiral derivatizing agent [8] followed by chromatography on achiral column (e.g., resolution of amino acids [9]).
2. Direct method via transient diastereomeric associations involving sep-aration of the enantiomers using either a CSP or a chiral selector in the mobile phase.

TLC and PC also are LC techniques. Same stationary phases are used in both LC and TLC [10]. The chiral separation mechanism that governs separation in LC has been extensively studied and is applicable to PC. Thus, it will be discussed here.

7.2 THE BINDING SITES: THE THREE-POINT CHIRAL RECOGNITION MODEL

Enantioselectivity is generated by the different interaction of a chiral selector with the two enantiomers. When a chiral selector is used, the primary retention mechanism is complexation with the chiral selector. Two different diastereomeric enantiomer–selector adducts are formed.

The concept of the "three-point fit" was proposed in 1933 [11]. In this model, stereochemical differences in pharmacological activities were due to the dif-ferential binding of enantiomers to a common site on a receptor surface. The "three-point" interaction model was revisited by Ogston [12]. However, the often-quoted paper is the one from Dalgliesh [13], who invoked a "three-point" interaction to explain the enantioselective separation of amino acids on cellulose paper. The three-point interaction rule differs from the three-point attachment rule as was pointed out by Davankov [14], who states that the condition for a chiral selector to recognize the enantiomers is that at least three configuration-dependent active points of the selector molecule should interact with three complementary

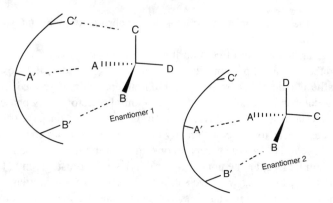

FIGURE 7.1 An illustration of the classical three-point interaction model. Enantiomer 2 provides a weaker association of the diastereomeric complex.

and configuration-dependent active points of the enantiomer molecules. A typical scheme of the three-point interaction model is shown in Figure 7.1. If three groups of a molecule can match three sites of the chiral selector, then its mirror image, after all possible rotations can only present two groups that can interact.

In Figure 7.1, the enantiomer labeled 1 interacts with the chiral selector on A–A′, B–B′, and C–C′. The enantiomer labeled 2 (the antipode) does not interact with the chiral selector on C–C′. In this mode, if the diastereomeric complex chiral selector–enantiomer 1 is stabilized by the interaction with C–C′, the enantiomer 1 is more retained by the chiral selector than the enantiomer 2. Conversely, if the interaction C–C′ is nonstabilizing the complex chiral selector–enantiomer 1, the enantiomer 2 is more retained.

The three interactions between the chiral selector and enantiomers considered do not need to be attractive, as some of them may be repulsive. Only the resulting sum for one (the favored diastereomeric combination) should produce a more favorable binding. Two interactions may be repulsive if the third one is strong enough to promote the formation of one of the two possible diastereomeric selector–ligand complexes.

The three-point attachment rule is largely qualitative and only valid with bimolecular processes (e.g., small Pirkle or ligand-exchange selectors). Another drawback of this model approach is that it cannot be applied to enantiomers with multiple chiral centers. Sundaresan and Abrol [15] proposed a novel chiral recognition model to explain stereoselectivity of substrates with two or three stereo centers requiring a minimum of four or five interaction points. In the same way, Davankov [16] pointed out that much more contact points are realized with chiral cavities of solids.

Table 7.1 lists the main intermolecular forces, which may occur between two pairs of enantiomers and a given chiral selector.

As shown in Table 7.1, high-performance liquid chromatography (HPLC) and TLC chiral separations involve different types of attractive or repulsive noncovalent-bonding interactions. These interactions combine both long-range

TABLE 7.1

Characteristics of the Common Types of Interactions Involved in Chiral Separation

Type of interaction	Chiral selector class	Relative strength	Comment
Electrostatic			
Ion–ion	Ion-pairing selectors (e.g., proteins and ligand exchange)	Very strong	Two oppositely charged species. Strongest interaction, nondirectional, and nonspecific
Ion–dipole	Ion-pairing selectors	Strong	One ion with a polar group
Coordinative (metal)	All selectors based on a metal complex (e.g., ligand exchange)	Strong	A particular ion–dipole interaction
Dipole–dipole	Selector with polar groups (e.g., polysaccharides)	Moderate	Alignment of permanent dipoles (e.g., carbonyls). Different orientations
Induction			
Ion-induced dipole	Ion-pairing selectors (e.g., proteins and ligand exchange)	Weak	Temporary-induced dipoles; depends on the number of polarizable electrons in molecules (polarizability) and size of the selector
Dipole-induced dipole	Any selector with a polar group (e.g., antibiotics)	Very weak	Induced dipole–induced dipole attraction. Nonpolar compounds
Dispersion (London forces)	All selectors. Driving force of hydrophobic and inclusion effects (e.g., CD)	Very weak	In aqueous system contribute to hydrophobic effects (e.g., alkyl chains)
Specific interactions (highly complementary)			
Hydrogen bonding	Native polysaccharides or CDs, antibiotics, etc.	Moderate to strong	Highly directional, strong unusual dipole–dipole force, weak with low electronegative atoms
π–π — charge transfer	Brush-type π — acceptor and donor (e.g., derivatized polysaccharides)	Weak to moderate	Electron rich/electron poor aromatics. Directional: face-to-face π–π stacking or edge-to-face
CH–π/XH–π	Mainly Pirkle selector	Weak	Comparable to a weak hydrogen bond

(electrostatic, induction, and dispersion) and short-range (repulsion and charge transfer) effects. The term *van der Waals force* covers attractive interactions (dipole, induced dipole, and London dispersion) as well as repulsive steric interactions. Hydrophobic interactions are also critical in chiral LC. They are not the result of an attractive force but a consequence of unfavorable interactions between water molecules and nonpolar molecules.

Intermolecular forces can be ranked according to their decreasing relative strength (from left to right):

ion–ion > ion–dipole > hydrogen bonds > dipole–dipole > π–π > dipole–induced dipole > dispersion

The strongest interaction comes from Coulomb force and produces the best resolution. This mode involves an ion as (e.g., in ligand exchange). The hydrogen bond interactions are involved in many schemes, the attractiveness can be strong or weak according to the distance between the negative site and the hydrogen atom. Hydrogen bond interactions are not solely responsible for chiral recognition; they are always associated with another mechanism such as inclusion (e.g., CD and polysaccharide) or π interactions. This is well illustrated by the few relative number of HPLC applications involving chiral selectors only based on hydrogen bonds when compared with other aromatic Pirkle phases.

Steric hindrance is always repulsive (Pauli exclusion) and mainly involved in tailor-made CSPs. High levels of chiral recognition can be attributed to the presence of bulky groups [17]. Many studies have shown that the steric effect can be the determinant interaction in chiral recognition [18–20].

π–π Interactions are associated with aromatic rings and play a major role in the development of Pirkle phases [21]. According to the nature and the number of the substituents, a π-donating or π-accepting moiety is produced. Electron-rich substituents such as NO_2 produce π-accepting moiety, whereas withdrawing substituents produce π-donating moiety. A pseudo two-point interaction is often supposed to be involved in chiral discrimination with π-complexes. These interactions are not popular in TLC due to the problems of detection since UV light is the popular mode of detection.

If a number of chiral recognition models have been proposed today, early models of the chiral recognition mechanism in chiral HPLC derive from the evolution of the first Pirkle CSP based on the use of a chiral solvating agent [22] followed by spectroscopic investigations [23].

Pirkle's [24] 1966 discovery of the nonequivalence of NMR signals arising from enantiomers in the presence of a chiral solvating agent had a dramatic effect on the course of enantiomer resolution. The proposed model supposes a two-point chelate-like complex with analytes containing two different electron-rich interaction sites. NMR spectroscopy is often used to confirm the model proposed from chromatographic results [25].

Today, the rational design of specific chiral selectors is one of the more challenging problems facing chiral separation. The current trend is moving away from "rational" to combinatorial approaches [26,27]. One possible explanation for this trend is the lack of a clear and informative classification scheme of chiral selectors that would make clear the relationship between them.

Wainer [28] has been able to introduce a first simple classification scheme for HPLC CSPs based on the mode of formation of the solute–CSP complexes:

1. Solute–CSP complexes are formed by attractive interactions such as hydrogen-bonding, π–π interactions, and dipole stacking (e.g., the Pirkle phase).
2. Solute–CSP complexes are formed by attractive interactions and through the inclusion into a chiral cavity (e.g., any cellulose or amylose-based CSPs).
3. The primary mechanism involves the formation of inclusion complexes (e.g., CDs).
4. The solute is involved in a ligand-exchange complex (Davankov chiral selectors).
5. The CSP is a protein and the solute–CSP complexes are based on the combination of hydrophobic and polar interactions.

Today, a more pertinent classification should be proposed in accordance with the great number of chiral selectors reported in LC (about 1500 CSPs have been collected in ChirBase database) and our new knowledge of chiral separation mechanisms. For instance, chromatographic results are clearly suggesting that bimodal and supramolecular systems should be gathered quite separately.

7.3 COMPLEXATION CONSTANTS AND RETENTION

7.3.1 CHIRAL SELECTOR IMMOBILIZED ON THE SOLID PHASE SUPPORT

When a chiral selector is used as the stationary phase and 1:1 solute–selector complexation occurs (the usual case), the equilibrium complexation constant K_c is written as

$$K_c = \frac{[\text{solute} - \text{selector}]}{[\text{solute}][\text{selector}]},$$

where [solute], [selector], and [solute − selector] represent the equilibrium concentrations of free solute, free selector, and solute–selector complex, respectively.

In LC, it can be written as

$$K_c = \frac{k}{\Phi[\text{selector}]},$$

where k is the retention factor and φ is the phase ratio. The retention factor k [29] is proportional to

$$\frac{1 - R_f}{R_f}.$$

Nevertheless, K_cs are difficult to estimate because many parameters are involved [30]. Determination of K_c is even more difficult from PC data since an accurate determination of φ is difficult.

It should be recognized that enantiomer separation by chromatography is caused by the difference in the Gibbs energy $-\Delta_{R,S}(\Delta G^0)$ of the diastereomeric association equilibrium between chiral selector and solute and is thus thermodynamic in nature. It should be pointed out that fast kinetics is required.

The Gibbs–Helmholtz equation is written as follows:

$$-\Delta_{R,S}(\Delta G^0) = RT \ln \left(\frac{K_{c_R}}{K_{c_S}} \right) = -\Delta_{R,S}(\Delta H) + T\Delta_{R,S}(\Delta S) = RT \ln \alpha,$$

where R refers arbitrarily to the second-eluted enantiomer and S to the first-eluted enantiomer, and α is the enantioselectivity.

The enthalpic and entropic terms have an opposite effect on the chiral recognition process. The unfavorable entropic term essentially arises from the loss of degree of freedom experienced by the most retained enantiomer.

The temperature dependence of the enantioselectivity α can be utilized to calculate the Gibbs–Helmholtz parameters $\Delta_{R,S}(\Delta H)$ and $\Delta_{R,S}(\Delta S)$ of chiral recognition according to the following equation:

$$\ln \alpha = \frac{-\Delta_{R,S}(\Delta H)}{RT} + \frac{\Delta_{R,S}(\Delta S)}{R}.$$

As indicated by these equations, both retention factor and separation factor are controlled by an enthalpic contribution, which decreases with the elevation of temperature, and an entropic contribution, which is independent of the temperature. The selectivity is a compromise between differences in enantiomeric binding enthalpy and disruptive entropic effects. The enthalpy term is a function of overall interactions between each enantiomer and the chiral selector. By plotting $\ln(\alpha)$ vs. $1/T$, all processes that do not contribute to the enantiomeric discrimination cancel out and the plot is linear, the slope being the difference between the enthalpy of association of the enantiomers with the stationary phase. The linear inverse relationship between $\ln \alpha$ and temperature demonstrates the enhancement of selectivity with a decrease in temperature. There exists a Tiso where $-\Delta_{R,S}(\Delta G^0) = 0$ owing to enthalpy/entropy compensation. However, many unusual effects of changes in temperature have been reported, and the van't Hoff plot may be nonlinear as it may occur with polysaccharides as chiral selector [31].

In spite of its usefulness, this procedure is seldom utilized in TLC. The presence of the gas phase and associated evaporation of the mobile phase precludes any variation in temperature. However, a decrease in temperature would be worth trying.

7.3.2 CHIRAL SELECTOR IN THE MOBILE PHASE

When the chiral selector is incorporated in the mobile phase, the solute retention is the result of competing interactions in the stationary and mobile phases. The R_f value is obviously in between R_{f0} when no selector is present and $R_{f\infty}$ when all the solute is present as solute–selector complex. Utilizing retention factor

$$\frac{1}{k} = \frac{1 + K_C[\text{selector}]}{k_0 + k_\infty K_C[\text{selector}]}$$

with

$$\frac{1}{k} = \frac{R_f}{1 - R_f},$$

where k_0 is the solute retention factor with no selector in the mobile phase and k_∞ is the retention factor when all solute is present as the solute–selector complex.

As an example, CDs are utilized as mobile phase additives in TLC.

If complexation occurs within the mobile phase, let us suppose that the stationary phase is hydrophobic, and the solute molecule is nonionic. The following equilibrium can be written as

$$(\text{S})_m + (\text{CD})_m \longleftrightarrow (\text{CD} - \text{S})_m$$

$$\uparrow\downarrow(1) \quad \uparrow\downarrow(2) \qquad\qquad \uparrow\downarrow(3)$$

$$(\text{S})_s \qquad (\text{CD})_s \qquad (\text{CD} - \text{S})_s$$

According to the affinity of the CD toward the stationary phase, two processes are possible: either chiral recognition within the mobile phase or formation of a dynamic CSP. The first process is involved in a system with a hydrophobic stationary phase in which equilibrium 2 can be neglected. Equilibrium in mobile phase are as follows:

$$(\text{S})_m + (\text{CD})_m \longleftrightarrow (\text{CD} - \text{S})_m$$

$$\uparrow\downarrow K_0 \quad K_M \uparrow\downarrow + M \qquad \uparrow\downarrow K_1$$

$$(\text{S})_s \qquad (\text{CD} - \text{M})_s \quad (\text{CD} - \text{S})_s$$

and

$$k_0 = (S)_g/(S)_m$$
$$k_D = (CD)_m(S)_m/(CD-S)_m$$
$$k_M = (CD-M)_m/(CD)_m(M)$$
$$k_1 = (CD-S)_g/(CD-S)_m$$

The retention factor is

$$k = \frac{[(S_g)+(CD-S)_g]}{[(S_m)+(CD-S)_m]} \cdot \varphi,$$

where φ is the phase ratio.

In the above equation, K_0 is neglected. This assumption is valid since CD as a solute is unretained on a RP18 phase. Furthermore, K_1 can be neglected as well, and the solute concentration is low (analytical chromatography). In this mode

$$\frac{1}{k} = \frac{1}{k_0} + \frac{(CD)_m}{K_D k_0}$$

and

$$k_0 = K_0 \varphi,$$

where k_0 is the retention factor in the system without CD.

One shortcoming of the chiral mobile phase approach is usually a lower selectivity than the one observed with CSP.

7.4 STOICHIOMETRY OF COMPLEXATION

Many solute–selector complexes arise from 1:1 association. However, some complexes exhibit multiple stoichiometries (see, e.g., 4-L-phenylalanyaminopyridine and methacrylic acid, or terpene-α-CD).

For the general case of m:n solute–selector complex with m:n solute–selector complexation, we have

$$K_C = \frac{[\text{solute}_m \cdot \text{selector}_n]}{[\text{solute}]^m [\text{selector}]^n}.$$

Stoichiometry is determined from the change in the equilibrium concentration of the complex when the selector concentration is varied. Adding selector to the mobile phase induces a change in the solute retention. The method is used with LC but not in TLC. NMR measurements provide a faster approach to stoichiometry but it requires the availability of pure solute.

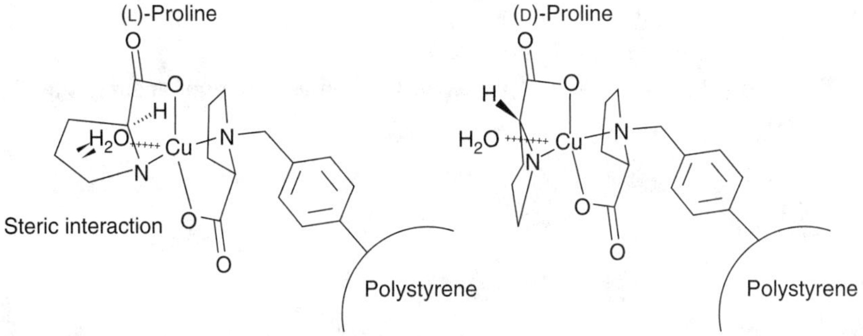

FIGURE 7.2 A representative ligand-exchange chiral recognition model based on (L)-proline chiral selector (adapted from Davankov, Y.A., Zolotarev, Y.A., and Kurganov, A.A., *J. Liquid Chromatogr.*, 2, 1191, 1979.). (L)-Proline binding to chiral selector is less favorable due to the unfavorable steric interaction of water with the pyrrolidine ring.

7.5 ENANTIOMERIC SEPARATION IN PLANAR CHROMATOGRAPHY

Two types of CSP are mostly utilized: those obtained via bonding an enantiopure ligand with localized chiral centers and potentially active moieties onto a support (e.g., silica) or those derived from natural (or chemically modified) molecules (e.g., cellulose) that exhibit chirality. Interactions are not as strong in the latter case as in the former.

7.5.1 LIGAND EXCHANGE

The technique was suggested by Helfferich in 1961 [32] and further developed by Rogozhin and Davankov [33]. Ligand-exchange chromatography (LEC) involves immobilized complexes of a ligand with a metal in a stationary phase. Such complexes are able to undergo very specific interaction with a certain solute. One ligand is a chiral selector with a complexing metal ion [generally copper (II)] bonded or loaded on a solid support (e.g., silica). The second ligand is one of the components of the sample mixture. In other words, the interaction between the chiral selector and the solute is mediated by a metal ion, which coordinates simultaneously to the chiral selector and the enantiomers to be separated [34].

Chiral separation mechanisms are often well explained, and models may offer reliable prediction of the elution orders. Figure 7.2 shows a LEC recognition model adapted from Davankov et al. [35].

LEC separation is based on the formation of labile ternary metal complexes in the stationary phase:

$$\text{Mobile phase} \quad A_\text{m}$$

$$\uparrow\downarrow$$

$$\text{Stationary phase} \quad A_\text{s} + ML_\text{s} \rightleftarrows AML_\text{s}$$

where M is the metal, L the ligand, and the subscripts "m" and "s" stands for mobile and stationary phases, respectively.

The two labile ternary complexes differ in their thermodynamic stability constants:

$$K_{A_R}ML = \frac{[A_RML]}{[A_R][M][L]}$$

and

$$K_{A_S}ML = \frac{[A_SML]}{[A_S][M][L]}.$$

Since it involves Coulombic forces, the method produces highest K_cs.

Davankov [16] observed that the stability of diastereomeric complexes formed in LEC is higher than the stability of the diastereomeric adducts formed by other chiral selectors. The consequence is that the association–dissociation cycle is not rapid and mass transfer is slow and thus generates low efficiency. This is a shortcoming in LC where both efficiency and selectivity are looked for, but not in TLC where efficiency is low.

The only commercially successful CSP for TLC is based on the ligand-exchange approach [36,37].

Gunther et al. [38] described a chiral phase made by treating octadecyl-modified TLC plates with a solution of copper acetate followed by a solution of $(2S,4R,2'RS)$-4-hydroxy-1-(2'-hydroxydodecyl)proline. These plates were commercially available from Macherey-Nagel. Similarly, the CHIR plates from Merck are based on a reversed-phase matrix impregnated with a copper salt and an optically active amino acid.

To investigate the interactions between the LEC-CSPs and the analytes, the effect of temperature on enantioselectivity (α) was examined in the range 25–45. The ligand-exchange approach is limited to the range of R_f values of separated amino acids, which is rather narrow ($0.4 \leq R_f \leq 0.73$). Polak et al. [39] investigated the effect of the pH* (pH* refers to the organic-aqueous mobile phase) and concluded that chromatography should be conducted at pH* 3–4 or 6–7 to obtain the highest enantioselectivity.

7.5.2 SUPRAMOLECULAR AND INCLUSION MECHANISMS

7.5.2.1 Cyclodextrins

Cyclodextrins are produced by the action of CD glycosyl transferase on starch. They are cyclic oligosaccharide molecules of α-(1,4)-linked glucose. Those containing six, seven, and eight glucose units (α-, β-, and γ-CDs, respectively) are most common. The native CD is a truncated cone having an axial cavity with primary hydroxyl groups [in the C(6) positions] around its narrower rim and, secondary hydroxyl groups [in the C(2) and C(3) positions] on the opposite wider rim

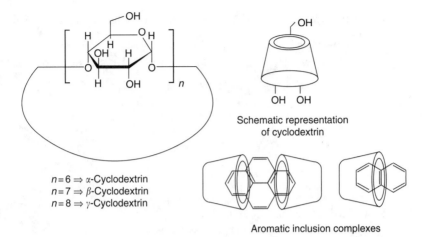

$n=6 \Rightarrow \alpha$-Cyclodextrin
$n=7 \Rightarrow \beta$-Cyclodextrin
$n=8 \Rightarrow \gamma$-Cyclodextrin

Schematic representation
of cyclodextrin

Aromatic inclusion complexes

FIGURE 7.3 Schematic representation of the structure of CDs and some inclusion complexes with aromatics.

(see Figure 7.3). Due to the high-electron density from the free electron pairs of glycosidyl oxygens, the internal cavity of the CD tends to be more hydrophobic than the exterior. The 2- and 3-hydroxyl groups at the edge of the cavity are oriented toward the outside of the molecule in a clockwise direction and are responsible for the aqueous solubility of the oligosaccharides. The depth of the cavity is the same (7.8 Å) for all CDs. CDs exhibit numerous chiral centers from the glucose units (each sugar provides five asymmetric centers). For instance, β-CD possesses 35 different chiral centers. The twisted shape explains why the CDs offer a broad recognition ability.

The chiral recognition mechanism of CD is understood fairly well and is based on several phenomena. In the beginning, the inclusion phenomenon was found to be a key interaction in chiral recognition by CDs. Small size molecule usually generate 1:1 inclusion complexes whereas with large molecule 2:1 complexes can be formed. However, data from GC and separations of enantiomers on different sizes of CDs evidenced that inclusion-complex formation can be excluded in some cases. Chiral recognition can involve hydrogen bonds with the spatially oriented hydroxyl groups at the rims of the cavity [40–42]. The hydrophobic part of the solute can fit the cavity while the other part of the enantiomer can interact (or not) with the glucose units as does a screw. However, the inclusion phenomenon is not always the only required chiral interaction as some racemates have been resolved with different sizes of CDs. CDs can change their shape to interact with solutes; it is the so-called "induced fit."

The host–guest inclusion process involves the hydrophobicity and the size of the solute:

$$CD + S \underset{K_S}{\Longleftrightarrow} CD - S,$$

where CD is the cyclodextrin, S the guest molecule (solute), and CD − S the inclusion complex.

The stability constant is

$$K_S = \frac{[CD - S]}{[CD][S]}$$

with

$$K_S = \frac{1}{K_D},$$

where K_D is the dissociation constant.

The parameters that influence the process are as follows:

- van der Waals interaction between the hydrophobic moiety of the solute molecule and the hydrophobic cavity
- Hydrogen bonding between the polar groups of the solute molecule and the hydroxyl groups of the CD
- Expelling the water molecules off the cavity.

The size of the solute is of paramount importance. A solute molecule with a phenyl ring fits the cavity of an α-CD pretty well. A molecule with two phenyl rings fits the β-CD cavity, and pyrene or similar length to breadth ratio molecules will better match the size of the γ-CD.

The complexation process can be improved by use of chemically modified CDs. Furthermore, partial or complete alkylation or acylation of hydroxyls can also increase the solubility of CDs in organic mobile phase.

Cyclodextrins were first used as mobile-phase additives in TLC to separate isomeric compounds. Later, they were immobilized or bonded onto chromatographic supports to form a highly effective CSP. CDs are well suited to chromatography since

- The inclusion process is stereoselective and reversible
- They are stable within a wide range of pH
- They do not absorb under usual UV detection wavelengths
- Complexation kinetics is of the same magnitude as the diffusion.

As a matter of fact, CSPs with bonded CDs are utilized in HPLC whereas CDs as mobile phase additives are mostly utilized in TLC. Separation between two enantiomers depends on the CD concentration in the mobile phase and the K_D values. K is a hyperbolic function of [CD]. Plot of $1/k$ vs. [CD] is linear with a slope of $1/K_D k_0$. The enantiomer forming the least stable complex will be preferentially retained.

7.5.2.2 Polysaccharides

Polysaccharides such as cellulose and amylose and to a much lesser extent chitin are among the most common naturally occurring chiral polymers. As quoted previously, the resolving ability of cellulose paper was recognized in the late 1940s.

Cellulose is a poly-β-D-1,4-glucoside. The β-D-glucose units exist in chair conformations with hydroxyl groups at the equatorial position. Chiral recognition ability of native cellulose is attributed to the crystalline array of parallel chains with a twofold screw axis of symmetry along the chain axes.

Amylose is similar to cellulose, but the acetal linkage is in an α configuration. This inversion of configuration at one asymmetric center is accompanied by an important conformational change of the structure. Native cellulose is mostly composed of linear chains whereas amylose chains tend to adopt a helical conformation.

Both polymers are very poor chiral selectors in their native state but exhibit enhanced selectivity when derivatized as esters or carbamates derivatives. Microcrystalline cellulose triacetate (MCTA) was the first derivative used in LC mostly in the reversed-phase mode. Chiral recognition by MCTA is mainly ascribed to its microcrystallinity as confirmed by Okamoto et al. [43,44] and then Francotte et al. [45], who have shown the influence of the crystalline arrangements on the enantioseparation.

The broad chiral recognition ability of cellulose derivatives result from the stereoregular structure of the polymer chains that give rise to a supramolecular organization. The importance of the supramolecular structure was identified for microcrystalline cellulose tribenzoate by Francotte and Wolff [46]. Inclusion of solute into the cavities of the supramolecular helical structure was assumed to mainly affect separations whereas dipole–dipole interactions and hydrogen bonding could play a minor role. However, data from the resolution of racemates bearing a carbonyl in α or β position from the stereogenic center emphasizes the dipole–dipole interaction with the carbonyl groups of MCTA.

The first diversely derivatized polysaccharide-based CSPs were introduced in chromatography by Okamoto and coworkers [44,47] in 1984. They prepared a series of tribenzoate and phenylcarbamate of cellulose. These derivatives were found to resolve a wide range of racemates having various functional groups.

One advantage of polysaccharide derivatization is that it introduces new interactions sites. However, mechanisms of chiral recognition are much more complex and still not well molecularly elucidated. Wainer and Alembik [48] proposed for benzoate cellulose a binding-steric fit formation, which involved hydrogen-bonding and dipole–dipole interactions rather than inclusion for amide derivatives. With aromatic alcohols, results suggest the existence of a determinant hydrogen-bonding interaction with the carbonyl of the benzoate.

As far as the cellulose triphenylcarbamate (CTPC) or amylose triphenylcarbamate (ATPC) is considered, the important sites are the polar carbamate groups. CTPC exhibits a left-handed threefold helix and glucose residues are regularly

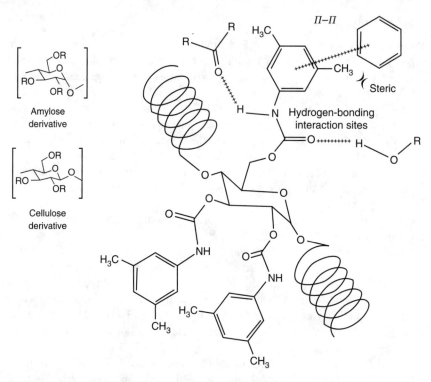

FIGURE 7.4 Potential interaction sites on a *tris*(3,5-dimethylphenylcarbamate) derivative of polysaccharide (e.g., amylose and cellulose).

arranged along the helical axis [49]. The helical structure is maintained in solution. The polar carbamate groups are located inside a chiral helical groove whereas hydrophobic aromatic groups are located outside. Enantiomers interact with the carbamates (Figure 7.4) via hydrogen bonding with the NH and C=O groups and dipole–dipole interactions on the C=O. π–π interaction as well as steric interactions of the solute should be considered. In addition, the introduction of substituents in the phenyl group greatly affects the chiral recognition mechanisms and thus the resolving ability of these derivatives.

Booth and Wainer [50] have demonstrated the role of multiple hydrogen-bonding interactions in the stability of the diastereomeric complex occurring throughout conformational changes of both ATPC chiral selector and solute. Authors pointed out in this study that chiral recognition does not fit a three-point interaction mechanism. Other aspects of chiral separation mechanisms on polysaccharide triphenylcarbamate derivatives have been recently reviewed by Yamamoto and Okamoto [51].

When polysaccharides are used as CSPs it is of importance to ensure that the analyte remains neutral, as there are no functional groups in the polysaccharide that are ionic in nature to interact with charged analytes. The retention of the analytes is also influenced by the polar solvent in the mobile phase [52] (apolar diluent/polar modifier) as the solvent molecule competes with the solute molecule for specific adsorption sites in the CSP. Solvation in the mobile phase will also increase with increased volume of polar modifier. The drawback of these polysaccharide derivatives is the solubility in a number of solvents. To overcome the problem, the polysaccharide can be bonded to a silica gel matrix via a spacer arm or photochemically cross-linked [53].

Aromatic carbamate derivatives of polysaccharides are today the most used CSP in chiral HPLC, but unfortunately, they are not ideal selectors for TLC because of the detection difficulty by UV light. However, this problem may be overcome in the future by the use of cycloalkylcarbamates [54].

7.5.2.3 Proteins

It seems that bovine serum albumin (BSA) is the only one protein considered in TLC. It is used as an additive in the mobile phase whereas it is used as immobilized on a solid anion exchanger or as an additive in HPLC.

BSA is highly soluble in water, which requires hydrophobic plates as stationary phase.

TLC enantioseparations were extensively studied by Lepri et al. [55–65]. Identification of the enantioselective domains and assessment of the noncovalent interaction sites contribution binding is governed by both electrostatic and hydrophobic interactions. In many cases, the contribution of hydrophobic forces is important.

On RP18 stationary phase with BSA in the mobile phase, the retention of a solute depends on the concentration of BSA according to the equation:

$$\frac{1}{k} = cC + C \sum n_i K_i [\text{BSA}].$$

Data from HPLC demonstrates that BSA is effective as chiral selector when the [BSA]/[solute] ratio is greater than 1.

7.5.2.4 Molecular Imprinted Polymers

Molecular imprinted polymers (MIPs) can be prepared according to a number of approaches that are different in the way the template is linked to the functional monomer and subsequently to the polymeric binding sites. The current technique makes use of noncovalent self-assembly of the template with functional monomers before polymerization, free radical polymerization with a cross-linking monomer, and then template extraction followed by rebinding by noncovalent interactions. The functional monomer is often methacrylic acid, the cross-linker is ethylene glycol dimethacrylate, the initiator is 2,2-azo-N,N'-bis-isobutyronitrile. They are mixed with template and the mixture is reacted at elevated temperature. The resultant rigid polymer is ground into a sieved powder and the template enantiomer washed off.

MIPs are very specific and consequently, the number of racemates resolved is roughly equal to the number of stationary phases as confirmed in the recent review of Ansell [66] reporting MIPs application to the separation of chiral drugs. The template interacts strongly with the MIP via a chelating ion-pair interaction and mobile phase with strong modifier is often required. According to Berthod [67], the mechanism is a "key-and-lock" association. Interactions are mainly steric. Mass transfer is usually poor with the result of elongated spots.

The reader will find a compilation of TLC enantiomer separation with MIPs in a chapter from Lepri et al. [68].

Sellergren [69] pointed out the factors to consider in the synthesis of MIP CSPs. The main limitations are as follows:

- Binding sites heterogeneity
- Extensive nonspecific binding
- Slow mass transfer
- Bleeding of template
- Low sample load capacity (but does not affect TLC)
- Impractical manufacturing procedure (that precludes release of manu-factured TLC plates)
- Poor recognition in aqueous systems
- Swelling–shrinkage (in HPLC or TLC mode) that may hamper the chiral resolution. To overcome the problem, covalent fixing on a matrix (e.g., silica) is performed
- Lack of recognition of a number of compound classes
- Preparative amounts of template required.

Sellergren also reviewed the factors affecting the recognition properties of MIPs related to the monomer template assemblies and those related to the polymer structure and morphology.

Lu et al. [70] studied the interactions between functional monomer and template by using UV-Vis and ^{1}H NMR spectroscopy. They wrote that the complexes formed between the print molecule and functional monomers is expressed as follows:

$$T + nM \underset{K}{\rightleftharpoons} TM_n,$$

where K is the association constant ($n = 1, 2, q$).

When increasing concentration b_0 of monomer are added to a concentration a_0 of template, a shift in UV absorbance ΔA occurs (the absorbance is measured at a wavelength where M does not absorb). A plot of $\Delta A/b_0^n$ vs. ΔA is linear and allows the determination of K.

Addition of the monomer to the template solution produces a NMR chemical shift and ΔAd changes in line widths. A plot of chemical shift vs. monomer concentration yields a straight line. The claimed mechanism is the preorganizing of the monomers around the template before polymerization.

The pH of the mobile phase is important since Sellergren found that average pK_a of carboxylic acid groups inside micro cavity is lower than that of carboxylic acid groups not inside micro cavity. Lu et al. [70] also observed a change of slope from positive to negative when changing the pH of the mobile phase.

7.6 CONCLUSION

In this overview, we have described the major classes of chiral selectors used in TLC and their different modes of separation mechanisms. Most of the available chiral recognition models deal with simple bimolecular systems for which very specific molecular mechanisms are often involved (e.g., ligand exchange). Other systems are dealing with macromolecule chiral selectors (e.g., polysaccharides) and implicate supramolecular and inclusion mechanisms combined with other different interactions (dipole–dipole and hydrogen-bonding). If these selectors clearly expand the application range of chiral chromatography, related chiral recognition mechanisms are not simple and depend in various ways on operating conditions (mobile phase composition, pH, and temperature). They are also strongly influenced by conformational changes and so any small change in the structure of the chiral selector results in dramatic shifts in enantioselectivity.

To understand the nature of ligand–selector interaction mechanism and build models, strategies based on systematic variation of the chemical structure of the ligand–selector complexes are very popular. Spectroscopic techniques are often necessary to confirm the models and discard the speculative features. However, information retrieved from the spectroscopic data on chiral selector–enantiomer complexes may be sometimes useless since the solvation of the solute and the selector by the mobile phase may hamper the prediction.

Although a number of enantioselective chiral selectors have been identified today, there is much that remains to be done to both improve them and understand the interaction processes. Furthermore, the user must know what chiral selector is involved (its chemical structure), what is selective for, and how it is used. When choosing a chiral method, the selectivity, efficiency, loadability, and reproducibility should also be considered.

REFERENCES

1. Cahn, R.S., Ingold, C., and Prelog, V., Specification of molecular chirality, *Angew. Chem. Int. Ed.*, 5, 385, 1966.
2. Maier, N.M., Franco, P., and Lindner, W., Separations of enantiomers: Needs, challenges, perspectives, *J. Chromatogr. A*, 906, 3, 2001.
3. Bojarski, J., Aboul-Enein, H.Y., and Ghanem, A., What's new in chromatographic enantioseparations, *Curr. Anal. Chem.*, 1, 59, 2005.
4. Gubitz, G. and Schmid, M.G., Chiral separation by chromatographic and electromigration techniques. A review, *Biopharm. Drug Dispos.*, 22, 291, 2001.

5. Dent, C.D., A study of the behavior of some sixty amino acids and other ninhydrin-reacting substances on phenol-'collidine' filter-paper chromatograms, with notes as to the occurrence of some of them in biological fluids, *Biochem. J.*, 43, 169, 1948.

6. Gil-Av, E., Feibush, B., and Charles-Sigler, R., Separation of enantiomers by gas liquid chromatography with an optically active stationary phase, *Tetrahedron Lett.*, 7, 1009, 1966.

7. Chirbase: A molecular data base for chiral chromatography, http://chirbase. u-3mrs.fr

8. Schulte, M., Chiral derivatization chromatography, in *Chiral Separation Techniques: A Practical Approach*, Subramanian, G., Ed., 2d rev. ed., Wiley-VCH, Weinheim, 2001, Chap. 7.

9. Nagata, I., Iida, T., and Sakai, M., Enantiomeric resolution of amino acids by thin-layer chromatography, *J. Mol. Catal. B Enzym.*, 12, 105, 2001.

10. Siouffi, A.M., Piras, P., and Roussel, C., Some aspects of chiral separations in planar chromatography compared with HPLC, *J. Planar Chromatogr.*, 18, 5, 2005.

11. Easson, L.M. and Steadman, E., Studies on the relationship between chemical constitution and physiological action, *Biochem. J.*, 27, 1257, 1933.

12. Ogston, A.G., Interpretation of experiments on metabolic processes, using isotopic tracer elements, *Nature*, 162, 963, 1948.

13. Dalgliesh, C.R., Optical resolution of aromatic amino acids on paper chromatograms, *J. Chem. Soc.*, 3940, 1952.

14. Davankov, V.A., The nature of chiral recognition: Is it a three-point interaction? *Chirality*, 9, 99, 1997.

15. Sundaresan, V. and Abrol, R., Towards a general model for protein–substrate stereoselectivity, *Protein Sci.*, 11, 1330, 2002.

16. Davankov, V.A., Chiral selectors with chelating properties in liquid chromatography: Fundamental reflections and selective review of recent development, *J. Chromatogr. A*, 666, 55, 1994.

17. Huang, J. and Li, T., Highly efficient chromatographic resolution of alpha-alpha'-dihydroxybiaryls, *Org. Lett.*, 7, 5821, 2005.

18. Pirkle, W.H. and Hyun, M.H., Alpha-arylalkylamine-derived chiral stationary phases: Evaluation of urea linkages, *J. Chromatogr.*, 322, 295, 1985.

19. Hyun, M.H., Na, M.S., and Jin, J.S., Chiral recognition models for the liquid chromatographic resolution of pi-acidic racemates on a chiral stationary phase derived from *N*-phenyl-*N*-alkylamide of (*S*-naproxen), *J. Chromatogr. A*, 752, 77, 1996.

20. Hyun, M.H., Ryoo, J.J., and Pirkle, W.H., Experimental support differentiating two proposed chiral recognition models for the resolution of *N*-(3,5-dinitrobenzoyl)-alpha-arylalkylamines on high-performance liquid chromatography chiral stationary phases, *J. Chromatogr. A*, 886, 47, 2000.

21. Pirkle, W.H. and Liu, Y., On the relevance of face-to-edge pi–pi interactions to chiral recognition, *J. Chromatogr. A*, 749, 19, 1996.

22. Pirkle, W.H. and House, D.W., Chiral high-performance liquid chromatographic stationary phases. 1. Separation of the enantiomers of sulfoxides, amines, amino acids, alcohols, hydroxy acids, lactones, and mercaptans, *J. Org. Chem.*, 44, 1957, 1979.

23. Pirkle, W.H., House, D.W., and Finn, J.M., Broad spectrum resolution of optical isomers using chiral high-performance liquid chromatographic bonded phases, *J. Chromatogr.*, 192, 143, 1980.

24. Pirkle, W.H., The nonequivalence of physical properties of enantiomers in optically active solvents. Differences in nuclear magnetic resonance spectra. I, *J. Am. Chem. Soc.*, 88, 1837, 1966.

25. Pirkle, W.H. and Welch, C.J., Chromatographic and ^{1}HNMR support for a proposed chiral recognition model, *J. Chromatogr. A*, 683, 347, 1994.

26. Murer, P. et al., On-bead combinatorial approach to the design of chiral stationary phases for HPLC, *Anal. Chem.*, 71, 1278, 1999.

27. Welch, C.J., Protopopova, M.N., and Bhat, G.A., Microscale synthesis and screening of chiral stationary phases, *Enantiomer*, 3, 471, 1998.

28. Wainer, I.W., Proposal for the classification of high-performance liquid chromatographic chiral stationary phases: How to choose the right column? *TrAC*, 6, 125, 1987.

29. Schlitt, H. and Geiss, F., Thin-layer chromatography as a pilot technique for rapid column chromatography, *J. Chromatogr.*, 67, 261, 1972.

30. Ferguson, P.D., Goodall, D.M., and Loran, J.S.J., Systematic approach to links between separations in capillary electrophoresis and liquid chromatography IV. Application of binding constant-retention factor relationship to the separation of 2-, 3- and 4-methylbenzoate anions using beta-cyclodextrin as selector, *J. Chromatogr. A*, 768, 29, 1997.

31. Persson, B.A. and Andersson, S., Unusual effects of separation conditions on chiral separations, *J. Chromatogr. A*, 906, 195, 2001.

32. Helfferich, F., "Ligand Exchange": A novel separation technique, *Nature*, 189, 1001, 1961.

33. Rogozhin, S. and Davankov, V.A., German patent, 1 932 19, 1970, CA 72, 90875c, 1970.

34. Kurganov, A., Chiral chromatographic separations based on ligand exchange, *J. Chromatogr. A*, 906, 51, 2001.

35. Davankov, V.A., Zolotarev, Y.A., and Kurganov, A.A., Ligand-exchange chromatography of racemates. XI. Complete resolution of some chelating compounds and nature of sorption enantioselectivity, *J. Liquid Chromatogr.*, 2, 1191, 1979.

36. Weinstein, S., Resolution of optical isomers by thin layer chromatography, *Tetrahedron Lett.*, 25, 985, 1984.

37. Gunther, K. and Moller, K., Part II: Applications of thin-layer, enantiomer separation, in *Handbook of Thin-Layer Chromatography*, Sherma, J. and Fried, B., Eds., 3d ed., Marcel Dekker, New York, 2003, pp. 471–535.

38. Gunther, K., Martens, J., and Schickedanz, M., Thin-layer chromatographic enantiomeric resolution via ligand exchange, *Angew. Chem. Int. Ed. Engl.*, 23, 506, 1984.

39. Polak, B., Golkiewicz, W., and Tuzimski, T., Effect of mobile phase pH* on chromatographic behavior in chiral ligand-exchange thin-layer chromatography (CLETLC) of amino acid enantiomers, *Chromatographia*, 63, 197, 2006.

40. Armstrong, D.W. and DeMond, W., Cyclodextrin bonded phases for the liquid chromatographic separation of optical, geometrical and structural isomers, *J. Chromatogr. Sci.*, 22, 411, 1984.

41. Armstrong, D.W., DeMond, W., and Czech, B.P., Separation of metallocene enantiomers by liquid chromatography: Chiral recognition via cyclodextrin bonded phases, *Anal. Chem.*, 57, 481, 1985.
42. Chang, S.C. et al., Evaluation of a new polar-organic high-performance liquid chromatographic mobile phase for cyclodextrin-bonded chiral stationary phases, *TrAc*, 12, 144, 1993.
43. Okamoto, Y. et al., Useful chiral packing materials for high-performance liquid chromatographic resolution. Cellulose triacetate and tribenzoate coated on macroporous silica gel, *Chem. Lett.*, 739, 1984.
44. Ichida, A. et al., Resolution of enantiomers by HPLC on cellulose derivatives, *Chromatographia*, 19, 280–284, 1984.
45. Francotte, E. et al., Chromatographic resolution of racemates on chiral stationary phases: I. Influence of the supramolecular structure of cellulose triacetate, *J. Chromatogr.*, 347, 25, 1985.
46. Francotte, E. and Wolff, R., Benzoyl cellulose beads in the pure polymeric form as a new powerful sorbent for the chromatographic resolution of racemates, *Chirality*, 3, 43, 1991.
47. Okamoto, Y., Kawashima, M., and Hatada, K., Chromatographic resolution. 7. Useful chiral packing materials for high-performance liquid chromatographic resolution of enantiomers: Phenylcarbamates of polysaccharides coated on silica gel, *J. Am. Chem. Soc.*, 106, 5357, 1984.
48. Wainer, I.W. and Alembik, M.C., Resolution of enantiomeric amides on a cellulose-based chiral stationary phase: Steric and electronic effects, *J. Chromatogr.*, 358, 85, 1986.
49. Yashima, E.J., Polysaccharide-based chiral stationary phases for high-performance liquid chromatographic enantioseparation, *J. Chromatogr. A*, 906, 105, 2001.
50. Booth, T.D. and Wainer, I.W., Investigation of the enantioselective separations of alpha-alkylarylcarboxylic acids on an amylose *tris*(3,5–dimethylphenylcarbamate) chiral stationary phase using quantitative structure-enantioselective retention relationships. Identification of a conformationally driven chiral recognition mechanism, *J. Chromatogr. A*, 737, 157, 1996.
51. Yamamoto, C. and Okamoto, Y., Optically active polymers for chiral separation, *Bull. Chem. Soc. Jpn.*, 77, 227, 2004.
52. Matthijsa, N., Maftouh, M., and Vander Heyden, Y., Screening approach for chiral separation of pharmaceuticals. IV. Polar organic solvent chromatography, *J. Chromatogr. A*, 1111, 48, 2006.
53. Francotte, E.R. and Huynh, D., Immobilized halogenophenyl carbamate derivatives of cellulose as novel stationary phases for enantioselective drug analysis, *J. Pharm. Biomed. Anal.*, 27, 421, 2002.
54. Kubota, T., Yamamoto, C., and Okamoto, Y., Tris(cyclohexylcarbamate)s of cellulose and amylose as potential chiral stationary phases for high-performance liquid chromatography and thin-layer chromatography, *J. Am. Chem. Soc.*, 122, 4056, 2000.
55. Lepri, L. et al., Reversed phase planar chromatography of racemic flavanones, *J. Liquid Chromatogr. Relat. Technol.*, 22, 105, 1999.
56. Lepri, L., Coas, V., and Desideri, P.G., Planar chromatography of isomers using beta-cyclodextrin solutions as mobile phases, *J. Planar Chromatogr.*, 3, 533, 1990.

57. Lepri, L., Coas, V., and Desideri, P.G., Reversed phase planar chromatography of isomers using alpha- and beta-cyclodextrin solutions as eluent, *J. Planar Chromatogr.*, 4, 338, 1991.

58. Lepri, L., Coas, V., and Desideri, P.G., Planar chromatography of optical isomers with bovine serum albumin in the mobile phase, *J. Planar Chromatogr.*, 5, 175, 1992.

59. Lepri, L., Coas, V., and Desideri, P.G., Reversed phase planar chromatography of optical active fluorenylmethoxycarbonyl amino acids with bovine serum albumin in the mobile phase, *J. Planar Chromatogr.*, 5, 294, 1992.

60. Lepri, L. et al., Reversed phase planar chromatography of enantiomeric tryptophans with bovine serum albumin in the mobile phase, *J. Planar Chromatogr.*, 5, 234, 1992.

61. Lepri, L. et al., Reversed phase planar chromatography of enantiomeric compounds with bovine serum albumin in the mobile phase, *J. Planar Chromatogr.*, 5, 364, 1992.

62. Lepri, L. et al., Reversed phase planar chromatography of dansyl DL amino acids with bovine serum albumin in the mobile phase, *Chromatographia*, 36, 297, 1993.

63. Lepri, L. et al., Thin layer chromatographic enantioseparation of miscellaneous compounds with bovine serum albumin in the eluent, *J. Planar Chromatogr.*, 6, 100, 1993.

64. Lepri, L., Coas, V., and Desideri, P.G., Planar chromatography of optical and structural isomers with eluents containing modified beta-cyclodextrins, *J. Planar Chromatogr.*, 7, 322, 1994.

65. Lepri, L. et al., Reversed phase planar chromatography of optical isomers with bovine serum albumin mobile phase additive, *J. Planar Chromatogr.*, 12, 221, 1999.

66. Ansell, R.J., Molecularly imprinted polymers for the enantioseparation of chiral drugs, *Adv. Drug Deliv. Res.*, 57, 1809, 2005.

67. Berthod, A., Chiral recognition mechanisms, *Anal. Chem.*, 78, 2093, 2006.

68. Lepri, L., Del Bubba, M., and Cincinelli, A., Chiral separations by TLC, in *Planar Chromatography. A Retrospective View for the Third Millenium*, Nyirdy, Sz., Ed., Springer, Budapest, 2001, Chap. 25.

69. Sellergren, B., Imprinted chiral stationary phases in high-performance liquid chromatography, *J. Chromatogr. A*, 906, 227, 2001.

70. Lu, Y. et al., Study on the mechanism of chiral recognition with molecularly imprinted polymers, *Anal. Chim. Acta*, 489, 33, 2003.

8 Separation of Diastereoisomers by Means of TLC

Virginia Coman

CONTENTS

8.1 INTRODUCTION

As is well known, isomers are compounds with the same molecular formula but differ in structure. The subdivision of isomers is according to the differences in constitution, configuration, or conformation. The isomers that differ only in configuration or conformation are recognized as stereoisomers that are divided into enantiomers and diastereoisomers (diastereomers) (Figure 8.1). The configurational isomers represent static spatial arrangements that are distinguished by different orientation modes of atoms vs. a rigid structural element (center or plane). The transformation of a stereoisomer into another one involves the change of the place of two substitutes by the formal cancellation of a covalent bond. The conformational isomers represent dynamic spatial arrangements, the stereoisomers being interconvertible by the simple torsion of substituted carbon atoms around a simple bond in the molecule. Some configurational isomers can have many conformations (i.e., cycloalkanes). Both enantiomers and diastereomers may differ in configuration or only in conformation [1–6].

Enantiomers are pairs of isomers in the mirror–image relationship (related as an object to its mirror image) that are not superposable. The diastereoisomers are

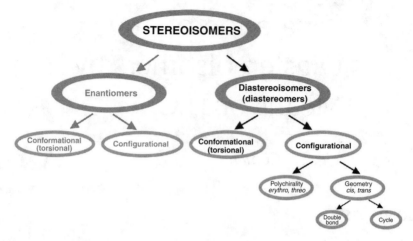

FIGURE 8.1   Classification of stereoisomers.

isomers with identical constitution that differ in 3D architecture and that are not related as an object to its mirror image [1–4].

It must be mentioned that this classification of stereoisomers is valuable only when it is taken into consideration the criterion of symmetry applied to the momentary geometry of the molecule [1].

8.2   DIASTEREOISOMERS

Diastereoisomerism (distance isomerism) refers to all the cases of stereoisomers that do not bear a mirror–image relationship with each other. In enantiomers, the distances between the atoms nonbonded directly are rigorously identical, whereas in diastereoisomers there are always differences between these distances. As a result, any molecular deformation of a chiral or an achiral substance consists in the appearance of a diastereoisomerical structure (Figure 8.2).

To observe this phenomenon by means of the experimental methods, it is necessary that the lifetime of the diastereoisomerical structure to be sufficiently long for its detection by the respective method [1,2,5,6].

Diastereoisomerism is caused by the differences in conformation or only in configuration at several sites in the molecule. In the first category, there is the torsion diastereoisomerism due to the magnitude of torsion angle, and in the second there is the diastereoisomerism due to the polychirality (two or more chirality centers) or the *cis–trans* geometry generated either by the existence of a double bond (C=C, N=N) or by a cycle (cycloalkanes, saturated heterocycles) and *syn–anti* isomerism, for example, at the compounds with C=N bond. Taking into consideration what was mentioned above, a classification of diastereoisomers is given in Figure 8.1 [1,2,5,6].

The diastereoisomers have scalar properties very well defined, thus they can be characterized experimentally without chiral testing. They can be easily

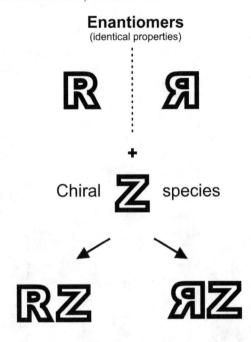

FIGURE 8.2 Interaction of a chiral species with a pair of enantiomers generating a pair of diastereoisomers.

differentiated by means of physical (melting or boiling point, density, refraction index, solubility, infrared, nuclear magnetic resonance, electronic or mass spectra, chromatographic behavior, etc.) (see Table 8.1 [3,7–9]) and chemical properties (oxidation or reduction reactions, etc.). In the case of enantiomers, these properties are identical, their characteristic for discrimination being the direction of the rotation of the plane of polarized light. (+) or D (dextro) indicates a rotation to right (clockwise), whereas (−) or L (levo) indicates a rotation to left (anticlockwise) [1–6].

Some examples of diastereoisomers are given in Figure 8.3 (tartaric acid and some monosaccharides) [1–3,10], Figure 8.4 (1,2-dimethylcyclohexane, decaline, 2-substituted bicyclo[2.2.1]heptane) [4], and Figure 8.5 (1,2-diphenylmethylglycol) [1,4].

8.3 NONCHIRAL TLC SYSTEMS USED FOR DIASTEREOISOMER SEPARATION

The TLC behavior of an organic compound is not only determined by the size of the molecule, the number, type and position of the functional groups, the presence and

TABLE 8.1
Some Physical Properties of Selected Stereoisomers (Enantiomers and Diastereoisomers)

Compound	Melting point (°C)	Specific optical rotation		Purity (%)	Reference
		$[\alpha]\ 20°/D$	Conditions		
Tartaric acids					
D-(−)-Tartaric acid	166–169	$-13.5 \pm 0.5°$	$c = 10$ in H_2O	>99	[7]
L-(+)-Tartaric acid	168–170	$+13.5 \pm 0.5°$	$c = 10$ in H_2O	purris >99.5	[7]
D,L-Tartaric acid	210–212			>99	[8]
Mesotartaric acid	140				[3]
Monosaccharides — Tetroses					
D-(−)-Erythrose	—	$-32.0 \pm 3°$	$c = 2$ in H_2O, 24 h	Syrup ~70	[9]
L-(+)-Erythrose	—	$+32.0 \pm 3°$	$c = 2$ in H_2O, 24 h	Syrup ~70	[9]
D-(−)-Threose	—	$-11.0 \pm 1°$	$c = 2$ in H_2O	Syrup (freeze dried)	[9]
L-(+)-Threose	—	$+11.0 \pm 1°$	$c = 2$ in H_2O	Syrup (freeze dried)	[9]
Monosaccharides — Pentoses					
D-(−)-Ribose	86–88	$-20 \pm 1°$	$c = 10$ in H_2O, 28 h		[7]
L-(+)-Ribose	84–86	$+20 \pm 1°$	$c = 10$ in H_2O, 28 h		[7]
D-(−)-Arabinose	156–160	$-104 \pm 1°$	$c = 10$ in H_2O, 24 h	>99	[7]

L-(+)-Arabinose	155–159	$+104 \pm 1°$	$c = 10$ in H_2O, 24 h	>99	[7]
D-(+)-Xylose	156–158	$+18.8°$	$c = 10$ in H_2O	>99	[8]
L-(−)-Xylose	150–152	$-18.7°$	$c = 4$ in H_2O, 24 h, $T = 24°C$	>99	[8]
D-(−)-Lyxose	105–107	$-14.0 \pm 0.5°$	$c = 6$ in H_2O, 1 h	>99	[7]
L-(+)-Lyxose	106–108	$+14.0 \pm 0.5°$	$c = 6$ in H_2O, 1 h	>99	[7]
Monosaccharides — Hexoses					
D-(+)-Glucose anh.	141–143[a]	$+53 \pm 3°$	$c = 10$ in H_2O, 3 h	>98	[7]
L-(−)-Glucose	135[a]	$-53 \pm 3°$	$c = 10$ in H_2O, 3 h	>99	[7,9]
D-(+)-Mannose	128–131[a]	$+13.8 \pm 0.5°$	$c = 10$ in H_2O, 24 h	>99	[7,9]
L-(−)-Mannose	143–146	$-13.5 \pm 0.5°$	$c = 10$ in H_2O, 24 h	>99	[7,9]
D-Allose	106–108	$+15.0 \pm 1°$	$c = 10$ in H_2O, 2 h	>99	[7]
L-Allose	103–105	$-14.0 \pm 0.5°$	$c = 2$ in H_2O	>99	[9]
D-Altrose	167–170	$+30.0 \pm 3°$	$c = 1$ in H_2O	>99	[7]
L-Altrose	166–167	$-32.0 \pm 1°$	$c = 2$ in H_2O	>99	[9]
D-(+)-Galactose	128–132	$+80.0 \pm 1°$	$c = 5$ in H_2O, 24 h	>99	[7]
L-(−)-Galactose	133	$-80.0 \pm 2°$	$c = 5$ in H_2O, 24 h	>99	[7]
D-(+)-Talose		$+19.0 \pm 0.5°$	$c = 2$ in H_2O	>99	[9]
L-(−)-Talose		$-19.0 \pm 0.5°$	$c = 5$ in H_2O	>99	[9]
D-Gulose		$-23.0 \pm 1°$	$c = 2$ in H_2O	Syrup (freeze dried)	[9]
L-Gulose		$+23.0 \pm 1°$	$c = 2$ in H_2O	Syrup (freeze dried)	[9]

[a]Values from the reference [9].

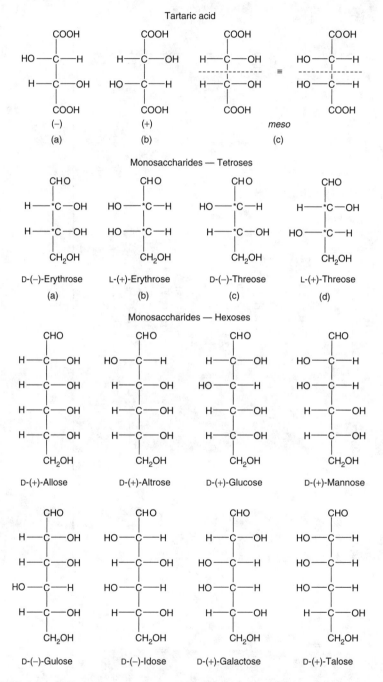

FIGURE 8.3 Stereoisomers of tartaric acid (a, b — enantiomers; a, c and b, c — diastereoisomers) and of monosaccharides with 4 and 6 carbon atoms. Tetroses: a, b and c, d — enantiomers; a, c, d and b, c, d — diastereoisomers.

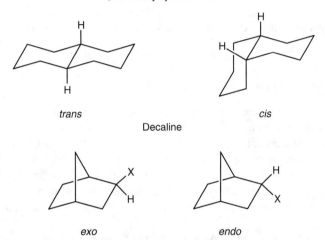

Conformation (e, e)   (a, a)   (e, a)
Configuration   trans   cis
(a)   (b)   (c)

1,2-Dimethylcyclohexane

trans   cis

Decaline

exo   endo

2-Substituted bicyclo[2.2.1]heptane

FIGURE 8.4 Examples of some diastereoisomers of different cyclic compounds. 1,2-Dimethylcyclohexane (a, b — enantiomers; a, c and b, c — diastereoisomers); (e, e) − (a, a) − (e, a) (e — equatorial; a — axial) — conformational diastereoisomers; *trans-cis* — configurational diastereoisomers; Decaline: *trans-cis* — configurational diastereoisomers; 2-Substituted bicyclo[2.2.1]heptane: *exo-endo* — configurational diastereoisomers.

the number of double bonds, but also is influenced by the geometric arrangement of the atoms, functional groups, and so forth, within the molecule. It is therefore possible to separate *cis–trans* and *erythro–threo* diastereoisomers from each other, compounds with a *syn–anti* configuration, compounds with *endo–exo* arrangements, and molecules with functional groups either in an equatorial (*e*) or axial (*a*) position [11,12].

The separation of diastereoisomeric compounds by TLC on polar adsorbents such as silica gel and alumina is widely used (References [12–30] and the references cited therein). The separation itself is the main problem as it enables isolation of each isomer in pure form. It is important that one isomer move faster than its diastereoisomer.

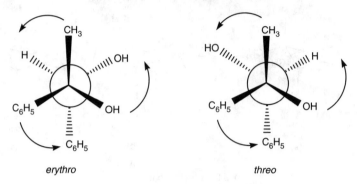

erythro	*threo*

FIGURE 8.5 Configurational diastereoisomers of *erythro* and *threo* 1,2-diphenylmethyl glycol. (Adapted from Palamareva, M.D., Kurtev, B.J., Mladenova, M.P., and Blagoev, B.M., *J. Chromatogr.*, 235, 299–308, 1982.)

8.3.1 INFLUENCE OF CONFIGURATION

The researches regarding the different chromatographic behavior of *cis–trans* isomeric compounds on thin layers show that the *trans* compounds always migrate faster than the *cis* ones, for example, fumaric acid > maleic [13,14], *trans*-aconitic acid > *cis*-aconitic acid [14], *trans*-azobenzene > *cis*-azobenzene [12], *trans*-stilbene > *cis*-stilbene [12]. In the case of the inorganic *cis–trans* isomeric cobalt complexes, the *cis* forms always migrate faster than the corresponding *trans* ones [15]. The *erythro* isomer of phenylmethylglycol has higher R_F value than the *threo* isomer [16].

Similar to the *cis–trans* isomerism related to the C=C double bond, the two possible stereoisomeric compounds with C=N double bonds are the *syn* and *anti* forms. These isomers (oximes) and related compounds obtained by the reaction between aldehydes and asymmetrical ketones and hydroxylamines can be separated by TLC. In the case of benzaldehyde and its derivatives, all *syn* (α-form) isomers had R_F values higher than the *anti* (β-form) isomers, but for the condensation derivatives of benzaldehyde, all *syn* isomers displayed R_F values smaller than the *anti* isomers [17].

The separation of α-form (I) and β-form (II) of isomers of *N*-allyl-DL-camphoramic acid demonstrates the importance of the carboxyl group regarding the adsorption affinity [18].

In the case of the β-isomer, a higher adsorption affinity $[R_F(\beta) = 0.58; R_F(\alpha) = 0.80]$ was observed, caused by the carboxyl group that has no other substituent on the carbon atom in the neighboring position.

Details on the TLC separations mentioned above are given in Table 8.2 [12–18].

Representative papers regarding the TLC separation of the diastereomeric pairs of tetrasubstituted ethane derivatives with two asymmetrical carbon atoms, such as the derivatives of 1,2-diarylethanes, have been reported extensively in a series of papers by Palamareva and coworkers (Table 8.3) [19–29]. The data obtained were analyzed based on the general Snyder theory [31–38] including the Soczewiński method [39]. They proposed the TLC method on silica gel to be used as a method for assignment of the relative configurations of diastereomeric pairs of 1,2-diarylethanes of the Ar—CH(X)—CH(Y)—Ar' type (type I), where Ar and Ar' are phenyl or m- or p-alkoxyl-substituted groups; X and Y are polar groups, such as NH_2, OH, and COOH, or their derivatives [19]. They demonstrated that the *erythro* configuration should be assigned to the diastereomer with a higher R_F value, whereas the *threo* configuration to that with a lower R_F value, irrespective of the polarity of the developing solvents as well as of the formation of an intramolecular hydrogen bond between X and Y. This fact is explained based on the preferred conformations of the diastereoisomers.

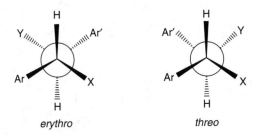

erythro   threo

The results presented in the papers [19,20] indicate a correlation of $R_{F(erythro)} > R_{F(threo)}$ for 50 of 52 diastereomeric pairs studied. The qualitative analysis of the experimental data by means of Snyder theory of linear adsorption chromatography [31] suggests that the aforementioned correlation between R_F values of *erythro* and *threo* diastereoisomers on silica gel can be used for the assignment of the relative configurations of other nonionic diastereomeric compounds of type I, which do not possess intramolecular hydrogen bonds or which have bonds of the types OH. . .OH or OH. . .N [20].

This group studied the behavior of different diastereomeric compounds of type RO_2C—$CH(Br)CHCO_2R$ on alumina [27] and silica [29] and the chromatographic separation of esters of Z- and E-2,3-diphenylpropenoic acids [28]. Based on Snyder theory [31–36] they used 20 computer-selected mobile phases. Palamareva and Kozekov [28] reported the measurement of the TLC retention of approximately 300 diastereoisomers on silica and alumina.

TABLE 8.2
Some Nonchiral TLC Systems Used for Diastereoisomer Separation

Diastereoisomer	Configuration	Stationary phase	Mobile phase (v/v)	R_F	Reference
Aldrin	endo–exo	Silica gel G	Cyclohexane	0.60	[12]
Isodrin	endo–endo			0.49	
Maleic acid	cis	Silica gel	Benzene–methanol–glacial acetic acid (90 + 10 + 4)	0.07	[13,14]
Fumaric acid	trans			0.23	
Aconitic acid	cis	Silica gel		0.03	[14]
	trans			0.12	
1-Phenyl-2-methylglycol	erythro	Kieselguhr	Methanol	0.68	[16]
	threo			0.47	
	erythro	Kieselguhr	Methanol–acetone (50 + 50)	0.64	[16]
	threo			0.18	
Cyclopentane-diol	trans	Kieselguhr	Methanol–acetone (50 + 50)	0.68	[16]
	cis			0.14	
Cobalt complexes [Co en₂Cl₂]Cl	cis	Silica gel	Methanol–0.5 N sodium acetate in methanol glacial acetic acid in methanol–H₂O (90 + 10 + 0.1 + 1)	$R_{F(cis)} > R_{F(trans)}$	[15]
	trans				
Benzaldoxime	α-form	Silica gel G	Benzene–ethyl acetate (50 + 10)	0.50	[17]
	β-form			0.32	
Benzoin oxime	α-form	Silica gel G	Benzene–ethyl acetate (50 + 10)	0.14	[17]
	β-form			0.37	
Anisoin oxime	α-form	Silica gel G	Benzene–ethyl acetate (50 + 10)	0.05	[17]
	β-form			0.23	
N-Allyl-DL-camphoramic acid	α-form	Silica gel G	n-Butanol-saturated with water (100 + 20)	0.80	[18]
	β-form			0.58	

TABLE 8.3

Data on the Chromatographic Behavior of the Diastereoisomeric Pairs of Tetrasubstituted Ethanes and Some Related Cyclic Compounds

No.	Diastereoisomers	General formula	Configuration	Number of separated pairs/used solvent system	Adsorbent	Reference
1	1,2-Disubstituted-1,2-diarylethanes	Ar—CH(X)—CH(Y)—Ar' X and Y = NH_2, OH, COOH, and their derivatives Ar, Ar' = phenyl or alkoxyphenyl	erythro–threo	37/17	Silica gel, Cellulose	[19]
2	1,2-Disubstituted-1,2-diarylethanes	Ar—CH(X)—CH(Y)—Ar' X = NH_2, NMe_2, OH, NHMe Y = COOMe, CH_2OH, etc. Ar, Ar' — phenyl or phenyl derivatives	erythro–threo	15/20	Silica gel	[20]
3	4-Substituted-6,7-dialkoxy-3-aryl-N-alkyltetra-hydroisoquinolines	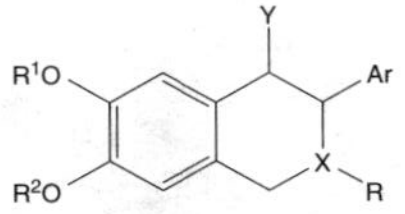 3. X = N; Y = COOMe, CH_2OH; R = alkyl 4. idem 3, except R = H 5. idem 3, except X = O; R = 0	cis–trans	13/8	Silica gel	[21]
4	4-Substituted-6,7-dialkoxy-3-aryltetrahydroisoquinolines		cis–trans	6/4	Silica gel	[21]
5	4-Substituted-6,7-dialkoxy-3-arylisochromans		cis–trans	4/3	Silica gel	[21]

(Continued)

TABLE 8.3
(Continued)

No.	Diastereoisomers	General formula	Configuration	Number of separated pairs/used solvent system	Adsorbent	Reference
6	Carbazole derivatives	Cbz—CH(OH)—CH(Y)—R 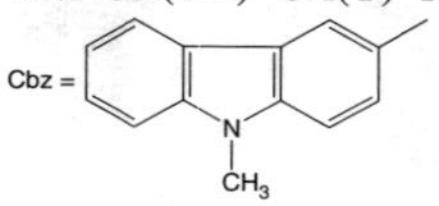 [carbazole structure, Cbz =, CH₃] Y = COOH, COOEt, CONHNH$_2$, NH$_2$, NHMe R = Me, Et	erythro–threo	8/3	Silica gel	[22]
7	Carbazole derivatives related to oxazolidones	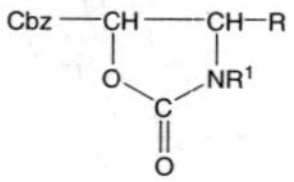 Cbz—CH—CH—R, NR^1, O—C=O [structure] R = idem 6; R^1 = H, Me	cis–trans	3/1	Silica gel	[22]
8	3-Hydroxy-2,3-diarylpropionates	Ar—CH(X)—CH(COOR)—Ar' Ar, Ar' = phenyl or phenyl derivatives X = OH, OAc; R = H, Me, i-Pr	erythro–threo	31/3	Silica gel	[23]
9	1,2-Disubstituted-1,2-diarylethanes	Ar—CH(X)—CH(Y)—Ar Ar = phenyl; X = CH$_2$COOMe, CH$_2$COO-t-Bu, CH$_2$CONMe$_2$ Y = CONMe$_2$, CONEt$_2$, CON-i-Pr$_2$, CON-n-Pr$_2$, CON(cyclohexyl)$_2$	erythro–threo	10/46	Silica gel	[24]

10	Aminoester	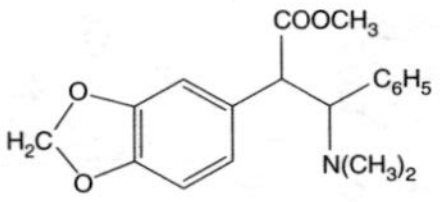	erythro–threo	1/21 ($\varepsilon = 0.354$)[a]	Silica gel	[25]
11	Isochroman	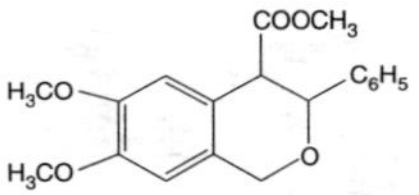	cis–trans	1/21 ($\varepsilon = 0.0354$)[a]	Silica gel	[25]
12	1,1-Disubstituted-2-arylethenes (oxazolones)	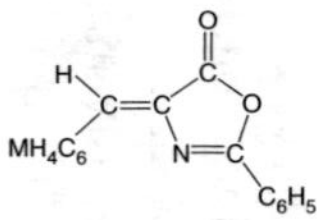 M = H, 4-Cl, 4-Me, 4-MeO, 3-MeO, 3-NO$_2$	Z–E	6/16 ($0.100 < \varepsilon < 0.250$)[a]	Silica gel	[26]
13	1,1-Disubstituted-2-arylethenes (cinnamates)	M–C$_6$H$_4$C(H) = CXY; M = H, 4-Cl, 4-CH$_3$O, 3-NO$_2$; X = NHCOC$_6$H$_5$; Y = CO$_2$Me	Z–E	4/24 ($0.290 < \varepsilon < 0.420$)[a]	Silica gel	[26]
14	Esters of 2,3-dibromobutane-1,4-dioic acids	ROOC–CH(Br)–CH(Br)–COOR; R = alkyl	erythro–threo	9/18 ($0.140 < \varepsilon < 0.180$)[a]	Alumina	[27]
				9/20 ($0.215 < \varepsilon < 0.305$)[a]	Silica gel	[29]
15	Esters of 2,3-diphenylpropenoic acids and similar compounds	Ar–CH = C(Ar)COOR; Ar = phenyl; R = Me, Et, n-Pr, n-Bu	Z–E	8/20 ($\varepsilon = 0.165$ or 0.250)[a]	Silica gel	[28]

[a] solvent strength value; Me = Methyl; Et = Ethyl; Pr = Propyl; Bu = Butyl

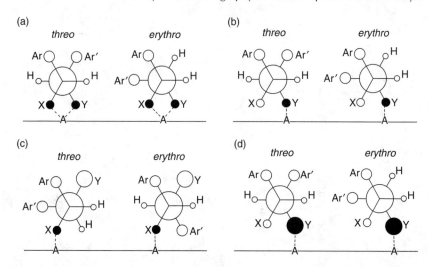

FIGURE 8.6 Illustrative representation of adsorption patterns for diastereoisomeric compounds of type I. The full circle denotes the strongest adsorbing groups of the molecule. A = active site comprising the different types of the adsorbent surface hydroxyl groups. (a) Two-point adsorption with X and Y; $R_{F(erythro)} > R_{F(threo)}$; (b) one-point adsorption with Y (or with X via conformations where X is between H and Y, not presented); $R_{F(erythro)} > R_{F(threo)}$; (c) one-point adsorption with X; $R_{F(threo)} > R_{F(erythro)}$; (d) one-point adsorption with Y; $R_{F(erythro)} > R_{F(threo)}$. In (a) and (b), X and Y are smaller than Ar and Ar'; in (c) and (d) X is smaller and Y bulkier than Ar and Ar'.
Source: Data from Palamareva, M.D., Kurtev, B.J., Mladenova, M.P., and Blagoev, B.M., *J. Chromatogr.*, 235, 299–308, 1982. With permission.

Four different patterns of adsorption regarding the retention of the diastereoisomeric compounds of type I (Figure 8.6) are discussed by Palamareva et al. [23]. The data and the conclusions of the study supported the previously elaborated criteria for TLC assessment of the relative configurations of other diastereoisomeric pairs of tetrasubstituted ethanes (type I) based on the relation $R_{F(erythro)} > R_{F(threo)}$. The main requirement is that the groups X and Y to be adsorbed more strongly, and to have smaller effective volumes, than the groups Ar and Ar'. The results also provide a further development of the criteria in the case where X is smaller and Y is bulkier than Ar and Ar'.

In the series of papers regarding the chromatographic behavior of diastereoisomers, Snyder et al. [24] studied the solvent strength values and the effects of solvent selectivity on silica for 20 diastereoisomeric compounds (ten *threo* and ten *erythro* pairs) as a function of mobile phase composition. They analyzed 46 mobile phases composed of 12 different solvents. The obtained data were compared with predictions from the model, solvent strength as a function of composition, and they showed a good agreement between experiment and theory. Palamareva [25] applied a microcomputer program [30] based on Snyder displacement model [31–35] to select 21 mobile phases composed of two to six solvents for the TLC separation of four diastereoisomeric amino esters and isocromans on silica. The experimental R_F

values (0.12–0.82 range) showed the applicability of the microcomputer programs for the choice of suitable mobile phases. The TLC behavior of some (E)- and (Z)-oxazolones and related cinnamates on silica showed a stronger retention of diastereoisomers in all instances [26]. The retentions were correlated with solvent selectivity effects or Hammett constants of the various substituents on the aryl group. Solvent selectivity effects were approximated based on Snyder theory by the localization parameter m, for experiments with mobile phases of constant strength, ε. The best separations were obtained with mobile phases of m with minimum values as predicted by the theory. The electronic effects related to the substituent M do not control the relative retentions of the studied diastereoisomers. The Hammett equation is of limited validity because of the deviation of m-CH$_3$ and m-NO$_2$ derivatives. The relative retentions were attributed to the adsorption models including two groups (nitrogen atom and ester carbonyl for the oxazolones and the amide carbonyl and ester carbonyl for cinnamates). Each adsorption model ensures less steric hindrance of the adsorbing groups for the stronger-retained isomer.

According to the obtained data, the model of adsorption is based on the electronic effects of any Z–E pair and the steric effects modify it by controlling the relative retention that is not affected by solvent selectivity effects and the nature of the aryl group.

TLC chromatographic retention on silica of some Z–E oxazolones and related cinnamates as a function of mobile phase are presented in Figure 8.7 (and Table 8.4) and Figure 8.8 (and Table 8.5).

Cooper [40] studied the TLC behavior of four *trans–cis* pairs of some geometric isomers, intermediates in the synthesis of some alicyclic analogues of mescaline on silica gel with different mobile phases without founding a definite correlation. Generally, the *cis*-isomers migrated faster than the *trans*-isomers.

Palamareva et al. [21] studied the chromatographic behavior of diastereomeric tetrahydroisoquinolines and isochromans (cyclic diastereoisomers of the type II).

The *trans–cis* retention sequence depends on the number of adsorbed groups. The conformational factors are more favorable for the *trans*-isomer under one-point adsorption and for the *cis*-isomer under two-point adsorption. The established relations of $R_{F(trans)} > R_{F(cis)}$ and $R_{F(cis)} > R_{F(trans)}$ have practical importance if the adsorption of the groups X and Y can be reliably predicted for other newly synthesized compounds of the type II without intramolecular hydrogen bonds or with bonds of the types OH. . .N or OH. . .O.

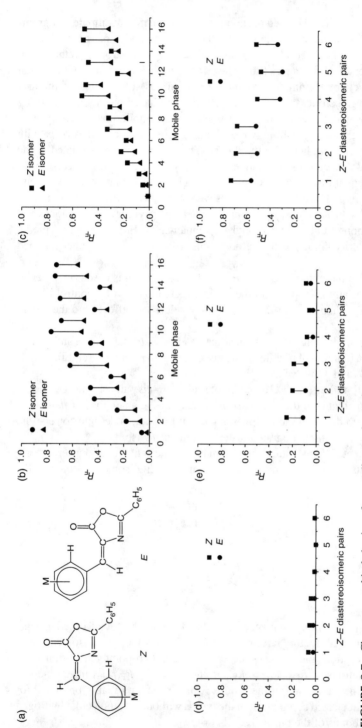

FIGURE 8.7   Chromatographic behavior of some diastereoisomeric pairs of oxazolones of type I using mobile phases with different values of strength, ε, according to Table 8.4: (a) Z–E isomers (M = 4-Cl, H, 4-CH$_3$, 3-CH$_3$, 4-CH$_3$O, 3-NO$_2$); (b) and (c) R_F values of Z–E diastereoisomers [(b) M = 4-Cl; (c) M = 3-CH$_3$O]; mobile phases arranged in increasing order of ε, according to data from Table 8.4; (d), (e), and (f) R_F values of Z–E diastereoisomers of [M: (1) 4-Cl; (2) H; (3) 4-CH$_3$; (4) 3-CH$_3$O; (5) 4-CH$_3$O; (6) 3-NO$_2$] eluted with three mobile phases of different values of ε [(d) $\varepsilon = 0.100$; (e) $\varepsilon = 0.150$; (f) $\varepsilon = 0.250$]. (Data from Palamareva, M.D., Kurtev, B.J., and Kavrakova, I., *J. Chromatogr.*, 545, 161–175, 1991.)

TABLE 8.4

Some Solvent Systems with Different Values of Strength, ε, in Increasing Order, Used as Mobile Phases in the Study of Chromatographic Behavior of Z–E Diastereoisomeric Pairs of Oxazolones Presented in Figure 8.7

No.	Mobile phase	Composition (v/v)	Strength (ε)
1	Cyclohexane–benzene	86.6 + 13.4	0.100
2	Cyclohexane–benzene	70.6 + 29.4	0.150
3	Tetrachloromethane–chloroform	84.7 + 15.3	0.150
4	Cyclohexane–benzene	44.0 + 56.0	0.200
5	Tetrachloromethane–chloroform	56.7 + 43.3	0.200
6	Cyclohexane–toluene–tetrahydrofuran	64.0 + 35.0 + 1.0	0.211
7	Cyclohexane–benzene	29.6 + 70.4	0.220
8	Tetrachloromethane–chloroform	41.2 + 58.8	0.220
9	Hexane–chloroform–diethyl ether	68.9 + 30.0 + 1.1	0.226
10	Hexane–benzene	20.0 + 80.0	0.233
11	Tetrachloromethane–diethyl ether	95.0 + 5.0	0.233
12	Hexane–1,2-dichloroethane	68.0 + 32.0	0.233
13	Toluene–methylene chloride	91.5 + 8.5	0.233
14	Tetrachloromethane–ethyl acetate	98.0 + 2.0	0.234
15	Cyclohexane–benzene	2.0 + 98.0	0.248
16	Tetrachloromethane–chloroform	12.5 + 87.5	0.250

Source: Data from Palamareva, M.D., Kurtev, B.J, and Kavrakova, I., *J. Chromatogr.*, 545, 161–175, 1991.

The chromatographic behavior of some different compounds is given in Figure 8.9 [41].

8.3.2 INFLUENCE OF CONFORMATION

The TLC behavior of the stereoisomeric menthols is strongly influenced by the position of the hydroxyl group, which is equatorial (*e*) in menthol and isomenthol and axial (*a*) in neomenthol and neoisomenthol [42].

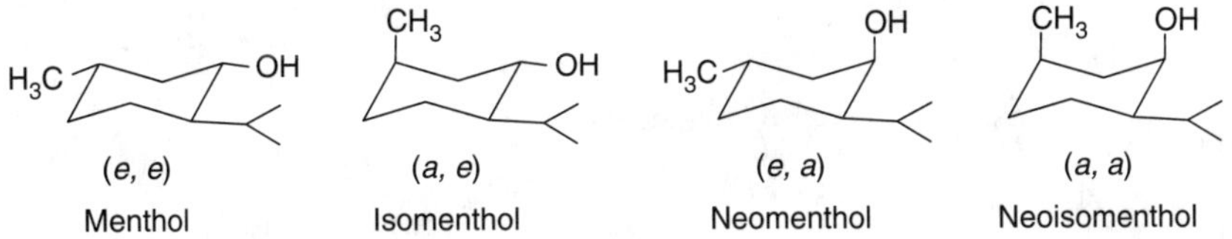

(e, e)	(a, e)	(e, a)	(a, a)
Menthol	Isomenthol	Neomenthol	Neoisomenthol

The molecules with equatorial hydroxyl groups showed a higher adsorption affinity with benzene as solvent (Figure 8.10). The effect of the methyl group, which is equatorial in menthol and neomenthol and axial in isomenthol and neo-isomenthol, is of less importance. The influence of the hydroxyl group can be

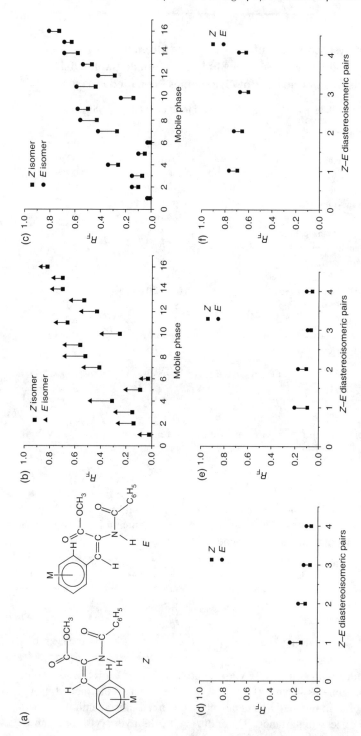

FIGURE 8.8 Chromatographic behavior of some diastereoisomeric pairs of cinnamates of type 1 using mobile phases with different values of strength, ε, according to Table 8.5: (a) Z–E isomers (M = 4-Cl, H, 4-CH$_3$O, 3-NO$_2$); (b) and (c) R_F values of Z–E diastereoisomers [(b) M = 4-Cl; (c) M = 3-NO$_2$]; mobile phases arranged in increasing order of ε, according to data from Table 8.5; (d), (e), and (f) R_F values of Z–E diastereoisomers [M: (1) 4-Cl; (2) H; (3) 4-CH$_3$O; (4) 3-NO$_2$] eluted with three mobile phases of different values of ε [(d) ε = 0.290; (e) ε = 0.378; (f) ε = 0.420]. (Data from Palamareva, M.D., Kurtev, B.J., and Kavrakova, I., *J. Chromatogr.*, 545, 161–175, 1991.)

TABLE 8.5
Some Solvent Systems with Different Values of Strength, ε, in Increasing Order, Used as Mobile Phases in the Study of the Chromatographic Behavior of Z–E Diastereoisomeric Pairs of Cinnamates Presented in Figure 8.8

No.	Mobile phase	Composition (v/v)	Strength (ε)
1	Hexane–diethyl ether	78.4 + 21.6	0.290
2	Benzene–acetone	97.8 + 2.2	0.290
3	Hexane–diethyl ether	57.6 + 42.4	0.330
4	Benzene–acetone	94.9 + 5.1	0.330
5	Hexane–chloroform–diisopropyl ether–acetonitrile	69.6 + 19.0 + 10.0 + 1.4	0.348
6	Hexane–chloroform–diisopropyl ether–ethyl acetate–acetonitrile	77.85 + 18.0 + 2.0 + 2.0 + 0.15	0.364
7	Hexane–diethyl ether	35.0 + 65.0	0.378
8	Toluene–diethyl ether	66.0 + 34.0	0.378
9	Benzene–acetone	89.8 + 10.2	0.378
10	Hexane–chloroform–diethyl ether	51.2 + 32.0 + 16.8	0.379
11	Cyclohexane–tetrachloromethane–benzene–tetrahydrofuran	26.6 + 25.0 + 25.0 + 23.4	0.379
12	Hexane–diethyl ether	34.0 + 66.0	0.380
13	Benzene–acetone	89.5 + 10.5	0.380
14	Hexane–diethyl ether	10.0 + 90.0	0.420
15	Benzene–acetone	81.6 + 18.4	0.420
16	Benzene–diethyl ether	8.0 + 92.0	0.420

Source: Data from Palamareva, M.D., Kurtev, B.J, and Kavrakova, I., *J. Chromatogr.*, 545, 161–175, 1991.

eliminated by its reaction with 3,5-dinitrobenzoic acid to form the respective dinitrobenzoate, which increases the influence of the methyl group on the chromatographic separation [12].

Due to their different conformations caused by the position of the methylene bridge, the chlorinated stereoisomeric hydrocarbons, aldrin and isodrin, present differences in the chromatographic behavior [12].

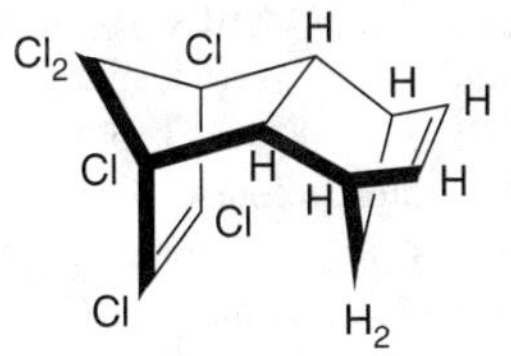

endo–exo
Aldrin

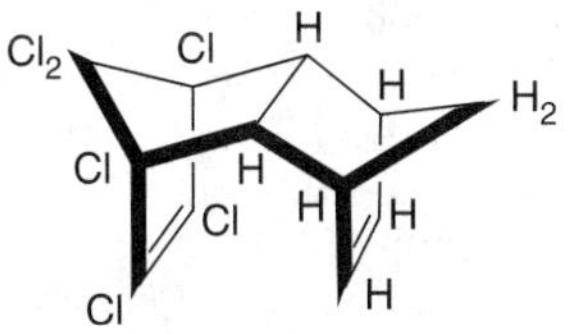

endo–endo
Isodrin

In aldrin, the C=C double bonds are in a *trans* position; thus, aldrin has a higher R_F value (0.60) than isodrin (0.49).

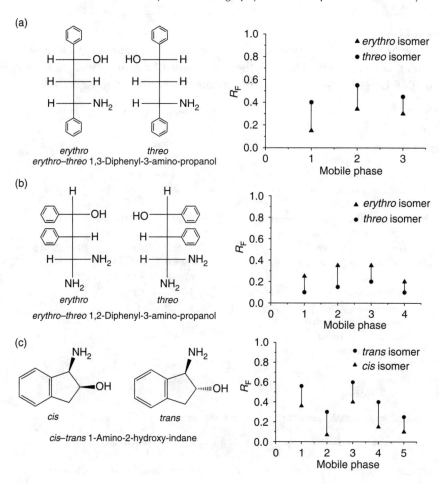

FIGURE 8.9 Chromatographic behavior of some diastereoisomeric pairs of *erythro–threo* and *cis–trans* isomers using different mixtures of dioxane–methanol with added boric acid as mobile phases. (a) R_F values of *erythro–threo* pairs of 1,3-diphenyl-3-amino-1-propanol; volume ratios of mobile phase: 1, $(17 + 3)$; 2, $(10 + 10)$; 3, $(8 + 12)$. (b) R_F values of *erythro–threo* pairs of 1,2-diphenyl-3-amino-1-propanol; mobile phase (v/v): 1, $(17 + 3)$; 2, $(10 + 10)$; 3, $(8 + 12)$; 4, $(6 + 14)$. (c) R_F values of *trans–cis* pairs of 1-amino-2-hydroxy-indane; mobile phase: 1^*, $(17 + 3)$; 2, $(17 + 3)$; 3^*, $(10 + 10)$; 4, $(10 + 10)$; 5, $(8 + 12)$. (d) R_F values of *trans–cis* pairs of 1-amino-2-hydroxy-tetraline; mobile phase: 1^*, $(17 + 3)$; 2, $(17 + 3)$; 3^*, $(10 + 10)$; 4, $(10 + 10)$; 5, $(8 + 12)$; 6, $(6 + 14)$. (e) R_F values of *trans–cis* pairs of 2-phenyl-2-hydroxy-1-amino-cyclohexane; mobile phase: 1, $(17 + 3)$; 2, $(10 + 10)$; 3, $(8 + 12)$; 4, $(6 + 14)$. Asterisks represent mobile phase without boric acid. (Data from Drefahl, G., Heublein, G., and Silbermann, K., *J. Chromatogr.*, 22, 460–464, 1966.)

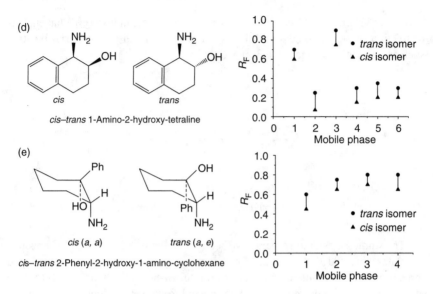

FIGURE 8.9   (Continued)

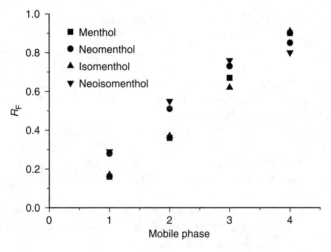

FIGURE 8.10   Chromatographic behavior of some diastereoisomers of menthols on silica gel. Mobile phases (v/v): 1, benzene; 2, benzene–methanol (95 + 5); 3, benzene–methanol (75 + 25); 4, methanol. (Data from Petrowitz, H.-J., *Angew. Chem.*, 23, 921, 1960.)

The physicochemical and analytical significance of topological indexes (I_t) lies in the possibility of using them to describe the chemical structures of groups of organic compounds. This index is not able to distinguish between stereoisomers.

Pyka [43] proposed a new stereoisomeric topological index (I_{STI}) that enables distinction of stereoisomers with hydroxyl group in axial or equatorial position such as menthol (see Section 8.3.2) and thujol.

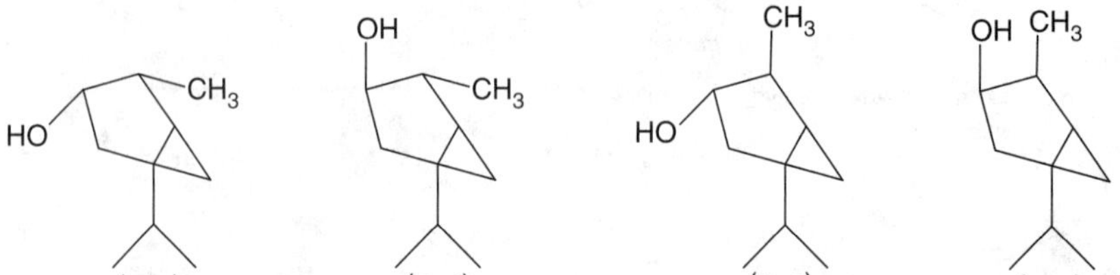

| (e, e) | (a, e) | (e, a) | (a, a) |
| Thujol | Neothujol | Isothujol | Neoisothujol |

The studied diastereoisomers have been separated by adsorption TLC on silica gel plates activated 30 min at 120°C before use. The spotted plates were developed three times, for menthols with the binary mixture n-hexane-ethanol (85 + 15, v/v) and for thujols with benzene. The experimental R_F and R_M values of investigated diastereoisomers and the numerical values of the proposed stereoisomeric topological indexes (I_{STI}) and the modified version of the index $^0B_{STI}$ are given by the author [43]. The data demonstrate close correlation between R_F and R_M values and the numerical values of the topological indexes.

The investigations demonstrated that the compounds with OH group in the equatorial position were adsorbed more strongly and the methyl group has no influence on the adsorption of separated compounds (Table 8.6).

Analyzing the studied parameters of pairs of diastereoisomers with OH group in the same position, but the methyl groups in the *trans* or *cis* configuration the correlations from Table 8.7 can be noted.

The calculated R_F (±0.030 deviation) and R_M values for the diastereoisomers of menthol and thujol are in good agreement with the experimental data that attest the physicochemical significance of the topological indexes proposed by the author. Such indexes can be used for analytical identification of diastereoisomers separated by TLC.

The silver nitrate impregnation of silica gel or other adsorbent layers has been used to separate unsaturated compounds. This technique has been applied in the field of lipids at the separation of *cis–trans* isomers and other compounds according to their number of double bonds [44]. This impregnation technique is very useful in the separation of unsaturated compounds; however, for maximum separation, the percentage impregnation is dependent on the types of compounds to be separated [45].

Marekov et al. [46] established a stand-alone argentation TLC procedure for the rapid estimation of the authenticity and densitometric quantification of fatty acid groups in milk fats. The unsaturated fatty acids contained in food are mostly of *cis* configuration. *Trans* fatty acids (TFAs) are formed during the hydrogenation (industrial or biological) of unsaturated fats and occur naturally in dairy

TABLE 8.6
General Correlations on the Chromatographic Behavior of Menthol and Thujol Diastereoisomers with Hydroxyl Group in Axial or Equatorial Position

Value of parameter	OH group position	Relation	Value of parameter	OH group position
R_F	Axial	>	R_F	Equatorial
I_{STI}	Axial	>	I_{STI}	Equatorial
R_M	Axial	<	R_M	Equatorial
$^0B_{STI}$	Axial	<	$^0B_{STI}$	Equatorial

Source: Data from Pyka, A., *J. Planar Chromatogr.*, 7, 389–393, 1994.

TABLE 8.7
Some Correlations on the Chromatographic Behavior of Diastereoisomers with Hydroxyl Group in the Same Position and Methyl Group in the *trans* or *cis* Configuration

Value of parameters	Diastereoisomer	Relation	Value of parameters	Diastereoisomer
R_F and I_{STI}	Menthol	>	R_F and I_{STI}	Isomenthol
R_F and I_{STI}	Neoisomenthol	>	R_F and I_{STI}	Neomenthol
R_F and I_{STI}	Thujol	>	R_F and I_{STI}	Isothujol
R_F and I_{STI}	Neoisothujol	>	R_F and I_{STI}	Neothujol
R_M and $^0B_{STI}$	Menthol	<	R_M and $^0B_{STI}$	Isomenthol
R_M and $^0B_{STI}$	Neoisomenthol	<	R_M and $^0B_{STI}$	Neomenthol
R_M and $^0B_{STI}$	Thujol	<	R_M and $^0B_{STI}$	Isothujol
R_M and $^0B_{STI}$	Neoisothujol	<	R_M and $^0B_{STI}$	Neothujol

Source: Data from Pyka, A., *J. Planar Chromatogr.*, 7, 389–393, 1994.

products derived from ruminant animals and in commercially hydrogenated fats, for example, margarines and shortenings. Special attention was paid to find the conditions for the correct measurement of short-chain saturated and low-level TFAs, which are very important in butter analysis.

Wall [47] has been achieved complete resolution of the *cis* and *trans* isomers of capsaicin based on the argentation HPTLC as an effective separation technique using a simple solvent system on reversed-phase plates. The separation is based on the known interaction of the silver ion (Ag^+) with the ethylenic π bonds. The method also gave good resolution from other capsaicin analogs present in chili extracts, enabling spectrodensitometric determination of the *trans*-capsaicin present. The stationary phases RP2, RP8, RP18, and RP18W were compared, the latter giving the optimum results.

TABLE 8.8

Comparison of R_F Values Obtained by Silver Nitrate Impregnation of Stationary Phase (Silica Gel 60 RP8) and by Addition of Silver Nitrate to the Mobile Phase (Methanol-Water: 80 + 20, v/v) at the Separation of cis-trans Isomers of Capsaicin

| | Impregnation | | Addition | |
| | R_F Value of isomer | | R_F Value of isomer | |
Silver nitrate %	cis	trans	cis	trans
0.5	—	—	0.40	0.42
1.0	—	—	0.54	0.46
1.5	—	—	0.56	0.48
2.0	0.46	0.43	0.66	0.57
3.0	—	—	0.69	0.60
4.0	0.73	0.68	—	—

Source: Wall, P.E., J. Planar Chromatogr., 10, 4–9, 1997. With permission.

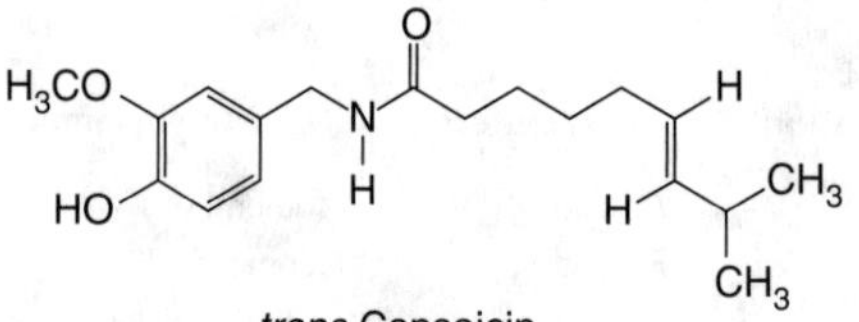

trans-Capsaicin 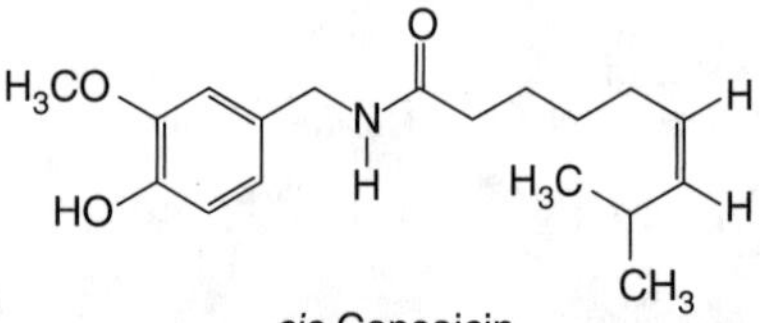   cis-Capsaicin

The use of silver nitrate as an impregnating agent on the layer was compared with results obtained by adding silver nitrate solution to the mobile phase (Table 8.8). A study was made with respect to the variation of ΔR_F with the concentration of silver nitrate, methanol, and orthoboric acid.

Kowalska et al. [48] have achieved a suitable TLC method for the separation of 19 pairs of the E–Z geometrical isomers of pyrazole, pyrimidine, and purine derivatives with potential cytokinin activity. These systems employed silica, silanized silica, silanized silica/Cu(II) cation, chemically bonded RP-8, and chemically bonded RP-18 as stationary phases, and a variety of binary (aqueous and nonaqueous) mobile phases. The quality of separations was evaluated by the use of three criteria — α, R_F, and ΔR_F. The average numerical values of these criteria, calculated for the whole population of E–Z isomer pairs, were used to rank the investigated TLC systems in order of decreasing separating power.

8.4 OPTIMIZATION OF DIASTEREOISOMER SEPARATION BY DERIVATIZATION

Separation of the enantiomers comprising the racemate is a common problem in stereochemical research as well as in the preparation of biologically active

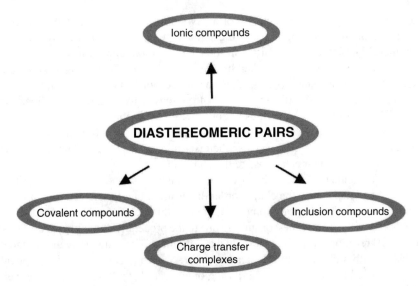

FIGURE 8.11 Types of diastereoisomers formed at the separation of enantiomers using different chiral agents.

compounds, particularly, drugs. The problem is that in contrast to diastereoisomers and all the types of isomeric species, in a nonchiral environment, the enantiomers display identical physical and chemical properties. In a racemate, the ratio of the two enantiomers is one and the sum of the optical rotations should be zero.

The chirality sense and optical activity of the enantiomers are determined by their absolute configuration (the spatial arrangement of the atoms in the molecule). In contrast to their conformation, the configuration of enantiomers can be modified only with the change in the connectivity of constituent atoms. Designation (notation) of the configuration of enantiomers should be made in accordance with R, S system of the Cahn–Ingold–Prelog (CIP) convention that describes the absolute configuration around the asymmetric carbon atom (the spatial arrangement of the substituents). The D, L designation that correlates the configuration of a molecule to the Fisher convention is predominantly restricted to amino acids and carbohydrates [1–6].

The chromatographic methods are considered to be most useful for chiral separations. Enantiomers can be separated by two methods: (a) indirect method that utilizes derivatizing agents and (b) direct method that uses chiral stationary phases (CSPs) or chiral mobile phase additives (CMPAs) [49–56].

The *indirect* separation method of a racemate into its enantiomers can be achieved by its derivatization with a chiral derivatizing agent before chromatography (Figure 8.2), resulting in a diastereomeric complex/salt (Figure 8.11) [11,56].

This fact is based on the formation of a covalent bond between the enantiomers and a pure chiral reagent (chiral derivatizing agent [54,56] and chiral selector [51,53]). The introduction of a new chiral center leads to the formation of a pair of isomeric compounds (diastereoisomers) that are not mirror images anymore. Diastereoisomers differ in their physical and chemical properties and they can be separated each other by any analytical method using a nonchiral separation mechanism [50–56]. This method enables the use of conventional stationary phases, a significant increase in the sensitivity of detection or location on the layers of some compounds not otherwise identifiable [51].

For a successful derivatization, the presence of a suitable functional group in the analyte is necessary. To increase the discrimination between enantiomers, the derivatization should occur close to the chiral atom.

The advantage of the indirect chromatographic method consists in the fact that it can predetermine the elution order that is important for the determination of optical purities, but there are some limitations to this technique. The derivatization method is tedious and time-consuming due to the procedure that involves different chemical reaction rates of the individual enantiomers. The chiral derivatizing agent must be with high-optical purity and stable in solution or solid state, since optical impurity will result in two more diastereomeric products [53,54,56]. Moreover, this method cannot be used easily with the environmental samples [56].

The derivatization can be due in the different reaction rates of the individual enantiomers, and for a proper interpretation of the analytical result, it is necessary to have the reaction completed before analysis. The formation of undesired should be avoided and the diastereomeric mixture must be chemically and stereochemically stable. For preparative purposes, the indirect chromatographic method involves an additional synthesis step, because the derivatizing agent has to be cleaved off the separated diastereoisomers after their resolution in a nonchiral environment. Hereby, the impurities can be introduced or even the racemization of the just resolved enantiomers could be caused. Still, once a suitable chiral derivatizing agent and procedure have been found, the indirect method offers some advantages for preparative separations over the direct preparative separations, mainly due to the fact that, in large-scale resolutions, the nonchiral media are better handled than, for example, the stationary phases, and the conditions can be adjusted more easily to obtain the desired resolution [54].

The *direct* chromatographic method involves the use of a chiral selector either in the stationary phase (CSP) or in the mobile phase (CMPA) in normal and reversed-phase chromatography. In the first case, the chiral selector is chemically bonded, coated, or otherwise attached to the solid support and in the second case the chiral selector is added to the mobile phase [11,49,50,54,56].

In the case of a CSP, the enantiomer that forms the more stable association with the chiral selector will be the more strongly retained species of the racemate. The enantioselectivity of system is expressed as the ratio of the retention factors of the two enantiomers. In the case of a chiral mobile phase, this reduces the retention of the solute enantiomer, which forms a stronger association with the chiral selector. The enantioselectivity of the latter system depends on the selector–solute association from the mobile phase, and it is proportional to the enantioselectivities of the association processes in the stationary and mobile phases [57].

The direct separation method of a racemate into its enantiomers is based upon the complex formation between the optical isomers of the solute and a chiral selector, resulting in the formation of labile diastereoisomers [50,53]. These differ in their thermodynamic stability, provided that at least three active points of the selector participate in the interaction with corresponding sites of the solute molecule. The rule of the three-point interaction model is generally valid for enantioselective chromatography, with the extension to the rule, starting that one of the required interactions may be mediated by the adsorption of the two components of the interacting pair onto the sorbent surface [50,55]. The separation of labile diastereoisomers can be accomplished if the complexes possess different stability constants. The major approaches in the formation of diastereomeric complexes are transition metal ion complexes, ion pairs, and inclusion complexes (diastereomeric complex/salt) (Figure 8.11). In this case, only the chiral purity of the selector influences the resolution [53].

A lot of number of racemic mixtures can be separated on conventional nonchiral stationary phases by using an appropriate CMPA (α-, β-, γ-cyclodextrins, bovine serum albumin, etc.) [11,49–52].

Enantiomeric separation by using CSP is based on the formation of labile (transient) diastereomeric complexes of solute–CSP between the enantiomers and the chiral molecule that is a part of the stationary phase. The five major CSP classes based on solute–CSP complexes are as follows [51,52,54]:

1. "Pirkle" phases that form solute–CSP complexes by π electron donor–acceptor mechanism (attractive–repulsive interactions).
2. Derivatized cellulose phases that form inclusion complexes by attractive interactions followed by inclusion into chiral cavities.
3. Cyclodextrins and crown ethers that form inclusion complexes.
4. Diastereoisomeric metal ion complexes (chiral ligand exchange chromatography).
5. Protein (bovine serum albumin) forms solute–CSP complexes based on hydrophobic and polar interactions.

A distinct class of CSPs is represented by the molecularly imprinted Polymers that form template-monomer complexes through reversible covalent or noncovalent bonds [55,56].

8.5 OVERVIEW ON DIASTEREOISOMER SEPARATION BY TLC

Application of TLC to determine the enantiomer purity via separation of diastereomeric derivatives is a relatively straightforward, though seemingly little used, technique. One example is the analysis of a nonracemic carboxylic acid synthetic intermediate as the amide formed with 1-(α-naphthyl)ethylamine (Reference [11] and references therein).

The TLC plates impregnated with nonracemic chiral selectors or coated with the latter bonded to nonchiral stationary phases (silica gel) have been developed for qualitative and quantitative analysis, especially for multicomponent mixture containing enantiomer pairs. The first report on this subject was of Wainer et al. [58], who demonstrated that the chiral alcohol TFAE could be resolved chromatographically on plates coated with (R)-N-(3,5-dinitrobenzoyl) phenylglycine covalently bonded to γ-aminopropyl silanized silica gel with a separation factor α estimated to be 1.50.

TLC plates coated with chiral ligand-exchange media have been applied to the analysis of amino acid mixture, as such (Reference [11] and literature therein) or either after derivatization [59]. The plates coated with β-cyclodextrin have been shown to fully resolve dansyl amino acids [60]. The quantitative determination of enantiomer purity by TLC is possible using densitometry or measurement of fluorescence or UV absorbance followed by the extraction of spots [61].

Cyclodextrins (α, β, γ) are frequently used in chromatography, added in the mobile phase, for the separation of enantiomers, diastereoisomers or structural isomers, and also in the routine analyses. In chromatography, the concentration of the CMPA is the critical parameter for obtaining enantiomeric separations.

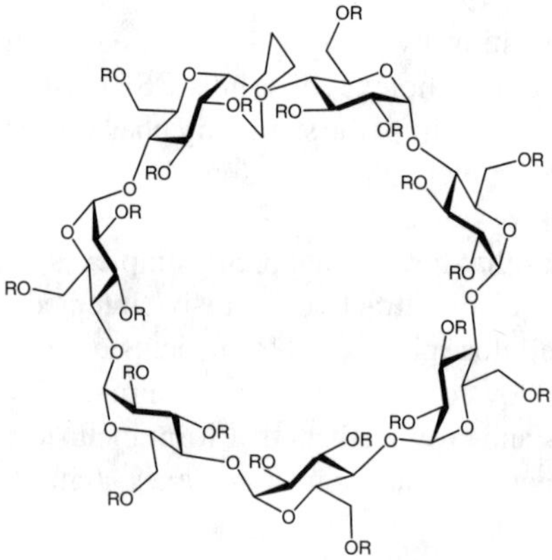

Armstrong et al. [62] used α-cyclodextrin (R = H) and β-cyclodextrin derivatives (R = hydroxypropyl) [63] as additives in the mobile phase for the separation of different diastereoisomers. Thus, they separated *syn*-azobenzene (R_F = 0.09) from *anti*-azobenzene (R_F = 0.53) on polyamide or polyamide with fluorescent indicator adding α-cyclodextrin to the mobile phase (0.1 M) [62] or some steroid epimers and alkaloids on plates coated with chemically bonded octadecylsilane reversed-phase [62]. The used mobile phases obtained by adding hydroxypropyl-β-cyclodextrin are presented in Table 8.9, beside the chromatographic results.

TLC separations of some diastereoisomers on plates with β-cyclodextrin bonded through a spacer to silica phase are presented in Table 8.10 [64]; one can observe that the *trans* isomer is stronger retained on the stationary phase than *cis* one $[R_{F(cis)} > R_{F(trans)}]$.

Chiral amines and amine alcohols are important building blocks in the synthesis of pharmaceutical compounds. A simple and fast method to perform the separation

TABLE 8.9
Chromatographic Behavior of Some Diastereoisomers

No.	Compound	R_F	R_S	Mobile phase (v/v)
1	Cinchonine	0.40	4.2	0.3 M HP-β-CD[a]
	Cinchonidine	0.23		Acetonitrile–water
				(35 + 65)
2	Quinidine	0.29	4.3	0.3 M HP-β-CD
	Quinine	0.15		Acetonitrile–water
				(35 + 65)
3	17α, 20α-Dihydroxy-4-pregnene-3-one	0.54	4.0	0.3 M HP-β-CD
	17α, 20β-Dihydroxy-4-pregnene-3-one	0.40		Acetonitrile–water
				(35 + 65)
4	20α-Hydroxy-4-pregnene-3-one	0.37	3.3	0.3 M HP-β-CD
	20β-Hydroxy-4-pregnene-3-one	0.16		Acetonitrile–water
				(35 + 65)
5	17α, 20α, 21-Trihydroxy-4-pregnene-3,11-dione	0.69	2.2	0.3 M HP-β-CD
	17α, 20β, 21-Trihydroxy-4-pregnene-3,11-dione	0.61		Acetonitrile–water
				(30 + 70)
6	11β, 17α, 20α, 21-Tetrahydroxy-4-pregnene-3-one	0.63	0.8	0.3 M HP-β-CD
	11β, 17α, 20β, 21-Tetrahydroxy-4-pregnene-3-one	0.00		Acetonitrile–water
				(35 + 65)
7	N'-(Menthoxycarbonyl)	0.02	0.8	0.3 M HP-β-CD
	Anabasine	0.04		Acetonitrile–water
				(35 + 65)
8	N'-(Menthoxycarbonyl)	0.11	2.2	0.3 M HP-β-CD
	3-Pyridyl-1-aminoethane	0.18		Acetonitrile–water
				(35 + 65)

[a] Hydroxypropyl-β-cyclodextrin.

Source: Armstrong, D.W., Faulkner, J.R., and Han, S.M., *J. Chromatogr.*, 452, 323–330, 1988. With permission.

TABLE 8.10
Chromatographic Behavior of Some Diastereoisomers on β-Cyclodextrin-Bonded Phase Plates

Compound	R_F	Mobile phase
Quinine (*erythro*)	0.38	25 + 75
Quinidine (*threo*)	0.46	
trans-Stilbene	0.38	80 + 20
cis-Stilbene	0.48	
Benzo[a]pyrene-*trans*-7,8-diol	0.46	80 + 20
Benzo[a]pyrene-*cis*-7,8-diol	0.52	

Mobile phase (v/v): methanol + 1% triethylammonium acetate (pH 4.1).

Source: Data from Alak, A. and Armstrong, D.W., *Anal. Chem.*, 58, 582–584, 1986.

of these compounds by HPTLC has been achieved after derivatization with Marfey's reagent (1-fluoro-2,4-dinitrophenyl-5-L-alanine amide) [65].

Solute   Marfey's reagent   Diastereomeric derivative

The derivatization of simple aliphatic chiral amines and amino alcohols with Marfey's reagent has resulted in the formation of diastereomeric compounds to enable separation on nonchiral HPTLC plates.

The tetrahydrophthalic ester of prednisolone, an anti-inflammatory drug, consists of two diastereoisomers. Olszewska et al. [66] describes the effect of addition of an optically active compound (amino acids, cyclodextrins, camphosulfonic acid) on the capacity ratios, k', and separation factors of these diastereoisomers by TLC. They used both mobile phases containing a chiral additive and stationary phases impregnated with an optically active compound, for example, silica gel plates impregnated with amino acids. The best resolution was obtained by impregnation of the stationary phase with D-camphosulfonic acid and copper acetate, and use of dichloromethane-isopropanol (90+10, v/v) as mobile phase. The addition of chiral compound to the mobile phase had less effect on the resolution of diastereoisomers.

The work of Lepri et al. [67–72] is concentrated on the enantioseparation of different compounds based on bovine serum albumin added in the mobile phase.

Researches regarding the separation of dansyl-DL-amino acids by normal-phase TLC on plates impregnated with a macrocyclic antibiotic [73] and amino acids from small peptides on reversed-phase TLC [74] are given in literature.

Enantiomers of β-blocking drugs have been separated on diol plate with dichloromethane and achiral counter ion, N-benzyloxycarbonylglycyl-L-proline as mobile phase additive [75].

The separation of pindolol enantiomers, a nonselective β-adrenergic antagonist, by chemical derivatization with sugar-based derivatizing agent is reported [76].

Molecular imprinting is a new methodology for the preparation of synthetic polymers with predicted selectivity for various substances [77]. Molecular imprinted polymers were prepared as layers on TLC plates to investigate the resolution of diastereomeric pairs of alkaloids (quinine–quinidine and cinconine–cinconidine) [78].

Both CSPs and CMPAs are based on the formation of transient diastereomeric complexes between the chiral selectors and enantiomers during chromatographic procedure. The diastereomeric complexes can be separated due to their different properties [79].

The current state of chiral separations by planar chromatography and the quantitative determination of resolved enantiomers with an advanced technique (molecular imprinting techniques) are discussed in detail by Günther and Möller (Reference [80] and references therein) and Lepri et al. (Reference [81] and references therein). TLC systems that include CSPs, chiral-coated phases, and chiral selectors as mobile phase additives on achiral supports are described. A detailed image of the various types of structurally related racemates that can be easily resolved into their optical antipodes by planar chromatography is reported. The mechanism of enantioresolution on hydrophobic supports eluted with mobile phases containing chiral cyclodextrins or bovine serum albumin is discussed. The future possibilities of chiral TLC development are critically advanced.

In the review of Maier et al. [49] with 259 references, we can find data on the separation of enantiomers: needs, challenges, and perspectives. They show that chiral drugs, agrochemicals, food additives, and fragrances represent classes of compounds with high economic and scientific potential. Owing to different biological activities on enantiomers of active ingredients, the preparation of highly enantiopure compounds is most important.

8.6 CONCLUSIONS

The importance of the spatial arrangements of the atoms in a molecule (the stereochemistry) is well put in evidence by the TLC separation of diastereoisomers.

The TLC separation of proper diastereoisomers using classical systems is well presented in the papers of Palamareva and coworkers [19–29], who showed that the separation is based on the adsorption mechanism. They demonstrated that TLC on silica gel can be considered as a method for the assignment of the relative configurations to some aliphatic diastereomeric compounds.

The potential of enantiomer purity determinations by TLC is evident via the separation of diastereoisomeric derivatives. The separation of enantiomers is extensively treated in this book.

The TLC separation of enantiomers is based on the following basic methods, where diastereoisomers have an important role [80,81]:

- Direct separation by using CSPs, performed by the formation of diastereomeric association complexes.
- Separation on common stationary phases by means of chiral additives in the eluent, which form diastereoisomeric complexes.
- Separation on nonchiral stationary phases via diastereoisomeric derivatives formed by reaction of the sample with a chiral reagent.

The largest number of recorded resolutions of enantiomers has been achieved by the conversion of a racemate to a mixture of diastereoisomers. The method consists in the treatment of the substrate to be resolved with one enantiomer of a chiral substance. The diastereoisomer pairs prepared in connection with resolutions may be ionic (diastereomeric salts), covalent, charge-transfer complexes, or inclusion compounds.

The diastereoisomers can be separated without using chiral reagents, but the resolution can be difficult and time-consuming.

Several chromatographic methods for the separation of racemates based on the addition of chiral compounds to the mobile phase or use of a CSP have been developed.

As a general conclusion, one can assert that the isomeric and stereoisomeric constitutions influence on the chromatographic migration; thus, the TLC method is very suitable for the separation of enantiomers via diastereoisomers.

REFERENCES

1. Mager, S., Grosu, I., and David, L., Diastereoizomeria (Diastereoisomerism), in *Stereochimia Compuşilor Organici (Organic Compound Stereochemistry)*, Dacia Publishing House, Cluj-Napoca, 2006, Chap. 5.
2. Eliel, E.L., Wilen, S.H., and Mander, L.N., Stereoisomers, in *Stereochemistry of Organic Compounds*, John Wiley & Sons, Inc., New York, 1994, Chap. 3.
3. Neniţescu, C.D., Stereochimie I (Stereochemistry I), in *Chimie Organică, Vol. I (Organic Chemistry, Vol. I)*, Didactic and Pedagogic Publishing House, Bucharest, 1980, Chap. 1.
4. Avram, M., Stereochimie I. Stereochimia compuşilor cu carbon hibridizat sp^3 (Stereochemistry I. Stereochemistry of compounds with sp^3 hybridized carbon), in *Chimie Organică (Organic Chemistry)*, Zecasin Publishing House, Bucharest, 1994, Chap. 6.
5. Neniţescu, C.D., Stereochimie II (Stereochemistry II), in *Chimie Organică, Vol. II (Organic Chemistry, Vol. II)*, Didactic and Pedagogic Publishing House, Bucharest, 1980, Chap. 3.
6. Avram, M., Stereochimie II. Chiralitate. Activitate optică (Stereochemistry II. Chirality. Optical activity), in *Chimie Organică (Organic Chemistry)*, Zecasin Publishing House, Bucharest, 1994, Chap. 16.
7. ***, Catalogue, Fluka — Riedel-deHaën, Scientific research, Fluka Chemie GmbH, 2003–2004.
8. ***, Catalogue Aldrich — *Handbook of Fine Chemicals*, Sigma-Aldrich Chemie GmbH, Aldrich Division, Steinheim, Germany, 1996–1997, 2005–2006.
9. www.Saccharides.net
10. Simiti, I. and Zaharia, V., Monozaharide (Monosaccharides), in *Produşi Naturali (Natural Products)*, Dacia Publishing House, Cluj-Napoca, 1996, pp. 13–42.
11. Eliel, E.L., Wilen, S.H., and Mander, L.N., Chromatographic and related separation methods based on diastereomeric interactions, in *Stereochemistry of Organic Compounds*, John Wiley & Sons, Inc., New York, 1994, pp. 240–265, 275–295.
12. Petrowitz, H.-J., *J. Chromatogr.*, 63, 9–14, 1971.
13. Petrowitz, H.-J. and Pastuska, G., *J. Chromatogr.*, 7, 128–130, 1962.
14. Pastuska, G. and Petrowitz, H.-J., *J. Chromatogr.*, 10, 517–518, 1963.
15. Seiler, H., Biebricher, Chr., and Erlenmeyer, H., *Helv. Chim. Acta*, 46, 2636–2638, 1963.
16. Fisher, F. and Koch, H., *J. Chromatogr.*, 16, 246–248, 1964.
17. Hranisavljević-Jakovljević, M., Pejković-Tadić, I., and Stojiljković, A., *J. Chromatogr.*, 12, 70–73, 1963.
18. Henein, R.G. and Dávid, Á., *J. Chromatogr.*, 36, 543–545, 1968.
19. Palamareva, M., Haimova, M., Stefanovski, J., Viteva, L., and Kurtev, B., *J. Chromatogr.*, 54, 383–391, 1971.

20. Palamareva, M.D. and Kurtev, B.J., *J. Chromatogr.*, 132, 61–72, 1977.
21. Palamareva, M.D., Kurtev, B.J., and Haimova, M.A., *J. Chromatogr.*, 132, 73–82, 1977.
22. Palamareva, M.D., Kurtev, B.J., Faitondzieva, K.B., and Zheljazkov, L.D., *J. Chromatogr.*, 178, 155–161, 1979.
23. Palamareva, M.D., Kurtev, B.J., Mladenova, M.P., and Blagoev, B.M., *J. Chromatogr.*, 235, 299–308, 1982.
24. Snyder, L.R., Palamareva, M.D., Kurtev, B.J., Viteva, L.Z., and Stefanovski, J.N., *J. Chromatogr.*, 354, 107–118, 1986.
25. Palamareva, M.D., *J. Chromatogr.*, 438, 219–224, 1988.
26. Palamareva, M.D., Kurtev, B.J., and Kavrakova, I., *J. Chromatogr.*, 545, 161–175, 1991.
27. Palamareva, M. and Kozekov, I., *J. Planar Chromatogr.*, 9, 439–444, 1996.
28. Palamareva, M. and Kozekov, I., *J. Chromatogr. A*, 758, 135–144, 1997.
29. Palamareva, M. and Kozekov, I., *J. Liquid Chromatogr. Rel. Technol.*, 20, 31–46, 1997.
30. Palamareva, M.D. and Palamarev, H.E., *J. Chromatogr.*, 477, 235–248, 1989.
31. Snyder, L.R., *Principles of Adsorption Chromatography*, Marcel Dekker, New York, 1968.
32. Snyder, L.R., *J. Chromatogr.*, 63, 15–44, 1971.
33. Snyder, L.R., *J. Chromatogr.*, 92, 223–230, 1974.
34. Snyder, L.R., Glajch, J.L., and Kirkland, J.J., *J. Chromatogr.*, 218, 299–326, 1981.
35. Snyder, L.R., *J. Chromatogr.*, 245, 165–176, 1982.
36. Snyder, L.R. and Glajch, J.L., *J. Chromatogr.*, 248, 165–182, 1982.
37. Glajch, J.L., Kirkland, J.J., and Snyder, L.R., *J. Chromatogr.*, 238, 269–280, 1982.
38. Glajch, J.L., Kirkland, J.J., Squire, K.M., and Minor, J.M., *J. Chromatogr.*, 199, 57–79, 1980.
39. Soczewiński, E., *J. Chromatogr.*, 388, 91–98, 1987.
40. Cooper, P.D., *J. Chromatogr.*, 67, 184–185, 1972.
41. Drefahl, G., Heublein, G., and Silbermann, K., *J. Chromatogr.*, 22, 460–464, 1966.
42. Petrowitz, H.-J., *Angew. Chem.*, 23, 921, 1960.
43. Pyka, A., *J. Planar Chromatogr.*, 7, 389–393, 1994.
44. Lawrence, B.M., *J. Chromatogr.*, 38, 535–537, 1968.
45. Nano, G.M. and Martelli, A., *J. Chromatogr.*, 21, 349, 1966.
46. Marekov, I., Tarandjiiska, R., Panayotova, S., and Nikolova, N., *J. Planar Chromatogr.*, 14, 384–390, 2001.
47. Wall, P.E., *J. Planar Chromatogr.*, 10, 4–9, 1997.
48. Kowalska, T., Sajewicz, M., Nishikawa, S., Kuś, P., Kashimura, N., Kołodziejczyk, M., and Inoue, T., *J. Planar Chromatogr.*, 11, 205–210, 1998.
49. Maier, N.M., Franco, P., and Lindner, W., *J. Chromatogr. A*, 906, 3–33, 2001.
50. Davankov, V.A., *Pure Appl. Chem.*, 69, 1469–1474, 1997.
51. Lepri, L., *J. Planar Chromatogr.*, 10, 320–331, 1997.
52. Armstrong, D.W. and Han, S.M., *CRC Crit. Rev. Anal. Chem.*, 19, 175–224, 1988.
53. Ingelse, B.A., *Chiral Separations Using Capillary Electrophoresis*, Eindhoven Technical University, Eindhoven, 1997, ISBN 90-386-0958-2.

54. Buga, L.A., *Enantioselective Enrichment of Selected Pesticides by Adsorptive Bubble Separation*, Dissertation Thesis, Technical University from München, 2005.

55. Gübitz, G. and Schmid, M.G., *Biopharm. Drug Dispos.*, 22, 291–366, 2001.

56. Ali, I. and Aboul-Enein, H.Y., *Chiral Pollutants*, John Wiley & Sons, Ltd, New York, 2004, Chap. 1, ISBN 0-470-86780-9.

57. Davankov, V.A., Kurganov, A.A., and Ponomareva, T.M., *J. Chromatogr.*, 452, 309–316, 1988.

58. Wainer, I.W., Brunner, C.A., and Doyle, T.D., *J. Chromatogr.*, 264, 154, 1983.

59. Grinberg, N. and Weinstein, S., *J. Chromatogr.*, 303, 251–255, 1984.

60. Ward, T.J. and Armstrong, D.W., *J. Liq. Chromatogr.*, 9, 407–423, 1986.

61. Brinkman, U.A.Th. and Kamminga, D., *J. Chromatogr.*, 330, 375–378, 1985.

62. Armstrong, D.W., Bui, K.H., and Barry, R.M., *J. Chem. Ed.*, 61, 457–458, 1984.

63. Armstrong, D.W., Faulkner, J.R., and Han, S.M., *J. Chromatogr.*, 452, 323–330, 1988.

64. Alak, A. and Armstrong, D.W., *Anal. Chem.*, 58, 582–584, 1986.

65. Heuser, D. and Meads, P., *J. Planar Chromatogr.*, 6, 324–325, 1993.

66. Olszewska, E., Kroszczyński, W., Łypacewicz, M., and Wasiak, T., *Acta Chromatographica*, 12, 219–225, 2002.

67. Lepri, L., Coas, V., and Desideri, P.G., *J. Planar Chromatogr.*, 5, 175–178, 1992.

68. Lepri, L., Coas, V., Desideri, P.G., and Pettini, L., *J. Planar Chromatogr.*, 6, 100–104, 1993.

69. Lepri, L., Coas, V., Desideri, P.G., and Zocchi, A., *J. Planar Chromatogr.*, 5, 234–238, 1992.

70. Lepri, L., Coas, V., Desideri, P.G., and Pettini, L., *J. Planar Chromatogr.*, 5, 264–367, 1992.

71. Lepri, L., Coas, V., Desideri, P.G., and Zocchi, A., *J. Planar Chromatogr.*, 7, 103–107, 1994.

72. Lepri, L., Coas, V., Del Bubba, M., and Cincinelli, A., *J. Planar Chromatogr.*, 12, 221–224, 1999.

73. Bhushan, R. and Thuku Thiong'o, G., *J. Planar Chromatogr.*, 13, 33–36, 2000.

74. LeFreve, J.W., Gublo, E.J., Botting, C., Wall, R., Nigro, A., Pham, M.-L.T., and Ganci, G., *J. Planar Chromatogr.*, 13, 160–165, 2000.

75. Tivert, A.-M. and Backman, A., *J. Planar Chromatogr.*, 3, 216–219, 1993.

76. Różyło, J.K., Siembida, R., and Jamrozek-Mańko, A., *J. Planar Chromatogr.*, 10, 225–228, 1997.

77. Kriz, D., Kriz, C.B., Anderson, L.I., and Mosbach, K., *Anal. Chem.*, 66, 2636–2639, 1994.

78. Suedee, R., Songkram, C., Petmoreekul, A., Sangkunakup, S., Sankasa, S., and Kongyarit, N., *J. Planar Chromatogr.*, 11, 272–276, 1998.

79. Zhu, Q., Yu, P., Deng, Q., and Zeng, L., *J. Planar Chromatogr.*, 14, 2137–2139, 2001.

80. Günther, K. and Möller, K., Enantiomer separations, in *Handbook of Thin-Layer Chromatography, 3rd edn.*, revised and expanded, Chromatographic Science Series, 89, Sherma, J. and Fried, B., Eds., Marcel Dekker, Inc., New York, 2003, Chap. 17.

81. Lepri, L., Del Bubba, M., and Cincinelli, A., Chiral separations by TLC, in *Planar Chromatography, A Retrospective View for the Third Millennium*, Nyiredy, Sz., Ed., Springer Scientific Publisher, Budapest, 2001, Chap. 25.

9 Selected Bottlenecks of Densitometric Detection with Chiral Analytes

Mieczysław Sajewicz and Teresa Kowalska

CONTENTS

9.1 INTRODUCTION

Modern applications of thin layer chromatography (TLC) should inevitably benefit from the best-standardized and best-performing materials and equipment. It means that the separations ought to be carried out on the commercially precoated chromatographic plates, the analyte samples ought to be spotted on to the adsorbent layer with the aid of an automatic sampling device, and the detection ought to be performed by means of densitometry. All these demands are particularly important in view of the fact that the TLC results tend to be less repeatable than in the fully automated liquid chromatographic techniques, and for multiple reasons. For example, compared with gas chromatography (GC), thin layer chromatograms are developed usually without a possibility to keep the chromatographic systems under the strict isothermal conditions. Compared with high-performance liquid chromatography (HPLC), thin layer chromatograms are in most cases developed in the so-called normal chambers, with a void space between the liquid surface of mobile phase and the chamber lid, which needs saturation with the mobile phase vapors. However, there is no possibility usually to control humidity inside the chromatographic chambers, which consequently affects saturation and makes it not exactly repeatable. Finally, in the column liquid chromatography the same packing can be utilized many times, whereas in TLC practically each analytical repetition is performed on a different sample of the adsorbent layer. In spite of all these shortcomings, TLC can still be considered as a relatively cheap, rapid, and versatile mode, well performing under the properly chosen separation conditions. Its separation performance is certainly sufficient in the case of enantioseparations, when the principal separation task is confined to resolution of the two antimers only. It can, however, happen that separation of the two different (although structurally very closely related) enantiomeric species encounters unforeseen and sometimes unsurpassable difficulties, mostly due to the phenomena of the physicochemical nature, which cannot be taken into account in advance. In this chapter, we are going to present a selection of such unpredictable effects, which sometimes accompany enantioseparations and in certain specific cases can even make such separations impossible.

9.2 2-ARYLPROPIONIC ACIDS AND THEIR SPONTANEOUS OSCILLATORY TRANSENANTIOMERIZATION

Let us start our presentation of the selected bottlenecks with enantioseparations — and consequently with the respective densitometric detection thereof — from a striking case of a spontaneous and oscillatory (i.e., repeated) transenantiomerization of the selected chiral 2-arylpropionic acids (APAs) reported in our papers [1–3].

9.2.1 TRACING OF OSCILLATORY TRANSENANTIOMERIZATION WITH THE SELECTED APAs BY MEANS OF TLC

Originally, we attempted to separate two pairs of enantiomers, namely, S,R-($\pm$)-ibuprofen and S,R-($\pm$)-2-phenylpropionic acid [1]. The two samples were dissolved in 70% ethanol and then the respective solutions were spotted with the aid of an automatic sampler on to the adsorbent layer. Thin layer chromatographic conditions that are best suited for separation of the APA enantiomers involve silica gel impregnated with L-arginine, which is kept in the cationic form, due to a properly fixed pH value (<4.8). The mobile phases used were the ternary mixtures of acetonitrile (ACN), methanol (MeOH), and H_2O (plus several drops of glacial acetic acid), in the volume proportions depending on a given pair of enantiomers. The mechanism of such separations can best be summarized with the aid of the following stoichiometric equations [4,5], which reflect the ion-pair formation between the cationic impregnant and the two APA enantiomers in the anionic form:

$$\text{L-arginine}^+ + S\text{-}(+)\text{-APA}^- \longleftrightarrow \text{L-arginine}^+ S\text{-}(+)\text{-APA}^- \quad (K_1) \qquad (9.1)$$

$$\text{L-arginine}^+ + R\text{-}(-)\text{-APA}^- \longleftrightarrow \text{L-arginine}^+ R\text{-}(-)\text{-APA}^- \quad (K_2) \qquad (9.2)$$

The separation success proved, however, rather limited, because considerably short-lasting. A prolonged (i.e., lasting several or more hours) storage of APAs dissolved in 70% ethanol resulted in disappearance of the two well-separated and symmetric chromatographic bands, each one representing a single separated anti-mer, and instead, in appearance of a single, broad, and skewed chromatographic band with a continually changing position (i.e., the R_F value) on the chromatogram. This striking and unexpected result is schematically presented in Figure 9.1.

We performed the storage experiment with $S\text{-}(+)$-naproxen as an optically pure enantiomer dissolved in 70% ethanol, and the obtained chromatographic result, in a sense, resembled those observed with the $S,R\text{-}(\pm)$-ibuprofen and the $S,R\text{-}(\pm)$-2-phenylpropionic acid enantiomer mixtures. At the beginning of the storage period, the concentration profile of $S\text{-}(+)$-naproxen appeared at a fixed and well-repeatable position (i.e., it showed a steady R_F value) and assumed a regular shape, witnessing to a homogenous sample, containing a single analyte. After a prolonged storage time, the respective R_F values of naproxen started continually

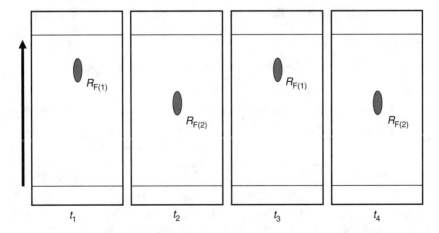

FIGURE 9.1   Schematic representation of the oscillatory changes of the R_F values, observed for each investigated APA, as a function of a prolonged storage time in 70% ethanol. (From Sajewicz, M., Piętka, R., Pieniak, A., and Kowalska, T., *Acta Chromatogr.*, 15, 131–149, 2005. With permission.)

changing in an oscillatory manner, and the concentration profile became irregular, changing with time.

To better illustrate the storage-induced and oscillatory changes in the concentration profiles (and also in the position on a chromatogram in terms of the respective R_F values) with three investigated APAs (i.e., with ibuprofen, naproxen, and 2-phenylpropionic acid), in Figure 9.2 we showed a time sequence of the densitograms in the form of characteristic "movie pictures." For the sake of brevity,

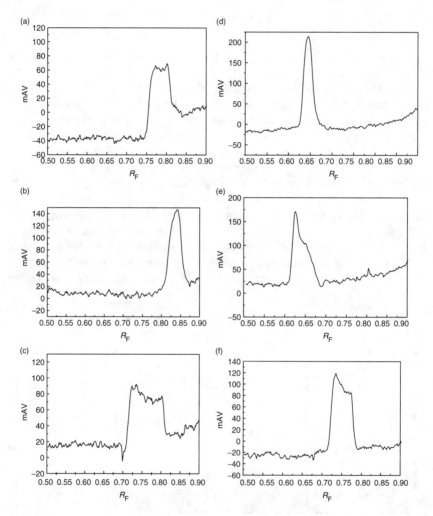

FIGURE 9.2 Sequence of densitometric concentration profiles of 2-phenylpropionic acid after storage at $22 \pm 2°C$ for (a) 0 h, (b) 22.5 h, (c) 27.5 h, (d) 46.5 h, (e) 51.5 h, and (f) 70.5 h. Changes of the consecutive peaks' concentration profiles are accompanied by the changing R_F values. (From Sajewicz, M., Piętka, R., Pieniak, A., and Kowalska, T., *Acta Chromatogr.*, 15, 131–149, 2005. With permission.)

we choose an example of 2-phenylpropionic acid only. However, it needs to be underlined that the remaining two APAs — ibuprofen and naproxen — behave in a fully analogous manner.

Summing up, an initial enantioseparation success with the two pairs of the S,R-($\pm$)-ibuprofen and the S,R-($\pm$)-2-phenylpropionic acid antimers under the applied working conditions, after a relatively short period of time, turned out to be a complete failure. However, it led to the discovery of a striking phenomenon of a moving position on the chromatograms and of the changing concentration profiles with all the employed test analytes. It was very carefully checked in the multiple additional investigations that none of these test analytes underwent a destructive process in the course of their storage. From chemical literature, it is also a well-established fact that APAs are practically indestructible, when stored in the nonreactive solvents (e.g., 70% ethanol) and under a mild temperature regime (like in the range from 6 to 22°C, as it was the case in our experiments).

To elucidate the real nature of the chromatographic phenomenon observed, we performed a thorough optical rotational study of the APA solutions in 70% ethanol by means of polarimetry, and this experiment fully confirmed our earlier TLC finding. Using the classical optical method (i.e., polarimetry), the oscillatory change of the specific rotation ($[\alpha]_D$) of ibuprofen, naproxen, and 2-phenylpropionic acid (stored for longer periods of time as solutions in 70% ethanol at two different storage temperatures) was clearly demonstrated. An example of such changes is shown for naproxen stored for 300 min at 6 ± 2°C (Figure 9.3). The remaining two APAs, that is, ibuprofen and 2-phenylpropionic acid, showed the analogous

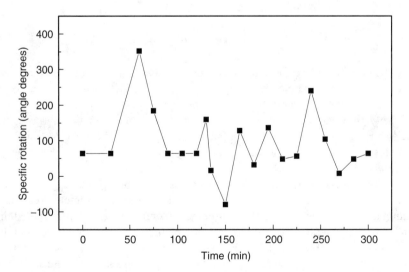

FIGURE 9.3 Dependence of the specific rotation $[\alpha]_D$ on the S-(+)-naproxen solution storage time ($[\alpha]_D = f(t)$) under refrigeration (6 ± 2°C) as a 70% solution in ethanol. (From Sajewicz, M., Piętka, R., Pieniak, A., and Kowalska, T., *Acta Chromatogr.*, 15, 131–149, 2005. With permission.)

oscillatory changes of their respective specific rotation ($[\alpha]_D$) values. The amplitudes of these oscillations were for each individual APA higher at 6°C than at 22°C. This characteristic trait can probably be related to the following two factors: (a) to the temperature-induced difference in kinetics of a process resulting in the oscillatory changes of the analytes' specific rotation and (b) to the difference in viscosity between the respective APA solutions kept at two different temperatures, which, with the most solutions, is expected to be lower at higher temperatures and higher at lower temperatures.

The analogous, although considerably less pronounced changes in the respective R_F values, the chromatographic peaks' concentration profiles, and the specific rotation ($[\alpha]_D$) values were established by means of polarimetry and TLC with densitometric detection in the case of ibuprofen, naproxen, and 2-phenylpropionic acid, when stored as solutions in dichloromethane and physiological salt [2,3]. Finally, we arrived at a conclusion that the true reason for all these oscillatory changes was a spontaneous and oscillatory transenantiomerization of the investigated APA solutions in the spirit of the classical Zhabotinskii–Belousov type oscillatory reactions.

Our next step was to reflect on the molecular mechanism of transenantiomerization of the APAs solutions, first observed in the TLC experiments, and then by means of polarimetry. In the literature [6], we found a report on a proven (and the base-catalyzed) ibuprofen racemization mechanism through the keto–enol tautomerism. We considered it as a convincing enough starting point for explanation of the oscillatory change of the R_F values and also of the concentration profiles with three chiral APAs, acting in our experiments as the test analytes. This molecular mechanism is given in the following scheme:

$$R\text{-}(-)\text{-APA} \longleftrightarrow \text{keto–enol} \longleftrightarrow S\text{-}(+)\text{-APA} \qquad (9.3)$$

We then decided to perform an independent experiment to additionally confirm the importance of the environment (and more precisely, of either its acidity or basicity) for the oscillatory transenantiomerization of ibuprofen, naproxen, and 2-phenylpropionic acid, perceived in our earlier studies. For this purpose, we selected S-(+)-naproxen as the best performing out of the three test species. The results of our experiment are extensively described in the paper cited under Reference [7].

In the classical literature on the mechanisms of the organic reactions, it is a well-established fact that the basic environment generally promotes keto–enol tautomerism, whereas the acidic environment hampers it. Thus, we prepared the two different solutions of S-(+)-naproxen, one in the mixture of ethanol and the pH 9 buffer (70:30, v/v) and the other one in the mixture of ethanol and the glacial acetic acid (70:30, v/v). Then these two S-(+)-naproxen solutions were stored for 5 h each. In the regular time intervals they were controlled by means of TLC and polarimetry, and finally the samples were chromatographed in two directions, using the ternary mobile phase ACN/MeOH/H$_2$O (5:1:1.5, v/v/v), containing several drops of glacial acetic acid to fix the pH at <4.8. The respective chromatograms

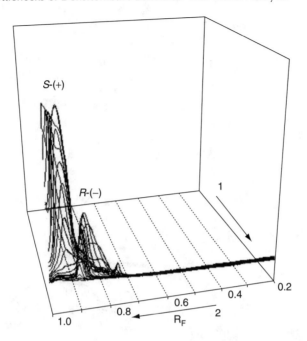

FIGURE 9.4   Three-dimensional presentation of the naproxen chromatogram with two development directions, 1 and 2, indicated. Densitometric scanning (at parallel 1.5-mm intervals) of the 35-mm-wide track in the second direction of the development was performed to better illustrate the separation performance and the skewed arrangement of S-(+)-naproxen relative to its R-(−) counterpart for the S-(+)-naproxen sample stored for 5 h in the EtOH–basic buffer mixture (pH 9; 70:30, v/v). (From Sajewicz, M., Piętka, R., Drabik, G., and Kowalska, T., *J. Liquid Chromatogr. Rel. Technol.*, 29, 2071–2082, 2006. With permission.)

were developed in two directions by using two mobile phases having exactly the same composition, the second development direction being perpendicular to the first one. Then the 35-mm-wide track, developed in the second direction, was scanned densitometrically in 1.5 mm intervals. The results obtained enabled presentation of the 3D chromatograms of naproxen, when stored in the basic and the acidic medium (Figures 9.4 and 9.5).

The asymmetric concentration profile of S-(+)-naproxen, when stored in the basic solution for 5 h and then chromatographed in the first direction signalized an efficient, base-catalyzed transenantiomerization, and also showed a possibility of attaining a complete separation of the S-(+)- and the R-(−)-antipode, if the samples were chromatographed in the second direction as well. In fact, full separation of S-(+)- and R-(−)-naproxen was attained in the 2D development of the chromatogram (as shown in Figure 9.4).

Contrary to the results obtained, when S-(+)-naproxen has been stored in the basic environment, the acidic mixed solvent caused no measurable transenantiomerization of the test analyte considered. In Figure 9.5, we gave the

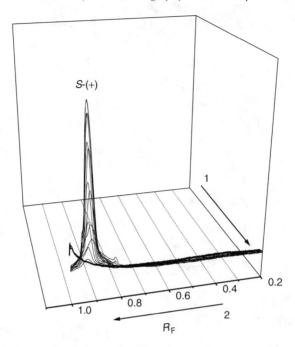

FIGURE 9.5 Three-dimensional presentation of the naproxen chromatogram with two development directions, 1 and 2, indicated. Densitometric scanning (at parallel 1.5-mm intervals) of the 35-mm-wide track in the second direction of the development was performed to better illustrate an absolute lack of transenantiomerization for the S-(+)-naproxen sample stored for 5 h in the EtOH/glacial acetic acid mixture (70:30, v/v). (From Sajewicz, M., Piętka, R., Drabik, G., and Kowalska, T., *J. Liquid Chromatogr. Rel. Technol.*, 29, 2071–2082, 2006. With permission.)

3D presentation of the chromatogram of S-(+)-naproxen dissolved and stored in the acidic medium (development in the second direction). In this figure, we can see the single 3D peak of S-(+)-naproxen only. In that way, we provided sufficient experimental evidence in favor of the base-catalyzed keto–enol tautomerism as a driving force for the configuration change of the chirally pure S-(+)-naproxen.

In the case of our initial and unsuccessful TLC attempts to enantioseparate the S,R-(±)-ibuprofen and the S,R-(±)-2-phenylpropionic acid antipodes [1–3], we kept our test samples for a longer period of time dissolved in 70% ethanol (and also in dichloromethane and physiological salt). Evidently, none of these solvents can be considered as a base or an acid, at least not in the spirit of the acid and base definitions introduced by Arrhenius. In other words, none of these solvents can catalyze or hamper transenantiomerization of the chiral APAs. However, the 70% ethanol solvent can easily be viewed as a weak ampholyte, able to simultaneously exert the catalytic and inhibiting effect on transenantiomerization of the chiral analytes considered. Perhaps, this perceptible ampholytic nature of 70% ethanol (combined with a change in viscosity of the APA solutions, as related to that of the pure solvent and reported in literature [8]) is responsible for the oscillatory transenantiomerization of our test analytes. In this case, it seems fully justified to expect that the molecular mechanism involving keto–enol tautomerism is responsible for the oscillatory transenantiomerization of the compounds discussed.

Bottleneck 1. The investigated pairs of enantiomers — as it is the case, for example, with the chiral APAs — can occasionally demonstrate a striking chromatographic behavior, which makes their enantioseparation unrepeatable in terms of the measured retention parameters, incomplete, or even impossible. Such effects are certainly not due to an insufficient separation performance of the TLC systems involved, but because of the chemical processes themselves (e.g., the oscillatory transenantiomerization, uninterruptedly running within the investigated samples). Sometimes, the lack of a good separation result can be counterbalanced by an evident applicability of TLC to the advanced physicochemical studies (even as a replacement for polarimetry, as was shown in this section).

9.2.2 Two-Dimensional Retention of the Selected APAs on Chiral Stationary Phase in the 1D Development Run

The phenomenon of a slight deviation of the analytes' migration tracks from strict verticality has been long recognized and well described in the TLC literature, and it is almost omnipresent in TLC. Usually, it occurs due to random distribution of sorbent particles and to the resulting randomness of the capillaries' arrangement in the solid bed of stationary phases, and it has absolutely nothing to do with the chirality of the analytes or other components of the chromatographic systems. It has been even more commonly observed in the past decades, when the home-coated glass plates were incomparably more common than they are now. Today this phenomenon is to a large extent extinguished or at least greatly suppressed by the better standardized commercial TLC plates. Deviations from verticality that are due to the random arrangement of sorbent particles are also random and they can differ from one plate to another. The difference can consist both in the direction of deviation (which can randomly be left- or right-handed) and in its magnitude (which is restricted to 1 or 2 mm the most).

On the contrary, in our study on thin layer chromatographic behavior of the selected pairs of the APA antipodes [i.e., of S,R-($\pm$)-ibuprofen, S,R-($\pm$)-naproxen, and S,R-($\pm$)-2-phenylpropionic acid] [9], we observed a "stereopeculiar," direction-wise deviation of the analytes' migration tracks from the strict verticality that was not random, but systematic. Moreover, the magnitudes of these deviations were perceptibly higher than those in the random cases. Our experiments were carried out in the chromatographic systems described in References [1–3,5] and also in Section 9.2.1. Thus the stationary phase, the best suited for the separation of APA antipodes, consisted of the silica gel impregnated with L-arginine and kept in the cationic form, due to a properly fixed pH value (<4.8). The applied mobile phases were the ternary mixtures of ACN, MeOH, and H_2O (plus several drops of glacial acetic acid), in the volume proportions depending on a given pair of antipodes. The mechanism of such separations can be best summarized with the aid of stoichiometric equations (9.1) and (9.2), which reflect the ion-pair formation between the cationic impregnant and the two APA antipodes in the anionic form.

The striking "stereopeculiar" specificity of the 2D separation of the selected pairs of the APA antipodes in the 1D development run had been confirmed by a wide number of thorough preliminary experiments that preceded proper and systematic investigations. At that preliminary stage, we have even checked the evenness and strict horizontality of the laboratory bench top with a water-level to eliminate the surprise external factors, which might negatively affect our results. Moreover, we saw to it that there was no disturbing air stream around the development chamber. In that manner, the influence of the nonchirality-based factors on an ultimate measuring result seems to have been entirely excluded.

The main results of our investigations on "stereopeculiar" migration tracks of the antimer pairs of ibuprofen, naproxen, and 2-phenylpropionic acid, when developed in the chiral chromatographic systems, are summarized in Table 9.1 and illustrated upon a selected example of the 2-phenylpropionic acid antimers in Figure 9.6.

TABLE 9.1

Deviation (in Terms of Handedness and Magnitude) from the Strict Verticality of the Enantiomer Migration with the Two Antipodes of Ibuprofen, Naproxen, and 2-Phenylpropionic Acid[a]

Analyte	Chiral configuration	R_F^b	Deviation from verticality[b] (mm)	Handedness of the deviation
Ibuprofen	S-(+)	0.91 (±0.02)	2 (±1)	Right
	R-(−)	0.88 (±0.02)	2 (±1)	Left
Naproxen	S-(+)	0.89 (±0.02)	3 (±1)	Left
	R-(−)	0.85 (±0.02)	3 (±1)	Right
2-Phenylpropionic acid	S-(+)	0.90 (±0.02)	5 (±1)	Right
	R-(−)	0.80 (±0.02)	2 (±1)	Left

[a] Stationary phase: silica gel 60 F_{254} (precoated plates, cat. no. 1.05715; Merck). Mobile phase: ACN/MeOH/H_2O; quantitative composition for ibuprofen (5:1:1, v/v/v); naproxen (5:1:1.5, v/v/v), and 2-phenylpropionic acid (5:1:0.75, v/v/v). Migration distance of mobile phase: 15 cm.

[b] The presented numerical results were derived from 27 individual enantiomer migration tracks (i.e., from three chromatographic plates, nine separate development lanes per plate).

Source: From Sajewicz, M., Pietka, R., Drabik, G., Namysło, E., and Kowalska, T., *J. Planar Chromatogr. — Mod. TLC*, 19, 273–277, 2006. With permission.

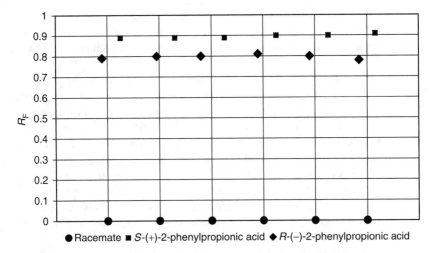

FIGURE 9.6 Schematic representation of the direction-wise deviation from verticality of the migration tracks with the antimers of 2-phenylpropionic acid. Stationary phase: silica gel 60 F_{254} (precoated plates, cat. no. 1.05715; Merck), impregnated with L-arginine. Mobile phase: ACN/MeOH/H$_2$O (5:1:0.75, v/v/v). (From Sajewicz, M., Piętka, R., Drabik, G., Namysło, E., and Kowalska, T., *J. Planar Chromatogr. — Mod. TLC*, 19, 273–277, 2006. With permission.)

Data shown in Table 9.1 well characterize the magnitude and the direction of the maximum deviations from verticality with the three-antimer pairs, originating from three different APAs considered. In the case of ibuprofen, vertical resolution of the two antimers is the lowest one ($\Delta R_F = 0.03$ R_F units only) and the maximum sum of the left- and the right-handed (i.e., horizontal) deviation was established as equal to 4 (± 2) mm. Paradoxically, a similar result had already been presented in the form of a photograph in the literature before (see Figure 9.1A in literature [4]), although at that time it has passed unnoticed, or at least not commented by the authors.

In the case of the remaining two APAs, the respective vertical resolution of the antimer pairs was higher ($\Delta R_F = 0.04$ for naproxen and $\Delta R_F = 0.10$ for 2-phenylpropionic acid). Accordingly, the observed maximum deviations of the migration tracks were also higher. With naproxen, the maximum sum of the left- and the right-handed (i.e., horizontal) deviation was equal to 6 (± 2) mm, and in the case of 2-phenylpropionic acid, the analogous value was equal to 7 (± 2) mm.

It is also noteworthy that with ibuprofen and 2-phenylpropionic acid (both having the asymmetric carbon atom substituted by the phenyl group), the S-$(+)$ antimers demonstrated the right-handed deviation from verticality and the tracks of their R-$(-)$ counterparts were left-handed. In the case of naproxen (having the asymmetric carbon atom substituted by the naphthyl group), the opposite regularity was observed. Namely, S-$(+)$-naproxen demonstrated the left-handed deviation of its migration track and R-$(-)$-naproxen deviated to the right. These regularities can prove quite important for the future molecular-level considerations regarding stereospecificity of interactions between L-arginine as the silica-gel-deposited chiral selector on the one hand and each individual APA in the anionic form on the other.

Taking advantage of the densitometric scans of all the chromatographic plates in 1 mm intervals, we managed to reconstruct the 3D concentration profiles of the two antimers of naproxen (Figure 9.7a and b) and of 2-phenylpropionic acid (Figure 9.8). We purposely made these reconstructions to better visualize the concentration profiles of the respective antimers that show the left- and the right-handed skewness, no doubt due to the horizontal deviation of the respective migration tracks. In the case of ibuprofen, 1D chromatographic development resulted in an insufficient vertical separation of the two antimers ($\Delta R_F = 0.03$) to draw an analogous 3D picture. Better resolution is only possible with the 2D development of the S,R-($\pm$)-ibuprofen sample, as reported in literature [1,4,5,10]. The 3D picture of the ibuprofen antipodes, better resolved by means of 2D TLC, is shown in the paper cited under Reference [1].

The results presented in this section provide a preliminary molecular-level insight into the retention mechanism of APAs in the chiral thin layer chromatographic systems with L-arginine as chiral selector. Namely, the experimental evidence was produced, showing that the impact of L-arginine on the separation of the three pairs of the APA antimers is 2D, that is, separating these pairs in the vertical direction of the mobile phase flow, and also in the horizontal direction, perpendicular to the former one. This two-dimensionality of the antimers' separations certainly enhances an overall separation effect.

It can rightfully be expected that further investigations on the detailed mechanism of the enantiomer retention in chiral systems can reveal some new and interesting features regarding the impact of the localized intermolecular interactions on separation of the antimer pairs. The preliminary hint is already contained in the observation reported in this section as to the systematic right-handed deviation from verticality with the S-(+) enantiomers of ibuprofen and 2-phenylpropionic acid, and to the systematic left-handed deviation from verticality with the S-(+) enantiomer of naproxen. From this observation, it can further be deduced that the type of the aryl substituent at the asymmetric carbon atom of APA — in our experiment either phenyl or naphthyl — plays an important role in the mechanism of retention, as considered from the molecular-level perspective.

Bottleneck 2. From the results discussed in this section it clearly comes out that densitometric detection applied to the thin layer chiral separations has to be performed in a very cautious and also very conscious manner, that is, anticipating the possibility of quite significant deviations of the chiral analytes' migration tracks from the straight-line direction. In this sense, the present section contains a troubleshooting advice addressed to all those, who use TLC with densitometric detection for chiral separations and — what is even more important — for quantification of the analyzed mixtures' composition. The recommended troubleshooting solution is to scan relatively broad bands on the respective chromatograms instead

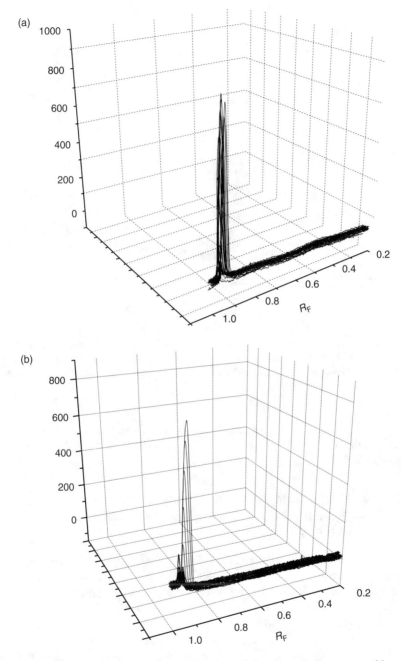

FIGURE 9.7 Three-dimensional representation of the skewed chromatographic peak shapes of (a) *S*-(+)-naproxen (left-handed) and (b) *R*-(−)-naproxen (right-handed). (From Sajewicz, M., Piętka, R., Drabik, G., Namysło, E., and Kowalska, T., *J. Planar Chromatogr. — Mod. TLC*, 19, 273–277, 2006. With permission.)

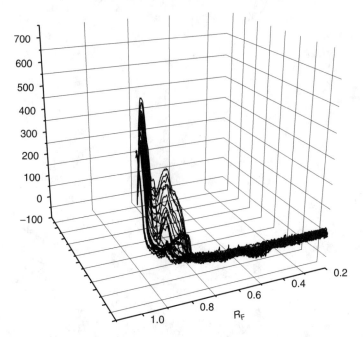

FIGURE 9.8 Three-dimensional presentation of the skewed chromatographic peak shapes of *S*-(+)-2-phenylpropionic acid (right-handed) and *R*-(−)-2-phenylpropionic acid (left-handed). (From Sajewicz, M., Piętka, R., Drabik, G., Namysło, E., and Kowalska, T., *J. Planar Chromatogr. — Mod. TLC*, 19, 273–277, 2006. With permission.)

of a narrow lane along the prospective straight-line migration track of the enantiomers to find the analyte's best pronounced (in terms of the height and the area) concentration profile.

9.2.3 OSCILLATORY CHANGES IN THE UV ABSORPTION SPECTRA AND THEIR IMPACT ON DENSITOMETRIC QUANTIFICATION OF THE SELECTED APAs

As mentioned in Sections 9.2.1 and 9.2.2, in the case of the selected APAs (i.e., of ibuprofen, naproxen, and 2-phenylpropionic acid) dissolved both in the aqueous and the nonaqueous solvents, the oscillatory transenantiomerization occurs, which can schematically be illustrated by equation (9.3) (see Section 9.2.1). The process is spontaneously carried out and it consists of a repeated and alternate transformation of the respective *S*-(+) enantiomers into their *R*-(−) antipodes and vice versa, via the keto–enol tautomerism. This process can be easily traced by means of TLC (Sections 9.2.1 and 9.2.2) and even easier by means of HPLC (which is beyond the scope of this book), and in the first instance with the aid of polarimetry, an optical technique of a paramount importance for any research in the field of the crystalline and molecular chirality.

Formation of keto–enols as intermediates in many organic reactions is known to be vigorously catalyzed in the basic environment and effectively hindered in the acidic one. The ethanol-aqueous environment, most often utilized in our experiments with the APAs, can rightfully be described as ampholytic. It seems that ampholytic environment combined with certain growth of viscosity of the APA solutions (in comparison with that of the pure solvents) [8] sufficiently fulfils the requirements for oscillatory transenantiomerization of the chiral analytes discussed.

In the course of our investigations, we found out that the UV absorption spectra of S-(+)-ibuprofen and S-(+)-naproxen, dissolved in several inert solvents that do not absorb light from the UV region above 200 nm [e.g., 70 and 95% ethanol, ACN, and tetrahydrofuran (THF)] showed a marked tendency to change both the position and the intensity of the respective absorption bands in the course of their prolonged storage [11]. It seems that this unexpected and even striking effect can only be attributed to the formation of measurable proportions of keto–enols in the investigated solutions, because both antimers from a given enantiomer pair are supposed to have the identical (or almost) UV spectra. In the two forthcoming subsections, the phenomenon of transformation of the spectral envelopes and the respective peak positions will be briefly illustrated by a comparison of the UV spectra of the selected APAs, measured after the different storage periods (Section 9.2.3.1). Then, we will discuss a possible negative impact of this phenomenon on the quality of densitometric quantification of the APA chromatograms (Section 9.2.3.2).

9.2.3.1 Oscillatory Changes in the UV Absorption Spectra of the Selected APAs

In this section, we are going to show — for the sake of brevity — a small choice only of the UV spectra measured for the samples of S-(+)-ibuprofen and S-(+)-naproxen dissolved in ethanol/water and in THF, and then stored for up to 4 h at $22 \pm 2°C$ (see Figures 9.9 through 9.11) [11]. In the 10-min intervals, the UV spectra were recorded by means of the UV-Vis spectrophotometer for the investigated solutions, which clearly showed the changes of the respective peak positions and of the intensities thereof. It seems rather evident to us that all these surprise changes and shifts — considerably surpassing the measurement error — are in a way due to oscillatory transenantiomerization of the investigated APAs in each of the selected solvents. With the compounds that are not chiral and apparently not undergoing any oscillatory structural transformation, the respective UV spectra repeatedly recorded by using our equipment remained practically unchanged. The most vigorous changes always took place during the first hour after dissolution of a given APA in a given solvent and then the situation slightly, although not completely, stabilized (because the less vigorous fluctuations could be observed all the time). At the moment, the only sensible explanation seems to be that within the first hour — in spite of the oscillatory transenantiomerization running in the entity of each investigated solution for many hours and even days — concentration

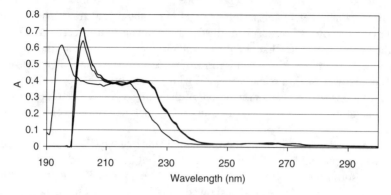

FIGURE 9.9   The changing UV spectra of the ibuprofen sample dissolved in 95% ethanol and stored at $22 \pm 2°C$, and repeatedly recorded for several hours. (Sajewicz, M. and Kowalska, T., Private communication.)

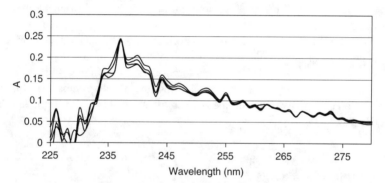

FIGURE 9.10   The changing UV spectra of the ibuprofen sample dissolved in THF and stored at $22 \pm 2°C$, and repeatedly recorded for several hours. (Sajewicz, M. and Kowalska, T., Private communication.)

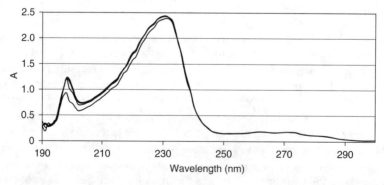

FIGURE 9.11   The changing UV spectra of the naproxen sample dissolved in 70% ethanol and stored at $22 \pm 2°C$, and repeatedly recorded for several hours. (Sajewicz, M. and Kowalska, T., Private communication.)

of an intermediate keto–enol structure attains a quasi-stationary level. The changes observed in the respective spectra might be due to superposition of the identical (or almost identical) UV spectra of the two antimers for a given APA with that of a corresponding keto–enol, which has to be markedly different.

9.2.3.2 The Impact of the Oscillatory Changes in the UV Absorption Spectra on Densitometric Detection of the Selected APAs

Oscillatory transenantiomerization of the three investigated APAs is a phenomenon that can most probably occur not only with ibuprofen, naproxen, and 2-phenylpropionic acid, but also with other APAs and with a number of different chiral compounds as well, provided their chemical structure permits such fluctuations. The quantification error resulting from the observed changes in the UV absorption spectra most probably adds to those (a) due to an oscillatory change of the quantitative proportions between the S- and the R-species (see Section 9.2.1), followed by (b) the deviation of the discussed analytes' migration tracks from the strict verticality, with a simultaneous horizontal split between the S- and the R-antimer (see Section 9.2.2), thus contributing to a further growth of an overall quantification error. The source of an error due to the fluctuation of the UV absorption spectra of the investigated compounds can briefly be characterized in the following way. Even if an intermediate keto–enol structure cannot be isolated and separately investigated, it seems indisputable that the molar extinction coefficient of a given APA (either of its S- or R-antimer) is markedly different from the molar extinction coefficient of its keto–enol counterpart, due to the change of the position of the double bond in a molecule between any given APA and its keto–enol derivative. Even if one cannot precisely differentiate among the three tightly interconnected sources of the densitometric quantification error with densitometry carried out in UV light, the error due to the difference between the molar extinction coefficients of the two APA antimers on the one hand and that of the respective keto–enol on the other hand seems — due to its physical nature — to be the most acute one. In the forthcoming paragraphs of this section, selected examples will be given of the quantification errors with ibuprofen and naproxen.

Before we present the results of the densitometric quantification error in TLC, let us start from showing the results of the UV detection of ibuprofen and naproxen in HPLC. In physicochemical terms, there is full analogy between the detection problems in HPLC and in TLC/densitometry, and a clear and neat presentation of these problems is even easier and more spectacular in the former case than in the latter one. In Figures 9.12 and 9.13, we showed selections of several superimposed UV spectra for the ibuprofen and naproxen samples, respectively, dissolved in 70% ethanol and then automatically injected to the C-18 chromatographic column by an auto sampler. The retention process was carried out in the RP-HPLC system employing the aqueous mobile phase. In the regular time intervals, the UV spectra were recorded for the peak maxima of the eluted ibuprofen and naproxen, with the

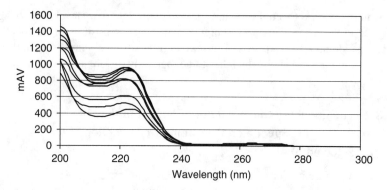

FIGURE 9.12 The changing UV spectra of the ten ibuprofen samples dissolved in 70% ethanol and then injected to the chromatographic system by means of auto sampler in equal quantities and at regular time intervals; the RP-HPLC system employed the C-18 type stationary phase and the ACN/H$_2$O (6:4, v/v) mobile phase. (Sajewicz, M. and Kowalska, T., Private communication.)

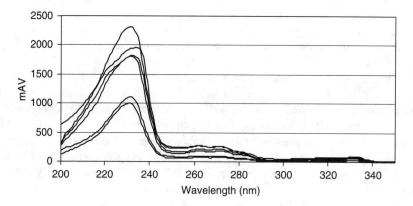

FIGURE 9.13 The changing UV spectra of the six naproxen samples dissolved in 70% ethanol and then injected to the chromatographic system by means of auto sampler in equal quantities and at regular time intervals; the RP-HPLC system employed the C-18 type stationary phase and the ACN/H$_2$O (6:4, v/v) mobile phase. (Sajewicz, M. and Kowalska, T., Private communication.)

equal concentrations and volumes of these analytes in each individual injection from a given series.

As it can be seen from the UV spectra of ibuprofen (Figure 9.12) and naproxen (Figure 9.13), recorded by means of the DAD detector in regular time intervals, again fluctuations can be observed in the peak positions and intensities, as it was first shown in Figures 9.9 through 9.11 with the selected UV spectra of ibuprofen and naproxen, recorded by means of a regular spectrophotometer. Obviously, such unstable UV spectra cannot provide a good enough basis for a reliable quantification of the APAs in the HPLC systems.

TABLE 9.2

The Average Peak Heights and the Respective Standard Deviations (SDs) for Ibuprofen, as Densitometrically Quantified at the Three Different Wavelengths λ = 200, 205, and 210 nm

| Plate number | The average peak height ($\pm$SD) (mAV) at the wavelength λ (nm) | | |
	200	205	210
1	465.58 ($\pm$28.04)	465.60 ($\pm$27.63)	465.69 ($\pm$26.46)
2	561.56 ($\pm$75.01)	562.08 ($\pm$75.22)	552.2 ($\pm$74.75)

Note: The measurements were carried out for the two chromatographic plates (Nos. 1 and 2), ten equal sample aliquots applied per plate (N = 10).

Source: Sajewicz, M. and Kowalska, T., Private communication.

An even worse situation is inherent of the quantitative TLC analysis, due to its generally lower precision than it is the case with the fully automated chromatographic techniques. To adequately illustrate this problem, let us quote the results of quantification of ibuprofen and naproxen each dissolved in 70% ethanol and then automatically spotted on to the adsorbent layers. We present the results obtained from the different chromatograms, with the equal sample aliquots spotted per plate. After the development of the chromatographic plates, each development lane on each chromatogram was densitometrically scanned at the following three wavelengths: λ = 200, 205, and 210 nm with ibuprofen, and λ = 202, 215, and 225 nm with naproxen. In the case of each lane, the peak height in its maximum was measured. The obtained average peak heights and the respective standard deviations (SD) of the measured values are shown in Table 9.2.

As it can be easily seen from the data collected in Tables 9.2 and 9.3, SD of the measured peak heights was in the case of each individual chromatographic plate and each UV light wavelength rather considerable.

In the case of ibuprofen, the numerical values of SD at λ = 200, 205, and 210 nm were — due to a relatively narrow region of the compared wavelengths — very similar. In the case of plate 1, the SD values were ca. $\pm$6% of the respective measured values, and in the case of plate 2, the analogical SD values were even higher, that is, ca. $\pm$13% of the respective measured values.

With naproxen, the relatively highest numerical values of SD were observed at λ = 202 nm (ranging up to ca. $\pm$9% of the measured value) and the lower SD values were characteristic of the densitograms recorded at the wavelengths λ = 215 and 225 nm (ranging up to ca. $\pm$6% of the measured values). These observations remain in good agreement with the pictures illustrating fluctuations of the UV spectra of ibuprofen and naproxen, as shown in Figures 9.9 and 9.11, respectively. From these figures, it clearly comes out that at lower wavelengths, these fluctuations are more strongly pronounced than in the higher wavelength regions.

TABLE 9.3

The Average Peak Heights and the Respective Standard Deviations (SDs) for Naproxen, as Densitometrically Quantified at the Three Different Wavelengths λ = 202, 215, and 225 nm

	The average peak height ($\pm$SD) (mAV) at the wavelength λ (nm)		
Plate number	202	215	225
1	177.53 ($\pm$10.78)	337.74 ($\pm$17.43)	285.18 ($\pm$13.03)
2	422.17 ($\pm$26.18)	653.99 ($\pm$24.75)	555.49 ($\pm$20.10)
3	234.97 ($\pm$21.10)	427.59 ($\pm$22.32)	354.50 ($\pm$20.12)
4	416.77 ($\pm$13.66)	661.47 ($\pm$18.29)	569.61 ($\pm$16.37)

Note: The measurements were carried out for the four chromatographic plates (Nos. 1–4), five equal sample aliquots applied per plate ($N = 5$).

Source: Sajewicz, M. and Kowalska, T., Private communication.

The analogical densitometric quantification of the nonchiral analytes (i.e., those that do not undergo an oscillatory transenantiomerization and, hence, show no resulting fluctuation of their respective UV absorption spectra) is relatively more accurate and the respective SD values per chromatographic plate are usually confined to the range up to $\pm$2% only.

Bottleneck 3. Oscillatory transenantiomerization of the investigated APA solutions in ethanol/water and also in the selected nonaqueous solvents is responsible for distinct changes in the UV absorption spectra of these compounds in the course of their prolonged storage. The observed changes are perceptible in terms of the peaks' intensity (the area and height) and position on the wavelength scale. As densitometric detection of the thin layer chromatograms is most often carried out in the UV wavelength region, any uncontrolled and spontaneous change of the analytes' UV spectra will negatively affect the respective calibration curves and more generally, the quality of densitometric quantification of the chromatograms.

9.3 TWO-DIMENSIONAL RETENTION OF THE SELECTED APAS ON SILICA GEL IN THE 1D DEVELOPMENT RUN AS EVIDENCE OF ITS CHIRALITY

Sajewicz et al. [12] managed to demonstrate yet another case of the deviation from verticality of the migration tracks with the two chiral APAs [i.e., with S-(+)-ibuprofen and S-(+)-naproxen dissolved in the acidified 70% ethanol solvent], this time when chromatographed on the ready-made chromatographic plates precoated with the plain silica gel. Now let us thoroughly discuss this case.

Keeping in mind that (a) the deviation of the analyte's migration track can only occur with a chiral analyte developed in an asymmetric chromatographic system

TABLE 9.4

Deviation (in Terms of Handedness and Magnitude) from the Strict Vertic- ality of the Migration Tracks with S-(+)-Ibuprofen and S-(+)-Naproxen[a]

Analyte	R_F^b	Deviation from verticality[b] (mm)	Handedness of deviation
S-(+)-ibuprofen	0.91 ($\pm$0.01)	5 ($\pm$1)	Right
S-(+)-naproxen	0.90 ($\pm$0.02)	3 ($\pm$1)	Left

[a] Stationary phase: silica gel 60 F_{254} (precoated plates, cat. no. 1.05715; Merck). Mobile phase: ACN/MeOH/H$_2$O; quantitative composition for S-(+)-ibuprofen (5:1:1, v/v/v) and for S-(+)- naproxen (5:1:1.5, v/v/v). Migration distance of mobile phase: 15 cm.

[b] The presented numerical results were derived from 27 individual migration tracks per analyte (i.e., from the three chromatographic plates, nine separate development lanes per plate).

Source: From Sajewicz, M., Hauck, H.-E., Drabik, G., Namysło, E., Głód, B., and Kowalska, T., *J. Planar Chromatogr. — Mod. TLC*, 19, 278–281, 2006. With permission.

and that (b) silicon dioxide (chemically identical with the chromatographic silica gel) can crystallize as quartz (i.e., as the rock crystal) in the two asymmetric — left-handed and right-handed — forms, we assumed the working hypothesis that the chromatographic silica gel layers can also appear as having an asymmetric microcrystalline structure, able to deviate the migration tracks of chiral analytes from verticality.

In our experiment, the direction-wise deviations of the migration tracks with S-(+)-ibuprofen and S-(+)-naproxen from the strict verticality were not random, but systematic; moreover, the magnitude of these deviations was higher than that might be expected in random cases. The respective numerical data are given in Table 9.4 and the phenomenon is schematically illustrated in Figure 9.14.

With S-(+)-ibuprofen, we observed the right-handed and well-pronounced horizontal deviation (5 $\pm$ 1 mm over the 15-cm-long developing distance). With S-(+)-naproxen, the horizontal deviation was to the contrary, left-handed, and again clearly above the sheer randomness level (3 $\pm$ 1 mm along the analogous developing distance). This specificity has been confirmed by the thorough addi- tional experiments that preceded proper and systematic investigations, reported in this study.

Systematic and well-pronounced deviation from verticality of the migration tracks both with S-(+)-ibuprofen and S-(+)-naproxen certainly needs some sort of explanation, even if it is not possible for the time being to furnish an exhaust- ive molecular-level insight. It seems rather obvious that one reason is structural asymmetry of the employed analytes. The second reason must be certain kind of asymmetry inherent of the chromatographic system involved. With the nonchiral components of the employed mobile phases, it can only be the case with the stationary phase.

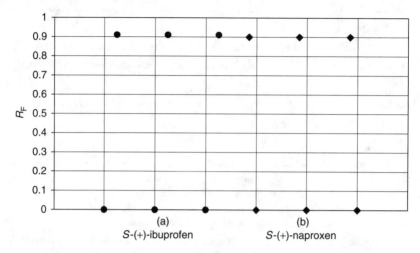

FIGURE 9.14 Schematic representation of the direction-wise deviation from vertical-ity of the migration tracks with (a) S-(+)-ibuprofen and (b) S-(+)-naproxen. Stationary phase: silica gel 60 F_{254} (precoated plates, cat. no. 1.05715; Merck). Mobile phase: ACN/MeOH/H$_2$O, (a) (5:1:1, v/v/v) and (b) (5:1:1.5, v/v/v), plus several drops of gla-cial acetic acid in each case. (From Sajewicz, M., Hauck, H.-E., Drabik, G., Namysło, E., Głód, B., and Kowalska, T., *J. Planar Chromatogr. — Mod. TLC*, 19, 278–281, 2006. With permission.)

To verify our presumption, we run the spectra of circular dichroism (CD) for the samples of the binder-free silica gel used for coating of the commercial chroma-tographic plates, and separately of the binder itself. As expected, the CD spectrum of the binder did not reveal any measurable Cotton effects, thus eliminating this particular substance from our further considerations. In the case of the investigated silica gel samples devoid of the binder, the two well-pronounced Cotton bands — one positive and one negative — were revealed in the spectra of each sample (as shown in Figure 9.15).

It is clear that silicon dioxide does not absorb in the UV light range and hence it cannot furnish the Cotton effect either. Moreover, it is not so easy to expect that precipitation of silica gel is stereospecific. It seems quite probable though that the microcrystalline silica gel matter is constituted of the right-handed and the left-handed particles, with quantitative predominance of one enantiomeric species over its mirror image. This thermodynamically possible predominance of one asymmetric form of silica gel over its antipode seems to be the cause of deviation of the migration track with our two APAs [i.e., with S-(+)-ibuprofen and S-(+)-naproxen], used as the test solutes.

How are we going to explain the two well-developed Cotton bands then (and of course, the UV absorption as a precondition for their appearance) in the invest-igated silica gel samples? The origin of these bands seems not very hard to explain. The silica gel samples can be contaminated with trace amounts of the adsorbed nonchiral organic compounds, having the chromophoric functionalities (e.g., the

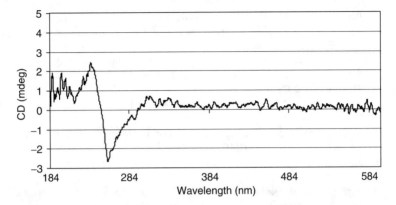

FIGURE 9.15   The CD spectrum of the binder-free silica gel for TLC sample recorded in the nujol suspension. (From Sajewicz, M., Hauck, H.-E., Drabik, G., Namysło, E., Głód, B., and Kowalska, T., *J. Planar Chromatogr. — Mod. TLC*, 19, 278–281, 2006. With permission.)

carbonyl groups). These compounds can be, for example, the adsorbed organic solvents from an ambient air in the course of the plate storage. Deposited on an asymmetric support, the electron orbitals of the adsorbed molecules are affected by its asymmetry. In fact, these molecules behave as the asymmetric species and in the case of quantitative predominance of one type of the asymmetric micro-crystalline variety, the Cotton bands are apt to appear in the CD spectrum of the investigated silica gel (Figure 9.15).

Bottleneck 4. This bottleneck is in a sense similar to bottleneck 2. Also in this particular case it clearly comes out that densitometric detection applied to separations carried out on the plain silica gel layers has to be performed in a cautious and also conscious manner, that is, anticipating the possibility of quite significant deviations of the chiral analytes' migration tracks from the straight-line direction. Such deviations can easily occur with any analytical separation involving one or more chiral analytes. In this sense, the present section is a troubleshooting advice addressed to all those, who use in their analytical work TLC with densitometric detection. The recommended solution is to scan the relatively broad bands on the respective chromatograms instead of a narrow lane along the supposed straight-line migration track of the analytes.

REFERENCES

1. Sajewicz, M., Piętka, R., Pieniak, A., and Kowalska, T., *Acta Chromatogr.*, 15, 131–149, 2005.
2. Sajewicz, M., Piętka, R., Pieniak, A., and Kowalska, T., *J. Chromatogr. Sci.*, 43, 542–548, 2005.
3. Sajewicz, M., Piętka, R., Pieniak, A., and Kowalska, T., *J. Liquid Chromatogr. Rel. Technol.*, 29, 2059–2069, 2006.

4. Bhushan, R. and Parshad, V., *J. Chromatogr. A*, 721, 369–372, 1996.
5. Sajewicz, M., Piętka, R., and Kowalska, T., *J. Planar Chromatogr. — Mod. TLC*, 17, 173–176, 2004.
6. Xie, Y., Liu, H., and Chen, J., *Int. J. Pharmaceut.*, 196, 21–26, 2000.
7. Sajewicz, M., Piętka, R., Drabik, G., and Kowalska, T., *J. Liq. Chromatogr. Rel. Technol.*, 29, 2071–2082, 2006.
8. Sajewicz, M., Piętka, R., Kuś, P., and Kowalska, T., *Acta Chromatogr.*, 16, 181–191, 2006.
9. Sajewicz, M., Piętka, R., Drabik, G., Namysło, E., and Kowalska, T., *J. Planar Chromatogr. — Mod. TLC*, 19, 273–277, 2006.
10. Sajewicz, M., Piętka, R., and Kowalska, T., *J. Liq. Chromatogr. Rel. Technol.*, 28, 2499–2513, 2005.
11. Sajewicz, M. and Kowalska, T., Private communication.
12. Sajewicz, M., Hauck, H.-E., Drabik, G., Namysło, E., Głód, B., and Kowalska, T., *J. Planar Chromatogr. — Mod. TLC*, 19, 278–281, 2006.

10 Chirality of Pharmaceutical Products

Jan Krzek, Jacek Bojarski, Irma Podolak, and Ewa Leciejewicz-Ziemecka

CONTENTS

10.1 INTRODUCTION

Chirality of pharmaceutical products is an important issue related to effective and safe therapy. Already more than 100 years ago, Louis Pasteur noticed the differences in the activity of compounds of natural origin [1]. The tragedy of thalidomide and development of stereospecific analytical techniques resulted in broadening of studies on drug quality and inclusion of monitoring of biological activity and pharmacokinetics of stereoisomers.

A growing interest in these subjects comes from the fact that many drugs used today have the chiral structure and it is well known that stereoisomers may differ in their pharmacokinetic and pharmacodynamic properties, efficacy, and undesired side effects thus having considerable influence on rational pharmacotherapy.

During the past two decades, chiral drugs constantly aroused particular concern of clinicians, drug companies, and legislative bodies.

Main emphasis was placed on pharmacological consequences of chiral drugs administration, technological processes leading to new stereoisomeric drugs and their analytics, as well as on requirements for their registration.

Some legal standards and regulations have already been issued by appropriate institutions for three important areas, that is, Europe [European Medicines Evaluation Agency (EMEA)], United States [Center for Drug Evaluation and Research (CDER)], and Japan [Pharmaceutical and Medicinal Safety Bureau (PMSB)].

Chiral medicinal products are also a matter of vital interest for International Conference on Harmonisation (ICH) and its parties [such as, among others, European Federation of Pharmaceutical Drug Manufacturers and Associations (EFPMA), U.S. Food and Drug Administration (FDA), and Japan Pharmaceutical Manufacturers Association (JPMA)], as well as International Federation of Pharmaceutical Drug Manufacturers and Associations (IFPMA), establishing and implementing guidelines for procedures and criteria for registration of new pharmaceuticals in terms of active substances and final products (ICH Guidelines Q3A, Q3B, and Q6A).

The majority of substances with chiral center was obtained as a racemate by chemical synthesis and, as such, drugs were used since many years in therapy. In fact, in the 1980s the problem of single stereoisomers emerged when it was found that particular stereoisomers might differ in their actions.

Therefore, broad studies on the methods of synthesis of pure enantiomers were performed, what may be justified from both scientific and clinical point of view, taking into consideration:

- An increase of safety of drug administration (therapeutic index) as a result of better selectivity of drug–receptor interaction and restriction of side effects.
- Prolongation or shortening of drug action time (modification of half-time) and determination of optimal dosage regimen.
- Diminution of interactions with other drugs.

An increase in bioavailability of drugs as homoisomers is a promising way of delivery of novel, safer, more efficacious, and better-tolerated pharmaceuticals. This trend is confirmed by many investigations on individual components of racemic drugs. Consequently, pharmaceutical industry, pharmacologists, and clinicians are more often interested in the development of stereoselective methods of drug synthesis and applications. These methods of asymmetric synthesis leading to products with distinct enantiomeric excess or in some cases to pure enantiomers require pure stereochemical forms of substrates and appropriate catalysts to yield the desired stereoisomeric product [2]. Efforts are also undertaken to use biotechnological methods of pure enantiomers production, due to the specific enzymatic mechanisms of bacterial origin to reach the same goal [3].

Closely associated with subject matter of asymmetric synthesis and clinical aspects of chiral drugs are important problems of their analytics dealing with control of identity, quantitative assay, and determination of enantiomeric purity of these pharmaceuticals. For safe and efficient therapy, methods and procedures of identification and quantitative estimation of undesired stereoisomeric form are of peculiar importance. At present, there are many advanced analytical methods, which enable appropriate control of enantiomeric purity. Among them, besides spectroscopic and thermal analysis, chromatographic methods, such as high-performance liquid chromatography (HPLC), gas chromatography (GC), and thin layer chromatography (TLC), as well as capillary electrophoresis (CE), are most extensively used [4].

10.2 STRUCTURE, FORMS, AND PHYSICOCHEMICAL PROPERTIES OF CHIRAL DRUGS

Chirality as a property of molecule is related to its chemical structure without any element of internal symmetry (point, axis, or plane).

In the case of drugs, this element of asymmetry is a chiral carbon atom with four different substituents and less often asymmetric atoms of nitrogen, sulfur, or phosphorus.

A common feature of enantiomeric forms of the drug is rotation of plane of polarized light with the same angle to the left or to the right. This rotation comes from absorption of one of the two components of linear polarization of light (circular polarization dextro- or levorotatory).

Dextro- or levorotation is indicated by $(+) = d = $ dextro or $(-) = l = $ levo signs, respectively, with preference of these "mathematical" notations. The spatial configuration of chiral centers in optical isomers is described by letters "d" and "l" (Fischer notation) [5] or by "R" and "S" according to Cahn et al. [6].

Almost 100% of chiral drugs may be classified as enantiomers, diastereoisomers, and epimers (Figure 10.1).

Chiral stereoisomers are compounds with the same structure, which differ by spatial arrangements of atoms or substituents. Their molecules are mirror images (enantiomers) or do not have that property (diastereoisomers and epimers) [5]. Pairs of enantiomers and diastereoisomers of the compound with two-carbon chain with four different substituents R_1-R_4 and their relations are presented in Figure 10.2.

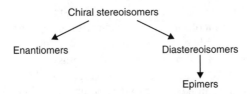

FIGURE 10.1 Types of chiral drugs.

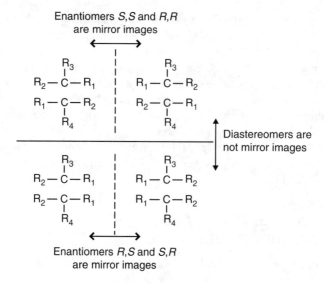

FIGURE 10.2   Pairs of enantiomers and diastereoisomers with substituents R_1–R_4.

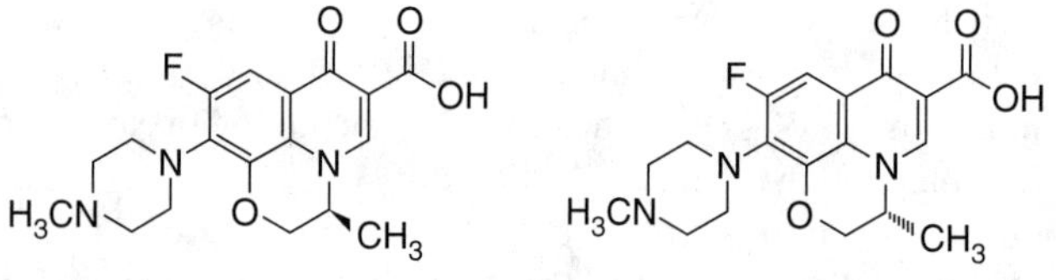

S-(−)-ofloxacin = Levofloxacin R-(+)-ofloxacin

FIGURE 10.3   Ofloxacin with one chiral center and two possible optical isomers: R-(+) and S-(−).

Among different types of three-dimensional structure of stereoisomeric compounds presented in Figure 10.1, enantiomers are of great interest. As was mentioned above, in each chiral molecule is at least one chiral center, which is mostly represented by asymmetric carbon atom bound to four different substituents. A number of possible stereoisomers, which may be ascribed to the given chiral drug, are directly related to number of chiral atoms in the molecule and for n such atoms amounts 2^n (or less, if *meso* forms exist).

For instance, in the molecule of ofloxacin with one asymmetric carbon atom (indicated by asterisk), there are two possible enantiomers: R-(+) and S-(−) (Figure 10.3), whereas for moxifloxacin with two asymmetric centers there are four stereoisomers: S,S-, R,R-, R,S-, and S,R- (Figure 10.4).

In a few drugs, the chiral center may be another atom, like in cytostatic drugs with chiral phosphorus atom (Figure 10.5).

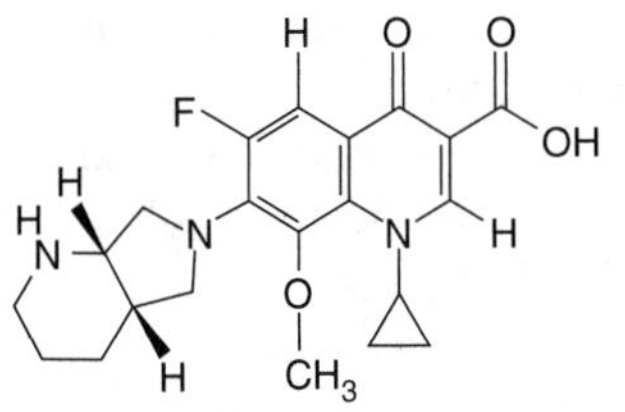

S,S-moxifloxacin

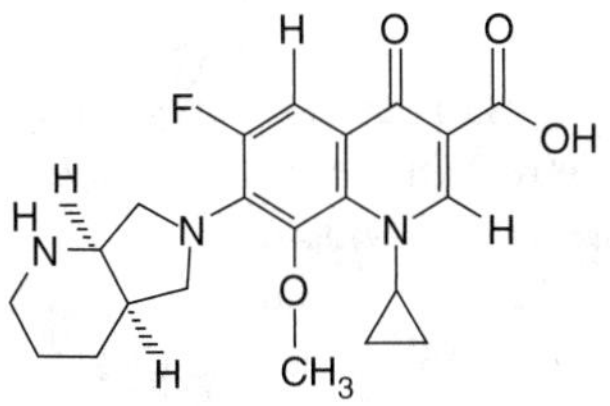

R,R-enantiomer

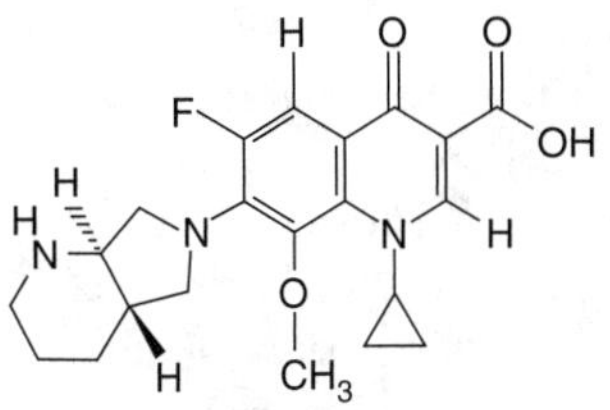

R,S-diastereomer

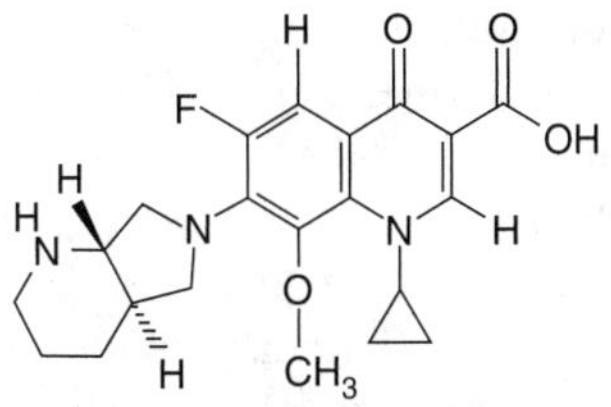

S,R-diastereomer

FIGURE 10.4 Moxifloxacin with two chiral centers and four possible optical isomers: *S,S*-, *R,R*-, *R,S*-, and *S,R*-.

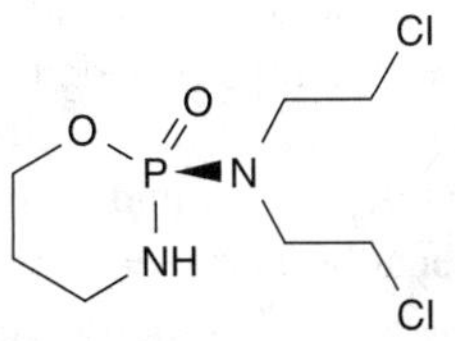

FIGURE 10.5 Structure of cyclophosphamide with chiral phosphorus atom.

The enantiomers have identical physicochemical properties, such as melting point, solubility, polarity, and hydrophobicity, in contrast to chiral diastereoisomers.

Epimers are diastereoisomers, which differ in configuration at only one stereogenic center, but of course, not all diastereoisomers are epimers.

Racemate or racemic mixture is an equimolecular mixture of enantiomers, which does not exhibit optical activity due to intermolecular compensation of rotation of opposite signs for each enantiomeric form.

10.3 METHODS OF DETERMINATION OF ENANTIOMERS

Different separation methods may be applied for enantioseparation of chiral drugs and their chiral precursors or metabolites. HPLC is considered the most important among them, but nowadays, CE becomes more and more popular. The use of

TLC method for this purpose is far less common, although it has some important advantages in comparison with those "column" techniques. One may assume that the progress in the production of chiral stationary phases, as well as automation and precision of quantitative assessment of chromatograms would result in higher interest in this technique.

Chromatographic enantioseparations may be performed using two methods called "direct" and "indirect." In this former method, chiral stationary phases (polymers of natural origin, protein phases, synthetic polymers with build-in chiral selectors, etc.) are used or appropriate chiral selectors (cyclodextrins, crown ethers, etc.) are added to the mobile phase [7].

In indirect method, the racemic mixture reacts with chiral reagent yielding diastereoisomeric pair of products [1,8], which may be separated by different methods due to their differences in physicochemical properties.

In both cases, the driving force of enantioseparation is formation of diastereoisomers, but in the direct method they are not covalently bound, but interactions are more discrete and may produce diastereoisomeric adducts between analytes and chiral selectors due to the formation of hydrogen bonds or inclusion compounds, interactions of π–π or dipole–dipole type, and so forth [4].

In the enantioseparation by electromigration methods different chiral selectors (such as cyclodextrins and their derivatives, cellulose, bile acids, or chiral surfactants, for instance, N-deoxycarbonylvaline) are added to the running buffer [9–11]. Cyclodextrins may also be incorporated in the silica bed or may be mixed with other selectors [10]. Many investigations revealed that the magnitude of enantioseparation depends not only on the concentration of chiral selector in the buffer but also on the addition of organic solvent and on its concentration.

Quantitative analysis of enantiomers separated by planar chromatography is usually accomplished by spectrophotometric methods (UV-Vis) either after extraction of spots or *in situ* on plates by densitometric measurements.

10.3.1 Qualitative Analysis

The identification of components separated by TLC can be performed by comparison with appropriate reference substances. Generally, a compound being examined and the reference substance are considered as identical if their R_f values are identical and if when applied as a mixture they produce a single spot on a chromatogram.

Spots on the chromatograms may be examined by:

- UV fluorescence detection
- Color reaction after treatment with appropriate spray reagent
- Comparison of UV spectra recorded directly from a chromatogram or after extracting the analytes from a TLC plate, in which case both UV and IR can be recorded and compared
- Radioactivity measurement
- Microbiological evaluation.

10.3.2 QUANTITATIVE ANALYSIS

The quantitative determination of compounds separated by TLC takes advantage of a plethora of available instrumental analysis methods following prior extraction of a compound from the coating layer or a direct densitometric scanning.

Densitometric analysis seems especially suitable since it allows rapid and reliable quantification, saving the need of additional sample preparation steps connected with the extraction of an analyte from the stationary phase.

Densitometric evaluation is based on measurements of light reflected by a spot or of its native fluorescence.

Today, modern computer-controlled densitometers allow quantitative determination of almost all chemicals, both colorless and colored, which have UV-Vis absorbance or are capable of measuring fluorescence. Densitometric detection can be performed in a wide wavelength range, from 200 to 800 nm, with high sensitivity and accuracy. The method is rapid and offers a possibility of analyzing simultaneously a large number of samples along with consecutive qualitative analysis (absorption spectra, R_f values).

Two types of factors affect the precision of densitometric analysis [12]:

1. Factors relative to instrumentation, such as choice of an analytical wavelength or a filter, the slit size, and the duration of the scanning process.
2. Factors relative to the quality of a chromatogram, such as preparation of the sorbent layer, development procedure (isocratic or gradient elution), and post-chromatographic derivatization.

The choice of the slit size (width and height) depends on the shape and size of the spot. Usually, for a spot whose round shape is sharply defined, the slit width is set at 120% of the spot diameter, when the spot is oval — at 120% of the spot length, and for a band — at 50–70% of its length [13].

Generally, the smaller the slit height the greater the precision and traceability. The slit size should be established experimentally on a series of standard spots.

The choice of scanning direction depends on the shape of the spot. For round spots, the scanning direction is irrelevant, whereas for oval-shaped spots, especially for quantitative work, the meander instead of the linear scan mode should be chosen and performed along the migration direction [14].

To assure the reproducibility of the results of quantitative densitometric evaluation, chromatographic plate movement with respect to the slit should remain constant during the analysis. Modern densitometers are equipped with automatic devices allowing constant and optimal for a given evaluation plate movement. The device can be programmed to scan each in its turn the individual paths.

Spots, which are suitable for densitometric evaluation, should be able to withstand exposure to light and the color intensity should be stable. For quantitative work, determination of a correlation between color intensity of a substance and its amount in a spot is required.

Results of densitometric scanning of spots separated by TLC are available as densitograms showing a series of peaks similar to chromatograms obtained in GC or HPLC analysis. Peak area or peak height measurement allows quantification.

Modern densitometry can also measure fluorescence, which enlarges the number of substances analyzed by this method. In such densitometers, mercury lamp used as a light source allows work in a monochromatic light.

This method requires TLC layers of well-defined own fluorescence because chromatographed compounds would quench fluorescence to a certain extent or even completely. The fluorescence quenching is proportional to the concentration of a compound in a spot [12].

Fluorescence quenching is a useful technique used for the quantification of substances lacking native fluorescence.

10.4 PURITY OF CHIRAL STEREOMERS

As numerous optical isomers have been shown to reveal major differences in pharmacological activity, metabolic fate, and excretion, the problem of stereochemical purity control seems to be widely appreciated and remains a topical issue, all the more so that it encompasses the problem of drug quality and aims at ensuring maximal safety and therapeutic efficacy. This issue has been investigated in more detail by the authors of other studies [15–17], who reviewed various possible modes of action of enantiomers present in drug products containing racemates.

The problem of stereochemical purity of a chiral active substance acquires even more importance along with licensing of new medicinal products obtained by chemical modification of an optical isomer, which results in a racemate containing two drug substances showing distinct differences in mechanisms of action. Examples of drugs developed in such a way include α- and β-adrenergic blockers, that is, labetalol, primidolol, or beta-adrenergic blockers, such as prizidol and sulfinalol [18]. Such products, despite their attractiveness, have fixed chemical composition that hinders modifications in dosage regimens, not to mention guaranteeing uniformity with respect to pharmacological profile. Further complications arise from the fact that the number of new components (drugs) in a diastereomeric mixture depends on the number of chiral centers what may lead to more isomers, as is the case for labetalol where two chiral centers give rise to four different stereoisomers.

Classification of enantiomeric forms as drug "impurities" is not generally accepted. The USP does not define them as impurities in the common sense of the term but as forms of the active substance [19] unless individual enantiomers differ in pharmacological activity, in which case they should be treated otherwise. The FDA's guidelines, published in 1992 [20], identify requirements relating to chiral compounds. If one of the enantiomers should appear more toxic, further investigations (including *in vivo* studies) would be required with the individual isomers. The development and use of racemic drugs is permitted in justified cases.

The ICH debated issues relating to chiral impurities but no consistent guidance has been produced. Even though the problem of classification and control of enantiomeric forms as impurities has been broadly discussed, the guideline Q3A does not address this issue clearly [21]. Some of the experts recommended that these forms be treated as impurities and included in the guidelines, whereas the majority held the contrary view arguing that analytical methods to resolve enantiomeric mixtures would require the use of specific, complicated, and expensive instrumentation. Such an approach cannot be fully justified as one may use widely available chromatographic methods, such as HPLC, GC, and TLC, or other analytical methods, that is, CE, modifying only the analytical conditions.

Other authors [22] agree that the presence of an unwanted enantiomer should be regarded as an impurity and should be subject to control according to specific recommendations in the guidelines. This opinion has been adopted by the ICH as reflected in the guideline Q6A, which came into operation in May 2000, in which chiral drug substances are included in the category of specific tests/criteria.

Decisions regarding the scope of tests for chiral drugs are made by regulatory bodies responsible for drug approval process. In this respect, an appropriate approach to analysis and results interpretation, dependent on whether the drug concerned is a racemate or a single enantiomer, is needed.

For example, the identity test for an individual enantiomer should be conducted in such a way so as to be able to discriminate between the isomer concerned and a racemate. When a given drug contains a single enantiomer, the presence of an antipode should be considered to be an impurity in concordance with the ICH guidelines Q3A and Q3B.

In the European Union, the regulatory requirements for investigation of chiral drugs are included in the guidelines prepared under the auspices of the Committee of Proprietary Medicinal Products (CPMP) issued in December 1993 [23], which recommend viewing an unwanted enantiomer as an impurity. These guidelines [24] are additional to other directives relating to the quality criteria for licensing of medicinal products in European Union and set out the scope of requirements with respect to chiral active substances.

Aspects relating to the control of enantiomeric impurities are still not addressed in pharmacopoeial monographs; however, it is now foreseen that information on their presence and chemical structure will be incorporated in the information section of the drug substance monographs in the European Pharmacopoeia.

In contrast to enantiomers, the European Pharmacopoeia differentiates qualitatively and quantitatively between epimeric chiral stereomers. This seems fully justified as individual epimers significantly differ in their pharmacological potency and may be present in a drug in various proportions.

10.5 PHARMACODYNAMIC AND PHARMACOKINETIC ASPECTS OF CHIRAL DRUGS

Differences in physicochemical properties of chiral diastereomers may result in inconsistencies with respect to the pharmacokinetic and pharmacodynamic behavior of individual isomers.

Enantiomers differ in their pharmacokinetic and pharmacodynamic properties provided a physiological partner involved in the reaction is chiral. Such chiral reaction partners include transport proteins, tissue and blood plasma proteins, enzymes, and receptors.

The mechanism of action of the majority of biologically active substances involves binding to specific receptors. These receptors have a specific three-dimensional stereochemistry; thus, the selectivity and binding affinity of highly active compounds, such as neurotransmitters, hormones, and some drugs, are governed by both the chemical and the three-dimensional molecular structures.

A more potent molecule is called an *eutomer*, whereas the less active, or even inactive, is called a *distomer*. An eutomer/distomer ratio, called the *eudismic ratio*, is a measure of stereoselectivity of a compound. The eudismic ratio increases with the potency of one of the isomers. This is especially true when the center of asymmetry is located in a part of the molecule, which is responsible for binding to a receptor (Pfeiffer's rule).

Most biochemical processes in nature are selective; therefore, the products that are generated in life-processes are stereospecific, that is, they occur in one optical form (−) or (+). This stereospecificity is a feature of plant- and animal-derived drugs.

Most synthetic drugs are available as isomeric mixtures. Individual enantiomers of a chiral drug may significantly differ not only with respect to selectivity of interaction with receptors or enzymes but also bioavailability, metabolic characteristics and excretion rate, therapeutic potency, and toxicity [25–27]. Therefore, the use of a desired stereoisomer of a drug may assure more selective pharmacokinetic profile, improvement of therapeutic indices, and, due to changed metabolic fate, a lower incidence of interactions with other drugs.

Stereoselective drug metabolism can be seen in close relation to the following issues:

- Generation of optically active drug metabolites and differences in the metabolic fate of drug enantiomers.
- Stereoselective drug interactions.
- Effect of stereoselectivity on drug genetic oxidative polymorphism.

Drug biotransformation may give rise to metabolites possessing chiral centers. This may be demonstrated on an example of a hypotensive drug debrisoquine oxidation to 4-hydroxydebrisoquine, which is excreted in urine exclusively as an L-isomer [28,29]. Similarly, reduction of achiral 5-HT$_2$ antagonist, ketanserin, gives rise to an optically pure D-isomer of its metabolite ketanserinol [29].

Another, commonly known example of stereoselective drug biotransformation is *in vitro* and *in vivo* conversion of inactive (R)-(−)-ibuprofen to the active (S)-(+)-enantiomer in drug formulations containing ibuprofen racemate [30–32].

Some chiral drugs available as racemates may interact with other drugs in such a way that biotransformation of only one enantiomer would be hindered. This may lead to major differences in the pharmacological potency and higher risk of unwanted side effects. Such an effect is observed, for example, with warfarin, which is used to prevent blood clots and is administered as a racemate. Increased anticoagulation is seen when some anti-inflammatory drugs, such as phenylbutazone, are added [33].

Pharmacokinetic and pharmacodynamic differences between drug isomers present another important issue relating to drug metabolism. Individual enantiomers of drugs administered as racemates show different pharmacokinetic profiles due to differences in metabolic clearance rates and binding affinities to blood plasma proteins [34].

Pharmacodynamic effects for individual enantiomers of drugs used as single enantiomers or as racemic mixtures are presented in the following sections/chapters.

10.6 PHARMACODYNAMIC DIFFERENCES AMONG STEREOISOMERS OF SELECTED CHIRAL DRUGS

According to Görög and Gazdag [4], ca. 56% of therapeutics administered today are chiral molecules. Compounds of natural origin or their semisynthetic derivatives most often occur enantiomerically pure, whereas as much as 88% of synthetic drugs are used as racemates.

Development of single drug enantiomers is justified not only clinically but also economically. This allows us to hope that new, more effective medicines will be introduced and overall treatment costs will be reduced.

As mentioned above, a majority of chiral drugs are stereoselective with respect to mode of activity, whereas only a small number lack this feature, for example, promethazine, flecainide, and liponic acid [35].

Despite high costs, studies on single enantiomeric development resulted in therapeutic and economic benefits, thus enabling pharmaceutical industry to submit patent applications on single optical isomers of long-used generic drugs [36].

10.6.1 SINGLE STEREOISOMERS

Chloramphenicol (R)(antibioticum)

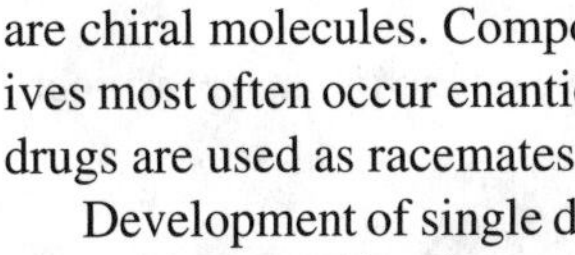

R-chloramphenicol is a potent antibacterial agent whereas S-isomer lacks this activity and is more toxic.

Levofloxacin (S)
(antibioticum)

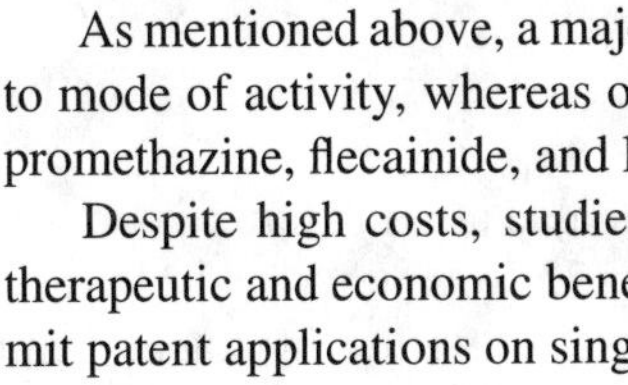

S-ofloxacin, a stereoisomer of racemic ofloxacin, is a more potent antibiotic than R-isomer [37].

Ibuprofen (*S*)
(analgeticum)

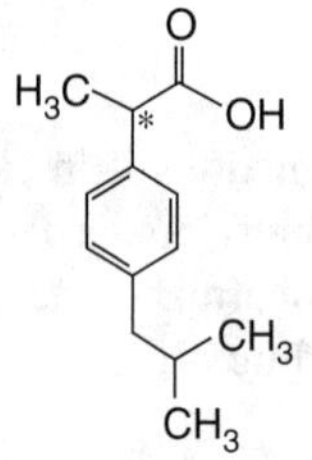

S-isomer is 100 times more potent as COX inhibitor than *R*-isomer [26,27,36]. Racemic mixture (*R*,*S*) is also used in medicine. Active *S*-ibuprofen is more rapidly absorbed than a racemate.

Selegiline (*R*)
(antiparkinsonicum)

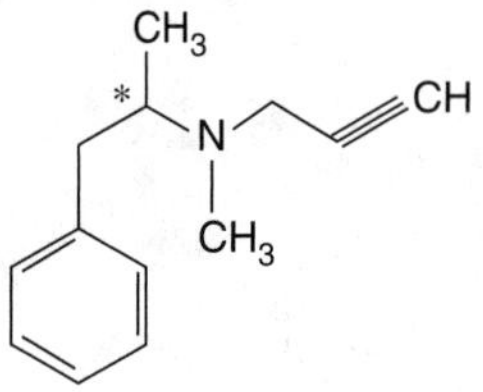

R-isomer is more selective to MAO-B, the enzyme responsible for deactivation of serotonin, as opposed to *S*-isomer, inhibits both MAO-A and MAO-B.

Penicillamine (*S*)
(antirheumaticum)

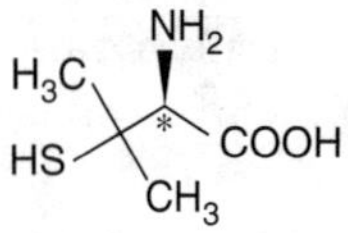

An anti-inflammatory agent, less toxic than *R*-isomer.

10.6.2 STEREOISOMER MIXTURES

Budesonide (*R*,*S*)
(hormonum)

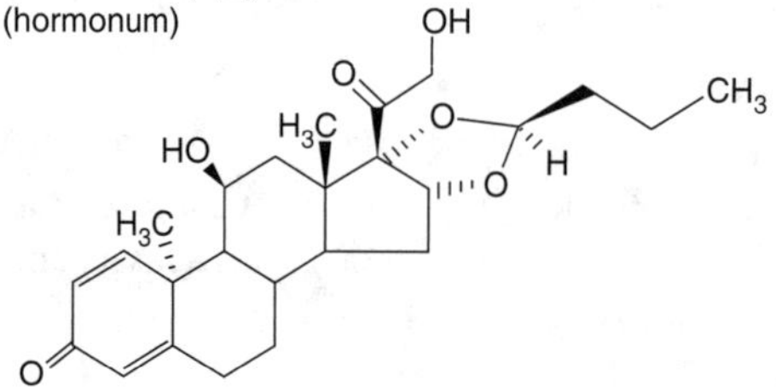

Available as a mixture of two isomers 22*S* (epimer A) and 22*R* (epimer B) differing in potency and pharmacokinetic profile. Epimer A is an anti-inflammatory agent, whose potency is twice that of epimer B and its volume of distribution and plasma clearance are also twice that of epimer B. In drugs the *R*:*S* ratio equals 55:45 or 90:10.

Fluoxetine (*R*,*S*)
(thymolepticum)

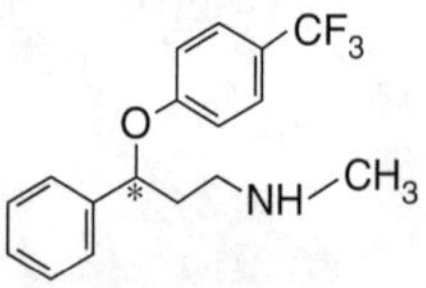

Available as a racemate (*R*,*S*). *S*-fluoxetine is a more effective antimigraine agent and shows fewer side effects than *R*-isomer. On the other hand, the *R*-enantiomer is a more effective antidepressant.

Ketamine (*R,S*)
(anaestheticum)

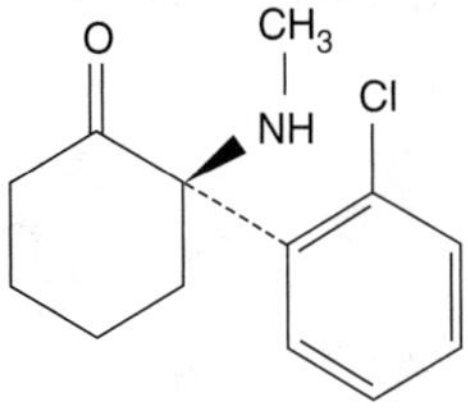

Both *R*- and *S*-enantiomers are analgesic, anesthetic, and sympathomimetic agents differing however in potency and toxicity. *S*-ketamine shows greater affinity to opioid receptor and more strongly inhibits catecholamine transport than *R*-ketamine [35].

Warfarin (*R,S*)
(anticoagulans)

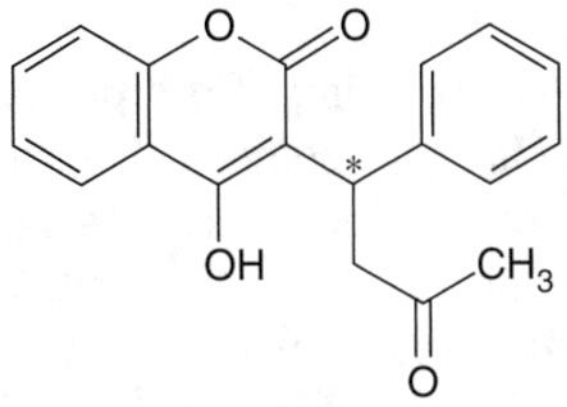

S-enantiomer is five times more potent than *R*-isomer.

Propafenone (*R,S*)
(antiarrhythmicum)

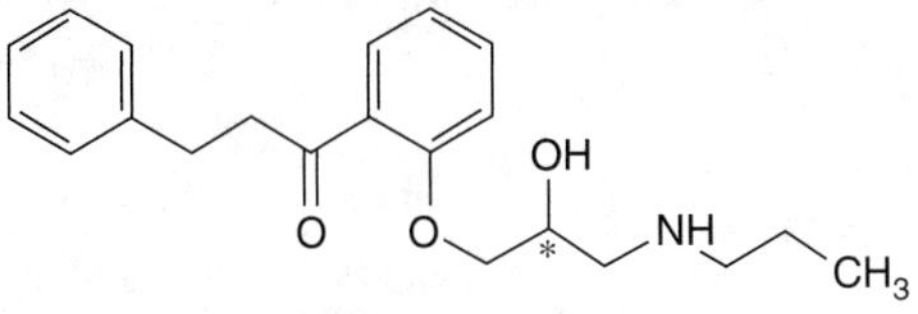

S-enantiomer is 100 times more potent *β*-adrenergic blocker than *R*-isomer [25].

Amlodipine (*R,S*)
(antihypertensivum)

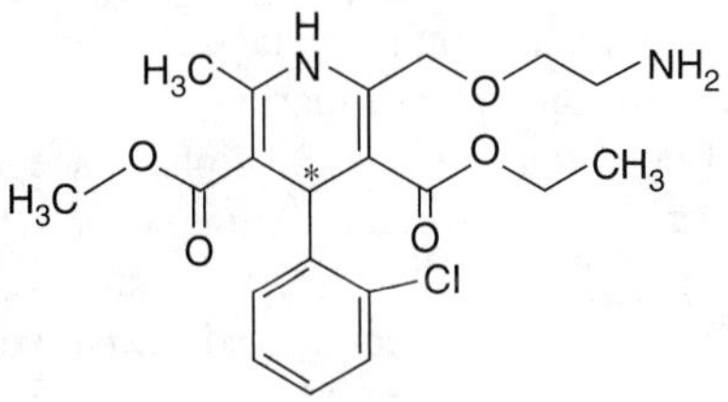

S-enantiomer is a selective calcium channel agonist as opposed to *R*-isomer, which is inactive [38,39].

Pentazocine (*R,S*)
(analgeticum)

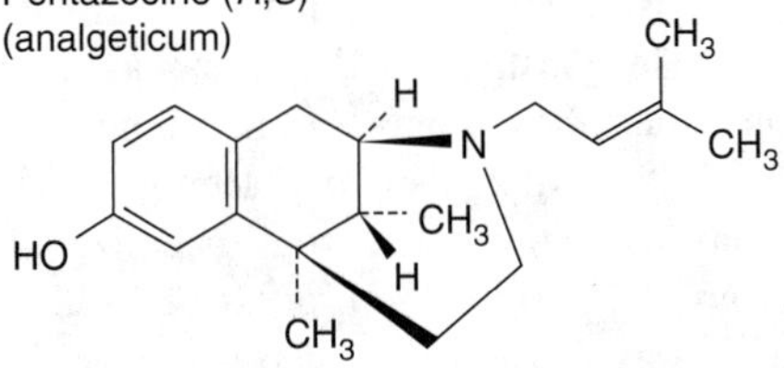

R-enantiomer is more potent analgesic causing stronger respiratory depression than *S*-form. *S*-enantiomer shows more psychomimetic side effects.

Pentobarbital (*R,S*)
(anaestheticum)

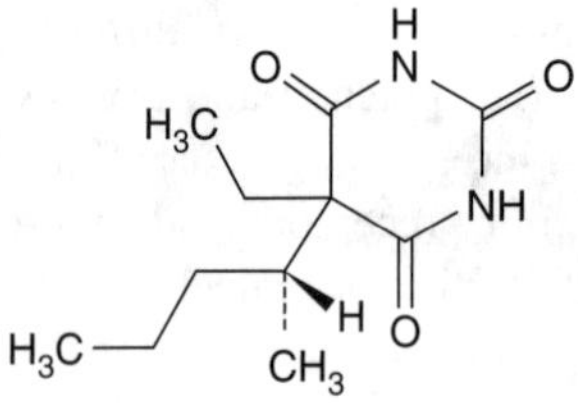

R-enantiomer induces sleep, whereas *S*-enantiomer is responsible for motor stimulation.

Verapamil (*R,S*)
(antiarrhythmicum)

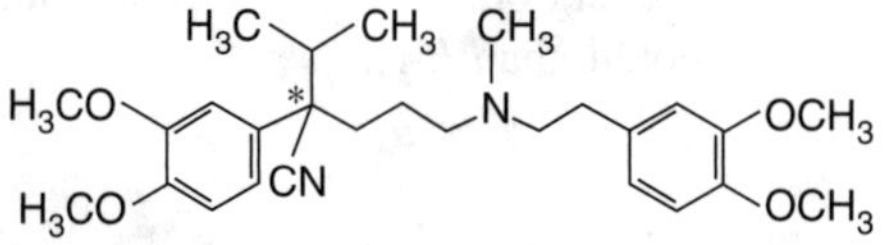

S-enantiomer blocks calcium channels more strongly than *R*-isomer. *R*-isomer is responsible for the negative chronotropic effect and its affinity for proteins is twice that of *S*-isomer. *R*-isomer is more rapidly metabolized in the liver therefore its bioavailability is over two times lower.

Procyclidine (*R,S*)
(antiparkinsonicum)

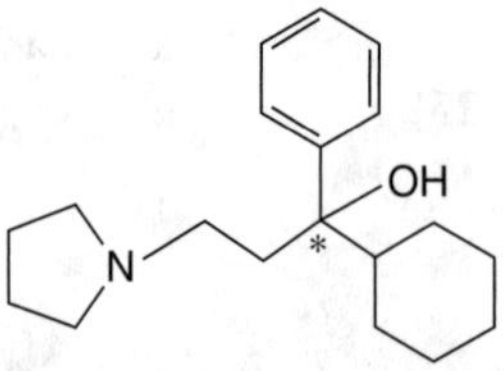

R-enantiomer shows twice higher affinity to muscarinic receptors and is more rapidly metabolized as compared with the inert *S*-isomer.

10.6.3 STEREOMERS OF NATURAL ORIGIN

Natural compounds, in opposition to most synthetic products, are biosynthesized enantiomerically pure. This stereospecificity is their great advantage in terms of seeking new drug sources. Compounds derived from plant or marine sources that exhibit pharmacological activity and are of potential interest to the pharmaceutical industry belong in a majority of cases to low molecular weight secondary metabolites (e.g., alkaloids, flavonoids, phenolic acids, coumarins, quinones, terpenoids, and steroids). They play an important physiological role being involved in, for example, insect–plant, insect–insect interactions, regulation of cell-to-cell signaling, and defense against pathogens [40,41].

Among natural substances there are classes of compounds that occur in only one of the enantiomeric forms, there are also constituents that occur as mixtures of enantiomers. One enantiomeric series may dominate and sometimes the optical antipode is present in a different type of organism. Racemic asymmetric natural products are much less common; however, partial racemization may occur in a plant or during the isolation and purification process [41,42]. In the case of tropane alkaloids, especially (−)-hyoscyamine, which is one of the principal constituents found in genus *Atropa*, *Datura*, *Hyoscyamus*, *Scopolia*, and *Duboisia* (*Solanaceae*), racemization to atropine occurs as a result of drying and extraction [43,44].

One should bear in mind that plant and animal secondary metabolites play many physiological roles; thus, the enantiomeric and diastereomeric composition of constituents has important implications for the observed bioactivity. Chiral pheromones, used by animals as communication substances, may be a good example. Activity is usually confined only to one enantiomer, whereas the other is either inactive or acts opposite. (*S*)-Japonilure, a female sex pheromone of the Japanese beetle *Popillia japonica* counteracts the (*R*)-form [45]. Research on the composition of sex pheromones of saw flies (*Neodiprion sertifer*) revealed that a slight change in the ratio of dipirinyl acetate diastereomers 2*S*,3*S*,7*S* and 2*S*,3*R*,7*S* resulted in the loss of activity (99.9% 2*S*,3*S*,7*S* and 0.1% 2*S*,3*R*,7*S*: very high activity; <99.5% 2*S*,3*S*,7*S* and >0.5% 2*S*,3*R*,7*S*: loss of activity) [41]. Another interesting example may be derived from studies on insect–host interactions. Enantiomeric composition of the common spruce *Picea abies* (L.) Karst was investigated. A weakened or dead tree produces α-pinene and ethanol, which are regarded as signal molecules for the primary aggregation of the spruce bark beetle *Ips typographus*. The active constituent of the aggregation pheromone is (*S*)-*cis*-verbenol, formed by microorganisms in the intestinal tract of the insect from (−)-α-pinene. When (+)-α-pinene is present in the plant (*S*)-*trans*-verbenol is produced, which was shown to inhibit aggregation, so trees rich in (+)-α-pinene would be more beetle resistant. Large variations in the enantiomeric composition of α-pinene among individual trees and different parts of a tree, from 90% of the (−)-form to 80% of the (+)-form were found [41].

As was mentioned earlier in this chapter, different drug isomers may account for a different range of activities and toxicities. The story of thalidomide is the most prominent example, but in the area of natural products, similar problems may be encountered as well, especially when many herbal products are available as OTC or dietary supplements.

Thujone, a monoterpene first extracted from yellow cedar — *Thuja occidentalis*, but found in high amounts in other sources such as volatile oils derived from wormwood (*Artemisia absinthium*), mugwort (*Artemisia vulgaris*), sage (*Salvia officinalis*), junipers (*Juniperus* sp.), or tansy (*Tanacetum vulgare*), is a well-known neurotoxic agent. It is the principal constituent responsible for the toxic effects observed in absinthe drinkers. Thujone containing plant oils are used in medicinal products indicated for loss of appetite and as flavoring substances in food and alcoholic drink industry. Its intakes are associated mainly with sage and sage-flavored products and alcoholic beverages such as vermouth. Two enantiomers of this compound: (+)-3-thujone known as β-thujone and (−)-3-isothujone known as α-thujone are present in plant sources in varying proportions and differ markedly in toxicity and convulsant activity. Most often, the content of β-thujone exceeds that of α-thujone. According to data cited in the EC Opinion of the Scientific Committee on Food on Thujone and the National Institutes of Health [46,47], its ratio in wormwood and tansy is higher, whereas volatile oils derived from sage and cedar leaf contain more α-thujone. α-Thujone is considered to be the more toxic isomer, which blocks brain receptors for γ-aminobutyric acid (GABA). The compound is also reported to have antinociceptive, insecticidal, and anthelmintic activity [46–49].

Another reason for concern is that some products may be adulterated or fortified with a cheaper synthetically produced enantiomer or racemate thus changing the pharmacological and toxicological profile of a drug. Dietary supplements containing ephedra, which are used as weight-loss enhancers and stimulants, were reported to contain (+)-ephedrine which does not occur naturally [50,51]. Moreover, herbal products claiming identity in the composition may not be equivalent in terms of enantiopurity. Sybilska and Asztemborska [52] reported that commercially available drugs of natural origin applied in liver and kidney diseases with similar chemical composition markedly differed in the content of enantiomers depending on the manufacturer.

Today, most phytomedicines constitute plant extracts containing complex mixtures of compounds. Isolated single plant constituents are usually not considered as phytomedicines. These compounds (especially alkaloids) are used widely in medicine as drugs, or serve as prototypes for new synthetic drugs.

10.6.3.1 Selected single compounds

Below, some selected examples of plant chiral compounds that are usually found in medicinal products as isolated ingredients are listed:

(−)-Ephedrine

An alkaloid present in *Ephedra sinica*, commonly known as *ma huang*; used as antiasthmatic, nasal decongestant, weight-reducing agent, and a stimulant. The plant is a component of many dietary supplements advertised as energy-boosters and calorie-burning. Ma huang also contains other alkaloids such as (+)-pseudoephedrine, (−)-norephedrine, (+)-norpseudoephedrine, (−)-N-methylephedrine, and (+)-N-methylpseudoephedrine. For each of these compounds an enantiomer is possible, which, however, does not occur in the plant. Therefore, detection of (+)-ephedrine, (−)-pseudoephedrine, or a racemic ephedrine suggests that a given extract or preparation might have been adulterated/fortified with less expensive and less active synthetic alkaloids [50,51,53].

(−)-Synephrine

An alkaloid similar in structure to ephedrine, present in the peel and the edible part of *Citrus aurantium* fruit. The (*S*)-(+)-enantiomer, not present in natural sources, is twice less active. The compound is most often administered clinically as a racemic mixture obtained by synthesis. The main indication is treatment of obesity.

Similar to ephedrine, serious side effects due to adrenergic stimulation may appear [54].

(−)-Paclitaxel

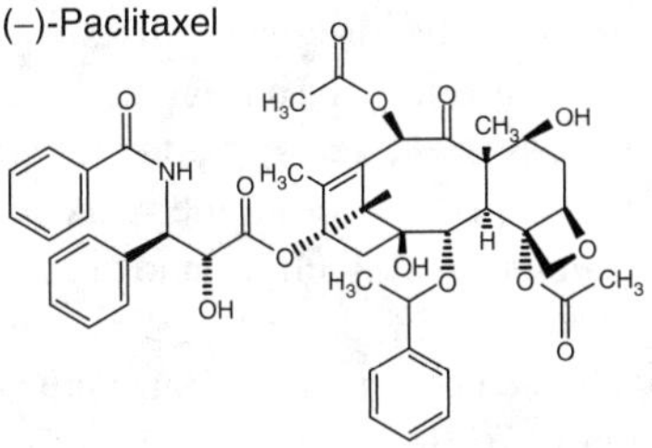

A diterpene alkaloid, initially isolated from the bark of the pacific yew *Taxus brevifolia*, has 11 chiral centers so the number of possible diastereoisomers exceeds 2000. An important chemotherapeutic agent. Lack of adequate supply from natural sources and chirality of the molecule were the reasons for the development of a semi-synthetic conversion from baccatins, which are relatively abundant in other yew species whereas full synthesis is multistep and costly [55,56].

(+)-*cis*-Vincamine

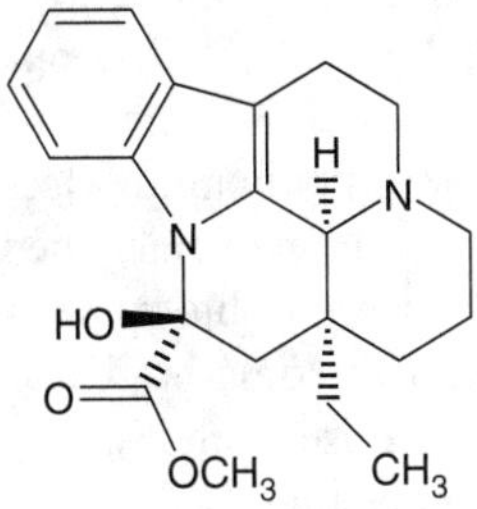

An indole alkaloid isolated from the common periwinkle *Vinca minor*, a valuable medicine used in cerebral insufficiency. The compound has three stereogenic centers, so eight stereoisomers in four enantiomeric pairs are possible [57].

(+)-Tubocurarine

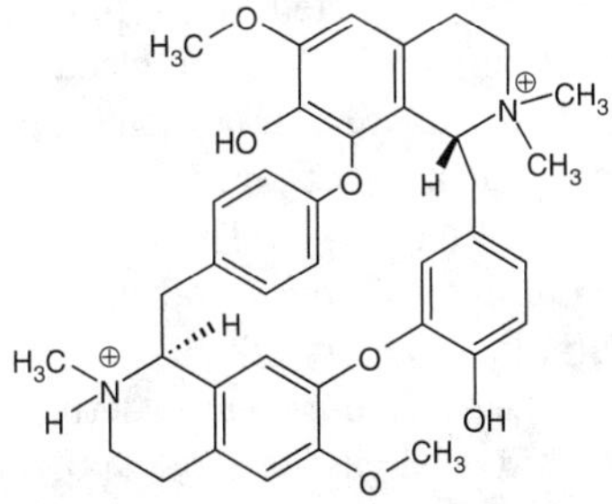

An isoquinoline alkaloid isolated from *Chondrodendron tomentosum*; causes muscular relaxation; more potent than its (−)-enantiomer; today synthetic alternatives such as atracurium are more often used.

(−)-Quinine

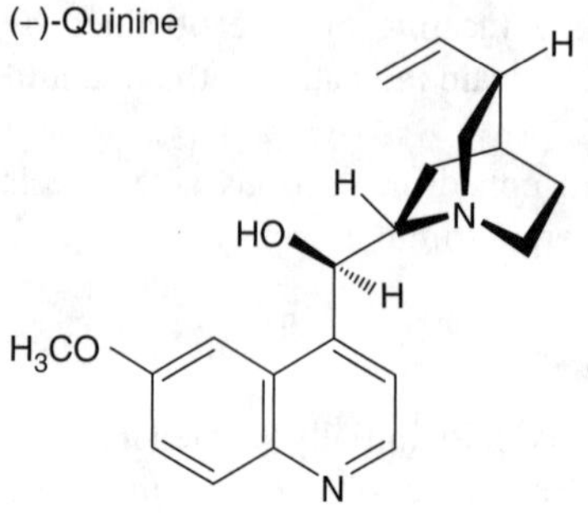

Quinoline alkaloid; its stereoisomer quinidine and their 6′ demethoxy derivatives, are the principal alkaloids of Peruvian bark (*Cinchona succirubra*). The quinine series has a configuration 8*S*,9*R* whereas quinidine 8*R*,9*S*. The ratios between alkaloids vary greatly dependent on the species or variety, environmental factors, and age of the tree. Before the advent of synthetic drugs, quinine was the most important antimalarial agent. Despite its serious adverse effects, it remains in use for the treatment of chloroquine-resistant *Plasmodium falciparum* infections and is used as a bitter flavoring in drinks. Quinidine, the (+)-isomer, is used as a cardiac antiarrhythmic agent [58]. Both enantiomers differ also in their effect on the activity of the CYP2D isoenzymes. Quinidine inhibits the human isozyme CYP2D6, whereas quinine has no effect on the metabolism of CYP2D6 substrates in humans [59].

Atropine

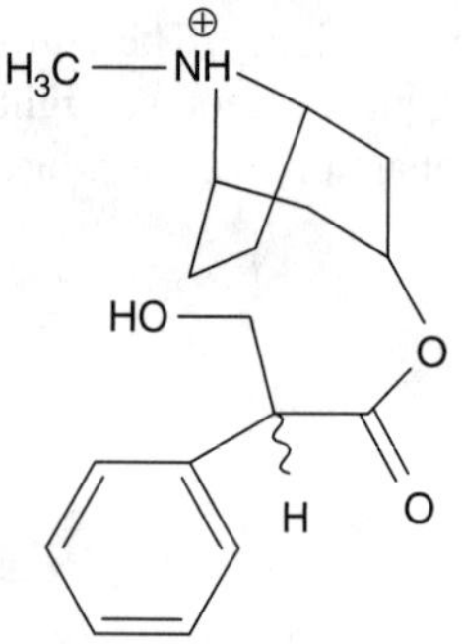

A racemic mixture of hyoscyamine, the principal tropane alkaloid of Belladonna herb (*Atropa belladonna*), henbane (*Hyoscyamus niger*) or stramonium leaf (*Datura stramonium*); used in medicine when temporary reversible muscarinic blockade is needed, for example, to overcome bradycardia, as an antidote for overdose of cholinergic drugs or cholinesterase poisoning.

Ergot alkaloids

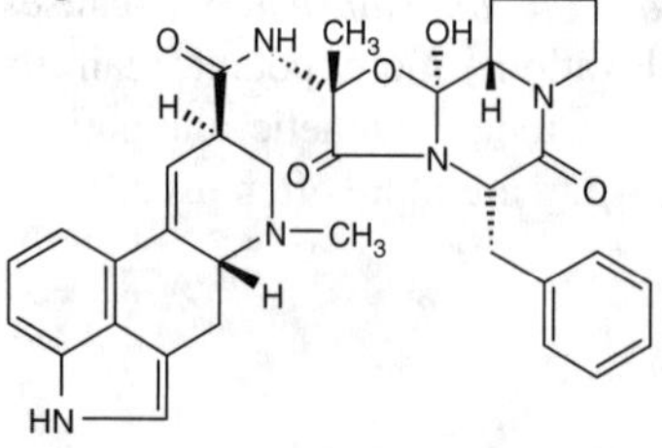

Produced by ergots collected from fungus (*Claviceps purpurea*) infected rye. Nowadays, artificial parasitic cultivation on rye and saprophytic growth techniques are employed to obtain ergot alkaloids. These compounds are present in pairs of stereoisomers. Natural levorotatory alkaloids, such as ergotamine (see Figure), are derived from lysergic acid whereas the corresponding dextrorotatory pairs from isolysergic acid.

The lysergic acid molecule possesses two centers of asymmetry with a stereoconfiguration 8R,5R. 5S isomers are not encountered in nature, however, such compounds were obtained by synthesis. Whether existing independently or forming part of the alkaloid molecule, lysergic acid readily undergoes epimerization, especially in the presence of alkalies, with respect to the center of symmetry C8 to isomeric isolysergic acid (8S,5R). Only levorotatory alkaloids have a powerful pharmacological activity, isolysergic acid derivatives have very weak activity [60].

(+)-Camphor

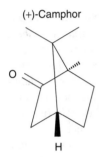

A monoterpene, constituent of an essential oil from the wood of *Cinnamomum camphora* which provides exclusively a dextrorotatory camphor; both isomers may be found in several *Chrysanthemum* species.

A rubefacient agent, constituent of many externally applied preparations [41,61–63].

(−)-Menthol

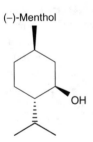

A monoterpene constituent of essential oil obtained from *Mentha piperita*; important agent in medicine (the cooling activity) and in perfumery — responsible for the typical peppermint aroma [63]. Its synthetic enantiomer shows significant pharmacokinetic differences: 69% of a dose of natural (−)-menthol was excreted in the bile in 24 h as compared with only 32% for the (+)-antipode [59]. Studies on the composition of chiral terpenes in *M. piperita* confirmed that the characteristic ratio, in which both enantiomers of a compound are naturally present, is independent of geographical origin and climatic conditions. Thus, the qualitative profile was identical even though quantitative differences were apparent depending on the origin of the plant. It was found that all major chiral constituents, such as menthyl acetate, menthol, menthone, neomenthol, and isomenthol, occurred as pure enantiomers in (−)-form except for (+)-isomenthol and neomenthol, for which both enantiomers were detected. Most of minor chiral constituents were pure (+)-enantiomers (α-thujene, sabinene, limonene, α-terpineol) whereas α- and β-pinene were present as both enantiomers in a characteristic ratio. The presence of (+)-menthyl acetate and (+)-menthol in peppermint oil indicates adulteration as these enantiomers do not occur naturally [63]. There is a strong evidence that hepatotoxicity associated with another constituent of peppermint oil — pulegone and its metabolite menthofuran (also present in pennyroyal oil from *Mentha pulegium*) depends mainly on (*R*)-(+)-isomers. The ratio of (*S*)-(−)-pulegone in these plant sources is much lower; however, it may isomerize and eventually give rise to (*R*)-(+)-menthofuran. (*S*)-(−)-pulegone is present in Buchu leaf oil derived from *Agathosma betulina* which is used as a flavoring ("cassis"-type aroma) [64].

(*R*)-(+)-Limonene

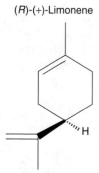

A monoterpene found in the volatile oil of bitter orange peel (*Citrus aurantium* L. subsp. *amara*) and lemon peel [*Citrus limon* (L.) Burm.]. (*R*)-(+)-isomer is the dominant enantiomer in most plant that produce it, however, both enantiomers (and the racemic form dipentene) are found in different ratios and amounts in many plant products, such as volatile oils of rosemary, lavender, lemon grass, eucalyptus, and others [65]. The use of this compound as a fragrance is widespread. The two optical forms of limonene, differ in the intensity of odor, which is stronger in case of (−)-limonene [the odor quality is more turpentine than orange as is for (*R*)-(+)-isomer] [66,67]. (*R*)-(+)-Limonene, although not allergenic itself, forms a number of oxidation products which are frequent skin sensitizers.

(+)-Usnic acid

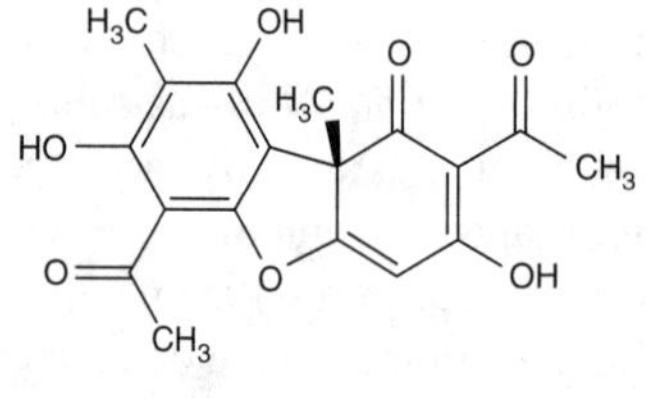

A compound typical of lichens, for example, genus *Cladonia* and *Usnea*. It occurs in two enantiomeric forms, which are effective against a large variety of Gram-positive bacterial strains. The (+)-enantiomer is selective against *Streptococcus mutans* and is included in some oral-care commercial preparations. (−)-Usnic acid, as opposed to (+)-form, is a selective herbicide, which irreversibly inhibits one of the important plant enzymes involved in carotenoid synthesis and bleaching — 4-hydroxyphenylpyruvate dioxygenase [68–71].

$(-)$-Δ^9-THC

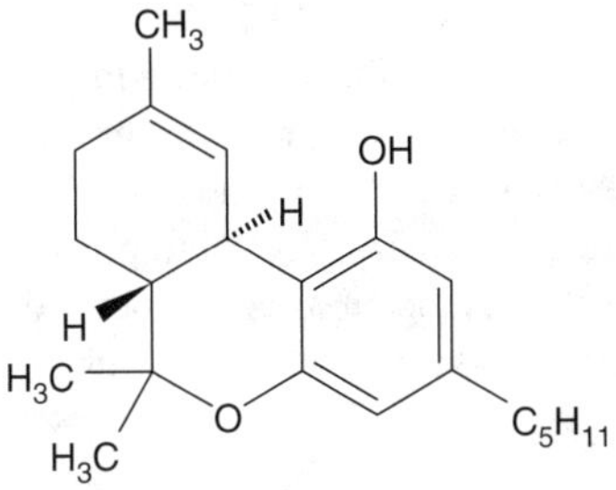

The C_{21} terpenophenolic compound present in *Cannabis sativa* L. among over 70 different cannabinoids differing in the skeleton type (11 types). $(-)$-Δ^9-*trans*-(6aR,10aR)-tetrahydrocannabinol (Δ^9-THC) is mainly responsible for the hallucinogenic properties of cannabis products such as marijuana. Recent clinical studies indicate that THC may be useful in stimulating appetite and preventing weight loss in cancer and AIDS patients. Promising results were also obtained with oral THC in the treatment of Tourette's syndrome [72].

Nine other Δ^9-THC-type cannabinoids were identified in *C. sativa*. The *cis*-isomer of Δ^9-THC was found as a major contaminant in samples of confiscated marijuana; however, its $(+)$-enantiomer or a racemate is not known. Cannabidiol CBD is the parent structure of six other representatives of this skeleton type having *trans*-(1R,6R) configuration and negative optical rotation. Cannabigerol CBG and CBG-type compounds lack CNS activity but were shown to have a marked antibacterial effect against Gram-positive bacterial strains. Except for cannabinerolic acid, which is the *trans*-isomer of cannabigerolic acid, other CBG-type compounds are all *cis*-isomers. Cannabitriol CBT is yet another compound worth mentioning, both $(+)$-*cis*-isomer and $(+)$-*trans*-cannabitriol were detected in the plant; however, the individual $(+)$- and $(-)$-*cis*-isomers have not been isolated separately [73].

10.6.3.2   Selected classes of compounds

Below, a few other selected examples of medicinally important classes of natural chiral compounds are given.

10.6.3.2.1   Monoterpenoids and sesquiterpenoids
Chiral chemodiversity is particularly displayed among monoterpenoids and sesquiterpenoids [41]. Some monoterpene compounds have already been described. The parent compound for monoterpenoids — geranyl pyrophospate (GPP) — gives rise to over 1000 compounds which may be acyclic, monocyclic, or bicyclic.

Tricyclic skeletons are rarely found. Research data indicate that specific cyclases are involved in the synthesis of individual enantiomers [41,63]. Monoterpenoids are most often the constituents of volatile oils, which are widely utilized in perfumery and in medicine as carminatives, sedatives, expectorants, and antimicrobial agents. Some most important essential oils are derived from peppermint leaf, lavender, coriander, caraway, lemon peel, turpentine, rosemary, sage, and others.

Sesquiterpenoids, similar to monoterpenes, very often occur as constituents of volatile oils, for example, zingiberene in ginger oil, (−)-*trans*-*β*-caryophyllene in clove leaf oil, and farnesol in rose oil.

(−)-*trans*-*β*-Caryophyllene

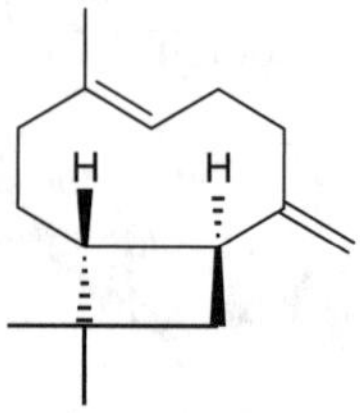

Acyclic, monocyclic, and bicyclic sesquiterpenoids are derived from 2*E*,6*E*-farnesyl pyrophosphate FPP or its 2*Z*,6*E* isomer. Similarly to monoterpenes, these compounds are also very abundant representatives of secondary metabolites; several thousands have been identified so far. Both marine and plant sesquiterpenes may have antifeedant properties. Sesquiterpene hydrocarbons from natural sources in most cases appear to be homochiral, sometimes enantiomeric mixtures are found; however, this is rather unusual as compared with monoterpene hydrocarbons for which the occurrence of both enantiomers of a single compound is considered common [63]. It is interesting to note that in many cases sesquiterpene enantiomers with opposite configuration as compared with higher plants are found in the volatile oils of liverworts [74]. Sesquiterpene lactones are characteristic of *Asteraceae*, but are also found in other families such as *Apiaceae* and *Magnoliaceae* [62,75]. Many of them have interesting pharmacological activities, antimigraine parthenolid (*Chrysanthemum parthenium*), bitter artabsine (*Artemisia absinthium*), or antimalarial artemisinine (*Artemisia annua*) are only some more prominent examples.

Artabsine

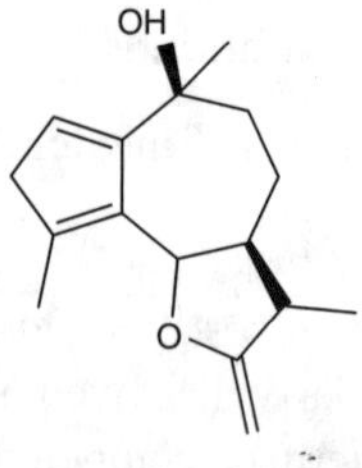

Gossypol is a dimeric sesquiterpene found in the immature flower buds and seed kernels of the cotton plant (*Gossypium* sp., *Malvaceae*). The compound has gained much attention as a potential male antifertility agent. In the plant it is present as a mixture, the enantiomer ratio is species-related. The (−)-isomer, which appears to be pharmacologically active principle, was found in excess in the seeds of various varieties of *Gossypium barbadense*, whereas the (+)-isomer was a predominant constituent in *G. arboreum*, *G. herbaceum*, and *G. hirsutum*. Unfortunately, the biological half-life of (−)-gossypol is very short [42].

10.6.3.2.2 Flavonoids

Flavonoids are phenolic compounds with a wide range of pharmacological activities, including antioxidant and vasoprotective properties. There are several groups of flavonoids, which have centers of asymmetry in their structures. Flavanones (e.g., pinocembrin, naringenin, and liquiritigenin), with one chiral carbon C2 may give rise to two isomeric forms, but all naturally occurring levorotatory (−)-flavanones have the (2S)-configuration. It seems that (−)-2S-naringenin, which is formed stereospecifically from the corresponding 4,2′,4′,6′-tetrahydroxychalcone, is a central intermediate in the biosynthesis of flavonoids [76]. Dihydroflavonols (e.g., pinobanksin and taxifolin), with two centers of asymmetry at C2 and C3 may have four isomers. Even though a few compounds with 2S,3S and 2R,3S configuration were reported the majority has a 2R,3R configuration. Another group of related compounds are flavan-3-ols, the precursor units in the biosynthesis of oligomeric procyanidins, which are cardiotonic agents, and of condensed tannins. Polyhydroxy-flavan-3-ols are considered to be important preventive agents in cardiovascular diseases and some forms of cancer. The basic structure of this group is catechin, which has two asymmetric carbons C2 and C3. There are four optical isomers possible: (+)-catechin, (−)-catechin, (−)-epicatechin, and (+)-epicatechin. Normally, in plants 2R flavan-3-ols and the corresponding procyanidins are biosynthesized so in most plant extracts only (+)-catechin and (−)-epicatechin are identified.

(+)-Catechin

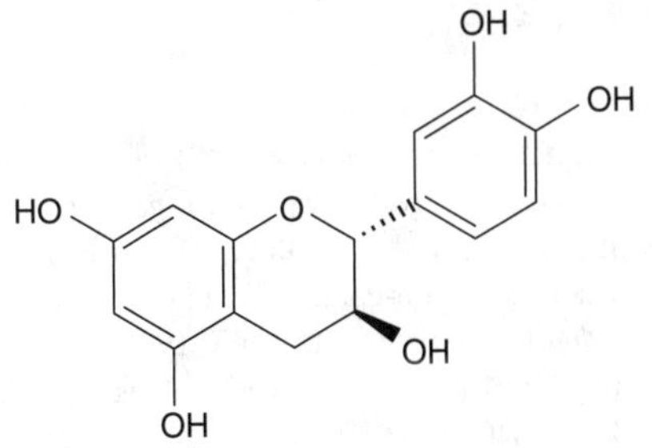

In certain monocotyledonous families, however, compounds with a 2S configuration were identified [66, 77, 78]. It is interesting to note that (−)-catechin, absent in cocoa beans, is predominant in chocolate, most probably as a result of epimerization of (−)-epicatechin during the manufacturing process [80].

10.6.3.2.3 Lignans

Lignans are yet another group of chiral plant metabolites. They are composed of two phenylpropane units coupled by a bond between the β-positions in the propane side chains. Dimers of two phenylpropanoids that are not linked by the bond between the central carbon atoms of the side chains are termed neolignans. In plants lignans occur mostly enantiomerically pure or in enantiomeric excess. They are present mostly in the free form, as dimers, trimers, or tetramers, but can also be glycosidized. They are widely distributed, especially in softwood species (angiosperms), for example, (+)-pinoresinol, and in many vascular plants.

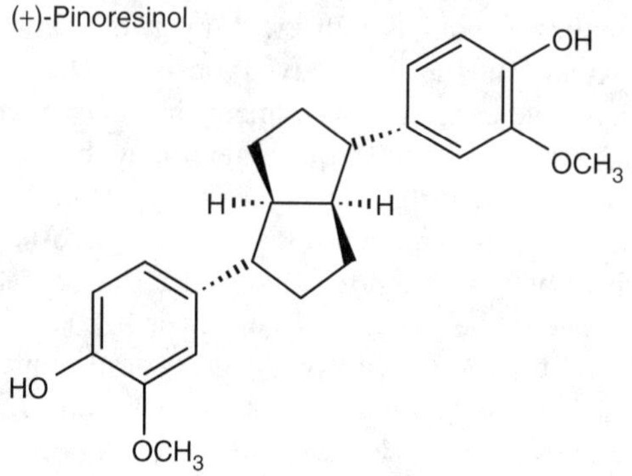

(+)-Pinoresinol

Considerable amounts of lignans are found in foods such as sesame seeds (*Sesamum indicum*), flaxseed (*Linum usitatissimum*), cereals, olive oil, and beverages such as beer, tea, coffee, and wine. Many of them possess interesting pharmacological activity: antitumor, antioxidant, cardiovascular, antihepatotoxic, adaptogenic, and other [80,81]. *Schizandra chinensis* fruit is one of the most important medicinal lignan plants. Some reports indicate a correlation between the presence of "mammalian" lignans enterolactone and enterodiol, which are produced in the colon from plant lignans, and a lower incidence rate of breast, prostate, and colon cancers, and reduced risk of cardiovascular heart disease [80,81].

REFERENCES

1. Drayer, D.E., Pharmacodynamic and pharmacokinetic differences between drug enantiomers in humans: An overview, *Clin. Pharmacol. Ther.*, 40, 125, 1986.
2. Dupau, P., Bruneau, C., and Dixneuf, P.H., New route to optically active amine derivatives: Ruthenium-catalyzed enantioselective hydrogenation of ene carbamates, *Tetrahedron Asymm.*, 10, 3467, 1999.
3. Snell, D. and Colby, J., Enantioselective hydrolysis of racemic ibuprofen amide to *S*-(+)-ibuprofen by *Rhodococcus* AJ270, *Enzym. Microb. Tech.*, 24, 160, 1999.
4. Görög, S. and Gazdag, M., Enantiomeric derivatization for biomedical chromatography, *J. Chromatogr. B Biomed. Appl.*, 659, 51, 1994.
5. Aboul-Enein, H.Y. and Wainer, I.W., *The Impact of Stereochemistry on Drug Development and Use*, John Wiley & Sons, Inc., New York, 1997, Chap. 1.
6. Cahn, R.S., Ingold, C.K., and Prelog, V., Specification of molecular chirality, *Angrew. Chem. Int. Ed.*, 5, 385, 1966.
7. Šubert, J. and Šlais, K., Progress in the separation of enantiomers of chiral drugs by TLC without their prior derivatization, *Pharmazie*, 56, 355, 2001.

8. Lai, F., Mayer, A., and Sheenhan, T., Chiral separation and detection enhancement of propranolol using automated precolumn derivatization, *J. Pharm. Biomed. Anal.*, 11, 117, 1993.

9. Fillet, M. et al., Enantiomeric purity determination of propranolol by cyclodextrin-modified capillary electrophoresis, *J. Chromatogr. A*, 717, 203, 1995.

10. Wren, S.A.C. and Rowe, R.C., Theoretical aspects of chiral separation in capillary electrophoresis. I. Initial evaluation of a model, *J. Chromatogr. A*, 603, 235, 1992.

11. Bojarski, J., Antibiotics as electrophoretic and chromatographic chiral selectors (in Polish), *Wiadomoœci Chemiczne*, 53, 236, 1999.

12. Matysik, G., *Problems of optimization of thin-layer chromatography (in Polish)*, AM, Lublin, 1997, Chap. 7.

13. Katz, E., *Quantitative Analysis using Chromatographic Techniques*, John Wiley & Sons, Chichester, New York, 1987.

14. Shellard, E.J., *Quantitative Thin-Layer Chromatography*, Academic Press, London, 1968.

15. Williams, K. and Lee, K., Importance of drug enantiomers in clinical pharmacology, *Drugs*, 30, 333, 1985.

16. Ariëns, E.J., Stereoselectivity in the action of drugs, *Pharmacol. Toxicol.*, 64, 319, 1989.

17. Ariëns, E.J., Nonchiral, homochiral and composite chiral drugs, *Trends Pharmacol. Sci.*, 14, 68, 1993.

18. Janicki, P.K., Rewerski, W., and Gomułka, W., Importance of drug stereochemistry in therapy (in Polish), *Pol. Tygodnik Lek.*, 45, 36, 1990.

19. *The United States Pharmacopeia (USP 24)*, Twinbrook, Parkway, Rockville, 2000.

20. *FDA's policy statement for the development of new stereoisomeric drugs*, U.S. Food and Drug Administration, May 1, 1992.

21. D'Arcy, P.F., Harron, D.W.G., *Proceedings of the Second International Conference on Harmonisation Orlando 1993*. The Queen's University, Belfast, UK.

22. Commission Europeenne de Pharmacopee, dokument PA/PH/SG (94) 123, 1994.

23. Ahuja, S. and Dong, M.W., *Handbook of Pharmaceutical Analysis by HPLC*, Elsevier Academic Press, 2005, Chap. 18.

24. Committee for Proprietary Medicinal Products Working Parties on Quality, Safety, and Efficacy of Medicinal Products. *Note for guidance; investigation of chiral active* III/3501/91, 1993.

25. Chen, X., Zhong, D., and Blum, H., Stereoselective pharmacokinetics of propafenone and its major metabolites in healthy Chinese volunteers, *Eur. J. Pharm. Sci.*, 10, 11, 2000.

26. Tracy, T.S., Role of cytochrome P450 2C9 and an allelic variant in the $4'$-hydroxylation of (R)- and (S)-flurbiprofen, *Biochem. Pharmacol.*, 49, 1269, 1995.

27. Panus, P.C. et al., Transdermal iontophoretic delivery of ketoprofen through human cadaver skin and in humans, *J. Controlled Rel.*, 44, 113, 1997.

28. Eichelbaum, M., Pharmacokinetic and pharmacodynamic consequence of stereoselective drug metabolism in man, *Biochem. Pharmacol.*, 37, 93, 1988.

29. Meese, C.O. and Eichelbaum, M., Stereoselective drug metabolism and pharmacodynamic consequence, in *Xenobiotic Metabolism and Disposition*, Kato, R., Estabrook, R.W., and Cayen, M.N., Eds., Taylor & Francis, London, 1989.

30. Evans, A.M., Pharmacodynamics and pharmacokinetics of the profens: Enantio-selectivity, clinical implication and special reference to $S(+)$-ibuprofen, *J. Clin. Pharmacol.*, 36, 7, 1996.

31. Sajewicz, M. et al., Application of thin-layer chromatography (TLC) to investig-ating oscillatory instability of the selected profen enantiomers, *Acta Chromatogr.*, 15, 131, 2005.

32. Sajewicz, M. et al., Application of thin-layer chromatography (TLC) to investig-ate oscillatory instability of the selected profen enantiomers in dichloromethane, *J. Chromatogr. Sci.*, 43, 542, 2005.

33. Podlewski, J.K. and Chwalibogowska, A., *Drugs for modern therapy (in Polish)*, Split Trading Sp. z o.o., Warszawa, 2000.

34. Drayer, D.E., Protein binding of drugs, drug enantiomers, active drug metabolites, and therapeutic drug monitoring, in *Drug–Protein Binding*, Reidenberg, M. and Erill, E., Eds., Praeger, New York, 1989.

35. Zajac, M. and Pawełczyk, E., *Chemistry of drugs (in Polish)*, AM, Poznań, 2002.

36. Tucker, G.T., Chiral switches, *Lancet*, 355, 1085, 2000.

37. Morissey, I. et al., Mechanism of differential activities of ofloxacin, *Antimicrob. Agents Chemother.*, 40, 1775, 1996.

38. Triggle, D.J., Stereoselectivity of drug action, *DDT*, 2, 138, 1997.

39. Lukša, J. et al., Pharmacokinetic behavior of R-$(+)$- and S-$(-)$-amlodipine after single enantiomer administration, *J. Chromatogr. B Biomed. Appl.*, 703, 185, 1997.

40. Fraternalli, F., Anselimi, C., and Temussi, P.A., Neurologically active plant com-pounds and peptide hormones: A chirality connection, *FEBS Lett.*, 448, 217, 1999.

41. Norin, T., Chiral chemodiversity and its role for biological activity. Some obser-vations from studies on insect/insect and insect/plant relationships, *Pure Appl. Chem.*, 68, 2043, 1996.

42. Cordell, G.A., Changing strategies in natural products chemistry, *Phytochemistry*, 40, 1585, 1995.

43. Mateus, L. et al., Enantioseparation of atropine by capillary electrophoresis using sulfated β-cyclodextrin: Application to a plant extract, *J. Chromatogr. A*, 868, 285, 2000.

44. Tahara, S. et al., Enantiomeric separation of atropine in Scopolia extract and Sco-polia rhizome by capillary electrophoresis using cyclodextrins as chiral selectors, *J. Chromatogr. A*, 848, 465, 1999.

45. Brenna, E., Fuganti, C., and Serra, S., Enantioselective perception of chiral odorants, *Tetrahedron Asymm.*, 14, 1, 2003.

46. EC Opinion of the Scientific Committee on Food on Thujone, 2002.

47. NIH Executive Summary: Thujone Information, 2006.

48. Patočka, J. and Plucar, B., Pharmacology and toxicology of absinthe, *J. Appl. Biomed.*, 1, 199, 2003.

49. Emmert, J. et al., Determination of α-/β-thujone and related terpenes in absinthe using solid phase extraction and gas chromatography, *Deut. Lebensm.-Rundsch.*, 9, 352, 2004.

50. Wang, M. et al., Enantiomeric separation and quantification of ephedrine-type alkaloids in herbal materials by comprehensive two-dimensional gas chromato-graphy, *J. Chromatogr. A*, 1112, 361, 2006.

51. Phinney, K.W., Ihara, T., and Sander, L.C., Determination of ephedrine alkaloid stereoisomers in dietary supplements by capillary electrophoresis, *J. Chromatogr. A*, 1077, 90, 2005.

52. Vansal, S.S. and Feller, D.R., Direct effects of ephedrine isomers on human β-adrenergic receptor subtypes, *Biochem. Pharmacol.*, 58, 807, 1999.

53. Sybilska, D. and Asztemborska, M., Chiral recognition of terpenoids in some pharmaceuticals derived from natural sources, *J. Biochem. Biophys. Methods*, 54, 187, 2002.

54. Pellati, F., Benvenuti, S., and Melegari, M., Enantioselective LC analysis of synephrine in natural products on a protein-based chiral stationary phase, *J. Pharm. Biomed. Anal.*, 37, 839, 2005.

55. De Smet, P.A.G.M., The role of plant-derived drugs and herbal medicines in health care, *Drugs*, 54, 801, 1997.

56. Cragg, G.M. et al., International collaboration in drug discovery and development: The NCI experience, *Pure Appl. Chem.*, 71, 1619, 1999.

57. Caccamese, S. and Prinzipato, G., Separation of the four pairs of enantiomers of vincamine alkaloids by enantioselective high-performance liquid chromatography, *J. Chromatogr. A*, 893, 47, 2000.

58. McCalley, D.V., Analysis of the Cinchona alkaloids by high-performance liquid chromatography and other separation techniques, *J. Chromatogr. A*, 967, 1, 2002.

59. Caldwell, J., Stereochemical determinants of the nature and consequences of drug metabolism, *J. Chromatogr. A*, 694, 39, 1995.

60. Komarova, E.L. and Tolkachev, O.N., The chemistry of peptide ergot alkaloids. Part I. Classification and chemistry of ergot peptides, *Pharm. Chem. J.*, 35, 504, 2001.

61. Wang, X. et al., Enantiomeric composition of monoterpenes in conifer resins, *Tetrahedron Asymm.*, 8, 3977, 1997.

62. Dey, P.M. and Harborne, J.B., *Methods in Plant Biochemistry*, vol. 7, Academic Press, London, 1991.

63. Ruiz del Castillo, M.L., Blanch, G.P., and Herraiz, M., Natural variability of the enantiomeric composition of biactive chiral terpenes in *Mentha piperita*, *J. Chromatogr. A*, 1054, 87, 2004.

64. EC Opinion of the Scientific Committee on Food on pulegone and menthofuran, 2002.

65. Matura, M. et al., Oxidized citrus oil (*R*-limonene): A frequent skin sensitizer in Europe, *J. Am. Acad. Dermatol.*, 47, 709, 2002.

66. Modnicki, D. and Klimek, B., Isomeric forms of naturally occurring compounds and their biological and pharmacological activities (in Polish), *Farm. Pol.*, 57, 620, 2001.

67. Laska, M., Liesen, A., and Teubner, P., Enantioselectivity of odor perception in squirrel monkeys and humans, *Am. J. Physiol. Regul. Integr. Comp. Physiol.*, 277, R1098, 1999.

68. Cocchietto, M. et al., A review on usnic acid, an interesting natural compound, *Naturwissenschaften*, 89, 137, 2002.

69. Tay, T. et al., Evaluation of the antimicrobial activity of the acetone extract of the lichen *Ramalina farinacea* and its (+)-usnic acid, norstictic acid and protocetraric acid constituents, *Zeitschr. Naturforsch.*, 59c, 384, 2004.

70. Yilmaz, M. et al., The antimicrobial activity of extracts of the lichen *Clado-nia foliacea* and its (−)-usnic acid, atranorin, and fumarprotocetraric acid constituents, *Zeitschr. Naturforsch.*, 59c, 249, 2004.

71. Lauterwein, M. et al., *In vitro* activities of the lichen secondary metabolites vulpinic acid, (+)-usnic acid and (−)-usnic acid against aerobic and anaerobic microorganisms, *Antimicrob. Agents Chemother.*, 39, 2541, 1995.

72. Ben Amar, M., Cannabinoids in medicine: A review of their therapeutic potential, *J. Ethnopharmacol.*, 105, 1, 2006.

73. ElSohly, M.A. and Slade, D., Chemical constituents of marijuana: The complex mixture of natural cannabinoids, *Life Sci.*, 78, 539, 2005.

74. Fricke, C. et al., Identification of (+)-β-caryophyllene in essential oils of liv-erworts by enantioselective gas chromatography, *Phytochemistry*, 39, 1119, 1995.

75. Merfort, I., Review of analytical techniques for sesquiterpenes and sesquiterpene lactones, *J. Chromatogr. A*, 967, 115, 2002.

76. Gel-Moreto, N., Streich, R., and Galensa, R., Chiral separation of six diaste-reomeric flavanone-7-*O*-glycosides by capillary electrophoresis and analysis of lemon juice, *J. Chromatogr. A*, 925, 279, 2001.

77. Dey, P.M. and Harborne, J.B., *Methods in Plant Biochemistry*, vol. 1, Academic Press, London, 1989.

78. Nowak, R. and Hawryl, M., Application of densitometry to the determination of catechin in rose-hip extracts, *J. Planar Chromatogr.*, 18, 217, 2005.

79. Gotti, R. et al., Analysis of catechins in *Theobroma cacao* beans by cyclodextrin-modified micellar electrokinetic chromatography, *J. Chromatogr. A*, 1112, 345, 2006.

80. Willför, S.M., Smeds, A.I., and Holmbom, B.R., Chromatographic analysis of lignans, *J. Chromatogr. A*, 1112, 64, 2006.

81. Slanina, J. and Glatz, Z., Separation procedures applicable to lignan analysis, *J. Chromatogr. B*, 812, 215, 2004.

11 Chiral Separation of β-Adrenergic Antagonists

Danica Agbaba and Branka Ivković

CONTENTS

11.1 INTRODUCTION

Stinson [1] reported that worldwide sales of chiral drugs in single-enantiomer dosage forms continued growing at more than 13% annual rate to $133 billion in 2000 and, following this progression rate, that sales might reach $200 billion in 2008. This could be partly a result of the fact that the U.S. Food and Drug Administration is giving new life to old products by approving the marketing of established drugs and offering benefits when one enantiomer is used instead of the racemic mixture [2].

Ours is an asymmetric world and life is not even-handed. Asymmetry plays an important role in the understanding and definition of biologically important interactions and as the consequence further in research, design, and development of drugs. Receptors are chiral entities and specific binding of either physiological or pharmacological ligands at these specific entities exhibits chirality of interaction [3].

Considering that different types and subtypes of adrenergic receptors are highly stereoselective, their chirality plays an important role in specific ligand–receptor interactions, as well as in pharmacokinetics, pharmacodynamics, and toxicology of adrenergic drugs acting as agonists, partial agonists, or antagonists. Drugs affecting adrenergic receptors as antagonists are mostly used in the treatment of different cardiovascular diseases, for example, hypertension, angina pectoris, cardiac arrhythmias, and glaucoma.

The group of drugs targeting β-adrenergic receptors as antagonists (β-blockers) exhibit a high stereoselectivity in provoking their β-blocking effects. At present, two structural classes of derivatives such as arylethanolamines and aryloxypropanolamines are clinically applied as β-blocking agents. Chemical structures of some commonly used β-blockers are presented in Figure 11.1. The configuration of the hydroxyl-bearing carbon in the side chain of both classes of the above derivatives plays a principal role in their interaction with β-receptors. First of all, for optimal activity at the β-receptor, the hydroxyl-bearing carbon in the side chain must be in either (R) or (S) configuration. Besides, relative positions of the four functional groups must have the same spatial arrangement in the two structures. Introduction of oxygen atom into the side chain of the aryloxypropanolamine changes the priority of two of the groups used in the nomenclature assignment. The structure of aryl moiety in both groups differs and substituents were shown to affect the selectivity of the molecules for different β-receptor subtypes, as well as lipophilicity, metabolism, and route of excretion of unchanged drug or the corresponding metabolites [4,5].

The stereoselectivity to different types of adrenergic receptors was observed in labetalol. This pharmaceutical has two chiral centers and exists as a mixture of four isomers. Different isomers, however, possess different α- and β-antagonistic activities. The β-blocking activity resides solely in the ($1R,1'R$) isomer while the α_1-antagonistic activity is seen in the ($1S,1'R$) and ($1S,1'S$) isomers, the ($1S,1'R$) isomer possessing the highest activity.

The β-blocking potency of $S(-)$ propranolol is ca. 40 times greater than that of $R(+)$ enantiomer. In humans, differences in the extent of binding to plasma proteins of different enantiomers have been observed. So, $R(+)$ propranolol binds to a lesser extent to plasma proteins than its $S(-)$ antipode and as the result, the bioavailability of $S(-)$ form exceeds that of the $R(+)$ isomer. No stereoselective plasma binding in humans was observed for pindolol enantiomers, but after oral administration of racemic pindolol, the $S(-)$ enantiomer achieved a higher plasma concentration than less active R-($+$)-pindolol. The renal clearance of the $S(-)$ enantiomer of pindolol was greater than that of its $R(+)$ counterpart. Metoprolol is a selective antagonist of β_1-adrenergic receptors and β-blocking capacity has been shown to reside predominantly in the $S(-)$ enantiomer, while $R(+)$ enantiomer does not contribute to this effect. Also, the bioavailability of $S(-)$ metoprolol was found to be higher than that of its $R(+)$ antipode. Although poorly metabolized, the

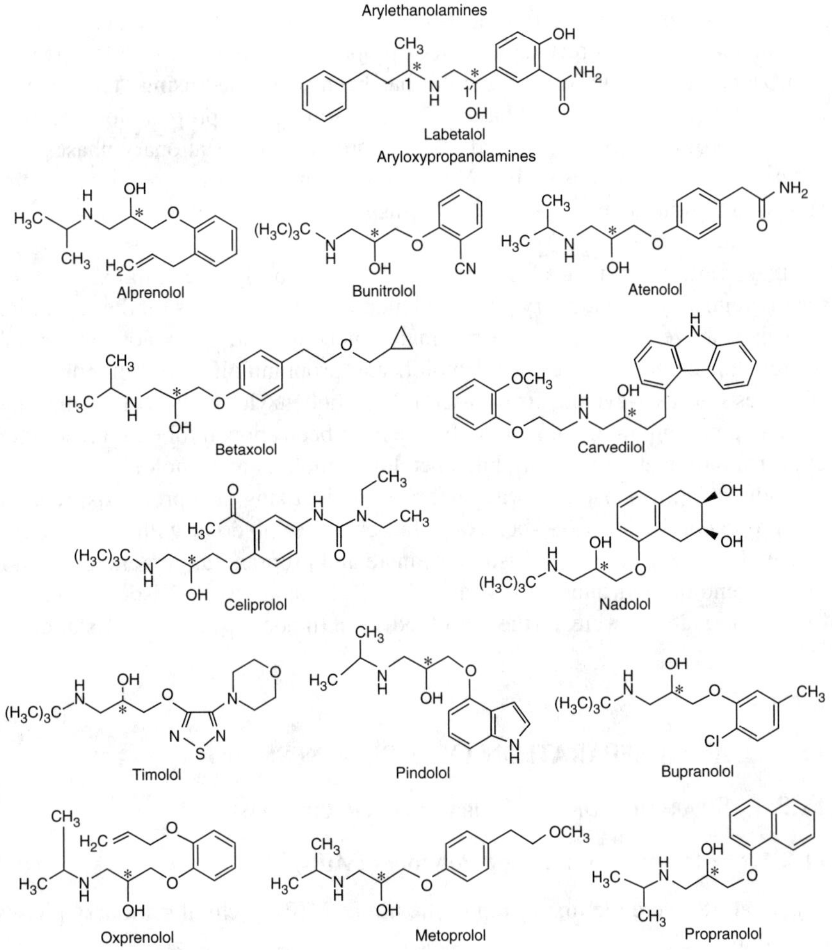

FIGURE 11.1 Chemical structure of selected β-blockers.

differences in the half-life of metoprolol enantiomers, as well as in their metabolic pathways have been observed [6].

In spite of the fact that nearly complete β-antagonistic activity resides in one enantiomer, propranolol and most of other β-blockers are applied in the therapy as racemic mixtures. The only exceptions are levobunolol, timolol, and penbutolol used as the (S) enantiomers.

To verify the enantiomeric purity and composition of pharmaceutical drugs either in bulk drug and dosage forms or in biological fluids, numerous separation techniques have been developed and employed, so far [7]. For that purpose, different chromatographic procedures such as HPLC, GC, and TLC have been used, although TLC has several advantages compared with other methods referring to simplicity, rapidity, low cost, and at the same time, detectability.

Chiral separations of different β-blockers have been performed using direct and indirect modes. Direct separation of propranolol, pindolol, nadolol, oxprenolol, atenolol, timolol, and bupranolol has been performed using "tailor-made" and native chiral stationary phases. Direct resolution of propranolol, atenolol, and metoprolol has been achieved using impregnation of stationary phases with optically pure compounds such as N-(3,5-dinitrobenzoyl)-R-(−)-α-phenylglycine and N-(3,5-dinitrobenzoyl)-L-leucine, L-aspartic acid, L-arginine, L-lysine, and D(−) tartaric acid.

D(−) Tartaric acid has also been used simultaneously as a chiral selector for the impregnation of stationary phase and mobile phase additives for the separation of a part above mentioned, the racemates of celiprolol, betaxolol, alprenolol, oxprenolol, carvedilol, atenolol, timolol, and propranolol. Chiral mobile phase additives such as β-cyclodextrin (β-CD), N-carbobenzyloxy-glycyl-L-proline and (R)-(−)-ammonium-10-camphorsulfonate have been applied for chiral resolution of propanolol, metoprolol, timolol, labetalol, pindolol, and atenolol.

Indirect chiral separation was performed derivatizing metoprolol, oxprenolol, and propranolol with S-(+)-benoxaprofen chloride, pindolol with 2,3,4,6-tetra-O-acetyl-β-D-glucopyranosyl-isothiocyanate and propranolol, alprenolol, pindolol, oxprenolol, and bunitrolol with R-(−)-1-(1-naphthyl)ethyl isocyanate. The formed derivatives were further resolved to antipodes using achiral stationary phases.

11.2 DIRECT SEPARATION OF β-BLOCKERS

11.2.1 SEPARATION ON THE CHIRAL STATIONARY PHASES

11.2.1.1 Molecular Imprinted Polymers (MIPs)

Kriz et al. [8] were the first to report the use of MIPs as chiral stationary phases (CSPs) in TLC.

Molecular imprinting is a technique for the preparation of "tailor-made" selectors, which are used for the separation of numerous classes of both achiral and chiral compounds. Imprinting of molecules occurs by the polymerization of functional and cross-linking monomers in the presence of a template/target ligand. Based on molecular recognition approach, the polymers/adsorbents are tailored to recognize molecules of interest, which are to be resolved [9]. Therefore, in the preparation of polymers for chiral discrimination of β-blockers, S-(−)-timolol, R-(+)-propranolol, R-(+)-atenolol, (+)-ephedrine, (+)-pseudoephedrine, and (+)-norephedrine were employed as the templates or imprint molecules. In fact, the template is initially allowed to establish bond formation with polymerizable functionality and the resulting complexes are subsequently copolymerized with cross-linkers into a rigid polymer. For that purpose, several functional monomers such as methacrylic acid (MAA) [10–12] and itraconic acid (ITA) [12] have been used so far, a much better thin-layer coating being obtained with MAA than with ITA. The functional monomer usually includes two functional groups. The compound binds by one end of the molecule with the template via noncovalent interactions (hydrogen bonding, van der Waals forces, or hydrophobic interactions) or reversible covalent interactions. At the other end of the monomer (i.e., the end that is not interacting with the template), there is a group that can bind covalently to the cross-linker. The cross-linker polymerizes the monomers around the template by covalent binding and holds them in place after the template is removed. In order to prevent the covering of the entire binding site with polymer, thus avoiding the difficulties in removing the template from the imprint, careful selection of an appropriate cross-linker and its concentration are of utmost importance [13]. Ethylene glycol dimethacrylate (EDMA) represents one of widely employed cross-linkers. The MIP can be prepared either by photo polymerization ($\lambda = 366$ nm, 4°C, 24 h, in tetrahydrofuran) [10], thermal polymerization (40°C, 16 h, in dichloromethane [11], or at 60°C for 6 h in chloroform, as reported by Suedee et al. [12]). These authors [12] emphasized that the type of solvent employed in the polymerization process significantly affects the characteristics of the MIP stationary layer. After the preparation, polymer has to be ground and sieved through a 100 μm sieve. In order to remove the template after sieving, the polymer particles were soaked either in the solvent mixture of acetic acid and methanol (1:9, v/v) for 24 h [11], or in the mixture of acetonitrile and acetic acid (9:1, v/v) for 24 h [10,12], washed with methanol and acetonitrile, respectively, and dried overnight *in vacuo*. Nonimprinted polymers used as a control were prepared identically, but in the absence of a print molecule. For the chiral separation, equal amounts of polymer and silica gel 6 GF_{254} [10], or polymer and calcium sulfate [11,12] should be mixed and applied as the coating material. The composition of the mobile phase used for the enantioseparation of β-blockers and the values of separation factor α are listed in Table 11.1. Visualization of spots was performed either under UV light ($\lambda = 254$ nm) [10] or spraying with anisaldehyde reagent [11,12]. The separation distance was up to 7.0 cm and separation time less than 7 min. Aboul-Enein et al. [10] found that mobile phase consisting of 5% acetic acid in acetonitrile or 1% acetic acid in methanol provided the best resolution of propranolol, timolol, and atenolol enantiomers. Higher concentration of acetic acid in the mobile phase, in most cases, increased the R_f values and reduced the tailing of the spots, as well as chiral separation factor. This could be ascribed to the competition of acetic acid with the analyte for the binding sites on MIP. The S-(−)-timolol-imprinted MIP showed enantiomeric selectivity for structurally related compounds such as pindolol and atenolol probably owing to the similarity of the side chain in this class of compounds that contain the chiral center. In resolutions of these compounds, the R_f values of S-(−) enantiomers were lower than those of their R-(+) counterparts, indicating a higher affinity of the S-(−)-enantiomers to the stationary phase comparing to that of their R-(+) antipodes.

From the results of Suedee et al. [11] listed in Table 11.2, it is clear that the geometry around the chiral carbon (β-C) in the print molecules of (+)-ephedrine, (+)-pseudoephedrine, and (+)-norephedrine was essential for enantiomeric recognition and separation of β-blocker racemates. Since (+)-antipodes of print molecules possess the same spatial arrangements as (−)-antipodes of β-blockers, this could explain a higher affinity and lower R_f values of the later. The substituents

TABLE 11.1

β-Blocker Enantiomers Separated by Imprinted CSPs

β-Blocker	Functional monomer	Imprint molecule	Mobile phase	α	Ref.
(±) Propranolol	Methacrylic acid	(−)-S-Timolol	5% Acetic acid in acetonitrile	1.52	[10]
			1% Acetic acid in methanol	1.47	
		R-(+)-Propranolol	5% Acetic acid in acetonitrile	2.4	[12]
		(+)-Ephedrine	5% Acetic acid in dichloromethane	1.30	[11]
		(+)-Pseudoephedrine	5% Acetic acid in dichloromethane	1.20	
		(+)-Norephedrine	5% Acetic acid in dichloromethane	1.25	
(±) Timolol	Methacrylic acid	(−)-S-Timolol	5% Acetic acid in acetonitrile	1.60	[10]
			1% Acetic acid in methanol	1.52	
(±) Atenolol	Methacrylic acid	(−)-S-Timolol	5% Acetic acid in acetonitrile	1.59	[10]
			1% Acetic acid in methanol	1.50	
(±) Oxprenolol	Methacrylic acid	(+)-Ephedrine	5% Acetic acid in dichloromethane	1.32	[11]
		(+)-Pseudoephedrine	5% Acetic acid in dichloromethane	1.32	
		(+)-Norephedrine	5% Acetic acid in dichloromethane	1.43	
(±) Pindolol	Methacrylic acid	(+)-Ephedrine	7% Acetic acid in dichloromethane	1.57	[11]
		(+)-Pseudoephedrine	7% Acetic acid in dichloromethane	1.50	
		(+)-Norephedrine	7% Acetic acid in dichloromethane	1.29	
(±) Nadolol	Methacrylic acid	(+)-Ephedrine	7% Acetic acid in dichloromethane	1	[11]
		(+)-Pseudoephedrine	7% Acetic acid in dichloromethane	1	
		(+)-Norephedrine	7% Acetic acid in dichloromethane	1	

TABLE 11.2
Retention and Resolution Data of Some β-Adrenergic Drugs

β-Blocker	Print molecule	hR_f value	α
(−)-Propranolol	(+)-Ephedrine	56	1.3 (−)
(+)-Propranolol		73	
(−)-Pindolol		14	1.57
(+)-Pindolol		22	
(−)-Oxprenolol		40	1.32
(+)-Oxprenolol		53	
(−)-Nadolol		4	1.00
(+)-Nadolol		4	
(−)-Propranolol	(+)-Pseudoephedrine	45	1.2 (−)
(+)-Propranolol		54	
(−)-Pindolol		24	1.5
(+)-Pindolol		36	
(−)-Oxprenolol		44	1.32
(+)-Oxprenolol		58	
(−)-Nadolol		33	1.0
(+)-Nadolol		33	
(−)-Propranolol	(+)-Norephedrine	57	1.25 (−)
(+)-Propranolol		71	
(−)-Pindolol		31	1.29
(+)-Pindolol		40	
(−)-Oxprenolol		42	1.43
(+)-Oxprenolol		60	
(−)-Nadolol		11	1.00
(+)-Nadolol		11	

in the aromatic region, particularly lipophilic groups, did not influence the chiral recognition of the MIPs. No chiral separation was achieved with nadolol. The observed correlation between structurally related compounds could also be useful in predicting whether it would be possible to determine, screen, or identify the enantiomers of optically active compounds on MIP.

11.2.1.2 Triphenylcarbamate Derivatives of Cellulose as CSP

In addition to its availability from natural sources and enantioselectivity of cellulose, as well as of its derivatives, this natural polymer has several important advantages such as the ease of substitution and functionalization of the glucose unit hydroxyl groups. The correlation of the chiral recognition availabilities and electron and structural properties between some cellulose derivatives have already been studied [14]. The chiral recognition of benzoate and phenylcarbamate cellulose derivatives greatly depends on the type and position of the substituents introduced into the phenyl group. Therefore, Suedee and Heard [15] investigated the effect of different cellulose triphenylcarbamate derivatives as stationary phases on enantiomer resolution of known β-blockers propranolol and bupranolol by TLC. These authors examined *tris*-phenylcarbamate-, *tris*-(2,3-chlorophenylcarbamate)-, *tris*-(2,4-dichlorophenylcarbamate)-, *tris*-(2,6-dichlorophenylcarbamate)-, *tris*-(3,4-dichlorophenylcarbamate)-, *tris*-(3,5-di-chlorophenylcarbamate)-, *tris*-(2,3-dimethylphenylcarbamate)-, and *tris*-(3,5-dimethylphenylcarbamate) cellulose derivatives. Some other investigators also have previously faced several weak points of these derivatives, for example, difficulties to prepare a stable thin layer on glass support owing to adhesion and cracking of the layer, the elution time of mobile phase was very long because of low polarity of solvents used for the separation, the visualization of the spots was limited due to an intense UV absorption of these derivatives containing strong chromophores, and the expense associated with the plate production. It was found that the addition of an equal proportion of cellulose to the derivative alleviated many of the above problems [15]. Also, adhesion to glass plates and physical properties were improved and elution time reduced (within 30 min). The spots were visualized after development of chromatograms with anisaldehyde reagent. A variety of mobile phases was used to achieve a satisfactory separation and the effects of solvent polarity were also discussed. The best resolution of propranolol and bupranolol racemates was obtained using cellulose *tris*-(3,5-dimethylphenylcarbamate) CSP ($R_f R = 0.26$, $R_f S = 0.06$, $\alpha = 4.3$) in mobile phase hexane/propan-2-ol (80:20, v/v), and cellulose *tris*(2,3-dimethylphenylcarbamate) CSP ($R_f R = 0.29$, $R_f S = 0.09$, $\alpha = 3.2$) in mobile phase hexane/propan-2-ol (80:20, v/v). It has been concluded that this can be a potentially useful method for the rapid determination of optical purity of propranolol and bupropranolol and, possibly, for the small-scale isolation of their enantiomers.

11.2.2 Separation on the Plates Impregnated with Pure Chiral Compounds

For enantioseparation of selected β-blockers, propranolol, metoprolol, and alprenolol, amino acids such as L-arginine [16,17], L-lysine [16], L-aspartic acid [18], or derivatives of amino acids, N-(3,5-dinitrobenzoyl)-R-(−)-α-phenyl glycine, and N-(3,5-dinitrobenzoyl)-L-leucine were the most frequently used chiral selectors impregnated in TLC layer [19]. Tartaric acid was employed not only as a chiral selector impregnated in the layer, but also as the CMA [20]. The structures of these chiral selectors are presented in Figure 11.2.

11.2.2.1 Separation on the Plates Impregnated with Chiral Selectors Preceded by Chemical Derivatization of the Selected β-Blockers

The separation of propranolol, atenolol, and metoprolol racemates was performed on ionically and covalently modified aminopropyl bonded HPTLC

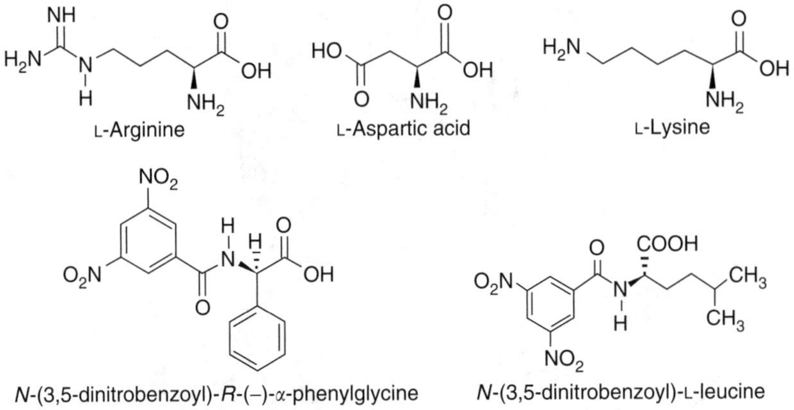

L-Arginine

L-Aspartic acid

L-Lysine

N-(3,5-dinitrobenzoyl)-R-(−)-α-phenylglycine

N-(3,5-dinitrobenzoyl)-L-leucine

FIGURE 11.2 The structures of some of chiral amino acids and their derivatives.

plates with N-(3,5-dinitrobenzoyl)-R-(−)-α-phenylglycine (CSP1) and N-(3,5-dinitobenzoyl)-L-leucine (CSP2). The results of separation of either chiral or achiral molecular species using HPLC or HPTLC on CSP1 and CSP2 revealed that the compounds subjected to optical resolution have to contain amides, carbonyl, or hydroxyl groups close to the chiral center. Hence, some classes of drugs, such as barbiturates and several benzodiazepines can be resolved without derivatization, but for the complete enantioseparation, β-blockers required the formation of naphthyl amides. So, the chromatographic resolution of β-blockers was preceded by derivatization with 1-isocyanatonaphthalene. The separated derivatives were visualized as yellow spots on a white background and no further treatment was necessary. The chromatogram representing separation of metoprolol antipodes on CSP1 is shown in Figure 11.3.

The proposed mechanism involved in chiral discrimination using CSP1 and CSP2 is based on hydrogen bonding, possible dipole stacking, and $\pi-\pi$ interactions between the aromatic groups of isomers and those in the stationary phases. The obtained chiral separation factors for propranolol, atenolol, and metoprolol were 1.1, 1.3, and 2.2, respectively.

11.2.2.2 Separation on the Plates Impregnated with Chiral Selector

Bhushan and Thiongo [16] used homemade plates, prepared by spreading of silica gel slurry in distilled water containing L-lysine or L-arginine and quite recently, Bhushan and Arora [18] employed L-aspartic acid. The best resolution of all tested β-blocker racemates was observed at 0.5% concentration of the above impregnating reagents. Changes in the concentrations of a chiral selector resulted in either poorer or no resolution of the selected racemates. So, decrease in the concentration of either of the above chiral selectors led to poorer resolution, while no resolution of atenolol and propranolol racemates were observed at increasing concentrations of

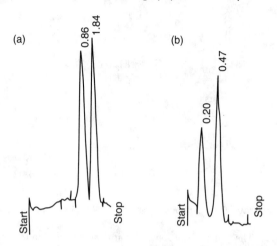

FIGURE 11.3 (a) 2,2,2-Trifluoro-1-(9-anthryl)ethanol separated on CSP1 (ionic) using hexane-2-propanol (80:20). Scanned at $\lambda = 380$ nm. (b) Metoprolol separated on CSP1 (ionic) using hexane-2-propanol–acetonitrile (80:15:5). Scanned at $\lambda = 410$ nm. (From Wall, P.E., *J. Planar Chromatogr.*, 2, 228, 1989. With permission.)

chiral selectors. Sajewicz et al. [17] also employed 0.5% L-arginine for the impregnation of commercial precoated glass TLC plates by a conventional dipping mode.

The mobile phases consisting of acetonitrile/methanol [16] or acetonitrile/methanol/water [18] in different ratios with a few drops of ammonia solution were found to discriminate between two enantiomers of these three β-blockers in one-dimensional (1D) ascending development mode. The mobile phase composition and values of chiral separation factor α are listed in Table 11.3. The zones of separated antipodes were detected using iodine vapor and the detection limit of both alprenolol and propranolol racemates was 2.6 μg, while that of metoprolol was 0.26 μg [16]. The effects of temperature on the separation of these three β-blockers were also investigated [16]. It was observed that no resolution of atenolol racemate was achieved at the temperatures higher than 15°C, as well as below 8°C. The best resolution of propranolol and metoprolol racemates was achieved at 22°C. Increase of the temperature above 22°C led to the tailing, while a decrease up to 6°C had little or no effect on the quality of resolution.

The effects of different temperatures on the separation of all three β-blockers using L-aspartic acid as a chiral selector were also investigated. It has been found that 17°C was the most suitable temperature for the resolution of the examined β-blockers, providing desired mobility to the diastereomeric ion pair formed anionic species of L-aspartic acid and protonated cations of amino moieties of the corresponding β-blockers. The presence of chiral selector *in situ* was established by treating the developed chromatograms with ninhydrin that produced a characteristic color with aspartic acid in both spots of the resolved enantiomers [18]. This method was very sensitive, enabling detection of 0.26 μg atenolol and 0.23 μg of each metoprolol and propranolol.

TABLE 11.3
Resolution of (±) β-Blockers on Plates Impregnated with Different Amino Acids

Compound	Impregnating agent	Mobile phase	α	Ref.
(±) Propranolol	30 mM L-arginine	Acetonitrile/methanol/aq. NH$_3$(15:4:0–1)	4–11	[17]
	0.5% L-arginine	Acetonitrile/methanol		[16]
		(15:3)	2.83	
		(15:4)	1.92	
	0.5% L-aspartic acid	Acetonitrile/methanol/water		[18]
		(16:5:0.5)	1.64	
	0.5% L-lysine	Acetonitrile/methanol		[16]
		(15:2)	2.83	
		(16:2)	1.92	
	N-(3,5-dinitrobenzoyl)- R-(−)-α-phenylglycine	Hexane/2-propanol (75:25)	1.08	[19]
(±) Atenolol	0.5% L-aspartic acid	Acetonitrile/methanol/water		[18]
		(18:4:2)	1.60	
	0.5% L-lysine	Acetonitrile/methanol		[16]
		(16:4)	3.33	
		(16:2)	2.00	
		(15:2)	1.75	
	0.5% L-arginine	Acetonitrile/methanol		[16]
		(15:5)	1.5	
		(14:6)	1.36	
	N-(3,5-dinitrobenzoyl)- R-(−)-α-phenylglycine	Hexane/2-propanol (75:25)	1.27	[19]
(±) Metoprolol	0.5% L-aspartic acid	Acetonitrile/methanol/water		[18]
		(10:4:1)	1.31	
	0.5% L-lysine	Acetonitrile/methanol		[16]
		(15:4)	1.36	
		(15:5)	1.67	
	0.5% L-arginine	Acetonitrile/methanol		[16]
		(15:3)	2.83	
		(16:4)	1.92	
	N-(3,5-dinitrobenzoyl)- R-(−)-α-phenylglycine	Hexane/2-propanol (75:25)	2.2	[19]

Two-dimensional mode of development was performed by Sajewicz et al. [17] who used acetonitrile/methanol (15:4, v/v) supplemented with 400 μl of aqueous ammonia for the separation of optical propanolol antipodes. The detection of chromatograms was performed by densitometric scanning at 210 nm. 3D representation of the TLC plate (Figure 11.4) with two development directions was done with an aim of getting a better insight into the enhanced chiral separation of S-(−)-propranolol from its R-(+) counterpart and into the "skewed" arrangement of the two species on the plate.

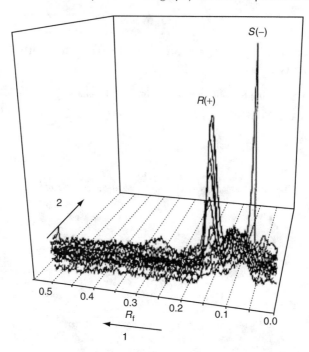

FIGURE 11.4 3D representation of part of the TLC plate with the two development directions, 1 and 2, indicated. Densitometric scanning as described in the text was performed to illustrate better the separation performance and the "skewed" arrangement of S-(−)-propranolol relative to its R-(+) counterpart. (From Sajewicz, M., Pietka, R., and Kowalska, T., *J. Liq. Chrom. Rel. Technol.*, 28, 2499, 2005. With permission.)

According to the results obtained, Sajewicz et al. [17] suggested the mechanism of ion-pair formation. These authors [17] hypothesized that L-arginine-derived anion interacts via hydrogen bonding with either of the two propranolol antipodes and affords the two pseudo-diastereoisomers with different values of their thermodynamic equilibrium constants.

Concerning the values of standard chemical adsorption potential ($\Delta\mu_a$), the S-propranolol antipode possesses a higher affinity for the adsorbent than its enantiomer but taking into account the results obtained with other chiral species such as ibuprofen, adsorption of propranolol on this stationary phase is governed not only by the moieties containing the chiral carbon, but also primarily by the aromatic moieties in the structure of chiral selector.

The results presented here gave a new prospect to TLC suggesting that this method could be used for thermodynamic assessment of the racemization process.

The most important characteristic of the chiral recognition is that the chiral selector, either present in mobile phase or impregnated in the stationary phase, has to be compatible in size and structure with the racemic species to be resolved. This approach has been used by Lučić et al. [20] who developed a method for the separation of (±)-metoprolol tartrate. In order to approach the separation, the effects of different chromatographic conditions were investigated (selection of a proper chiral selector, choice of chromatographic support, selection of organic modifier, the impregnation solution, and composition of the mobile phase). These authors [20] found that D-(−) tartaric acid was a better chiral selector than its L(+) antipode. Also, among several examined chromatographic supports (silica gel, octadecyl silica gel, precoated silica gel with concentration zone, LiChrospher Si60, and aluminum oxide), the best results were obtained when precoated silica gel with concentration zone was used for the impregnation. In selecting the most suitable organic modifier, they also tested ethanol vs. methanol and n-propanol and concluded that 70% ethanol provided the most satisfactory enantioseparation. Also, optimal resolution of optical isomers was achieved when 11.6 mM D-(−) tartaric acid in ethanol/water mixture (70:30, v/v), refluxed for 15 min at 70°C was applied as the impregnation solution and mobile phase. Concentrations of D-(−) tartaric acid below 5.8 mM resulted in an unsatisfactory resolution of enantiomers, with visible tailing. On the other hand, although increasing D-(−) tartaric acid concentration to 23 mM in both the mobile phase and the impregnation solution provided a good resolution, the time for the chiral separation was substantially longer than at 11.6 mM tartaric acid. The mechanism of separation was proposed by taking into account that tartaric acid (pK_a 2.9) in the presence of a large excess of ethanol, which shifts the reaction to the right, formed monoethyl ester, that is, monoethyl tartrate that might play the role of a real chiral selector (Figure 11.5). Silica gel with free silanol groups could form hydrogen bonds with monoethyl tartrate and thus, reduce the possibility of bond formation between metoprolol base and silanol groups of the support in favor of forming diastereoisomers (−)-metoprolol-(−)ethyl tartrate or (+)-metoprolol-(−)-ethyl tartrate which differ in chromatographic mobility.

The separated zones on the chromatogram were observed either in an iodine vapor chamber or under UV light at λ = 254 nm. The chromatograms were also scanned at λ = 230 nm with a TLC scanner.

The scanned profiles of chromatograms obtained applying different (±)-metoprolol tartrate concentrations on the plate are presented in Figure 11.6. It was observed that S-(−) enantiomer had a higher mobility than its R-(+) counterpart. The applicability of this chromatographic system for the enantioseparation of some other β-blockers is illustrated in Table 11.4. Racemates of atenolol, alprenolol, betaxolol, celiprolol, carvedilol, metoprolol tartrate, oxprenolol, propranol, and timolol were quite well separated, while the spots of carvedilol were elongated [21,22].

11.2.3 Separation Using Mobile Phase Additives (CMPAs)

Antipodes of β-blockers were discriminated using only in one case β-CD [23] and more often chiral counter-ions such as (1R)-(−)-ammonium-10-camphorsulfonate (CSA), N-carbobenzyloxy-glycyl-L-proline (ZGP), and various

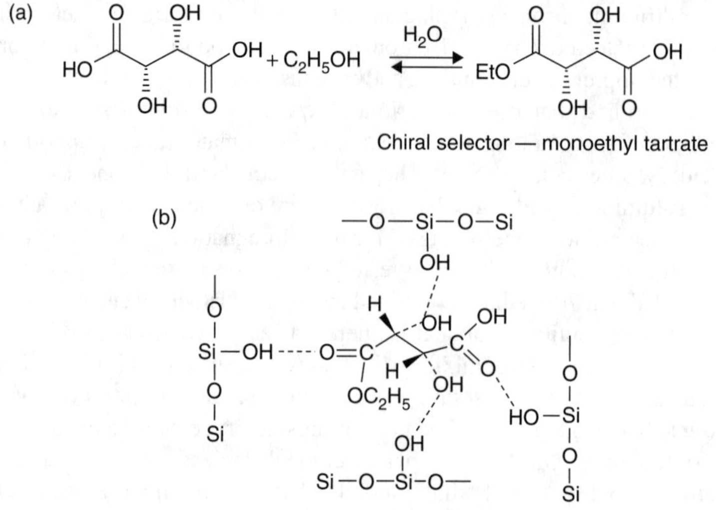

(a)

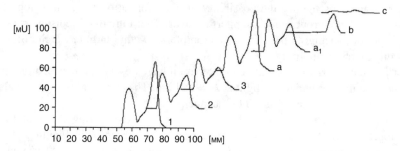

Chiral selector — monoethyl tartrate

(b)

FIGURE 11.5 Reaction leading to formation of monoethyl tartrate (a) and possible interaction of D-(−)-tartaric acid as chiral selector with free silanol groups of the silica gel support (b). (From Lučić, B. et al., *J. Planar Chromatogr.*, 18, 294, 2005. With permission.)

FIGURE 11.6 Densitogram obtained after separation of different concentrations of the enantiomers of ±MeT standard substance (1, 2, 3), ±MeT extracted from Presolol tablets (a, a_1), and the pure enantiomers S-(−)-alprenolol (b) and D-(−)-tartaric acid (c). (From Lučić, B. et al., *J. Planar Chromatogr.*, 18, 294, 2005. With permission.)

N-carbobenzyloxy)-amino acid derivatives (*N*-CBZ), for example, *N*-CBZ-isoleucyl-L-proline (ZIP), *N*-CBZ-alanyl-L-proline (ZAP), and *N*-CBZ-proline (ZP). Different β-blockers have been enantioseparated so far (propranolol, pindolol, timolol, atenolol, alprenolol, and metoprolol) using CSA as a chiral counter-ion. There is only one example in the available literature for the use of methanol/0.262 M β-CD (35:65, v/w) for the chiral separation of labetalol and the obtained separation factor was 1.07.

TABLE 11.4
Resolution of ($\pm$) β-Blockers on Plates Impregnated with
D-($-$)-Tartaric Acid

Compound	Mobile phase	Enantiomer	Migration distance (mm)	α	Ref.
($\pm$) Atenolol	11.6 mM tartaric acid in	$S(-)$	62	1.26	[21]
	ethanol/water (70:30)	$R(+)$	49		
($\pm$) Alprenolol	11.6 mM tartaric acid in	$S(-)$	68	1.26	[21]
	ethanol/water (70:30)	$R(+)$	54		
($\pm$) Betaxolol	11.6 mM tartaric acid in	$S(-)$	68	1.23	[21]
	ethanol/water (70:30)	$R(+)$	55		
($\pm$) Celiprolol	11.6 mM tartaric acid in	$S(-)$	63	1.24	[21]
	ethanol/water (70:30)	$R(+)$	51		
($\pm$) Carvedilol	11.6 mM tartaric acid in	$S(-)$	51	1[a]	[21]
	ethanol/water (70:30)	$R(+)$	51		
($\pm$) Metoprolol tartrate	11.6 mM tartaric acid in	$S(-)$	57	1.24	[20,21]
	ethanol/water (70:30)	$R(+)$	46		
($\pm$) Oxprenolol	11.6 mM tartaric acid in	$S(-)$	68	1.24	[21]
	ethanol/water (70:30)	$R(+)$	55		
($\pm$) Propranolol	11.6 mM tartaric acid in	$S(-)$	69	1.23	[21]
	ethanol/water (70:30)	$R(+)$	56		
($\pm$) Timolol	11.6 mM tartaric acid in	$S(-)$	67	1.22	[21]
	ethanol/water (70:30)	$R(+)$	55		

[a] Tailing spots were obtained.

11.2.3.1 Separation Using Counter-Ions as CMPA

The formation of diastereomeric ion pairs using chiral-ion interaction agents represents the basis for the resolution of enantiomers. Dissimilar distribution properties of the formed diastereomers enable their discrimination on nonchiral stationary phases [24]. Precoated HPTLC and polar diol-HPTLC are the most commonly used stationary phases [24–27]. Besides these conventional homemade stationary phases, silica gel GF$_{254}$ plates can be employed for chiral separation [28]. Some preliminary results showed that aminopropylsiloxane silica layer could be also successfully used for the chiral separation of β-blockers [29]. Eluent composition, type, and concentration of counter-ions and temperature play a crucial role in the chiral discrimination of β-blockers. Considering the structure of β-blockers, as well as the properties of stationary phases and solubility of selected counter-ions used for chiral discrimination, it was established that methylene chloride is the major component of the mobile phase. For the chiral separation, methanol or 2-propanol have been used as organic modifiers. The content of organic modifiers in mobile phase was determined mostly depending on the type of stationary phase whenever CSA or ZGP were used. So, the highest concentrations of organic modifiers (30–50%) were applied for conventional silica gel chromatography, somewhat lower (10–25%) for HPTLC and the lowest ones (5–10%) for polar diol-HPTLC. Triethylamine (TEA) [24] or ethanolamine [26,27] were applied to provide a higher selectivity but no general trend was noticed, so far. The concentration of ZGP varied from 6.5 to 16.8 mM [25] or from 1.0 to 7.0 mM [27], but it was observed that the capacity factor of the enantiomers was usually decreasing parallel to an increase in the concentration of counter-ions. This could be ascribed to an increased competition between the counter-ions and the diastereomeric ion-pair for adsorption sites on the stationary phase. Owing to a limited solubility of CSA in dry dichloromethane (up to 10 mM), the effects of concentrations over 10 mM were not studied, but it was observed that low-polarity alcohols act improving the solubility of CSA. The details on counter-ion concentrations and mobile phase composition for each of the separated β-blocker racemates are summarized in Table 11.5. The presence of water in chromatographic system was found to play an important role in the chiral separation; however, this phenomenon is still far from being fully understood. Duncan [25] used either HPTLC or diol-HPTLC dried plates and CSA, ZIP, ZAP, ZG, or ZGP as solvents for the separation of pindolol, propranolol, metoprolol, and timolol. However, Tivert and Backman [26] revealed that a higher relative humidity of stationary phases is required for optimal resolution of antipodes. Separation of metoprolol, propranolol, and alprenolol enantiomers lasting for 27 h and performed on moisture-equilibrated plates in the absence of ethanolamine is presented in Figure 11.7. Satisfactory and reproducible results for migration distance and chromatographic performance were obtained after keeping the chromatographic plates at a high relative humidity for about 20 h or longer-time period. The separation of metoprolol racemate from control-release tablets using 5 mM ZGP in dichloromethane and moisture-equilibrated plates for 12 days is shown in Figure 11.8.

The reason for these quite opposite results is not still fully understood, even though water should express a negative influence on the formation of the neutral diastereomeric association. Temperature was also reported to be a very important factor for chiral discrimination. Huang et al. [28] performed chiral separation of pindolol and propranolol at 5°C in a refrigerator, using homemade silica gel plates and CSA as a chiral-ion agent. However, enantioseparation of atenolol required temperature of 30°C. This could be explained in terms of a stronger binding of polar amido group of atenolol to the adsorbent comparing with other examined β-blocker racemates.

11.3 INDIRECT SEPARATION OF β-BLOCKERS

Indirect separation of some β-blockers has been performed after derivatization with several chiral reagents, such as R-($-$)-1-(1-naphthyl)ethyl isocyanate (R-($-$)-NEIC) [30], 2,3,4,6-tetra-O-acetyl-β-glucopyranosylisothiocyanate (GITC) [31], and S(+) benoxaprofen chloride (S-($-$)-BOP-Cl) [32]. The derivatization was based on the formation of the corresponding diastereomers, which were further resolved by TLC on achiral stationary phases.

TABLE 11.5
Chromatographic and Resolution Data of Some β-Adrenergic Blocking Agents Using Counter-Ion as CMPA

β-Blocker	Plate	Counter-ion	Mobile phase	α	Ref.
(±) Propranolol	HPTLC precoated	9.3 mM CSA	Dichloromethane/methanol (9:1)	4	[24]
	Homemade silica gel	6.8 mM CSA	Dichloromethane/methanol (5:5)	1.3	[28]
	HPTLC precoated	6.7 mM ZGP + 5 mM TEA	Dichloromethane/methanol (9:1)	2.5	[24]
	HPTLC precoated	6.3 mM ZGP + 5 mM TEA	Dichloromethane/methanol (7.5:2.5)	3	[24]
	Diol-HPTLC precoated	5 mM ZGP	Dichloromethane + 0.4 mM ethanolamine	1.3–1.4	[27]
(±) Timolol	Diol-HPTLC precoated	13.9 mM CSA	Dichloromethane/2-propanol (9.5:0.5)	1.4	[24]
	HPTLC precoated	5.8 mM ZGP + 5 mM TEA	Dichloromethane/methanol (7.5:2.5)	2	[24]
(±) Atenolol	Homemade silica gel	6.8 mM CSA	Dichloromethane/methanol (7:3)	2	[28]
(±) Alprenolol	Diol-HPTLC precoated	5 mM ZGP	Dichloromethane	1.3–1.4	[27]
(±) Pindolol	Homemade silica gel	6.8 mM CSA	Dichloromethane/methanol (6:4)	1.3	[28]
	Diol-HPTLC precoated	6.7 mM ZGP + 5 mM TEA	Dichloromethane/2-propanol (9.5:0.5)	1.7	[24]
(±) Metoprolol	Diol-HPTLC precoated	10.7 mM CSA	Dichloromethane/2-propanol (9.5:0.5)	1.5	[24]
	Diol-HPTLC precoated	6.9 mM ZIP	Dichloromethane/2-propanol (9.5:0.5)	1.6	[24]
	Diol-HPTLC precoated	7 mM ZAP	Dichloromethane/2-propanol (9.5:0.5)	1.6	[24]
	Diol-HPTLC precoated	7.8 mM ZP	Dichloromethane/2-propanol (9.5:0.5)	1.6	[24]
	Diol-HPTLC precoated	5 mM ZGP	Dichloromethane	1.3–1.4	[27]

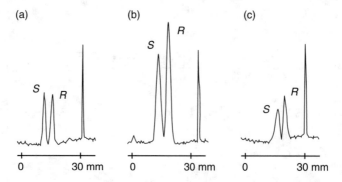

FIGURE 11.7 Separation of the enantiomers of metoprolol (a), propranolol (b), and alprenolol (c) on moisture-equilibrated plates (b) 27 h, (a) and (c) 12 days: mobile phase, 5 m*M* ZGP in dichloromethane; amount of sample, (a) 1 μg, (b) 0.5 μg, (c) 1 μg; scanning wavelength, 280 nm. (From Tivert, A.M. and Backman, Å., *J. Planar Chromatogr.*, 6, 216, 1993. With permission.)

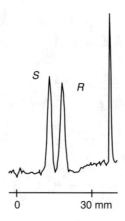

FIGURE 11.8 Enantiomeric separation of *R*- and *S*-metoprolol from controlled release tablet: mobile phase, 5 m*M* ZGP in dichloromethane; amount of sample, 1 μg of metoprolol succinate; length of time the plate was equilibrated with moisture, 12 days; scanning wavelength, 280 nm. (From Tivert, A.M. and Backman, Å., *J. Planar Chromatogr.*, 6, 216, 1993. With permission.)

R-(−)-NEIC reacts with the amino group of the investigated 1-aryloxy-3-isopropylamino-2-propanol derivatives, antipodes of propranolol, oxprenolol, metoprolol, pindolol, alprenolol, and bunitrolol forming the corresponding diastereomeric ureas. For that purpose, to 1.0–50 μmol of the free bases or their salts dissolved in the mixture of dry chloroform and dimethylformamide (8:2, v/v), an equimolar amount of TEA and approximately twofold molar excess of *R*-(−)-NEIC were added. After 20 min reaction time, the excess of the reagent was destroyed by adding diethylamine and 15 min later, an aliquot of the reaction mixture was applied to the HPTLC plates. No racemization was observed during the reaction. The chromatograms were developed in benzene/ether/acetone mixture (88:10:5, v/v) over the distance of 5.0 cm and visualized under a UV lamp. It was observed that the retention of S-($-$)-antipodes was higher than that of their R-($+$)-counterparts. The chiral separation factors for the investigated β-blockers ranged from 1.15 to 1.24. It was also established that nanogram amounts of one enantiomer derivative with NEIC were detectable in the presence of a 100-fold excess of the other stereoisomer. This method can be used for quantification of the enantiomers as well. The derivatization reaction of chiral, sugar-based derivatization reagent (GITC) with secondary amines of pindolol has been used for the separation of its optical antipodes. In this case, a certain volume of each R- and S-pindolol and racemic pindolol were mixed with GITC solution in acetonitrile and the reaction carried out for 35 min at 35–40°C on a stirrer-hotplate by gentle stirring. The chromatography was performed on reversed-phase thin layer C8 F_{254} silica gel plates preconditioned by development in methanol/water (30:70, v/v), air-dried and activated for 20 min at 120°C. After the application of derivatized R-($+$) and S-($-$) pindolol, the chromatogram was developed by ascending chromatography using the mixture of distilled water and 2-propanol (70:30, v/v) as the mobile phase. The development distance was about 75 mm and development time 75 min. The separation zones were detected by scanning the plates at 256 nm, and a typical HPTLC chromatogram of the separation of the GITC derivatives of R-($+$) and S-($-$) pindolol is shown in Figure 11.9. The HPTLC method was further evaluated by linearity and accuracy. This approach was found to be applicable for quantitative assessment of each of optical antipodes in a racemic pindolol mixture.

Pflugmann et al. [32] developed a sensitive method for determination of optical antipodes of propranolol, oxprenolol, and metoprolol in urine samples after derivatization with S-($+$)-BOP-Cl and TLC separation. The derivatization of the above β-blockers with S-($+$) BOP-Cl in methylene chloride was performed overnight at room temperature. The reaction was terminated by adding methanol and the

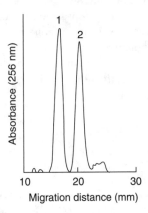

FIGURE 11.9 Chromatogram of (1) R-($+$)-pindolol and (2) S-($-$)-pindolol. (From Spell, J.C. and Stewart, J.T., *J. Planar Chromatogr.*, 10, 222, 1997. With permission.)

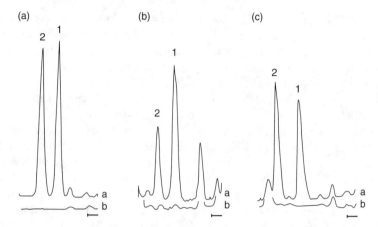

FIGURE 11.10 Thin-layer chromatograms of metoprolol (a; 9.54 μg/ml), oxprenolol (b; 1.81 μg/ml) and propranolol (c; 1.67 μg/ml) after extraction from urine (2 h after oral administration of a commercial preparation) and derivatization with S-(+)-BOP-Cl. Peaks: 1 = R-(+)-enantiomer; 2 = S-(−)-enantiomer. Traces: a = 0–2 h sample; b = blank urine. The concentrations given for a, b, and c are total concentrations ($R + S$). (From Pflugmann, G., Spahn, H., and Mutschler, E., *J. Chromatogr.*, 416, 331, 1987. With permission.)

solution was evaporated to dryness under the vacuum. The residue was redissolved in cyclohexane and a certain volume of the solution subjected to TLC. The chromatogram was developed in a glass tank in freshly prepared toluene/acetone mixture (100:10, v/v) as a mobile phase and ammonia-saturated atmosphere, produced by two open 50 ml beakers filled with ammonia (33%) and inserted in the chromatographic tank at room temperature. The separated optical antipodes of all three β-blockers were detected by measuring the fluorescence (λ_{ex} = 313 nm, λ_{em} = 365 nm) *in situ*. The R_f values of the examined β-blocker derivatives with S-(+)-BOP-Cl were 0.24 (R)/0.28 (S), 0.32 (R)/0.38 (S), and 0.32 (R)/0.39 (S) for metoprolol, oxprenolol, and propranolol, respectively. The detection limit of all tested β-blockers was ca. 1.6 ng per zone after extraction from urine and derivatization. The corresponding chromatograms are presented in Figure 11.10. The method was evaluated by selectivity in the presence of all other metabolites contained in urine samples, as well as by linearity, accuracy, reproducibility, detection, and quantification limits.

REFERENCES

1. Stinson, S.C., Chiral pharmaceuticals, *C & E News*, 79, 79, 2001.
2. Ranade, V.V. and Somberg, J.C., Chiral cardiovascular drugs, *Am. J. Ther.*, 12, 439, 2005.
3. Triggle, D.J., The transition from agonist to antagonist activity: Symmetry and other considerations, in *The Practice of Medicinal Chemistry*, Wermuth, C.G., Ed., Academic Press, 2003, Chap. 28.

4. Griffith, R., Drugs affecting adrenergic neurotransmission, in *Foy's Principles of Medicinal Chemistry*, Williams, D.A. and Lemke, T.L., Eds., Lippincott Williams & Wilkins, 2002, 292.

5. Johnson, R.L., Adrenergic agents, in *Wilson and Gisvold's Textbook of Organic and Pharmaceutical Chemistry*, Block, H.J. and Beale, J.M., Jr., Eds., Lippincott Williams & Wilkins, 2004, 524.

6. Kulig, K., Nowicki, P., and Malawska, B., Influence of the absolute configuration on pharmacological activity of antihypertensive and antiarrhythmic drugs, *Pol. J. Pharmacol.*, 56, 499, 2004.

7. Zhang, Y. et al., Enantioselective chromatography in drug discovery, *Drug Discov. Today*, 10, 571, 2005.

8. Kriz, D. et al., Thin-layer chromatography based on the molecular imprinting technique, *Anal. Chem.*, 66, 2636, 1994.

9. Ansell, R.J., Molecularly imprinted polymers for the enantioseparation of chiral drugs, *Adv. Drug Deliv. Rev.*, 57, 1809, 2005.

10. Aboul-Enein, H.Y., El-Awady, M.I., and Heard, C.M., Direct enantiomeric resolution of some cardiovascular agents using synthetic polymers imprinted with (−)-*S*-timolol as chiral stationary phase by thin layer chromatography, *Pharmazie*, 57, 169, 2002.

11. Suedee, R. et al., Chiral determination of various adrenergic drugs by thin-layer chromatography using molecularly imprinted chiral stationary phases prepared with α-agonists, *Analyst*, 124, 1003, 1999.

12. Suedee, R. et al., Thin-layer chromatographic separation of chiral drugs on molecularly imprinted chiral stationary phases, *J. Planar Chromatogr.*, 14, 194, 2001.

13. Kandimalla, V.B. and Ju, H., Molecular imprinting: A dynamic technique for diverse applications in analytical chemistry, *Anal. Bioanal. Chem.*, 380, 587, 2004.

14. Aboul-Enein, H.Y. et al., Application of thin-layer chromatography in enantiomeric chiral analysis — An overview, *Biomed. Chromatogr.*, 13, 531, 1999.

15. Suedee, R. and Heard, C.M., Direct resolution of propranolol and bupranolol by thin-layer chromatography using cellulose derivatives as stationary phase, *Chirality*, 9, 139, 1997.

16. Bhushan, R. and Thiongo, T.G., Direct enantioseparation of some β-adrenergic blocking agents using impregnated thin-layer chromatography, *J. Chromatogr. B*, 708, 330, 1998.

17. Sajewicz, M., Pietka, R., and Kowalska, T., Chiral separations of ibuprofen and propranolol by TLC. A study of the mechanism and thermodynamics of retention, *J. Liq. Chrom. Rel. Technol.*, 28, 2499, 2005.

18. Bhushan, R. and Arora, M., Direct enantiomeric resolution of (±)-atenolol, (±)-metoprolol, and (±)-propranolol by impregnated TLC using L-aspartic acid as chiral selector, *Biomed. Chromatogr.*, 17, 226, 2003.

19. Wall, P.E., Preparation and application of HPTLC plates for enantiomer separation, *J. Planar Chromatogr.*, 2, 228, 1989.

20. Lučić, B. et al., Direct separation of the enantiomers of (±)-metoprolol tartarate on impregnated TLC plates with D-(−)-tartaric acid as a chiral selector, *J. Planar Chromatogr.*, 18, 294, 2005.

21. Ivković, B.L. et al., Chiral separation of beta blocker enantiomers by thin-layer chromatography using D(−) tartaric acid as a chiral additive in the mobile phase, Presented at *10th Int. Symp. on Separation Sciences*, Opatija, Croatia, October 12–15, 2004, 74.

22. Ivković, B.L. et al., Direct separation of beta-blockers enantiomers by thin-layer chromatography using D(-) tartaric acid as a chiral mobile phase aditive, Presented at *4th Symposium of Serbian Pharmaciest with International Participation*, Belgrade, Serbia, Novermber 28–October 2, 2006, 738.

23. Armstrong, D.W., He, F.Y., and Han, S.M., Planar chromatographic separation of enantiomers and diastereoisomers with cyclodextrin mobile phase additives, *J. Chromatogr.*, 448, 345, 1988.

24. Duncan, J.D., Armstrong, D.W., and Stalcup, A.M., Normal phase TLC separation of enantiomers using chiral ion interaction agents, *J. Liq. Chromatogr.*, 13, 1091, 1990.

25. Duncan, J.D., Chiral separations: A comparison of HPLC and TLC, *J. Liq. Chromatogr.*, 13, 2737, 1990.

26. Tivert, A.M. and Backman, Å., Enantiomeric separation of aminoalcohols by TLC using a chiral counter-ion in the mobile phase, *J. Planar Chromatogr.*, 2, 472, 1989.

27. Tivert, A.M. and Backman, Å., Separation of the enantiomers of β-blocking drugs by TLC with a chiral mobile phase additive, *J. Planar Chromatogr.*, 6, 216, 1993.

28. Huang, M.B. et al., Enantiomeric separation of aromatic amino alcohol drugs by chiral ion-pair chromatography on a silica gel plate, *J. Liq. Chrom. Rel. Technol.*, 20, 1507, 1997.

29. Bazylak, G. and Aboul-Enein, H.Y., Enantioseparation in series of beta-blockers by normal-phase planar chromatography systems with bonded aminopropyl-siloxane silica layer and chiral counter ion, Presented at 22nd Int. Symp. on Chromatography, Rome, Italy, September 13–18, 1998.

30. Gubitz, G. and Mihellyes, S., Optical resolution of β-blocking agents by thin-layer chromatography and high-performance liquid chromatography as diastereomeric R-(−)-1-(1-naphthyl)ethylureas, *J. Chromatogr.*, 314, 462, 1984.

31. Spell, J.C. and Stewart, J.T., A high-performance thin-layer chromatographic assay of pindolol enantiomers by chemical derivatization, *J. Planar Chromatogr.*, 10, 222, 1997.

32. Pflugmann, G., Spahn, H., and Mutschler, E., Rapid determination of the enantiomers of metoprolol, oxprenolol and propranolol in urine, *J. Chromatogr.*, 416, 331, 1987.

12 Chiral Separation of Amino Acid Enantiomers

Władysław Gołkiewicz and Beata Polak

CONTENTS

12.1 INTRODUCTION

About 20 L-amino acids are the building blocks of proteins, so there is no need to use the chiral chromatographic techniques to resolve such mixtures; there are many different chromatographic techniques that allow to separate mixture of L-amino acids (for a review see [1]).

On the other hand, the optically active amino acids can be converted into a racemic mixture by racemization reaction [2]. In reality, racemization reaction, especially when occurring in fossil or teeth [3], takes a lot of time, but it has also become established that racemization takes place with higher reaction rates, at natural pH as well as in dilute acid or base [2].

Racemization in the metabolically stable proteins has also been detected [2,4–6].

Supplementation of livestock feeds with certain amino acid enantiomers has generated interest in analyses of D- and L-amino acids. Nutritional studies on utilization of amino acid enantiomers have shown that e.g. D-phenylalanine has better growth-promoting activity for chicks and rats than D-histidine, whereas D-arginine has no growth-promoting activity in either chicks or rats.

305

The two enantiomers of a given amino acid have identical chemical and physical properties in a symmetrical environment. To resolve such a pair of amino acid by chromatography, diastereomers must be formed. Diastereomers can be formed if a chiral reagent (selector) is introduced to either the mobile or the stationary phase. In case of thin-layer chromatography (TLC), latter manner has largely been used for resolution of amino acids, their PTH-, dansyl, and other derivatives [2,4–6].

In TLC, it is possible to utilize one of four basic techniques for the separation of enantiomeric compounds, including amino acids:

1. Direct separation of racemate by using natural chiral stationary phases (e.g., cellulose and β-cyclodextrin [β-CD]) effected by the formation of diastereomeric association complexes.
2. Separation of enantiomers on TLC plates coated witch chiral reagent.
3. Separation of racemate on ordinary stationary phase (silica gel or hydrophobic RP-18 plates) by means of chiral additives in mobile phase, which form diastereomeric complexes with D- and L-enantiomer.
4. Separation of enantiomers using chiral derivatizing reagent which form diastereomeric derivatives.

Many reviews [1,7–12] were published in books and journals, which describe all techniques listed here.

12.2 SEPARATION OF AMINO ACID ENANTIOMERS ON THE COMMERCIAL CHIRAL PLATES

At present, the only ready-to-use chiral plates are manufactured by Macherey-Nagel. Separation of the enantiomers on these plates is based on ligand-exchange mechanism. Although ligand-exchange chromatography (LEC) is not the only technique used for separation of amino acid enantiomers, using of LEC is easy to perform and for many racemates gives good resolution. LEC can also be combined with scanning densitometry for quantification and may be an alternative to high-performance liquid chromatography (HPLC) for certain applications.

Ligand-exchange chromatography is one of the most typical cases of complexation chromatography; diastereomeric complexes of different stability are formed during separation process. Such complexes are formed between metal cations (usually Cu^{2+}) and enantiomers (ligands) which are able to donate a lone pair of electrons. The formation of five-membered ring (Figure 12.1) results in the best chance to obtain a ternary complex of sufficient stability.

Such requirements very well meet especially when α-amino acids and α-hydroxy acids are separated. It is interesting that resolution of amino acids racemates could be easily obtained for α-amino acids possessing a free carboxyl group. In case of esterification, amidation, elimination, reduction, or substitution of carboxylic group, resolution was poorer or there was no resolution.

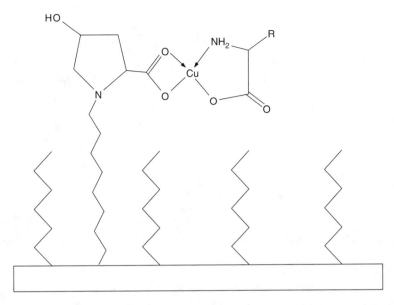

FIGURE 12.1 Schematic representation of the ternary complex of chiral selector, copper, and D- or L-amino acid.

Many different selectors have been used for TLC by LEC, but the new selectors proposed in addition to $(2S,4R,2'RS)$-4-hydroxyl-1-($2'$-hydroxydodecyl)-proline (Figure 12.2) do not have any specific advantage.

As seen in Figure 12.1, the stationary phase used in LEC is in general, silica-based hydrocarbonaceous phase; therefore, long hydrocarbonaceous chains of the chiral selector are probably oriented parallel to the C_{18} chains of the plate, according to the strong hydrophobic effect in water-organic mobile phase. This is the reason why it is not easy to remove chiral selector by mobile phase from the stationary phase during chromatographic process.

Thin-layer ligand-exchange chromatographic separation of enantiomers were independently published by Günther et al. [13,14] and Weinstein [15]. Although the procedure used for obtaining homemade chromatographic plates was very similar, the chiral selector used was quite different. Weinstein [15] used N,N-di-n-propyl-L-alanine as the selector and resolved all proteinogenic enantiomers of amino acids (with the exception of proline) as dansyl derivatives.

Günther et al. [13,14] used a chiral selector whose chemical formula and name are given in Figure 12.2.

The typical impregnation procedure was as follows: a commercial glass plate (RP-18, TLC) was dipped into a 0.25% copper (II) acetate solution (methanol/water 1:9, v/v) and dried. Next, the plate was immersed in a 0.8% methanolic solution of chiral selector for 1 min. After drying, the plates were ready for enantiomeric separation [7].

As mentioned earlier, hydrophobic RP-18 stationary phase is covered by chiral selector possessing long hydrocarbonaceous chain. Such stationary phase is very

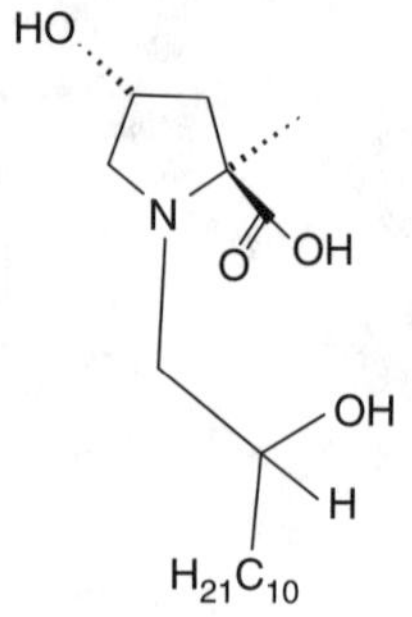

FIGURE 12.2 Chemical structure of (2S, 4R, 2′RS)-4-hydroxyl-1-(2′-hydroxy dodecyl)-proline.

similar to typical reversed-phase system, especially when water–organic mobile phase is used.

Careful inspection of the mobile phases, used for the separation of different amino acid enantiomers, show that in most cases mobile phases composed of methanol/acetonitrile/water was used [1,7,8,11]. In a few cases instead of acetonitrile, acetone or other solvents were used. It is also interesting that for resolving of α-hydroxy acids enantiomers, mobile phases without water were used.

Gołkiewicz and Polak [16] have shown that the chromatographic behavior of ternary mixed ligand complexes of α-amino acids can be described by an equation used in reversed phase:

$$R_M = \log k_w - m\varphi, \qquad (12.1)$$

where $R_M = \log(1 - R_F)/R_F$; k_w is the retention of a solute in pure water as developing solvent, m is the slope of the R_M against volume fraction φ.

This statement (linear relationship between the R_M and φ) is true only for mobile phase composed of phosphate buffer (water)/methanol but not for phosphate buffer (water)/acetonitrile or phosphate buffer (water)/dioxane; for those mobile phases, nonlinear relationships between the R_M and volume fraction, φ, of acetonitrile or dioxane was obtained, as shown in Figure 12.3a,b.

It was also stated that if concentration of acetonitrile or dioxane increased, the R_M values of α-amino acids decreased. Such chromatographic behavior of ternary complexes is not in agreement with general chromatographic experience, but such behavior is probably a result of coordination of one or two acetonitrile molecules by the ternary mixed ligand copper complex, which increases its hydrophobicity.

Amino acids have basic amino, acidic carboxyl groups; these affect their chromatographic behavior. Changing the pH of the mobile phase changes the amount of protonation or deprotonation of amino groups and the extent of dissociation of carboxyl group. It should be stressed that the chiral stationary phase also contains one enantiomer of an amino acid molecule, so variation of the pH of the mobile phase can also change the properties of the chiral stationary phase.

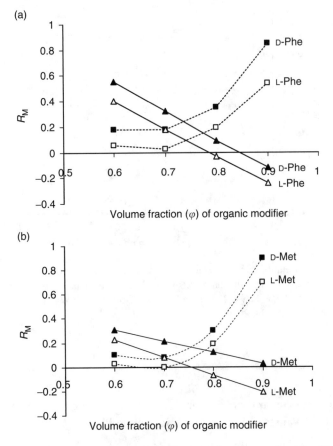

FIGURE 12.3 Relationship between R_M and volume fraction, φ, of methanol (solid lines) and acetonitrile (dashed lines) in phosphate buffer (pH 7) for D,L-phenylalanine (a) and for D,L-methionine (b). (From Gołkiewicz, W. and Polak, B., *J. Planar Chromatogr.*, 7, 453, 1994. With permission.)

The effect of mobile phase pH* (pH* means that pH was measured in water–organic solution) is presented in Figure 12.4 as relationships between R_F and mobile phase pH [17].

It is apparent from Figure 12.4, that the R_F increases slowly as pH* increases from 3.5 to 4.5, then decreases rapidly at pH* 5.0–5.5. The lowest R_F values are obtained at pH close to the isoelectric points of the investigated amino acids.

Variation of separation factor, α, as a function of mobile phase pH* was determined with the buffer and modifier (methanol and acetonitrile) concentrations kept constant (Table 12.1).

Usually, as it is seen from Figure 12.4 (although with some exceptions), the value of separation factor, α, first decreases quite sharply with increasing pH*, next passes through a minimum and then moderately increases. In described TLC experiments [17], the lowest enantioselectivity was observed at pH* 5.0–5.5 and

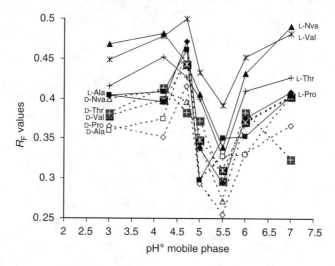

FIGURE 12.4 Relationship between the R_F values of the enantiomers of the amino acids and the pH* of the mobile phase consisted of: acetonitrile/methanol/water/phosphate buffer, 70:10:10:10 (%, v/v). (From Polak, B., Gołkiewicz, W., and Tuzimski, T., *Chromatographia*, 63, 197, 2006. With permission.)

TABLE 12.1
Values of the Separation Factor, α, for Investigated Amino Acids at Different pH Values of Water–Organic Mobile Phase

Amino acid	pH of mobile phase						
	3.0	**4.2**	**4.7**	**5.0**	**5.5**	**6.0**	**7.0**
Proline	1.31	1.21	1.26	1.23	1.22	1.21	1.19
Tryptophane	2.07	1.63	1.64	1.58	1.53	1.60	1.64
Phenylalanine	1.69	1.42	1.55	1.43	1.46	1.46	1.46
Norvaline	1.55	1.34	1.22	1.31	1.38	1.30	1.44
Leucine	1.51	1.38	1.23	1.35	1.33	1.39	1.42
Threonine	1.25	1.17	1.21	1.13	1.15	1.12	1.55
Valine	1.59	1.38	1.26	1.43	1.43	1.40	1.38
Methionine	1.15	1.14	1.20	1.21	1.19	1.19	1.22
Dopa	1.77	1.41	1.64	1.57	1.54	1.48	1.62
Alanine	1.36	1.15	1.00	1.00	1.10	1.11	1.00
Tyrosine	1.00	1.00	1.00	1.00	1.00	1.00	1.00

Values of the separation factors were calculated from $\alpha = k_2/k_1$ and $k = (1 - R_F)/R_F$.

Source: From Polak, B., Gołkiewicz, W., and Tuzimski, T., *Chromatographia*, 63, 197, 2006. With permission.

the highest at pH* in range 3–4 or 6–7. Contrary to these results, Veig and Lindner [18] showed that the highest enantioselectivity in HPLC LEC experiments was obtained at 5.5 (with some exceptions).

The hR_F values and conditions of separations of racemates on commercial chiral plates by LEC are collected in reviews by different authors: amino acids [1,7,8,11], α-substituted of the amino acids [7,8,11], dipeptides [7,8], β-methyl derivatives of amino acids [7,11,19], α-dialkyl amino acids [1], other derivatives — fluoro, bromo, benzyl, etc. [1,7,8,11].

12.3 SEPARATION OF AMINO ACID ENANTIOMERS ON THE IMPREGNATED PLATES WITH CHIRAL REAGENT

Incorporating a chiral reagent with the adsorbent, prior to application of the samples and development of chromatogram, cause that new type of modified stationary phase is obtained. Such type of stationary phase in TLC is termed *impregnated TLC*.

In reality, impregnated chiral reagent plates do not differ from commercial plates (Section 12.2), if preparation method of such plates is taken into consideration. The only difference is in the kind of chiral selector used and the separation mechanism (ligand exchange, formation of inclusion complexes, ion pair formation, etc.).

Different chiral selectors have been used for TLC resolution of variety of racemates of amino acids into enantiomers. Bhushan and Martens [20] in a review summarized various chiral selectors used as impregnating reagents, along with the mobile phases used for resolution of DL-amino acids (Table 12.2).

Mixing of an appropriate chiral selector with adsorbent, for example, silica gel during chromatographic plate impregnation, results in the formation of two diastereomers: L-amino acid-chiral selector and D-amino acid-chiral selector molecules. It is known that two enantiomers have the same physicochemical properties in achiral environment but not diastereomers. Two diastereomers of given amino acids have different properties (solubility, diastereomeric complex stability, etc.), so can be separated on achiral, conventional phases.

Bhushan and Ali [21] used optically pure (−) brucine as a chiral selector; they selected DL-amino acids which have isoelectric point below pH 7 and carried out separation on stationary phase where pH was 7.1–7.2 (during preparation of the plates they added the silica gel slurry 0.1 N NaOH). In such a case, amino acids existed as an anion and interacted with the basic brucine. Selectivity of the used system for some amino acids (isoleucine, leucine, and alanine) enantiomers was very high; hR_F for D,L-serine was 12 and 50, respectively.

Use of the optically pure (−) quinine [22] as chiral selector for the separation of amino acid enantiomers was not as successful as in the case of (−) brucine. Only for methionine, very good results were obtained; hR_F for D- and L-methionine were 25 and 50, respectively.

TABLE 12.2

Chiral Selectors and Solvent Systems Used for the Separation of Enantiomers of DL-Amino Acids and Their Derivatives on Impregnated Plates

Chiral selector/plate	Solvent system	Cited references
Amino acids		
(−) Brucine	Butanol/acetic acid/chloroform (3:1:4)	[21]
(−) Quinine	Butanol/chloroform/acetic acid (3:7:5) for methionine	[22]
	Butanol/chloroform/acetic acid (10:1:4) for threonine	
	Butanol/chloroform/acetic acid (6:8:4) for alanine	
	Ethyl acetate/carbon tetrachloride/propionic acid (10.5:6.5:3.5) for valine	
Cooper (II)–L-proline complex	(A) n-Butanol/methanol/water (6:2:3)	[20]
	(B) Chloroform/methanol/propionic acid (15:6:4)	
	(C) Acetonitrile/methanol/H$_2$O (2:2:1)	
$(1R,3R,5R)$-2-azabicyclo-[3.3.0]octan-3-carboxylic acid	Acetonitrile/methanol/water (different ratio)	[20]
$(2S,4R,2'RS)$-4-hydroxy-1-$(2'$-hydroxydodecyl)-proline	Acetonitrile/methanol/water (4:1:1 or 3:5:5)	[13]
Chiral plate	Acetonitrile/methanol/water (4:1:1)	[20]
Chitin or chitosan + Cu(II)	Methanol/water/acetonitrile (1:1:x)	[20]
Cellulose (0.5%) with silica gel, immersed in Cu(II) acetate followed by L-arginine	Methanol/acetonitrile/tetrahydrofuran/water, pH 2	[20]
PTH amino acids		
(+)-Tartaric acid	Chloroform/ethyl acetate/water (28:1:1)	[21]
(+)-Ascorbic acid	n-Butyl acetate/chloroform (1:5)	[21]
Microcrystalline cellulose triacetate + silica gel (3:1)	Ethanol/water (4:1) for PTH Phe	[26]
	2-Propanol/water (4:1) for PTH Tyr	

Dansyl-DL-amino acids

Erythromycin	0.5 M aqueous NaCl/acetonitrile/methanol (different ratio)	[23]
(1R,3R,5R)-2-azabicyclo-[3,3,0]octan-3-carboxylic acid	Acetonitrile/methanol/water (different ratio)	[20]
Vancomycin	Acetonitrile/0.5 M aqueous NaCl (10:4; 14:3)	[24]
β-Cyclodextrin	Methanol/1% aqueous triethyl ammonium acetate (1:1)	[25]
Poly-L-phenylalaninamide	Acetonitrile/water (1:2 or 9:11)	[20]
N,N-di-n-propyl-L-alanine, followed by cupric acetate (on RP-TLC plates)	0.3 M sodium acetate in 40% acetonitrile and 60% H_2O adjusted to pH 7 by acetic acid	[15]

α-Dialkylamino acids

(2S,4R,2′RS)-4-hydroxyl-1-92′-hydroxydodecylproline (Chiralplate)	Acetonitrile/methanol/water (4:1:1 or 3:5.5)	[13]

Halo-, methoxy-, methyl-, formyl-, and benzyl-, substituted

(2S,4R,2′RS)-4-hydroxyl-1-92′-hydroxydodecylproline (Chiralplate)	Acetonitrile/methanol/water (4:1:1 or 3:5.5)	[13]

α-Hydroxy amino acids

(2S,4R,2′RS)-4-hydroxyl-1-92′-hydroxydodecylproline (Chiralplate)	Dichloromethane/methanol (9:1)	[20]

Meta-tyrosine, diiodo-, and dimethyl tyrosine

Chiralplate	Acetonitrile/methanol/water (4:1:1 or 4:1:2 or 4:1:2 + 0.1 M diisopropylethylamine)	[20]

N-Benzyloxycarbonyl-phenylalanine-nitrophenyl ester and tert-butyloxycarbonyl-phenylalanine-nitrophenyl ester (N-CBZ-Phe-ONp and N-t-Boc-Phe-ONp)

Microcrystalline cellulose triacetate/silica gel (3:1)	Ethanol/water (4:1) for PTH-Phe / 2-Propanol/water for PTH-Tyr	[26]

Source: From Bhushan, R. and Martens, J., *Biomed. Chromatogr.*, 15, 155, 2001. With permission.

Also macrocyclic antibiotics, erythromycin and vancomycin, were successfully employed as the chiral selectors [23,24] for separation of dansyl derivatives of DL-amino acids. It was shown that such chromatographic system is very sensitive especially to changes of pH and temperature, owing to possibility of ionization of functional groups or changing conformational structure of the antibiotic molecule.

Erythromycin and vancomycin possess multiple stereogenic centers and a variety of functional groups, which can interact with functional groups of solute to create diastereomeric complex. Layers impregnated with antibiotics show good selectivity for acidic solutes, where interaction between the carboxylate group of the solute and amino group of chiral selector can occur.

Cyclodextrins have been used mainly as mobile phase additives for chiral resolutions of a number of racemates, including amino acids and their derivatives, but Alak and Armstrong [25] used β-CD bounded to silica gel for resolution of dansyl derivatives of DL-amino acid enantiomers, by forming a reversible inclusion complex of different stability. Dansyl DL-amino acids were better separated on β-CDs layers than nonderivatized amino acids, because they have additionally two or more carbohydrogen rings.

Lepri et al. [26] have reported separation of PTH-amino acids on silica gel mixed with microcrystalline triacetate cellulose. The authors [26] used 2-propanol or ethanol/water as mobile phase and have shown that retention of amino acid enantiomers increased as concentration of organic solvent decreased, in accordance with the behavior of chromatographed substances in reversed-phase systems.

12.4 SEPARATION OF AMINO ACID ENANTIOMERS ON THE COMMERCIAL PLATES WITH CHIRAL REAGENT IN MOBILE PHASE

In the previous section, the use of the β-CD-bonded phase [25] to achieve separation of racemic mixture of DL-amino acids was described. Alternatively, a great variety of chiral separations can be done using β-CD as chiral selector is added to mobile phase. The advantage of using β-CD as chiral mobile phase additive is that separation of enantiomers can be achieved on a less expensive and more durable nonchiral stationary phase.

Cyclodextrins are cyclic polysaccharides produced by bacterial enzymatic degradation of starch [27]. These cyclic polysaccharides contain 6–12 D-(+)-glucopyranose units connected through α-(1,4)-glycosidic linkages. Though, chromatographically important CDs can be α-CD (6 glucopyranose units) and γ-CD (8 units), nevertheless β-CD (7 glucopyranose units) and its derivatives (e.g., hydroxypropyl β-CD) have been used for successful separation of amino acids.

Cyclodextrins are shaped like open-ended truncated cones that can form inclusion complexes. The size of mouth cavity of β-CD is about 6–8 Å (600–800 pm), so that the moieties of the size of naphthalene (and smaller) can easily penetrate into CD cavity.

Cyclodextrins have a hydrophilic shell because each of glucopyranose unit has 2- or 3-hydroxyl groups at the mouth of CD cavity and hydrophobic cavity formed due to two rings of C—H groups and glucosidic oxygen groups. Therefore, the hydrophobic part of the solute can be included into the nonpolar cavity of the CD, while the polar part of the solute may form hydrogen bonds with the polar cavity shell.

The first application of β-CD as a chiral selector added to mobile phase, for separation of DL-amino acids on reversed phase TLC plates, was described by Armstrong et al. [28].

The success of separation of dansyl derivatives of amino acids was strongly depended on the concentration of β-CD. One of the reasons of difficulties is the low solubility of β-CD in water, that can be improved by addition of urea. It should be noted that if the concentration of β-CD in mobile phase is less than 0.04 M then there is no resolution of racemic mixture. Optimum of enantiomeric resolution occurs when the concentration of β-CD in mobile phase is in range 0.08–0.12 [28] (Figure 12.5).

Enantiomeric resolution depends also on the concentration and type of organic modifier (methanol or acetonitrile) in mobile phase. Optimum enantiomeric resolution occurs in a narrow range 20–30% of organic modifier and the R_F values of enantiomers do not change in this region [28] (Figure 12.6).

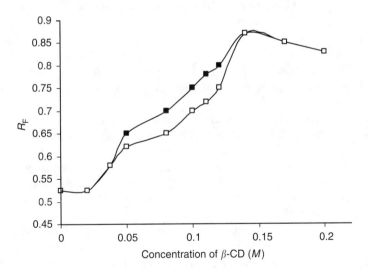

FIGURE 12.5 Plot showing the effect of β-CD concentration in the mobile phase on the R_F values of dansyl-D-glutamic acid (■) and dansyl-L-glutamic acid (□). In addition to the indicated levels of β-CD, the mobile phase consisted of acetonitrile/water (30:70) (saturated with urea). (From Armstrong, D.W., He, F., and Han, S.M., *J. Chromatogr.*, 448, 345, 1988. With permission.)

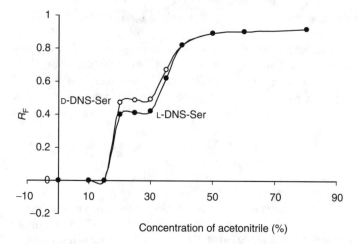

FIGURE 12.6 Plot showing the effect of % acetonitrile in the mobile phase on the TLC separation of dansyl-D-serine (open circles) from dansyl-L-serine (closed circles). The concentration of β-CD is 0.106 M. (From Armstrong, D.W., He, F., and Han, S.M., *J. Chromatogr.*, 448, 345, 1988. With permission.)

Solubility of the β-CDs can be improved by derivatization; partially substituted hydroxypropyl-β-CD and hydroxyethyl-CD show better solubility in water and aqueous-organic solvents [29]. The solubility of hydroxypropyl- and hydroxyethyl-β-CDs increases as the degree of substitution of hydroxyl group increases. Solutions exceeding 0.4 M CD can be made without additives enhancing solubility.

Authors of reviews [7,9,11] inserted tables containing DL-amino acids resolved with β-CD or chemically modified β-CD as chiral mobile phase additives, along with the R_F's and composition of mobile phase concentration, and type of stationary phases.

Another chiral selector used as the mobile phase additive was bovine serum albumin (BSA). The first application of BSA in TLC was described by Lepri et al. [9,30–33] for separation of various enantiomers, including amino acids derivatives.

Various enantiomers were separated on RP-18W/F$_{254}$ commercial plates using mobile phase prepared by dissolving 4–6% of BSA in buffer and then adding the desired amount of 2-propanol (generally 2–6%, v/v). There is no general suggestion which of the pH of mobile phase is the best. The results obtained suggest using acidic eluents for separation of N-derivatives of DL-amino acids, but in few cases, use of neutral or basic mobile phase gave better results. For names and structures of chiral solutes resolved on TLC plates and BSA as chiral selector added to mobile phase, see Reference [9].

It is very interesting to find out that chromatographic method of separation of amino acid enantiomers was used for educational purposes [34]. Students realizing project of complete analysis of a biological active tetrapeptide, D-Ala-Gly-L-Phe-D-Leu (D- and L-amino acids!) carried out overnight hydrolysis, derivatization using dansyl chloride (DNS-), identification of DNS-derivatives of amino acids by normal phase TLC, and determined the stereochemistry of the individual DNS-amino acid enantiomers, using β-CD as a chiral selector added to mobile phase, according to the method described by Armstrong et al. [28].

12.5 MOST POPULAR CHIRAL REAGENTS USED FOR DERIVATIZATION OF AMINO ACID ENANTIOMERS

The success of a chiral separation depends upon the formation of diastereo-meric pair between each enantiomer and a chiral reagent. There are two types of diastereomeric complex: long-lived and short-lived diastereomers. Short-lived diastereomers were discussed in the previous sections; such diastereomers occur between a pair of enantiomers and a chiral environment. Long-lived diastereomers are formed by chemical reaction between a pair of enantiomers and a chiral reagent.

The chromatographist fixed attention on reaction between racemic amino acids with $-NH_2$ and $-COOH$ functional groups and with the auxiliaries known from HPLC, especially ready-to-use reagents.

Marfey [35] resolved 1-fluoro-2,4-dinitro-5-L-alanine amide (FDAA) deriv-atives of DL-amino acids through HPLC, using a reversed-phase C_{18} stationary phase and a linear gradient of mobile phase, from 10 to 40% of acetonitrile in triethylamine-phosphate buffer (pH 3.5).

Commercially available Marfey's [35] reagent FDAA is frequently used for derivatization of racemic amino acids separated by TLC. The structure of Marfey's reagent and a pair of FDAA derivatives of serine are shown in Figure 12.7.

Ruterbories and Nurok [36] have shown that it is possible to separate FDAA derivatives of each of the 22 racemic amino acids. The values of ΔR_F of 22 pairs of, chromatographed in reversed-phase mode enantiomers are in the range of 0.06 (arginine) to 0.22 (aspartic acid and tyrosine). In all cases, L-enantiomer had the higher R_F value, which was ascribed to greater intramolecular hydrogen bonding in D- than in L-diastereomer. When FDAA derivatives of amino acids are separated on silica gel layers in the normal phase mode, it is the D-diastereomer that has the higher R_F.

Nagata et al. [37] also separated FDAA derivatives of 17 DL-amino acids, using reversed-phase RP-18 F_{254S} plates and acetonitrile/triethylamine/phosphate buffer (pH 5.5) as mobile phase. Separation of FDAA derivatives obtained by Nagata et al. [37] was not as good as in the case of Ruterbories and Nurok [36]; the values of ΔR_F of 17 pairs of enantiomers are in range from 0.03 to 0.09. It seems that such differences are caused by using in described experiments different mobile phases as well as pH.

Brückner and Wachsmann [38] synthesized 11 chiral derivatizing reagents by nucleophilic replacement of one chloride atom in 2,4,6-trichloro-1,3,5-triazine by L-alanine amide and a second chloride by L-alanine amide or methoxy, phen-oxy, and other substituents. In fact, the obtained chiral derivatizing reagent shows

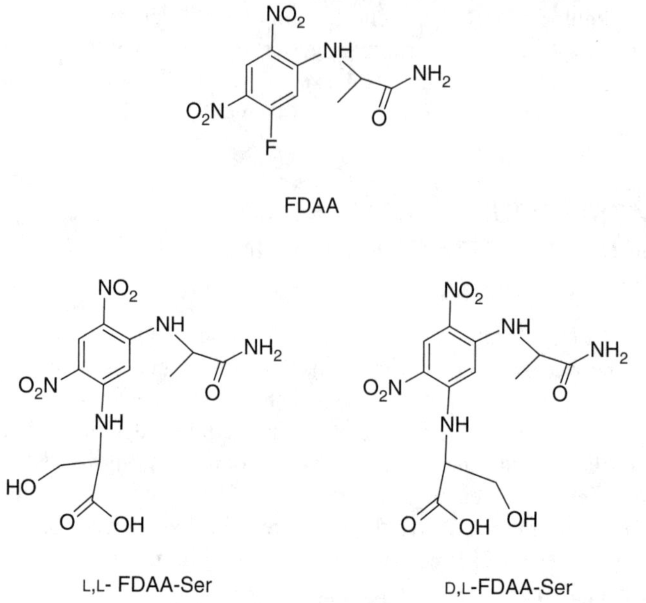

FDAA

L,L- FDAA-Ser

D,L-FDAA-Ser

FIGURE 12.7   Chemical structures of Marfey's reagent and FDAA derivatives of serine.

structural similarity to Marfey's reagent, but trifunctionality of triazine and the possibility of controlled sequential replacement of the chlorine atoms by nucleophiles, make the design of many new chiral reagents possible.

The structure of two chiral derivatizing agent obtained by Brückner and Wachsmann, for which the highest selectivity was obtained, are depicted in Table 12.3. After derivatization, selected DL-amino acids were separated by HPLC, but undoubtedly they would also be separated by TLC, because separation factor, α, was in the range of 1.14 to 2.31.

Nishi et al. [39] synthesized a new derivatizing reagent, $S(-)$-N-1-(2-naphthylsulphonyl)-2-pyrrolidinecarbonyl chloride (NSP-Cl) and separated the obtained diastereomeric derivatives by normal phase TLC and HPLC (Kieselgel 60F$_{254}$) and chloroform/methanol (98:2) as mobile phase. For the structure of NSP-Cl, see Table 12.3.

Unfortunately, under the reaction conditions, diastereomeric derivatives were formed only for 17 proteinogenic DL-amino acids but not for DL-cysteine and DL-histidine. Authors presented only the picture of TLC plates without the R_F values of separated diastereomeric derivatives, so there was no possibility of estimating selectivity in TLC; however, the presented HPLC chromatograms were of high quality.

Kleidernigg and Lindner [40] described application of a new fluorescence tagging chiral derivatizing agent (CDITC), for the separation of DL-amino acids and chiral amines by HPLC; 18 amino acids racemates were completely resolved, on a standard ODS reversed-phase column, with separation factor in range of 1.14–3.16.

TABLE 12.3
Structures of Chemical Reagents Used for Derivatization of Racemic Amino Acids

Structure of derivatizing reagents	References
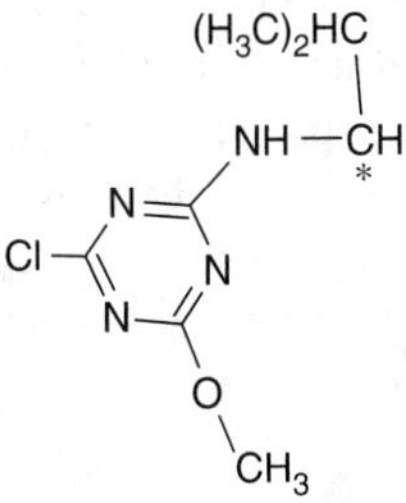 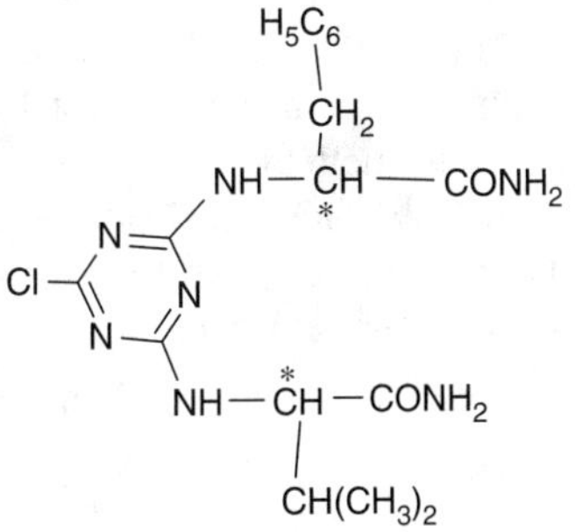	[38]

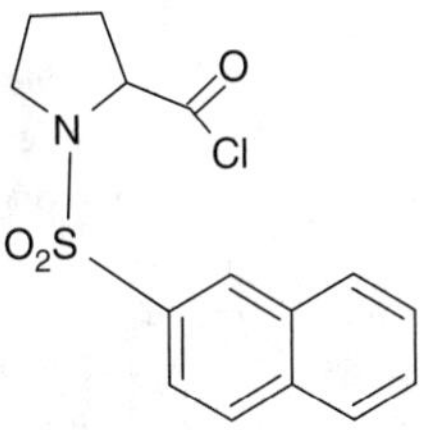

However, separation of DL-tyrosine was not achieved. Chemical structure of CDITC is shown in Table 12.3.

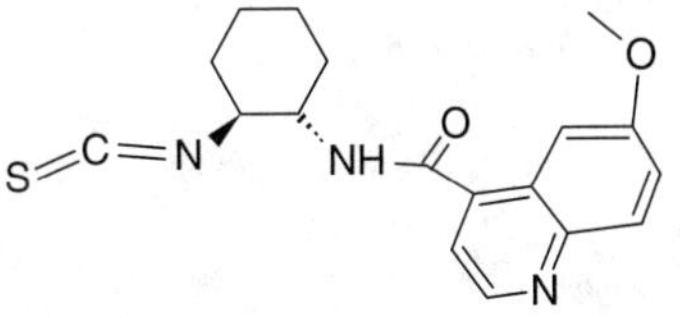

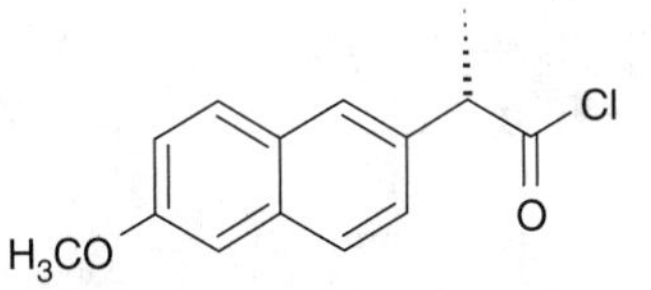

Popular drug (S)-(+)-naproxen was also used as the derivatizing reagent [41] for DL-amino acids. Commercially available (S)-(+)-naproxen was converted to (S)-(+)-naproxen chloride and used for derivatizing of methyl esters of DL-amino acids. Diastereomeric derivatives were separated in normal phase TLC; mobile phase consisted of toluene/dichloromethane/tetrahydrofuran (5:1:2; v/v; ammonia atmosphere). Maximum of ΔR_F is in the range of 0.03–0.1.

12.6 CONCLUSIONS

It was not the aim of this chapter to describe all chromatographic separations of racemic amino acids accomplished so far, but rather demonstrate applicability of these methods to separation of DL-amino acids. Chiral separations, using TLC, enables rapid and inexpensive testing both of optical enantiomers, their derivatives, peptides, and control of enantiomeric purity.

The separation of enantiomers is a challenge for analytical chemists, because separation conditions are, in most cases, established using trial-and-error method. We hope that different methods of separation of enantiomers presented in this chapter will help to solve analytical problems. It should be stressed that chiral TLC separations are very sensitive to changes in pH, polarity of solvents and concentrations of mobile phase, type of chiral selector, and temperature. A small change in these parameters can cause drastic changes in the selectivity.

REFERENCES

1. Bhushan, R. and Martens, J., Amino acids and their derivatives, in *Handbook of Thin-Layer Chromatography*, 3rd edn., Chromatographic Science Series, Sherma, J. and Fried, B., Eds., Taylor and Francis former Marcel Dekker, New York, Basel, 2003, Chap. 15, pp. 373–415.
2. Bada, J.L., Racemization of amino acids in nature, *Interdiscip. Sci. Rev.*, 7, 30, 1982.
3. Bada, J.L., Paleoanthropological applications of amino acids racemization dating of fossil bones and teeth, *Anthrop. Anz.*, 45, 1, 1987.
4. Helfman, P.M. and Bada, J.L., Aspartic acid racemization in dentine as a measure of ageing, *Nature*, 262, 279, 1976.
5. Masters, P.M., Bada, J.L., and Zigler, J.S., Aspartic acid racemization in heavy molecular weight crystallines and water insoluble protein from normal human lenses and cataracts, *Nature*, 268, 71, 1977.
6. Man, E.H. et al., Accumulation of D-aspartic acid with age in the human brain, *Science*, 220, 1407, 1983.
7. Günther, K. and Möller, K., Enantiomer separations, in *Handbook of Thin-Layer Chromatography*, 2nd edn., Sherma, J. and Fried, B., Eds., Taylor and Francis Marcel Dekker, New York, Basel, 2000, Chap. 20, pp. 621–681.
8. Martens, J. and Bhushan, R., TLC enantiomeric separation of amino acids, *Int. J. Peptide Protein Res.*, 34, 433, 1989.
9. Lepri, L., Enantiomer separation by TLC, *J. Planar Chromatogr.*, 10, 320, 1997.
10. Abdoul-Einen, H.Y. et al., Application of thin-layer chromatography in enantiomeric chiral analysis — An overview, *Biomed. Chromatogr.*, 13, 531, 1999.
11. Lepri, L., Del Bubba, M., and Cincinelli, A., Chiral separations by TLC in *Planar Chromatography, A Retrospective View for the Third Millennium*, Nyiredy, Sz., Ed., Springer Scientific Publisher, Budapest, 2001, p. 517.
12. Bereznitski, Y. et al., Thin-layer chromatography — A useful technique for the separation of enantiomers, *J. AOAC Int.*, 84, 1242, 2001.
13. Günther, K., Thin-layer chromatographic enantiomeric resolution via ligand exchange, *J. Chromatogr.*, 448, 11, 1988.

14. Günther, K., Martens, J., and Schckedanz, M., Thin layer chromatographic enantiomeric separation via ligand exchange, *Chem. Int. Ed. Engl.*, 23, 506, 1984.

15. Weinstein, S., Isolation of optical isomers by thin-layer chromatography, *Tetrahedron Lett.*, 25, 985, 1984.

16. Gołkiewicz, W. and Polak, B., Relationship between capacity factors and composition of a mobile phase in thin-layer liquid exchange chromatography of enantiomers of some amino acids, *J. Planar Chromatogr.*, 7, 453, 1994.

17. Polak, B., Gołkiewicz, W., and Tuzimski, T., Effect of the mobile phase pH* on the chromatographic behaviour in chiral ligand exchange thin-layer chromatography (CLETLC) of some amino acids enantiomers, *Chromatographia*, 63, 197, 2006.

18. Veig, E. and Lindner, W., Epimeric N-substituted L-proline derivatives as chiral selectors for ligand-exchange chromatography, *J. Chromatogr. A*, 660, 255, 1994.

19. Peter, A. and Toth, G., Chromatographic methods for the separation of enantiomers and epimers of β-alkyl amino acids and peptides containing them, *Anal. Chim. Acta*, 352, 335, 1997.

20. Bhushan, R. and Martens, J., Separation of amino acids, their derivatives and enantiomers by impregnated TLC, *Biomed. Chromatogr.*, 15, 155, 2001.

21. Bhushan, R. and Ali, I., TLC resolution of enantiomeric mixtures of amino acids, *Chromatographia*, 23, 141, 1987.

22. Bhushan, R. and Arora, M., Resolution of enantiomers of DL-amino acids on silica gel plates impregnated with optically pure (−)-quinine, *Biomed. Chromatogr.*, 15, 433, 2001.

23. Bhushan, R. and Parshad, V., Thin-layer chromatographic separation of enantiomeric dansyloamino acids using a macrocyclic antibiotic as a chiral selector, *J. Chromatogr. A.*, 736, 235, 1996.

24. Bhushan, R. and Thiong'o, G.T., Separation of the enantiomers of dansyl-DL-amino acids by normal-phase TLC on plates impregnated with a macrocyclic antibiotic, *J. Planar Chromatogr.*, 13, 33, 2000.

25. Alak, A. and Armstrong, D.W., Thin-layer chromatographic separation of optical, geometrical, and structural isomers, *Anal. Chem.*, 58, 582, 1986.

26. Lepri, L. et al., Reversed-phase planar chromatography of racemic flavanones, *J. Liquid Chromatogr. Rel. Technol.*, 22, 105, 1999.

27. Szejtli, J., *Cyclodextrins and Their Inclusion Complexes*, Akademai Kiado, Budapest, 1982.

28. Armstrong, D.W., He, F., and Han, S.M., Planar chromatographic separation of enantiomers and diastereomers with cyclodextrin mobile phase, *J. Chromatogr.*, 448, 345, 1988.

29. Armstrong, D.W., Faulkner, R., Jr., and Han, S.M., Use of hydroxypropyl- and hydroxyethyl-derivatized square-cyclodextrin for the TLC separation of enantiomers and diastereomers, *J. Chromatogr.*, 452, 323, 1988.

30. Lepri, L. et al., Planar chromatography of optical isomers with bovine serum albumin in the mobile phase, *J. Planar Chromatogr.*, 5, 175, 1992.

31. Lepri, L. et al., Reversed phase planar chromatography of enantiomeric tryptophans with bovine serum albumin in the mobile phase, *J. Planar Chromatogr.*, 5, 234, 1992.

32. Lepri, L. et al., Reversed phase planar chromatography of optically active fluorenylmethoxycarbonyl amino acids with bovine serum albumin in the mobile phase, *J. Planar Chromatogr.*, 5, 294, 1992.

33. Lepri, L. et al., Reversed phase planar chromatography of dansyl DL amino acids with bovine serum albumin in the mobile phase, *Chromatographia*, 36, 297, 1993.

34. LeFevre, J.W. and Dodsworth, D.W., Complete analysis of a biologically active tetrapeptide: A project utilizing thin-layer chromatography and mass spectrometry, *J. Chem. Educ.*, 77, 503, 2000.

35. Marfey, P., Determination of D-amino acids. II. Use of a bifunctional reagent, 1,5-difluoro-2,4-dinitrobenzene, *Carlsberg Res. Commun.*, 49, 591, 1984.

36. Ruterbories, K.J. and Nurok, D., Thin-layer chromatographic separation of diastereomeric amino acid derivatives prepared with Marfey's reagent, *Anal. Chem.*, 59, 2735, 1987.

37. Nagata, Y., Iida, T., and Sakai, M., Enantiomeric resolution of amino acids by thin-layer chromatography, *J. Mol. Cat. B Enzym.*, 12, 105, 2001.

38. Brückner, H. and Wachsmann, M., Design of chiral monochloro-s-triazine reagents for the liquid chromatographic separation of amino acid enantiomers, *J. Chromatogr. A*, 998, 73, 2004.

39. Nishi, H. et al., New Chiral derivatization reagent for resolution of amino acids as diastereomers by TLC and HPLC, *Chromatographia*, 27, 301, 1989.

40. Kleidernigg, O.P. and Lindner, W., Indirect separation of proteinogenic α-amino acids, using the fluorescence active $(1R,2R) - N$-[(2-isothiocyanato)cyclohexyl]-6-methoxy-4-quinolinylamide as chiral derivatizing agent. A comparison, *J. Chromatogr. A*, 795, 251, 1998.

41. Büyüktimkin, N. and Buschaner, A., Separation and determination of some amino acid ester enantiomers by thin-layer chromatography after derivatization with (S)-(+)-naproxen, *J. Chromatogr.*, 450, 281, 1988.

13 Chiral Separation of Nonsteroidal Anti-Inflammatory Drugs

Ravi Bhushan and Jürgen Martens

CONTENTS

13.1 INTRODUCTION

Nonsteroidal anti-inflammatory drugs (NSAIDs) are widely used for the treatment of minor pain and for the management of edema and tissue damage resulting from inflammatory joint disease (arthritis). A number of these drugs possess antipyretic property in addition to having analgesic and anti-inflammatory actions, and thus have utility in the treatment of fever.

13.1.1 GENERAL STRUCTURE AND PROPERTIES OF NSAIDS

In general, NSAIDs structurally consist of an acidic moiety (carboxylic acid and enols) attached to a planar aromatic functionality. Some of them contain a polar linking group, which attaches the planar moiety to an additional lipophilic group. This is represented in Figure 13.1.

Nonsteroidal anti-inflammatory drugs are characterized by the following chemical/pharmacological properties:

- All are relatively strong organic acids with pK_a in the range of 3–5. Most, but not all, are carboxylic acids. Thus, salt forms can be generated upon treatment with bases, and all of these compounds are extensively ionized at physiologic pH. The acidic group is essential for cyclooxygenase (COX) inhibitory activity.
- NSAIDs differ in their lipophilicities based on the lipophilic character of their aryl groups and additional lipophilic moieties and substituents.
- The acidic group in these compounds serves a major binding group (ionic binding) with plasma proteins. Thus, all NSAIDs are highly bound by plasma proteins (drug interactions).
- The acidic group also serves as a major site of metabolism by conjugation. Thus, a major pathway of clearance for many NSAIDs is glucuronidation (and inactivation) followed by renal elimination.

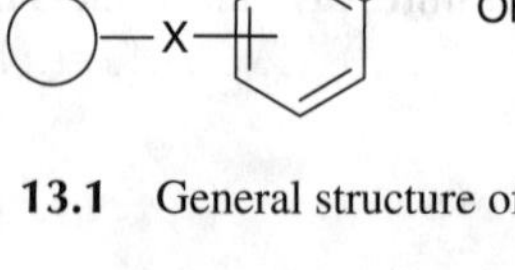

FIGURE 13.1   General structure of NSAIDs.

13.1.2 Classification of NSAIDs

The most prominent NSAID is aspirin. Nonaspirin NSAIDs can be classified based on the chemical structure as follows:

- Salicylates
- Propionic acids (profens)
- Aryl and heteroarylacetic acids
- Anthranilates (fenamates)
- Oxicams (enol acids)
- Phenylpyrazolones
- Anilides

13.1.2.1 Salicylates

Salicylates are derivatives of 2-hydroxybenzoic acid (salicylic acid). They were discovered in 1838, following the extraction of salicylic acid from willow bark. Salicylic acid was used medicinally as the sodium salt but replaced therapeutically in the late 1800s by the acetylated derivative, acetylsalicylic acid (ASA), or aspirin (Figure 13.2). Therapeutic utility is enhanced by esterification of the phenolic (hydroxyl) group as in aspirin, and by substitution of a hydrophobic/lipophilic group at C-5 as in diflunisal.

13.1.2.2 Propionic Acid Derivatives (Profens)

Some of the most useful NSAIDs are structurally derived from arylacetic acids. These compounds are often referred to as the "profens" based on the suffix of the prototype member, ibuprofen. As with the salicylates these agents are all strong organic acids (pK_a = 3–5) and thus form water-soluble salts with alkaline reagents. The aryl propionic acids are characterized by the general structure Ar—CH(CH$_3$)—COOH which conforms to the required general structure (Figure 13.3). All of these compounds are predominantly ionized at physiologic pH and are more lipophilic than ASA or salicylic acid.

The α-CH$_3$ substituent present in the profens, increases COX inhibitory activity and reduces toxicity of the profens. The α-carbon in these compounds is

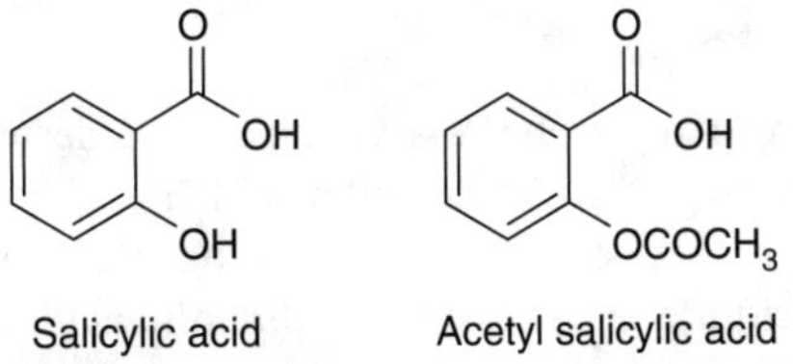

Salicylic acid　　　　　Acetyl salicylic acid

FIGURE 13.2 Structures of salicylates.

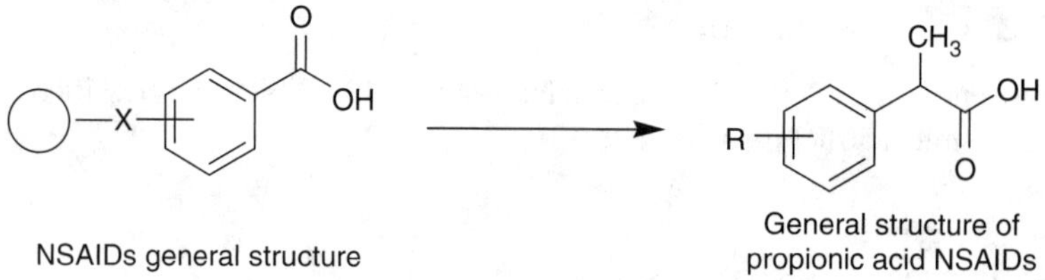

FIGURE 13.3 General structure of propionic acid NSAID.

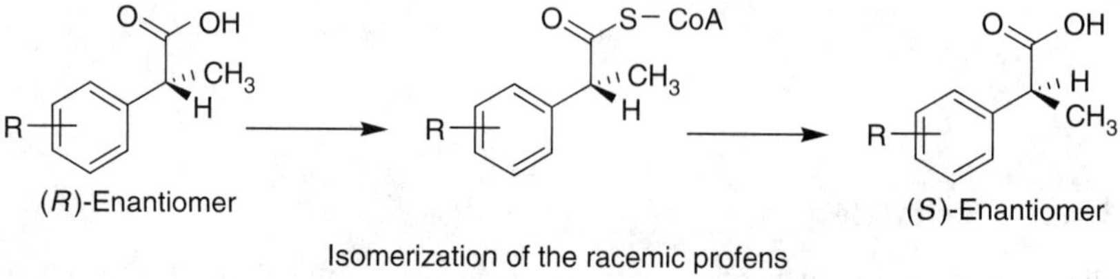

FIGURE 13.4 Isomerization of profens.

chiral, and the (*S*)-(+)-enantiomer of the profens is the more potent COX inhibitor. Most profen products, except naproxen, are marketed as the racemates. In addition to the metabolism described below, the profens undergo a metabolic inversion at the chiral carbon involving stereospecific transformation of the inactive (*R*)-enantiomer to the active (*S*)-enantiomer. This is believed to proceed through an activated (more acidic α-carbon) thioester intermediate (Figure 13.4). Normally, only the (*S*)-(+)-isomer is present in plasma. The chemical structures of a few commonly used profens are shown in Figure 13.5.

13.1.2.3 Aryl and Heteroarylacetic Acids

These compounds are also derivatives of acetic acid, but in this case the substituent at the 2-position is a heterocycle or related carbon cycle (Figure 13.6). This does not significantly affect the acidic properties of these compounds. The heteroarylacetic acid NSAIDs marketed can be further subclassified into indene/indoles, the pyrroles, and the oxazoles, as shown in Figures 13.7 through 13.9, respectively.

13.1.2.4 Anthranilates

These agents are considered to be *N*-aryl-substituted derivatives of anthranilic acid, which itself is a bioisostere of salicylic acid. These agents retain the acidic properties that are characteristic of this class of agents; though mefenamic acid and meclofenamic acid are derivatives of anthranilic acid, diclofenac is derived from 2-arylacetic acid (Figure 13.10). The most active fenamates have small alkyl or halogen substituents at the 2′, 3′, or 6′ position of the *N*-aryl moiety

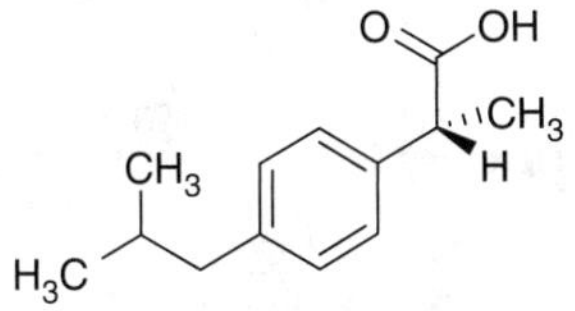

(R)-Ibuprofen

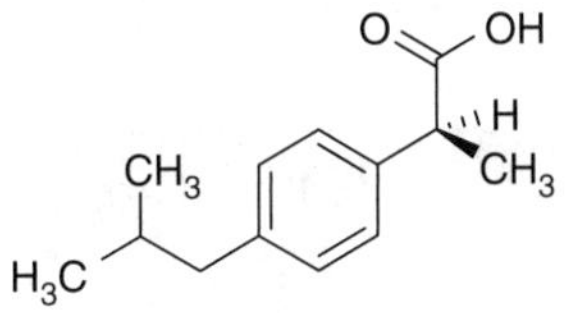

(S)-Ibuprofen

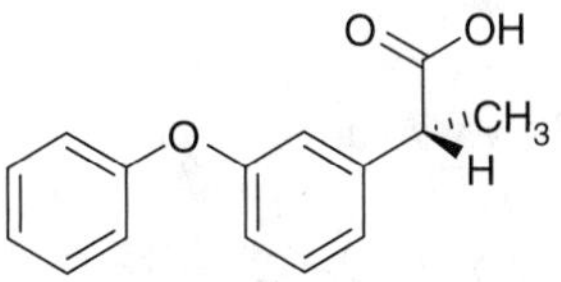

(R)-Fenoprofen

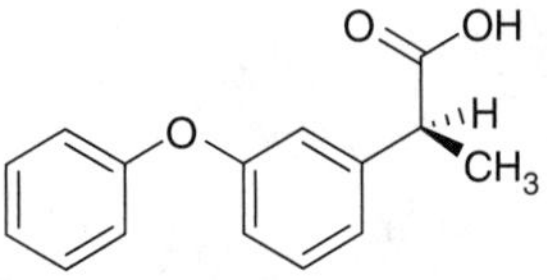

(S)-Fenoprofen

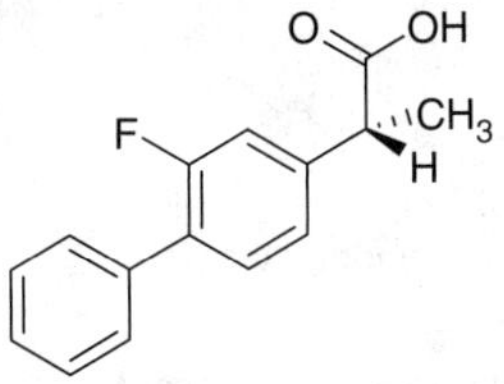

(R)-Flurbiprofen

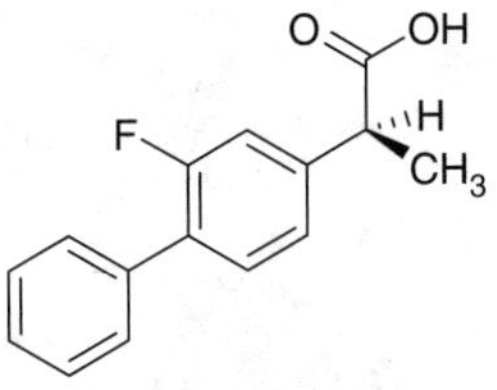

(S)-Flurbiprofen

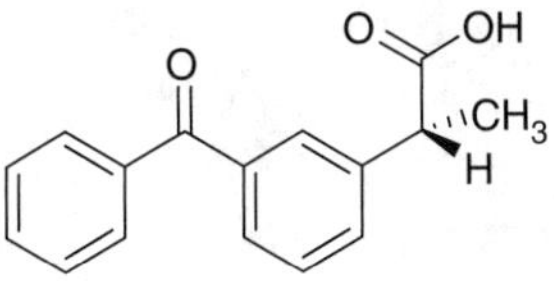

(R)-Ketoprofen

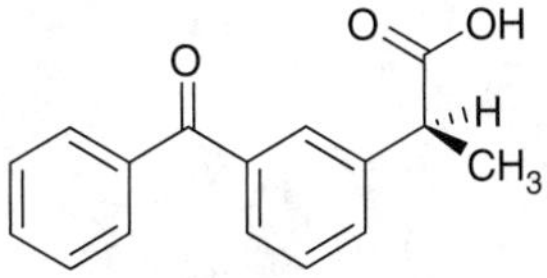

(S)-Ketoprofen

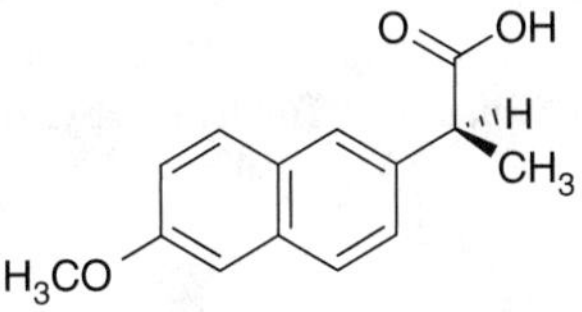

(S)-Naproxen

FIGURE 13.5 Structures of a few commonly used profens.

(meclofenamate is reported to be 25 times more potent than mefenamate). Among the disubstituted N-aryl fenamates, the 2′,3′-derivatives are most active suggesting that the substituents at the 2′,3′-positions serve to force the N-aryl ring out of coplanarity with the anthranilic acid. Hence, this steric effect is proposed to be important in the effective interaction of the fenamates at their inhibitory site on COX.

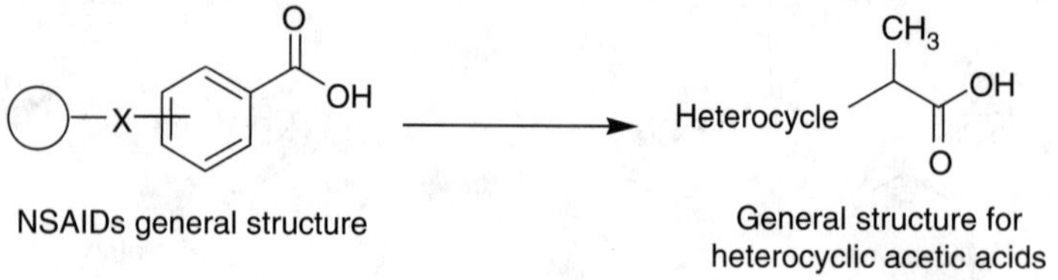

NSAIDs general structure   General structure for heterocyclic acetic acids

FIGURE 13.6   General structure for heterocyclic acetic acids.

Sulindac   Indomethacin

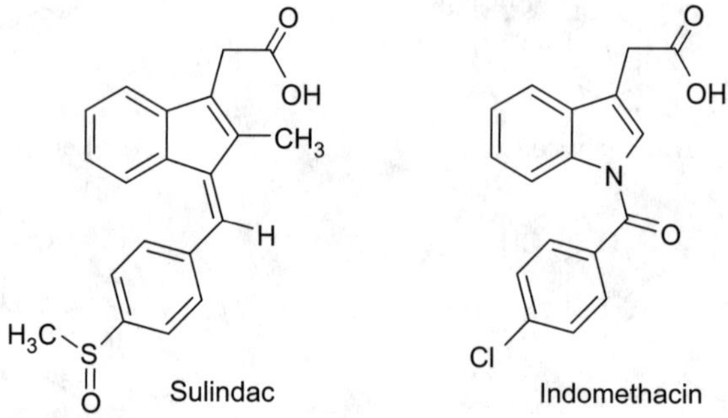

FIGURE 13.7   Structures for indene and indole acetic acid.

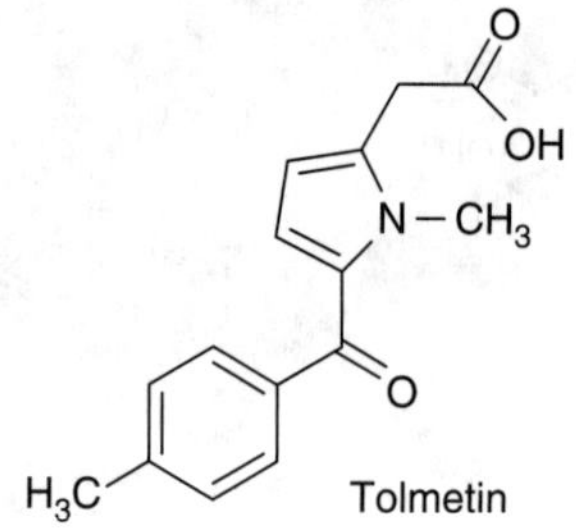

Tolmetin

FIGURE 13.8   Structure of Tolmetin, a pyrrole acetic acid.

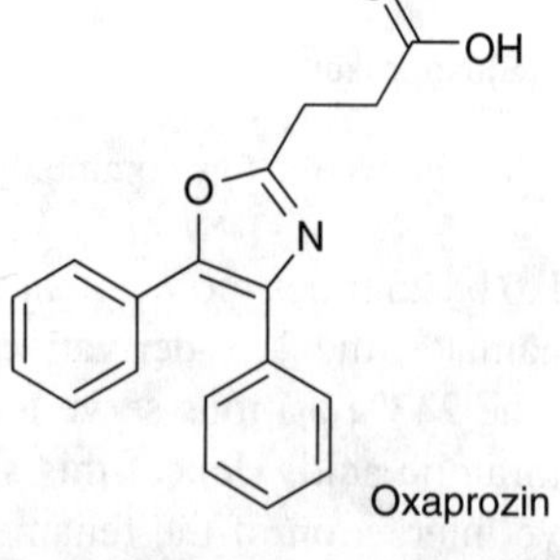

Oxaprozin

FIGURE 13.9   Structure of Oxaprozin, an oxazole acetic acid.

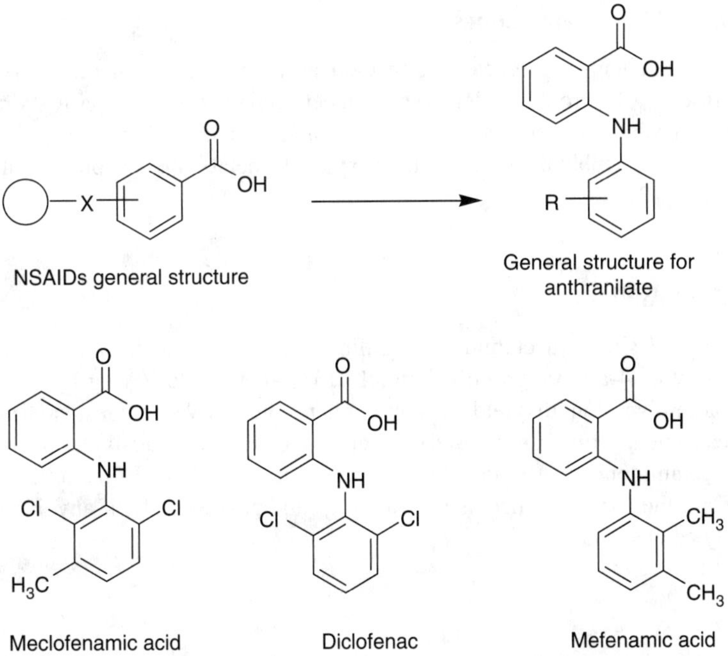

NSAIDs general structure

General structure for anthranilate

Meclofenamic acid

Diclofenac

Mefenamic acid

FIGURE 13.10 General structure of anthranilates and some typical examples.

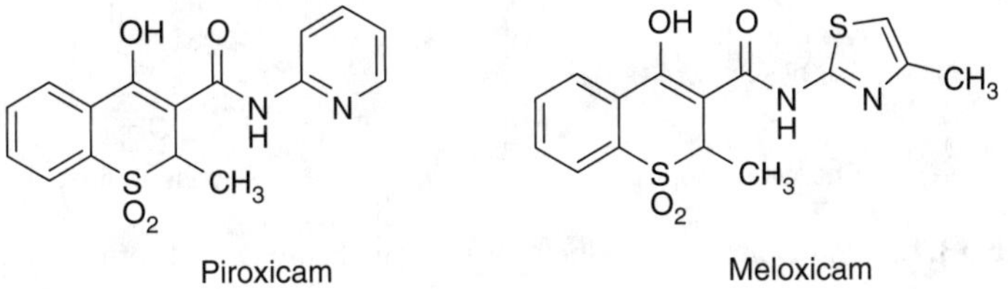

Piroxicam

Meloxicam

FIGURE 13.11 Structures of piroxicam and meloxicam (the general class of oxicams).

13.1.2.5 Oxicams (Enolic Acids)

Oxicams (piroxicam and meloxicam) are characterized by the 4-hydroxybenzothi-azine heterocycle (Figure 13.11). The acidity of the oxicams is attributed to the 4-OH with the enolate anion being stabilized by intramolecular H-bonding to the amide N—H group. Also, the presence of the carboxamide substituent at the 3-position of the benzothiazine ring contributes toward acidity by stabilizing the negative charge formed during ionization (resonance stabilization). Although these compounds are acidic ($pK_a = 6.3$), they are somewhat less acidic than carboxylic acid NSAIDs. Yet, the oxicams are primarily ionized at physiologic pH, and acidity is required for COX inhibitory activity.

13.1.2.6 Phenylpyrazolones

This class of agents is characterized by the 1-aryl-3,5-pyrazolidinedione structure. The presence of a proton, which is situated between the two electron-withdrawing carbonyl groups, renders these compounds acidic. The pK_a for phenylbutazone is 4.5. Oxyphenbutazone is a hydroxylated metabolite of phenylbutazone (Figure 13.12).

13.1.2.7 Anilides

Anilides are simple acetamides of aniline, which may or may not contain a 4-hydroxy or 4-alkoxy group (Figure 13.13). Acetanilide is ring hydroxylated after administration to yield acetaminophen, the active analgesic/antipyretic, whereas phenacetin (rarely used) undergoes oxidative-O-dealkylation to produce acetaminophen. Anilides do not possess carboxylic acid functionality and, therefore, they are classified as neutral drugs and possess little, if any, inhibitory activity against COX.

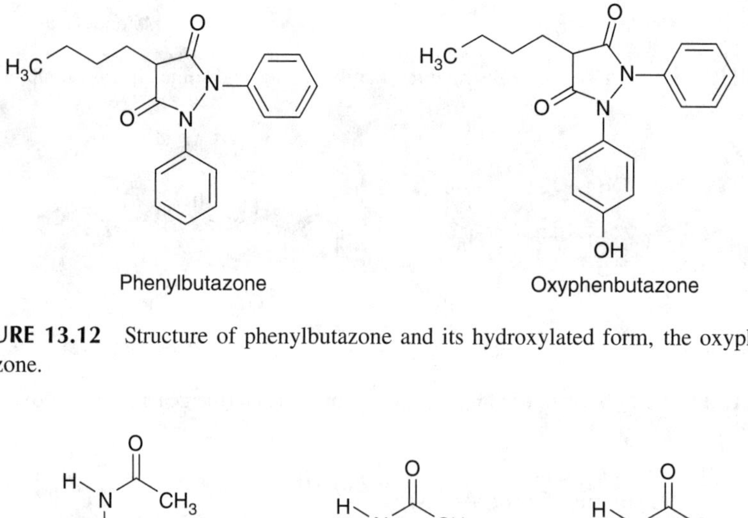

Phenylbutazone   Oxyphenbutazone

FIGURE 13.12   Structure of phenylbutazone and its hydroxylated form, the oxyphenbutazone.

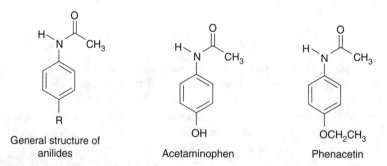

General structure of anilides

Acetaminophen

Phenacetin

FIGURE 13.13   General structure of anilides and the characteristic compounds acetaminophen and phenacetin.

13.2 APPROACH TO ENANTIOSEPARATION

During the past two decades, chirality and stereochemistry became very important topics in pharmacology and analytical chemistry. The chiral nature of living systems has evident implications on biologically active compounds interacting with them.

In drug development, analytical methods are required to evaluate the enantiomeric purity of starting materials, reagents, and catalysts, because the quality of these compounds limits the enantiomeric purity of the resulting products. Therefore, the goal of the analytical laboratory in the pharmaceutical industry is to develop as much separations as possible in a minimum time. In this context, it is useful to have general screening strategies based on, for instance, chromatographic techniques.

13.2.1 CHROMATOGRAPHIC METHODS OF SEPARATION

The most commonly used approach to chromatographic resolution of enantiomers has been the formation of diastereomers either transiently or covalently. As diastereomers have different physicochemical properties, there is a differential retention in a chromatographic system.

Chromatographic resolution of racemic or scalemic mixtures (e.g., NSAIDs) is generally accomplished by the following methods:

1. Derivatization of the enantiomers (or the racemic mixture) with a suitable chiral reagent, followed by separation of the resulting diastereomers on an achiral phase. This is the indirect approach. Since NSAIDs possess a carboxyl or a similar easily derivatizable functional group, the method is or has largely been successful. Following basic considerations are to be taken into account for indirect enantioseparation:
   a. The chiral derivatizing agent should be readily available in high and known optical purity and should not racemize during storage.
   b. The chiral derivatizing agent should react quantitatively or at least the reaction rates with the compounds to be resolved must be kinetically equal; in fact, the rates are normally different.
   c. The racemization of the stereogenic center is negligible under the reaction conditions.
2. The second broad approach is the direct separation. No chemical derivatization of the analyte is required prior to chromatography. It may adopt any of the following:
   a. Use of chiral stationary phase that may be either due to chirality of the material as such, like cellulose, or due to some sort of synthetic modification.
   b. Addition of chiral discriminating agents to the mobile phase.

    c. Use of a suitable chiral selector to impregnate the solid inert sup-
port, in thin-layer chromatography (TLC) particularly during the
plate making or at a stage before developing the chromatogram.

13.3 THIN-LAYER CHROMATOGRAPHY

13.3.1 ADVANTAGES OF TLC

The inert character of the thin-layer material, such as silica gel, alumina, and
cellulose, enhances its versatility and suits for use even with stronger corrosive
reagents, and many kinds of chemical reactions can be performed on the plate.
Certain groups of interest can be bonded by chemical modifications of the support.
Stationary phases for TLC have not improved substantially with time in contrast to
HPLC. However, the reasons for using TLC include parallel separation of samples,
high-throughput screening, static and sequential detection for identification, and
integrity of the total sample. It is evident that simultaneous procedure, in principle,
yields a higher precision. Probably one of the most advantageous features of TLC,
as opposed to other chromatographic methods, is that a number of samples can be
handled simultaneously.

    Moreover, future prospects for improved separation performance in TLC
include the use of zone refocusing, forced flow, and electro-osmotic flow methods, the use of multiple chemical and biological visualization, transfer from silica
gel layer onto polymer membrane, and instrumental spectrometric procedures for
detection and identification. TLC being simple and inexpensive facilitates the use
of chiral mobile phase additives or impregnation of the inert support with a suitable
(chiral) reagent during plate making or prior to applying the samples and at a stage
before developing the chromatogram and thus formation of *in situ* diastereomers,
leading to enantiomeric resolution.

    Sometimes, TLC and HPLC are unjustly looked at as competitive methods.
Each of these has its own advantages. In HPLC, finding suitable separation parameters is frequently expensive in time and materials, therefore, a combination of
the two by first optimizing the particular separation parameter with TLC would
be a step leading to a considerable saving in time and expenses for an analysis.
TLC is suitable as a pilot technique for the investigation of appropriate separation
conditions, particularly because with TLC various phase systems can be checked
at the same time without expensive apparatus. The mobile and stationary phases
are also very much comparable in TLC and HPLC as a large accordance in the
retention mechanism exists.

13.3.2 IMPREGNATED TLC

The technique of incorporating a suitable reagent with the adsorbent, prior to
applying the samples on the plates and development of chromatogram is termed
as impregnation. It originated from simple TLC. The impregnation thus means
a modification to a turning point where the simple TLC is used successfully

with greater improvements for all analytical purposes. The adsorption characteristics are changed without covalently affecting the inert character of the adsorbent. Multicomponent or complex mixtures and the enantiomers of a variety of compounds, such as amino acids and their derivatives and compounds of pharmaceutical importance, have been separated on TLC plates impregnated with different reagents or chiral selectors as the case may be. Direct resolution of enantiomers, by impregnated TLC, has been reviewed for amino acids and derivatives by Bhushan and Martens [1].

The reagents and methods used for impregnation are not to be mistaken with locating or spray reagents, which of course are required for the purpose of identification even on impregnated plates. Addition of reagents to the mobile phase is excluded from the basic definition of impregnation.

13.3.3 METHODS FOR IMPREGNATION

Mixing of the impregnating reagent, say a chiral selector, with the silica gel prior to plate making, is one of the various methods used for impregnation and has successfully been used in the author's laboratory.

A second approach is the immersion of plain plates into an appropriate solution of the impregnating reagent, but the whole process should be done carefully and slowly so as not to disturb the thin layer; peeling off of the layers has been experienced with the commercially available plates too. Alternatively, a solution of the impregnating material is allowed to ascend or descend the plate in a normal manner of development; this method is least apt to damage the thin layer. Furthermore, exposing the layers to the vapors of the impregnating reagent or spraying the impregnating reagent (or its solution) onto the plate has also been employed; spraying provides a less uniform dispersion than by development or immersion.

Another approach is to have a chemical reaction between the inert support and a suitable reagent, that is, the support is chemically modified before making the plate, the moieties of interest are bonded to the reactive groups of the packing material of the layers.

13.3.4 SEPARATION MECHANISM

Three main/basic mechanisms of resolution of enantiomers on impregnated TLC could be considered, which would largely be related to the approach of impregnation. These are ligand exchange, inclusion complex formation, that is, guest–host interaction, and ion exchange. For NSAIDs, there are reports of enantiomeric separation only via the last named approach.

13.3.4.1 Role of Impregnation

The chemical nature of the impregnating agent and the structure of the compounds being separated, along with the influence of the impregnating reagent on the adsorption/partition, are together held responsible for the overall results of resolution. Certain additives that improve the detection due to fluorescence and also allow detection at picomolar level do not influence the chromatographic properties or resolution efficiency and are not included under this category:

1. *Ion pairing*: Ion-pair formation between the ion of the analyte and the oppositely charged counter ion (from the impregnating reagent) is considered to be responsible for a change in retention. An appropriate pH is required to ensure that the analyte and the ion-pairing agent are in a charged form. Large cations are employed for the separation of anions, whereas for the separation of cations hydrophobic or hydrophilic ion-pairing reagents are used. Use of different ion-pairing reagents suggest that, in general, such reagents cause variation in retention, reduction in tailing, improvement in selectivity, particularly in the presence of oppositely charged counter ions. Ion exchange in the surrounding medium leading to separation of ions or molecules with ionic or polar properties has also been effective in many cases.

2. *Complex formation*: Formation of different types of complexes, such as charge transfer complex due to interaction between the analytes and the modified stationary phase or the formation of complexes between a metal ion and the donor molecule, having at least one pair of electrons or formation of molecular inclusion complex due to steric interactions have also been considered. There are no reports on TLC resolution of enantiomers of NSAIDs using this approach.

13.3.4.2 Ligand Exchange

The first successful separation using ligand exchange, via the technique of impregnation, on TLC was reported for enantiomers of amino acids by Günther et al. [2–4]; the commercial RP-18 silica gel plates were first immersed in a 0.25% copper (II) acetate solution prepared in methanol/water (1:9), dried and then immersed for 1 min in a 0.8% methanolic solution of the chiral selector (2S,4R,2$'RS$)-4-hydroxy-1-(2$'$-hydroxydodecyl)-proline. The impregnated plate, thus prepared, provided resolution of a number of enantiomeric amino acids, including substituted ones, when developed in methanol/water/acetonitrile (1:1:4 or 5:5:3).

13.3.4.3 Inclusion Complex (Guest–Host Steric Interaction)

The most common and the best example in this category comes from the application of cyclodextrins (CDs). Alak and Armstrong [5] mixed β-CD-bonded silica gel with a binder in 50% aqueous methanol, whereas methanol/1% aqueous triethyl ammonium acetate (pH 4.1) was used as mobile phase for the resolution of enantiomers of dansyl amino acids. Because, then, CDs have been used in different manners, that is, as impregnating agent as well as chiral mobile phase additive for

TLC resolution of enantiomers of amino acids and a variety of other compounds. No reports are available on TLC resolution of enantiomers of NSAIDs using CDs.

Cyclodextrins are cyclic oligoglucose molecules having a structure similar to a truncated cone with both ends open. The large opening of the cone is rimmed with secondary 2-hydroxy groups of the glucose units, all rotated to the right, and the smaller opening is rimmed with the polar primary hydroxyl groups and, therefore, is relatively hydrophobic. A variety of water-soluble as well as insoluble compounds can fit into the hydrophobic cavity of the CD molecule, thereby forming reversible inclusion complex of different stability. If a chiral molecule fits exactly into the cavity with its less polar side, a separation into the enantiomers can be expected, and the better the fit of the guest molecule, the better the separation. If the guest molecules are small, they are completely enclosed by the CD and cannot be separated. On the other hand, if the molecules are larger than the CD cavity, there may be little or no interaction.

Macrocyclic antibiotics, for example, erythromycin and vancomycin, have also been used as impregnating agents for TLC resolution of enantiomers of dansyl amino acids by Bhushan and Parshad [6] and Bhushan and Thiong'o [7]. Separation in this case is due to chirality of macrocyclic antibiotic molecule, which is ionizable and contains hydrophobic and hydrophilic moieties as well providing enantioseparation via $\pi-\pi$ complexation, H-bonding, inclusion in a hydrophobic pocket, dipole stacking, steric interactions, or combination thereof. In view of the scope of the chapter and nonavailability of reports related to NSAIDs, these are not being discussed here.

13.3.5 PREPARATION OF PLATES IMPREGNATED WITH CHIRAL SELECTOR

Impregnation of thin-layer material with different suitable chiral selectors provides an inexpensive wider choice of separation conditions for direct enantiomeric resolution, that is, the analyte can be applied to the plates in the form of racemic or scalemic mixture as such, without resorting to derivatization of the enantiomeric mixture with one or the other chiral reagent that would otherwise involve several experimental stages of synthesis and purification, etc. A general protocol for impregnated TLC successfully applied in author's laboratory [8] is described in the following text.

A slurry of silica gel G (50 g) is prepared in distilled water (100 ml) containing the chiral selector as the impregnating agent (e.g., L-arginine and (−)-brucine), spread over glass plates with a Stahl type applicator to give plates of 20 × 20 cm × 0.5 mm. The plates are activated overnight at 60°C. The solutions of the enantiomeric mixture and standards are prepared in appropriate concentrations and applied suitably. The chromatograms are developed and dried generally at 60°C for a few minutes and spots are located using a suitable reagent, for example, iodine vapors, spray with ninhydrin, etc.

Impregnation has also been carried out in the following manner. Precoated silica gel TLC plates, without fluorescent indicator, were dipped into a large dish, for 60 min, containing a solution of the chiral selector. Varying amounts of selector were initially dissolved in 5 ml distilled water then made up to 100 ml with ethanol. The plates were then removed from the solution and left to dry completely in a fume hood for 24 h. These were then activated in an oven at 100°C for 20 min [9].

In another approach, commercial TLC plates of silica gel 60 F_{254} were first carefully washed, by predevelopment, with methanol/water (9:1, v/v), then dried at ambient temperature for 3 h, and these were then impregnated with a 3×10^{-2} mol/l solution of L-arginine in methanol by conventional dipping for 2 s [10,11]. The concentration of the impregnating solution was calculated as that depositing 0.5 g L-arginine per 50 g of the dry silica gel adsorbent layer.

13.3.6 APPLICATIONS (TLC RESOLUTION OF ENANTIOMERS)

A little attention has been paid to TLC resolution of enantiomers of NSAIDs though in terms of equipment cost and cost of running/maintaining the HPLC equipment is simpler and less expensive and provides a direct resolution.

13.3.6.1 Ibuprofen

Ibuprofen is one of the most effective NSAID and analgesic with fewer side effects, and is extensively used worldwide. The current demand for ibuprofen is around 10,000 TPA, which is expected to grow by around 5% per year mainly led by its new applications in Alzheimer's disease. It is marketed as a racemic mixture. Ibuprofen has a weak chromophore and UV detection of nanogram amounts is difficult, and to detect low levels of ibuprofen in biological samples a moiety with a high-UV molar absorptivity or a fluorescent label is added by derivatization. The direct resolution of racemic ibuprofen on impregnated TLC has an advantage over general indirect methods involving chiral derivatization prior to chromatography.

Impregnated TLC has been, for the first time, reported by Bhushan and Parshad [8] for the resolution of (±)-ibuprofen into its enantiomers. Two-dimensional TLC on silica gel plates impregnated with optically pure L-arginine (0.5%) as chiral selector, using acetonitrile/methanol/water (5:1:1, v/v) as the solvent system, has been successful (Table 13.1). As L-arginine has a pI of 10.8 a few drops of acetic acid were added. This allowed to maintain pH below the isoelectric point and to keep the amino acid in the cationic form. Separate plates were run for one- and two-dimensional modes. In the one-dimensional mode the spots of (±)-ibuprofen and the (+)-isomer were applied side by side on the same plate. In the two-dimensional mode, first the spot of (±)-ibuprofen was applied to the L-arginine impregnated plate and then the spot of the (+)-isomer was applied after the first run at the side of the former spot. The developed plates were dried and the spots were located in iodine chamber.

L-Arginine, the impregnating reagent in the cationic form interacted with the two enantiomers of ibuprofen to give two diastereomeric salts, (+)(−) and (−)(−), leading to enantiomeric separation. The presence of L-arginine in the two spots was established by spraying the chromatogram with diluted HCl, after iodine

TABLE 13.1

Separation of (±)-Ibuprofen Using (−)-Brucine or L-Arginine as Chiral Selector

Compound	Solvent system	hR_f Pure (+)	From mixture (−)	From mixture (+)	Temperature	pH	References
Ibuprofen	(i) MeCN/MeOH (16:3, v/v) (ii) MeCN/MeOH/H$_2$O (16:3:0.4, v/v)	93	86	93	30 ± 2°C	6–7	[13]
Flurbiprofen	As above	96	90	96	24 ± 2°C	6–7	[13]
Ibuprofen	MeCN/MeOH (5:1)	85	71	85	28 ± 2°C	6–7	[15]
Ibuprofen[a]	MeCN/MeOH/H$_2$O (5:1:1, v/v)	80	77	80	32 ± 2°C	<10.8	[8]

(i) and (ii): Solvents for first and second dimension TLC run, respectively.

[a] Plates impregnated, by mixing, with L-arginine, same solvent system in both dimensions.

Source: From Bhushan, R. and Parshad, V., *J. Chromatogr. A*, 721, 369, 1996; Bhushan, R. and Thiong'o, G.T., *Biomed. Chromatogr.*, 13, 276, 1999; Bhushan, R. and Gupta, D., *Biomed. Chromatogr.*, 18, 838, 2004.

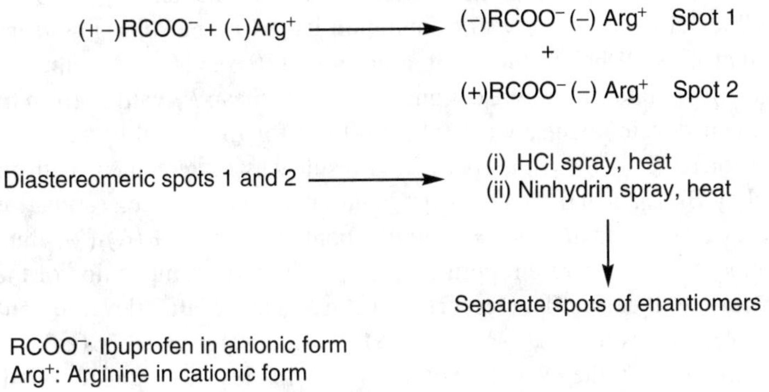

$$(+−)RCOO^− + (−)Arg^+ \longrightarrow (−)RCOO^− (−) Arg^+ \quad \text{Spot 1}$$
$$+$$
$$(+)RCOO^− (−) Arg^+ \quad \text{Spot 2}$$

Diastereomeric spots 1 and 2 ⟶ (i) HCl spray, heat
(ii) Ninhydrin spray, heat

↓

Separate spots of enantiomers

RCOO⁻: Ibuprofen in anionic form
Arg⁺: Arginine in cationic form

SCHEME 13.1　Reaction on the chromatogram.

was allowed to evaporate off, followed by heating and spraying with ninhydrin (0.2% in acetone). It showed that arginine formed diastereomeric salts *in situ* with the components of the racemic mixture. The reaction, taking place on the chromatogram can be represented as shown in Scheme 13.1. The method was successful in resolving as little as 0.1 μg of the enantiomeric mixture (i.e., 0.05 μg of each enantiomer), which is much lower than the reported minimum quantifiable concentration of 0.2 μg/ml by RP-HPLC using fluorescence detection [12].

In a similar manner, Bhushan and Thiong'o [13] used (−)-brucine as a chiral impregnating reagent for two-dimensional TLC resolution of (±)-ibuprofen. The chromatograms were developed in solvent systems acetonitrile/methanol (16:3, v/v) for the first dimension for 20 min and acetonitrile/methanol/water (16:3:0.4, v/v) for the second dimension for 20 min and the spots were located in an iodine chamber. The method was successful in resolving as little as 0.1 μg of the racemate. Experiments confirmed that (−)-brucine impregnated on thin layer plates was uniformly immobilized on the stationary phase.

The experiment included in-house coating of glass plates with silica gel slurry containing an accurately known amount of L-arginine as chiral selector. This direct approach to the separation of enantiomers of ibuprofen was later extended to TLC systems with commercially available silica-coated glass plates, again impregnated with L-arginine; successful separation of ibuprofen enantiomers was shown as chromatographic bands visualized densitometrically [10].

13.3.6.1.1   Reinvestigation and Densitometry

The resolution studies on L-arginine impregnated silica gel plates prepared in-house [8] were reinvestigated in both one- and two-dimensional modes on commercially available precoated plates which were impregnated with L-arginine by immersion followed by densitometry (210 nm) [10,11] of the separated enantiomers. The results obtained with the modified procedures were in agreement with the original reports; these are briefly described below.

Trace amounts of acetic acid were added to the ternary mobile phase (MeCN/MeOH/H$_2$O, 5:1:1, v/v) to maintain its pH at the solid–liquid interface at 4.8, that is, well below the pI of L-arginine (10.8), thereby maintaining the impregnating amino acid in the cationic form. The mean R_f values, from the one-dimensional developments, were 0.82 (+0.02) for (S)-(+)-ibuprofen and 0.79 (+0.02) for its (R)-(−) counterpart. This result was in agreement with the data reported by Bhushan and Parshad [8]. Quantitative analysis (i.e., estimation from relative areas) showed that the (±)-mixture contained ca. 10% (R)-(+)-ibuprofen, in agreement with literature reports of the enantiomeric composition of the commercial form of the drug. The respective mean R_f values, after development in the second direction, were 0.83 (+0.02) for (S)-(+)-ibuprofen and 0.76 (+0.02) for its (R)-(−) counterpart; the resolution of the two antipodes was found to be enhanced to $\Delta R_f = 0.07$. The in situ identification of the resolved chromatographic bands, by acquiring their UV absorption spectra, was the most persuasive proof of successful separation of the two enantiomers. In such circumstances, one can justifiably expect two identical spectra. The only difference between the two was the different intensity of the absorption bands because of the different amounts of the two separated species on the adsorbent layer. Thus, it was clearly proven that the two resolved chromatographic bands were of two enantiomers present in different quantitative proportions. This result provided up-to-date instrumental confirmation of successful complete TLC separation of the two ibuprofen antipodes.

The procedure resulted in enhancement of the separation. In the original procedure, two-dimensional development of the chromatogram resulted in resolution of the two enantiomers such that ΔR_f was 0.03. In the modified procedure, exactly the same separation (i.e., $\Delta R_f = 0.03$) was obtained by one-dimensional development. Two-dimensional development resulted in enhancement of the resolution of the enantiomers so that $\Delta R_f = 0.07$. Further, in the modified procedure, resolution of the two enantiomers was documented directly by *in situ* acquisition of identical UV absorption spectra from the two resolved species whereas in the original procedure visualization was achieved with iodine vapor only.

13.3.6.1.2 *Peculiar Two-dimensional Separation*
The first two-dimensional resolution of ($\pm$)-ibuprofen was reported and documented by Bhushan and Parshad [8] but did not comment on the stereochemically peculiar migration tracks of the enantiomers migrating in two mutually perpendicular directions.

Experiments were performed with (S,R)-($\pm$)-ibuprofen, (S)-($+$)-naproxen, and (S,R)-($\pm$)-2-phenylpropionic acid as test analytes to investigate the stereochemically peculiar two-dimensional separation of 2-arylpropionic acids by chiral TLC [14] using commercial precoated silica gel plates that were impregnated with L-arginine by immersion [10,11] before chromatography. The results showed that the effect of L-arginine on the separation of the three pairs of enantiomers was two-dimensional, that is, separation of the pairs of enantiomers was both in the direction of mobile phase flow and in the direction perpendicular to this. Scanning of the chromatographic plates at 1 mm intervals revealed that the tracks of the (S)-($+$) and the (R)-($-$) APA deviated markedly from the vertical in mutually opposite directions. This striking effect in the two-dimensional separation of the enantiomers was most probably caused by stereo-specific intermolecular interactions, which occur during ion-pair formation between L-arginine in the cationic form, deposited on the silica gel layer as chiral selector, and each individual 2-APA in the anionic form. The main results on the stereochemically peculiar migration tracks of the enantiomers of the three NSAIDs are summarized in Table 13.2.

The results in Table 13.2 show the magnitude and direction of the maximum deviations from the vertical for the three pairs of enantiomers of the three NSAIDs. For ibuprofen, resolution of the two enantiomers is the least ($\Delta R_f = 0.03\ R_f$ units only) and, consequently, for this compound the maximum sum of the left- and right-handed deviation was established as 4 ($\pm$2) mm. For the other two 2-APA, resolution of the enantiomer pairs was greater ($\Delta R_f = 0.04$ for naproxen and 0.10 for 2-phenylpropionic acid). Accordingly, the observed maximum deviations of the migration tracks were also greater. For naproxen, the maximum sum of the left- and the right-handed deviations was 6 ($\pm$2) mm; for 2-phenylpropionic acid it was 7 ($\pm$2) mm. It is also worthy note that for ibuprofen and 2-phenylpropionic acid (both of which have the asymmetric carbon atom substituted by the phenyl group), the (S)-($+$) enantiomers deviated to the right of the vertical and the tracks of their (R)-($-$) counterparts deviated to the left. For naproxen (in which the asymmetric carbon atom substituted by a naphthyl group) the opposite regularity was observed, that is, (S)-($+$)-naproxen deviated to the left and (R)-($-$)-naproxen deviated to the right. These regularities may prove quite important for future molecular-level

TABLE 13.2
Data Showing Migration of the Enantiomers of Ibuprofen, Naproxen and 2-Phenylpropionic Acid from Strict Vertical

Analyte	Configuration	$R_f^{(a)}$	MeCN-MeOH-H$_2$O (v/v)	Handedness of the deviation	Deviation in min[b]
Ibuprofen	(S)	0.91	5:1:1	Right	2
	(R)	0.88		Left	2
Naproxen	(S)	0.89	5:1:1.5	Left	3
	(R)	0.85		Right	3
2-Phenylpropionic	(S)	0.90	5:1:0.75	Right	5
acid	(R)	0.80		Left	2

Stationary phase: Silica gel 60 F$_{254}$, impregnated with L-arginine

Distance: 15 cm

Numerical values were derived from 27 individual enantiomer migration tracks.

[a] ±0.02; [b] ±1 mm.

Source: Adapted from Sajewicz, M., Piętka, R., Drabik, G., Namysło, E. and Kowalska, T., On the stereochemically peculiar two-dimensional seperation of 2-arylpropionic acids by chiral TLC, *J. Planar Chromatogr.*, 19, 273, 2006.

consideration of the stereospecificity of interactions between L-arginine as chiral selector and each individual 2-APA in the anionic form.

13.3.6.1.3 Thermodynamics of Retention

Further, the chiral discrimination model via the formation of ionic diastereomers, as proposed by Bhushan and Parshad [8] in Scheme 13.1, was viewed by Kowalska and coworkers [11] in terms of the energy difference for ion-pair formation (i.e., in the formation of the two diastereomeric salts). Ibuprofen is a carboxylic acid and, therefore, apt to dissociate (and form an organic anion) and the separation of the two enantiomers of ibuprofen can be achieved only because the thermodynamic equilibrium constants (K) for the ion-pair formation process for the two enantiomers $(K_1$ and K_2, respectively) have different numerical values. From the theory of adsorption liquid chromatography, it is well known [15] that the thermodynamic equilibrium constant of adsorption, K, can be defined as follows:

$$\log K = \frac{-\Delta\mu_a}{(2{:}303\ RT)},$$

where $\Delta\mu_a$ is the standard chemical potential for adsorption of the analyte on the adsorbent surface, R is the universal gas constant, and T is the temperature of the experiment. From the chromatographic results (i.e., R_f values) obtained for the two ibuprofen enantiomers and keeping in mind another fundamental relationship of adsorption liquid chromatography [3], namely, $R_f = 1/(1 + K\phi)$ where ϕ is the so-called phase ratio (i.e., the ratio of the volume of the adsorbed mobile phase to the volume of the flowing mobile phase), the thermodynamic magnitudes of $\Delta\mu_a$ for (S)-$(+)$- and (R)-$(-)$-ibuprofen were calculated. These are given in Table 13.3. In the calculation, it was assumed that ϕ is approximately equal to 0.1 and T was measured as 295 K. From the calculated values of $\Delta\mu_a$ given in Table 13.3, it is clearly evident that the affinity of (R)-$(+)$-ibuprofen for the adsorbent layer is considerably greater than that of its enantiomeric antipode. This readily apparent difference between the standard chemical potentials of adsorption of the two antipodes was considered due to their different ability to form ion pairs, although both the energetics of ion-pair formation and the energetic of adsorption of ibuprofen's phenyl ring, on active sites of silica, contribute to the numerical value of $\Delta\mu_a$.

13.3.6.1.4 Treatment with Chiral Selector Prior to Chromatography

In another approach, optically pure $(-)$-brucine was mixed with $(\pm)$-ibuprofen while preparing the solution prior to its application onto TLC plates [16]. The premixing resulted in the formation of diastereomeric salts of the type $(+)$-ibuprofen–$(-)$-brucine and $(-)$-ibuprofen–$(-)$-brucine without resorting to any covalent linkage. It was the movement of these ionic diastereomers on TLC plates that resulted in separation. Chromatograms were developed at $28 \pm 2°C$ for 20 min in acetonitrile/methanol (5:1, v/v), and the spots were located with iodine vapors.

TABLE 13.3

Numerical Values of $\Delta\mu_a$ Estimated for Enantiomers of Ibuprofen Separated in Impregnated Silica Gel Plates

Development direction	Enantiomer	R_f	$\Delta\mu_a$ kJ mol^{-1}	$\Delta\Delta\mu_a$ kJ mol^{-1}
First	(S)	0.82	−1.9	0.4
	(R)	0.79	−2.3	
Second	(S)	0.83	−1.7	1.1
	(R)	0.76	−2.8	

Mobile phase: MeCN-MeOH-H$_2$O (5:1:1, v/v) adjusted to pH 4.8 by addition of trace amount of acetic acid.

TLC on commercial pre coated plates impregnated with L-arginine by immersion.

Source: Adapted from, Sajewicz, M., Pietka, R. and Kowalska, T., Chiral separations of ibuprofen and propranolol by TLC, a study of the mechanism and thermodynamics of retention, *J. Liq. Chromatogr. and Rel. Tech.*, 28, 2499, 2005.

The best resolution conditions include mixing of 0.12 mmol of chiral selector with 0.24 mmol of ibuprofen, TLC at $28 \pm 2°C$, and pH between 6 and 7. The minimum detection limit was found to be 2.45 μg of each enantiomer. Pure separated enantiomers could be obtained by cutting the two spots from the TLC plate followed by extraction in ethanol and column chromatography for each to remove (−)-brucine from the ibuprofen enantiomer. The hR_f and other chromatographic conditions are summarized in Table 13.1.

13.3.6.1.5   TLC of 3,5-dinitroanilide Derivatives on Naphthylethyl Urea TLC-CSP

Using the impregnation approach, Brunner and Wainer [17] allowed to react (R)-(−)-1-(1-naphthylethyl)ethyl isocyanate with a commercially available amino propyl HPTLC plate to form a naphthylethyl urea TLC-CSP. For this purpose, 1 g of isocyanate was dissolved in 100 ml of methylene chloride and the aminopropyl HPTLC plate was soaked in 20 ml of the solution for 5 min. The plate was removed from the solution and air-dried. It was then washed by immersion in methylene chloride (twice) and air-dried. The amount of isocyanate bound to the plate was calculated from the loss of UV absorbance of the reagent solution, measured at 281 nm; it was found to be 86 mg isocyanate reagent per 10×10 cm^2 plate. The chromatograms were developed in hexane/isopropanol/acetonitrile (20:8:1, v/v) and detection was carried out at both short (254 nm) and long (360 nm) UV wavelengths. Racemic ibuprofen was converted to 3,5-dinitroanilide, by reacting it first with an acid chloride followed by condensation with nitroaniline, and the derivative was then chromatographed on the naphthylethyl urea TLC-CSP. The lower detection limit for the 3,5-dinitroanilide derivative of ibuprofen was 0.5 μg. The R_f values and chromatographic conditions are given in Table 13.4. Independent chromatography for the pure single isomer revealed that the (S)-isomer of ibuprofen was more retained than the (R)-isomer. Further investigations

TABLE 13.4

R_f Values and Experimental Conditions for Resolution of Certain NSAIDs on Naphthylethyl Urea TLC-CSP

Sl no.	Compound	R_f	α	Order
1	Ibuprofen	0.45, 0.28	2.10	R,S
2	Naproxen	0.24, 0.15	1.79	R,S
3	Fenoprofen	0.33, 0.23	1.65	ND
4	Flurbiprofen	0.33, 0.23	1.65	ND
5	Benoxaprofen	0.30, 0.20	1.71	ND

The compound chromatographed as 3,5-dinitroanilide. ND = not determined. Mobile phase: hexane/isopropanol/acetonitrile (20:8:1, v/v). Detection: 254 or 360 nm.

Source: From Brunner, C.A., and Wainer, I., *J. Chromatogr.*, 472, 277, 1989.

suggested that $\pi-\pi$ interactions between a π-acidic moiety on the solute and the π-basic naphthyl moiety on the CSP are the key aspects in chiral recognition. Thus, the potential solutes for the naphthylethyl urea TLC-CSP which contain an amine or a carboxylic acid moiety should be converted into the corresponding 3,5-dinitrobenzoyl amides or 3,5-dinitroanilides before chromatography.

TLC using silica gel plates has successfully been applied for qualitative and quantitative separation of ibuprofen from commercial formulations using acetonitrile/methanol (10:6 or 2:1, v/v). The detection limit was found to be 4.1 μg while the recoveries through preparative layer were not very good [18]. Assay of ibuprofen in commercial tablets has been reported using silica gel 60 HPTLC layers developed for 15 min up to 50 mm with toluene/acetone/formic acid (70:28:2, v/v); quantitative evaluation was done by measurement of fluorescence quenched zones by UV absorption scanning densitometry at 230 nm [19].

Racemic ibuprofen could not be resolved into its enantiomers using TLC plates impregnated, by immersion, with L-serine or L-threonine, or with both of them as the chiral selectors and combinations of acetonitrile/methanol/water as the mobile phase [9]. The same impregnated plates were, however, successful in resolving certain other 2-arlypropionic acids, as described in the following section.

13.3.6.2 Flurbiprofen

Flurbiprofen is marketed as a racemic mixture of (S)-$(+)$- and (R)-$(-)$-enantiomers. Flurbiprofen is a white or slightly yellow crystalline powder. It is slightly soluble in water at pH 7.0 and readily soluble in most polar solvents.

Impregnation of the stationary phase with the chiral selectors is the highly promising approach. Racemic flurbiprofen has been resolved into its enantiomers on plates impregnated with $(-)$-brucine in a manner similar to resolution of racemic ibuprofen on brucine impregnated plates [13] by two-dimensional TLC using acetonitrile/methanol (16:3, v/v) for first dimension and acetonitrile/methanol/water (16:3:0.4, v/v) for the second dimension.

Impregnation by dipping the precoated TLC plates of silica gel in the solution of chiral selector for the resolution of enantiomers of flurbiprofen and a few other NSAIDs has been reported by Aboul-Enien et al. [9]. Precoated TLC plates without fluorescent indicator were dipped in the solution of L-(−)-serine, L-(−)-threonine, and a mixture of the two optically pure amino acids for 60 min at ambient temperature. The plates were dried for 24 h in a fume hood and then activated at 100°C for 20 min. (±)-Flurbiprofen resolved into its enantiomers only on the plates impregnated with the mixture of L-serine and L-threonine (0.05 g each) using acetonitrile/methanol/water (16:4:0.5, v/v), when the (S)-(−)-enantiomer eluted first. The investigations showed that the best resolution was achieved at 25 ± 2°C, and between pH 6 and 7, and interestingly there was no resolution on the plates impregnated with L-serine or with L-threonine.

13.3.6.3 Other NSAIDs

Some of the other NSAIDs investigated for their enantiomeric separation include (±)-ibuproxam, (±)-benoxaprofen, (±)-ketoprofen, (±)-pranoprofen, and (±)-tiaprofenic acid. All of these have been investigated on plates impregnated with L-(−)-serine, L-(−)-threonine, and the mixture of both the L-amino acids as the chiral selectors. The TLC plates were prepared by dipping the precoated silica gel plates into a large dish, containing the solution of the selector, for 60 min at ambient temperature. TLC plates impregnated with L-serine or L-threonine provided the separation of (±)-ibuproxam, (±)-ketoprofen, and (±)-pranoprofen into their enantiomers; the chiral separation factors were 1.04, 1.44, and 1.38, respectively, on plates impregnated with L-serine, while these were 1.09, 1.52, and 1.56, respectively, on plates impregnated with L-threonine. TLC plates impregnated with both L-threonine and L-serine were able to resolve the racemic ibuproxam, benoxaprofen, flurbiprofen, ketoprofen, pranoprofen, and tiaprofenic acid. The results and experimental conditions are summarized in Table 13.5.

Impregnation of L-(−)-serine by mixing with silica gel, while making the plates, has also been used as a chiral selector for the TLC resolution of (±)-ibuproxam, (±)-ketoprofen, and (±)-tiaprofenic acid [20]. The successful solvent systems were acetonitrile/ethanol/water in the ratio of 16:4:0.5 (v/v) for the resolution of (±)-ibuproxam and (±)-ketoprofen and in the ratio of 16:3:0.5 (v/v) for (±)-tiaprofenic acid. The hR_f values reported are shown in Table 13.6. When an impregnated plate was developed without spotting any of the test compounds, a uniform staining of the entire surface of the plate was visible that indicated that L-(−)-serine was immobilized uniformly on the silica gel.

The naphthylethyl urea TLC-CSP, as described above [17] for ibuprofen, was also used for racemic naproxen, fenoprofen, flurbiprofen, and benoxaprofen. These were also converted to corresponding 3,5-dinitroanilide before chromatography. The R_f and other conditions are summarized in Table 13.4.

Kowalska and coworkers [21] studied two-dimensional TLC separation of the (S)-(+)-naproxen from its (R)-(−)-antipode, generated in the (S)-(+)-naproxen solution when stored in the basic and the acidic solutions. They adopted the

TABLE 13.5
Summary of Results and Experimental Conditions for Resolution of Certain Other NSAIDs on Impregnated Plates

	Impregnation								
	A			B			C		
Compound	hR_{f1}	hR_{f2}	Ratio	hR_{f1}	hR_{f2}	Ratio	hR_{f1}	hR_{f2}	Ratio
Ibuproxam	13.0	12.5	1.04	12.5	11.5	1.09	13	12	1.08
Benoxaprofen	13.0	13.0	1.00	11.0	11.0	1.00	13	12	1.08
Flurbiprofen	6.0	6.0	1.00	8.0	8.0	1.00	13.2	11.7	1.13
Ketoprofen	14.0	9.7	1.44	12.5	8.2	1.52	12.8	8.3	1.54
Pranoprofen	13.8	10.0	1.38	12.5	8.0	1.56	12.5	8.0	1.56
Tiaprofenic acid	14.0	14.00	1.00	13.7	13.7	1.00	13.0	8.5	1.53

Impregnation: (A) L-(−)-serine (0.1 g); (B) L-(−)-threonine (0.1 g); (C) mixture of L-serine and L-threonine (0.05 g each). Mobile phase: acetonitrile/methanol/water (16:4:0.5, v/v); temperature: 25°C; pH 6–7; time: 45 min for each run; solvent front: 15 cm; detection: iodine vapors.

Source: From Aboul-Enien, H.Y., El-Awady, M.I., and Heard, C.M., *Biomed. Chromatogr.*, 17, 325, 2003.

TABLE 13.6
TLC Resolution of Three Racemic 2-Arylpropionic Acids Using L-(−)-Serine as Chiral Selector at 25±2°C

		hR_f (R_f × 100)		
		From (±)-mixture		Pure
No.	Compound resolved	(+)	(−)	(−)
1	(±)-Tiaprofenic acid	53	93	93
2	(±)-Ketoprofen	57	83	83
3	(±)-Ibuproxam	28	95	95

Mobile phase: CH_3CN-CH_3OH-H_2O (16:3:0.5 v/v/v) for (±)-tiaprofenic acid while (16:4:0.50 v/v) for (±)-ibuproxam and (±)-ketoprofen.
Solvent front, 15 cm in 45 min; Spots located with iodine vapours.

Source: Adapted from Aboul-Enien, H.Y., El-Awady, M.I. and Heard, C.M., Enantiomeric resolution of some 2-arylpropionic acids using L-(−)-serine–impregnated silica as stationary phase by thin layer chromatography, *J. Pharma. Biomed. Analysis*, 32, 1055, 2003.

separation procedure first proposed for separation of the two enantiomers of ibuprofen by Bhushan and Parshad [8] and then successfully adapted to commercial precoated plates and densitometric detection [10]. Full separation of (S)-(+)- and (R)-(−)-naproxen was attained in the two-dimensional development of the chromatograms. The numerical values of R_f were 0.90 ($\pm$0.02) and 0.82 ($\pm$0.02) for the (S)-(+)- and the (R)-(−)-species, respectively, and the $\Delta R_f = 0.08$.

Detection limits for various NSAIDs with different chiral impregnating reagents are summarized in Table 13.7.

13.3.7 EFFECT OF TEMPERATURE, pH, AND CONCENTRATION OF CHIRAL SELECTOR

Systematic studies to investigate the effect of concentration of impregnating reagent, temperature, and pH to optimize the separation conditions have been reported by different workers. These showed that the best resolution for ($\pm$)-ibuprofen was at 0.1 g of (−)-brucine for every preparation of slurry of silica gel (30 g), at 24 $\pm$ 2°C and between pH 6 and 7 [13].

Studies to investigate the effect of concentration of the impregnating chiral selector on resolution of enantiomers by Aboul-Enien et al. [20] showed that the best resolution was achieved at a specific concentration of the chiral impregnating reagent, that is, 0.1 g of L-serine and L-threonine or 0.1 g of the two (0.05 g each) when used as a mixture. The relationship of the concentration of the chiral impregnating reagent with the enantioselectivity has never been linear. It is not established as to why doubling the amount of chiral selector fails to demonstrate enantioselectivity. Possibly, at the higher concentration the chiral selector was less dispersed and impregnated into the silica particles. To achieve enantioselectivity, interactions between the analytes and chiral selector, the process should take place within the confines of the pores of the silica gel. This indicates a steric parameter for the recognition process. The presence of excess chiral selector could give rise to interactions taking place outside the silica pores and in the absence of the steric component, eliminate or inhibit enantioselectivity [20].

Chiral discrimination is reported to decrease with increase in temperature and there has been a different temperature for different compounds that was found to be successful. Similarly, the pH of the solvent system affects chiral recognition. All such studies favor the formation of the ionic diastereomer leading to separation of enantiomers. The specific temperatures at which the enantiomeric resolution of ($\pm$)-ibuprofen or ($\pm$)-flurbiprofen occurred on the plates impregnated with L-arginine or (−)-brucine are shown in Table 13.1.

According to Dalgliesh [21], for enantiomer separation to occur, at least three factors among the π–π interactions, H-bonding, and steric and dipole–dipole interactions would be involved. Bhushan and Thiong'o [13] did not observe resolution of ($\pm$)-ibuprofen and ($\pm$)-flurbiprofen at pH 4 and 5, because within that pH range these (acids) were in unionized form and therefore no electrostatic interactions could occur between the chiral selector and the components of the racemates. Failure of resolution in the absence of water suggested the possibility of a lack of participation of some kind of H-bonding. Thus, the enantioseparation of both ($\pm$)-ibuprofen and ($\pm$)-flurbiprofen was due to electrostatic interaction between COO^- of these compounds and ^+NH of (−)-brucine, π–π interactions between phenyl moieties and H-bonding. These interactions provided

TABLE 13.7
Summary of Detection Limits for Various NSAIDs Using Different Chiral Impregnating Reagents

Sl. no.	(±)-NSAID	Chiral selector	Mobile phase	Detection limit for the enantiomer	References
1	Ibuprofen	(−)-Brucine	Acetonitrile/methanol (5:1, v/v)	2.45 μg	[16]
	Ibuprofen	(−)-Brucine	(i) CH_3CN/MeOH (16:3, v/v)	0.1 μg	[13]
	Flurbiprofen		(ii) CH_3CN/MeOH/H_2O (16:3:0.4, v/v)		
2	Ibuproxam, ketoprofen, pranoprofen, benoxaprofen, flurbiprofen, tiaprofenic acid	L-(−)-Serine, L-(−)-threonine, and a mixture of L-(−)-serine and L-(−)-threonine	Acetonitrile/methanol/water (16:4:0.5, v/v)	Between 0.25 and 0.5 μg	[9]
3	Ibuprofen	L-Arginine	MeCN/H_2O/MeOH (5:1:1, v/v)	0.05 μg	[8]
4	Ibuproxam, ketoprofen, tiaprofenic acid	L-(−)-Serine	Acetonitrile/methanol/water (16:4:0.5, v/v) and (16:3:0.5, v/v)	Between 0.25 and 0.50 μg	[20]

TABLE 13.8

R_f Values at $25 \pm 2°C$ due to Change in Solvent Composition Using L-Serine as the Chiral Selector

No.	Compound under study	Solvent system*	hR$_f$ (R$_f$ × 100) From racemic mixture (+)	(−)	Ratio S to R
1	(±)-Tiaprofenic acid	A	35	39	1.11
		B	68	78	1.15
		C	53	93	1.76
2	(±)-Ketoprofen	A	32	37	1.16
		B	57	83	1.46
		C	43	47	1.09
3	(±)-Ibuproxam	A	30	35	1.17
		B	28	95	3.39
		C	26	38	1.46

* Solvent system: CH_3CN-CH_3OH-H_2O (v/v), in the ratio, (A), 14:4:0.5; (B), 16:4:0.5; (C), 16:3:0.5 Solvent front, 15 cm in 45 min; Spots located with iodine vapors; Sample concentration, 250 μg/mL.

Source: Adapted from Aboul-Enien, H.Y., El-Awady, M.I. and Heard, C.M., Enantiomeric resolution of some 2-arylpropionic acids using L-(−)-serine–impregnated silica as stationary phase by thin layer chromatography, *J. Pharma. Biomed. Analysis*, 32, 1055, 2003.

in situ formation of diastereomers on the impregnated plates and consequently the resolution.

The effect of mobile phase system on enantiomeric resolution of certain NSAIDs is shown in Table 13.8 and the results showing influence of temperature and pH on L-(−)-serine impregnated plates are summarized in Table 13.9. Resolution in the presence of water assumes the possibility of participation of some kind of H-bonding. It was observed that resolution obtained for both ketoprofen and tiaprofenic acid was smaller than ibuproxam. It has been suggested that the ketone group between the two ring structures of ketoprofen and tiaprofenic acid interact with the terminal hydroxyl group of serine through hydrogen bonding, while the stereogenic centers of NSAID and serine formed the necessary transient diastereomeric pair. The formation of differential transient complexes may have been hindered within the sterically confined environment of the pore of the silica gel, resulting in relatively greater resolution for ibuproxam, which does not possess the same ketone ring configuration.

13.3.8 DETECTION USING IMPREGNATION

Compounds separated by TLC that absorb in the UV region but without emission in the visible region can be detected using thin layers impregnated with luminophores.

TABLE 13.9
R_f Values due to Change in Temperature and pH Using L-(−)-Serine as the Chiral Selector

Temperature	pH*	(±) Ibuproxam R_f from racemic mixture			(±) Ketoprofen R_f from racemic mixture			(±) Tiaprofenic acid R_f from racemic mixture		
		R	S	Ratio, S to R	R	S	Ratio, S to R	R	S	Ratio, S to R
15±2°C	4–5	25	25	1.00	25	29	1.16	35	37	1.66
	6–7	32	35	1.09	55	65	1.18	36	43	1.19
	8–9	16	18	1.13	24	28	1.17	38	41	1.08
25±2°C	4–5	22	27	1.23	24	27	1.13	55	66	1.20
	6–7	28	95	3.39	57	83	1.46	53	93	1.76
	8–9	55	63	1.15	36	38	1.06	48	52	1.08
35±2°C	4–5	22	22	1.00	20	22	1.10	41	41	1.00
	6–7	25	27	1.08	50	57	1.14	43	54	1.26
	8–9	14	14	1.00	16	19	1.19	43	46	1.07

*pH adjusted by addition of *dil* phosphoric acid and/or dil NaOH.
Mobile phase: Acetonitrile–methanol–water 16:3:0.5 for (±)-tiaprofenic acid while (16:4:0.5, v/v) for (±)-Ibuproxam and (±)-Ketoprofen.
Sample concentration: 250 µg/mL.

Source: Adapted from Aboul-Enien, H.Y., El-Awady, M.I. and Heard, C.M., Enantiomeric resolution of some 2-arylpropionic acids using L-(−)-serine–impregnated silica as stationary phase by thin layer chromatography. *J. Pharma. Biomed. Analysis,* 32, 1055, 2003.

Such layers show UV fluorescence and the luminophore used determines the color; the separated compounds appear as dark spots, usually at 254 nm. Silica gel F$_{254}$ precoated plates available commercially containing zinc orthosilicate (green fluorescence at 254 nm) are among the most popular ones. The luminophores used in TLC offer a poor selectivity because the substances appear as colorless spots, but they are efficient in locating the separated compounds.

Sârbu [23] compared the results for the detection of some NSAIDs on plates with a luminophore mixture of magnesium fluorogermanate, zinc orthosilicate, and zinc sulfide, and on plates with a mixture of magnesium fluorogermanate activated by manganese (red fluorescence), zinc–magnesium germanate activated by manganese (green fluorescence) and calcium tungstate (blue fluorescence) at 254 nm on a nearly white background. The NSAIDs examined included ibuprofen, piroxicam, diclofenac, indomethacin, fenbufene, aspirin, and phenylbutazone. Thin layers of 0.5 mm × 20 × 20 cm were prepared using silica gel R with 1% starch/agar-agar (1:1) as binder to which the mixture of magnesium fluorogermanate/zinc orthosilicate, and zinc sulfide (7:8:2) was added. Chromatograms were developed with chloroform/methanol/25% ammonia (18:15:5). The results obtained on the two types of plates in comparison with plates precoated with silica gel F$_{254}$ (Merck) are shown in Table 13.10.

TABLE 13.10
hR$_f$ and Detection of Some NSAIDs on Silica Gel Places with Single and Mixed Luminophores

| Sl. No. | NSAID | hR$_f^*$ | Colour at 254 nm | |
			A*	B
1	Ibuprofen	27	Violet	—
2	Oxyphenbutazone	12	Violet	Red-brown
3	Piroxicam	34	Green	Bluish
4	Diclofenac	44	Grey	Mauve
5	Indomethacin	49	Brown	Brown
6	Fenbufene	37	Blue	Bluish
7	Aspirin	16	Bluish	Violet

*Mobile phase: Chloroform-methanol-25% ammonia (18:15:5, v/v), Distance, 10 cm in 45 min
(A): Plate with magnesium fluorogermanate, zinc orthosilicate and zinc sulphide.
(B): Plate with magnesium fluorogermanate, zinc-magnesium germanate and calcium tungstate.

- Plates were first dried in air followed by drying in a chamber to 100–120°C.
- Detection limit 1 μg per spot at 254 nm except on pre-coated silica gel F$_{254}$ plate (Merck).
- All compounds, except ibuprofen, showed brown spots on pre-coated silica gel F$_{254}$ plate (Merck).

Source: Adapted from Sârbu, C. Detection of some non-steroidal anti-inflammatory agents on thin-layer chromatographic plates coated with fluorescent mixtures, *J. Chromatogr.*, 367, 286, 1986.

All the compounds could be detected, after separation, with a visual limit of 1 μg per spot at 254 nm on these impregnated plates while detection at this concentration was difficult on Merck plates.

13.3.9 INVESTIGATION OF OSCILLATORY INSTABILITY OF ENANTIOMERS

The usefulness of TLC has been demonstrated in physiological studies of the oscillatory instability of profens when stored for long periods of time in aqueous media. The oscillatory transenantiomerization of the profens from the (S)- to the (R)-form, and the vice versa, during storage has been shown using TLC. The TLC assessment of the structural oscillation of enantiomeric profens was also confirmed by polarimetry [24].

Solutions of (S)-(+)-ibuprofen, (S)-(+)-naproxen, and (S,R)-(±)-2-phenyl propionic acid in EtOH/H$_2$O (7:3, v/v) were stored at two different temperatures (6 ± 2°C and 22 ± 2°C) for prolonged period of time (12 days for ibuprofen and 5 days for the other two). TLC was performed on commercial glass plates precoated with 0.25 mm layers of silica gel 60 F$_{254}$. Before use, the plates were carefully washed by predevelopment with methanol/water (9:1, v/v) and then dried at ambient temperature for 3 h. The washed and dried plates were then impregnated with a solution of L-arginine in methanol by conventional dipping for 2 s. Finally, the washed, impregnated, and dried adsorbent layers were ready for chromatography. Chromatograms for (S)-(+)-ibuprofen solution were developed to a distance of 15 cm using the ternary mobile phase acetonitrile/methanol/H$_2$O (5:1:1, v/v); the solvent ratio were (5:1:1.5, v/v) and (5:1:0.75, v/v) for development of the naproxen and 2-phenylpropionic acid samples, respectively. The developing solvent contained several drops of acetic acid to fix the pH at 4.8. After development of the chromatograms, the plates were dried at ambient temperature for 3 h, and the three lanes were scanned densitometrically.

Oscillation of the respective numerical values of R_f at both measurement temperatures was apparent and the amplitude of these oscillations was higher for samples stored under refrigeration than for those kept at room temperature, for all the three profens. The specific rotation of each individual profen was measured and the results were in agreement with TLC experiments.

These oscillatory changes were observed not only for the optically pure enantiomers but also with the racemic 2-phenyl propionic acid sample. The molecular mechanism was attributed to keto–enol tautomerism that is acid catalyzed as well. All the profens are carboxylic acids with relatively well-pronounced electrolytic dissociation. Thus, the keto–enol transenantiomerization of profens in aqueous media was considered justified because of the self-catalytic effect of the protons originating from the dissociated carboxyl groups. Since reducing the temperature stabilizes the short-lived tautomers there was greater amplitude of oscillation at 6°C than at 22°C.

The oscillatory transenantiomerization of the selected test profens, that is, (S)-(+)-ibuprofen, (S)-(+)-naproxen, and (R,S)-(±)-2-phenylpropionic acid, was also assessed when dissolved in physiological salt to investigate whether profen drugs dispensed as physiological salt solutions can undergo oscillatory transenantiomerization in this particular medium [25]. Samples of the three compounds, dissolved in physiological salt, were stored at two different temperatures ($6 + 2°C$ and $22 + 2°C$) for the period of 5 days each and R_f values were measured for each compound and each experimental series twice a day at 5 h intervals. Each storage experiment was repeated twice, thus providing the two series of data.

TLC was performed as described earlier [24]. The results confirmed an occurrence of oscillations of the investigated profens when stored for a longer period of time as solutions in physiological salt. Polarimetry was not applied as a parallel measuring technique to assess transenantiomerization due to low solubility of profens in physiological salt.

Chromatography at ambient temperature ($22 + 2°C$), for (S)-(+)-ibuprofen stored in physiological salt showed the amplitude of R_f oscillations equal to $0.06 R_f$ units. For the sake of comparison, the same compound stored in dichloromethane had amplitude of $0.12 R_f$ units, and in 70% ethanol the amplitude was $0.11 R_f$ units [24]. For (S)-(+)-naproxen stored in physiological salt and developed at ambient temperature, the amplitude of R_f oscillations was equal to $0.04 R_f$ units. The same compound stored in dichloromethane and in 70% ethanol had the amplitudes of R_f oscillations equal to 0.05 and $0.07 R_f$ units [24], respectively.

Chromatography at $6 + 2°C$ for (S)-(+)-ibuprofen stored in physiological salt showed the amplitude of R_f oscillations equal to $0.12 R_f$ units. The same compound, when dissolved in dichloromethane had amplitude of $0.05 R_f$ units and, when dissolved in 70% ethanol, the value of amplitude was $0.18 R_f$ units [24]. For (S)-(+)-naproxen stored in physiological salt and developed at $6 + 2°C$, the amplitude was equal to $0.07 R_f$ units. The same compound dissolved in dichloromethane and in 70% ethanol showed the amplitudes of R_f oscillations equal to 0.04 and to $0.06 R_f$ units [24], respectively.

Further support to the hypothesis that the observed oscillations of the R_f and $[\alpha]_D$ values are due to keto–enol tautomerism [24] was obtained with (S)-(+)-naproxen investigated in the basic medium. Kowalska and coworkers [21] studied two-dimensional TLC separation of the (S)-(+)-naproxen from its (R)-(−)-antipode, generated in the (S)-(+)-naproxen solution when stored in the basic and the acidic medium. The separation was carried out using silica gel plates impregnated with L-arginine [8,10]. Full separation of (S)-(+)- and (R)-(−)-naproxen was attained in the two-dimensional development of the chromatograms. The result presented a very persuasive proof of transenantiomerization of (S)-(+)-naproxen to the (R)-(−)-species in the course of a prolonged storage in the basic mixed solvent. The more probable explanation accorded was that in the basic solvent partial racemization of (S)-(+)-naproxen to its (R)-(−)-counterpart took place, which resulted in lowering of the overall specific rotation value measured for the mixture of the two antipodes. Thus, keto–enol tautomerism was responsible for oscillations of the numerical R_f values for (S)-(+)-naproxen on prolonged storage in the basic solution. This evidence was furnished with the aid of polarimetry and TLC.

It has been concluded by these studies that there occurs an oscillatory change of configuration with the three selected profens, namely, (S)-(+)-naproxen, (S)-(+)-ibuprofen, and (S,R)-(±)-2-phenylpropionic acid, when stored for a long enough period of time in certain solvents and it occurred via keto–enol tautomerism. Thus, it was deduced that profens stored both in the aqueous and the nonaqueous media oscillate; the only difference is the frequency and the amplitude of these oscillations, which depends on the solvent and the storage temperature applied. However, TLC performed excellently and allowed tracing oscillatory transenantiomerization of profens in its own independent way.

13.4 CONCLUSION

Although the analysis of enantiomers of NSAIDs has been carried out for about 20 years, there often remains a deficit within the pharmaceutical and medical sciences to address this issue. The importance of stereoselectivity in pharmacokinetic, pharmacodynamic, and pharmacological activity is yet to be fully established and practiced. There is a steady demand for simple, inexpensive, and successful chiral separation methods; these are preferably executed with TLC. With the increasing application of HPLC (and GC), as revealed by the number of papers published, TLC still seems efficient enough to ensure even difficult separation of pairs of analytes. From well-documented experiments [8,10,11], the possibility of analytes migrating in two mutually perpendicular directions, that is, in the direction of mobile phase flow and in the direction perpendicular to this, becomes an evident advantage of chiral separations by TLC. By this method enhanced separation is undoubtedly achieved, and can easily be monitored by densitometric detection. Two-dimensional separation of the enantiomers certainly enhances overall resolution. It can reasonably be expected that further investigation of the detailed mechanism of enantiomer retention in chiral systems will reveal new and interesting features of the effect of localized intermolecular interactions on separation of pairs of enantiomers.

ACKNOWLEDGMENTS

The authors wish to thank Alexander von Humboldt-Stiftung, Bonn, Germany for the award of fellowship, and to the Indian Institute of Technology Roorkee, for granting the leave of absence to Ravi Bhushan.

REFERENCES

1. Bhushan, R., and Martens, J., Direct resolution of enantiomers by impregnated TLC, *Biomed. Chromatogr.*, 11, 280, 1997.
2. Günther, K., Martens, J., and Schickendanz, M., Dünnschichtchromatographische enantiomerentrennung mittels ligandenaustausch, *Angew. Chem.*, 96, 514, 1984.
3. Günther, K., Martens, J., and Schickendanz, M., Thin layer chromatographic enantiomeric resolution by ligand exchange, *Angew. Chem. Int. Ed. Engl.*, 23, 506, 1984.

4. Günther, K., Martens, J., and Schickendanz, M., Resolution of optical isomers by thin layer chromatography (TLC). Enantiomeric purity of L-DOPA, *Fresenius Z. Anal. Chem.*, 322, 513, 1986.

5. Alak, A., and Armstrong, D.W., Thin-layer chromatographic separation of optical, geometrical, and structural isomers, *Anal. Chem.*, 58, 582, 1986.

6. Bhushan, R., and Parshad, V., TLC separation of enantiomeric dansyl amino acids using macrocyclic antibiotic as a chiral selector, *J. Chromatogr. A*, 736, 235, 1996.

7. Bhushan, R., and Thiong'o, G.T., Separation of enantiomers of dansyl-DL-amino acids by normal phase TLC on plates impregnated with macrocyclic antibiotic, *J. Planar Chromatogr.*, 13, 33, 2000.

8. Bhushan, R., and Parshad, V., Resolution of (±)-ibuprofen using L-arginine-impregnated thin layer chromatography, *J. Chromatogr. A*, 721, 369, 1996.

9. Aboul-Enien, H.Y., El-Awady, M.I., and Heard, C.M., Thin layer chromatographic resolution of some 2-arylpropionic acid enantiomers using L-(−)-serine, L-(−)-threonine and a mixture of L-(−)-serine, L-(−)-threonine-impregnated silica gel as stationary phases, *Biomed. Chromatogr.*, 17, 325, 2003.

10. Sajewicz, M., Pietka, R., and Kowalska, T., Chiral separation of (S)-(+)- and (R)-(−)-ibuprofen by thin layer chromatography — an improved analytical procedure, *J. Planar Chromatogr.*, 17, 173, 2004.

11. Sajewicz, M., Pietka, R., and Kowalska, T., Chiral separations of ibuprofen and propranolol by TLC, a study of the mechanism and thermodynamics of retention, *J. Liq. Chromatogr. Rel. Tech.*, 28, 2499, 2005.

12. Kondo, J., Suzuki, N., Naganuma, H., Imaoka, T., Kawasaki, T., Nakanishi, A., and Kawahara, Y., Enantiospecific determination of ibuprofen in rat plasma using chiral fluorescence derivatization reagent, (−)-2-[4-(1-aminoethyl)phenyl]-6-methoxybenzoxazole, *Biomed. Chromatogr.*, 8, 170, 1994.

13. Bhushan, R., and Thiong'o, G.T., Direct enantiomeric resolution of some 2-arylpropionic acids using (−)-brucine impregnated thin layer chromatography, *Biomed. Chromatogr.*, 13, 276, 1999.

14. Sajewicz, M., Piêtka, R., Drabik, G., Namyslo, E. and Kowalska, T., On the stereochemically peculiar two-dimensional separation of 2-arylpropionic acids by chiral TLC, *J. Planar Chromatogr.*, 19, 273, 2006.

15. Snyder, L.R., *Principles of Adsorption Chromatography*, Marcel Dekker, Inc., New York, 1968.

16. Bhushan, R., and Gupta, D., Resolution of (±)-ibuprofen using (−)-brucine as a chiral selector by thin layer chromatography, *Biomed. Chromatogr.*, 18, 838, 2004.

17. Brunner, C.A., and Wainer, I., Direct stereochemical resolution of enantiomeric amides via thin layer chromatography on a covalently bonded chiral stationary phase, *J. Chromatogr.*, 472, 277, 1989.

18. Bhushan, R., Joshi, S., Arora, M., and Gupta, M., Study of the liquid chromatographic separation and determination of NSAID, *J. Planar Chromatogr.*, 18, 164, 2005.

19. Jamshidi, A., Adjvadi, M., Shahmiri, S., and Massoumi, A., HPTLC assay for ibuprofen in newly formulated liquorice-containing tablets, *J. Planar Chromatogr.*, 13, 45, 2000.

20. Aboul-Enien, H.Y., El-Awady, M.I., and Heard, C.M., Enantiomeric resolution of some 2-arylpropionic acids using L-(−)-serine-impregnated silica as stationary phase by thin layer chromatography, *J. Pharm. Biomed. Anal.*, 32, 1055, 2003.

21. Sajewicz, M., Piêtka, R., Drabik, G., and Kowalska, T., On the mechanism of oscillatory changes of the retardation factor (R_f) and the specific rotation $[\alpha]_D$ with selected solutions of (S)-(+)-naproxen, *J. Liq. Chromatogr. Rel. Tech.*, 29, 2071, 2006.
22. Dalgliesh, C.E., The optical resolution of aromatic amino acids on paper chromatograms, *J. Chem. Soc.*, 137, 3940, 1952.
23. Sârbu, C., Detection of some non-steroidal anti-inflammatory agents on thin-layer chromatographic plates coated with fluorescent mixtures, *J. Chromatogr.*, 367, 286, 1986.
24. Sajewicz, M., Pietka, R., Pieniak, A., and Kowalska, T., Application of thin-layer chromatography to investigating oscillatory instability of the selected profen enantiomers, *Acta Chromatogr.*, 15, 131, 2005.
25. Sajewicz, M., Piêtka, R., Pieniak, A., and Kowalska, T., Application of thin-layer chromatography to the investigation of oscillatory instability of selected profen enantiomers in physiological salt, *J. Liq. Chromatogr. Rel. Tech.*, 29, 2059, 2006.

14 Determination of Components in Selected Chiral Drugs

Jan Krzek, Irma Podolak, Urszula Hubicka, and Anna Kwiecień

CONTENTS

14.1 INTRODUCTION

In the available literature concerning the determination of chiral products by separation techniques, consistent opinion of the authors about the application of thin-layer chromatography (TLC) for the analysis of optical isomers is observed in many cases when column techniques such as high-performance liquid chromatography (HPLC), gas chromatography (GC), and capillary electrophoresis (CE) are in competition [1–3].

TLC has some advantages that may be preferred for the analysis of chiral drugs:

- Chromatographic separation and detection takes place separately, which enables carrying out the analysis at a different time and making full use of the detection techniques for the analysis of constituents.
- TLC makes simultaneous analysis of greater number of samples possible; it is more economical in comparison with column techniques.
- The fact that the chromatographic plate is disposable enables us to reduce a series of operations connected with the preparation of columns for HPLC.
- An advantage of TLC is the possibility of localization of all constituents on chromatograms, which always appear between the start line and the solvent front, contrary to HPLC, where lack of detection may be observed in case of the retention of a constituent on the column.
- The low resolution and higher detection limits are relative weak points of TLC in comparison with HPLC.

In spite of the aforementioned advantages, TLC is more rarely applied for the determination of enantiomers than HPLC or CE.

The present state of knowledge indicates the huge progress in column techniques which very often surpasses planar chromatography in relation to quantitative analysis.

It seems that modernization of TLC in the scope of production of modern equipments such as video densitometers, densitometers, sample applicators, chromatographic chambers, and the application of more effective separation techniques, together with modern computer programs, may contribute to greater interest in this method accordingly to the advantages described above.

14.2 SYNTHETIC DRUGS

14.2.1 QUINOLONES

The majority of quinolones in clinical use belong to the subset of fluoroquino-lones, which have a fluoro group attached to the central ring system. Among fluoroquinolones, there are molecules with one or two chiral centers. The position of a chiral center in relation to the quinolone ring influences antibacterial activity and pharmacokinetic parameters of isomers [4].

Enantiomers of compounds, in which chiral center is distant from 4-quinolone ring, such as lomefloxacin, clinafloxacin, gatifloxacin, and tosufloxacin, do not show significant differences in antibacterial activity and pharmacokinetic properties [5–9].

In therapeutics, racemic mixtures (ofloxacin, gemifloxacin, and clinafloxa-cin), active enantiomers (moxifloxacin and levofloxacin), and diastereoisomers (sparfloxacin) are applied [10].

TLC has been applied for the separation of ofloxacin enantiomers under conditions described below [11].

14.2.1.1 Ofloxacin

Synthetic chemotherapeutic agent of a broad spectrum of antibacterial activity.

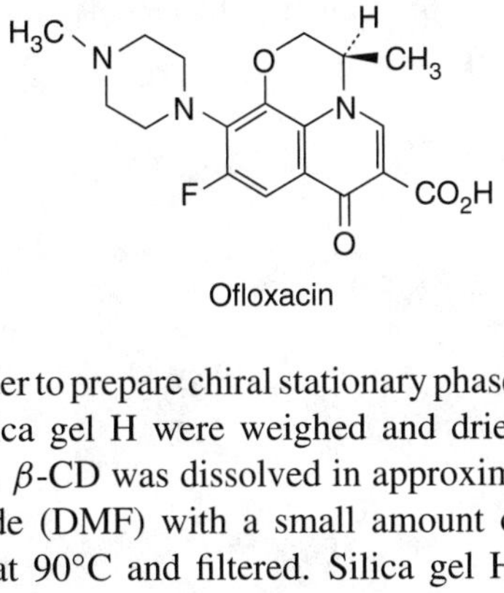

Ofloxacin

Stationary phase: In order to prepare chiral stationary phase, 10 g of β-cyclodextrin (β-CD) and 5 g of silica gel H were weighed and dried at 110°C in an oven under vacuum for 12 h. β-CD was dissolved in approximately 120 ml of anhyd-rous dimethylformamide (DMF) with a small amount of sodium; the mixture was stirred for 2–3 h at 90°C and filtered. Silica gel H and 2.5 g of KH-560 (3-glycidoxypropyltrimethoxysilane) were added to the filtrate and the mixture was stirred for 12–18 h at 90°C. The suspension was filtered and washed with DMF, toluene, methanol, water, again with methanol, and then dried in air. To 2 g of obtained β-CD-bonded silica gel, 6 ml of 0.3% aqueous solution of carboxy-methyl cellulose (CMC) sodium were added, and the slurry obtained was spread as a layer 0.2–0.4 mm thickness on glass plates 2.5 × 7.5 cm in size. Plates were dried in air and activated before use in an oven at 80°C for 1 h.

Mobile phase: Methanol/1% aqueous triethylammonium acetate at pH 4.1 (2:2, v/v).

Ofloxacin solution (1 μl) in n-butanol at a concentration of 1 mg/ml was applied to the prepared plate 10 mm from the edge. Chromatograms were developed to a distance 50 mm in a glass chamber 40 × 40 × 80 mm in size at room temperature.

Detection: The spots were identified by UV illumination at λ = 365 nm or by use of iodine vapor.

Under established conditions, good separation of ofloxacin enantiomers was achieved, the R_F values were 0.46 and 0.54 and the selectivity factor was α = 1.38. The authors revealed that separation of ofloxacin enantiomers may be carried out by use of 1% aqueous triethylammonium acetate at pH 4.1 without methanol.

For the separation of ofloxacin enantiomers by TLC, some other stationary phases were used such as cellulose *tris*(4-methylbenzoate), cellulose *tris*(4-nitrobenzoate) [12], and cellulose (3,5-dinitrobenzoate) [13].

14.2.2   1,4-DIHYDROPYRIDINE DERIVATIVES

1,4-Dihydropyridine (DHP) derivatives are chiral compounds, except for nifedipine and lacidipine. The center of chirality is the tetrahedral carbon atom C4 in the DHP ring.

14.2.2.1   Felodipine

Felodipine, first-vessel selective calcium channel blocking drug, is being used to control high blood pressure [14].

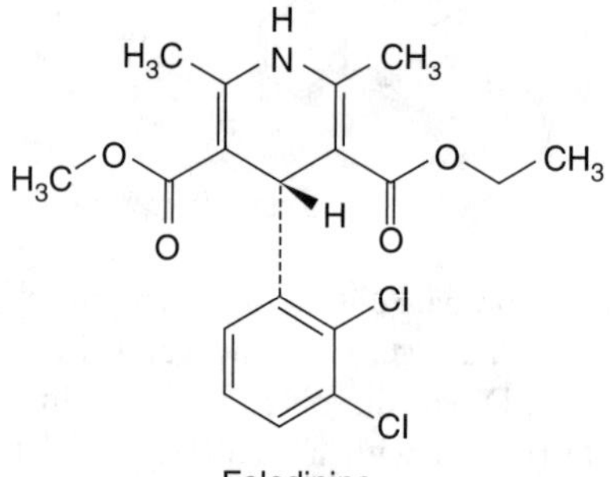

Felodipine

The separation of enantiomers was carried out by TLC with the application of chiral selectors present in the stationary phase [15]. This method is called ligand-exchange chromatography. According to the author, the formation of diastereoisomeric complexes composed of a DHP derivative enantiomer, selector–proline and Cu(II) ions occurred. The mechanism of separation of enantiomers is based on different stability of the complexes enantiomer–selector–Cu(II) ion and on the adsorption of the three-component diastereoisomer complexes on a chemically bonded stationary phase.

Stationary phase: Chromatographic plates (Chiralplate), Macherey-Nagel.

Mobile phase: The separation was carried out by the application of two mobile phases with the following components, containing methanol as an organic modifier (φ):

Phase 1 — chloroform/25% ammonia/methanol (10:0.02:φ).

Phase 2 — acetonitrile/0.1% triethylamine/methanol (5:3:φ).

On chromatographic plates 10×3 cm in size, $0.3\ \mu l$ of solutions at a concentration of 3.039×10^{-4} mol/l were applied. Chromatograms were developed to a distance 9 cm in a saturated chamber in the dark.

Detection: The spots were visualized under UV light (300–400 nm).

The best separation of enantiomers in Phase 1 was obtained at $\varphi = 1.9570 \times 10^{-2}$, the R_F values were 0.6200 for R-isomer and 0.6400 for S-isomer. The best separation of enantiomers in Phase 2 was obtained at $\varphi = 1.11 \times 10^{-1}$, the R_F values were 0.3713 for R-isomer and 0.3550 for S-isomer.

14.2.2.2 Nimodipine

Nimodipine blocks calcium flow at smooth muscles of vessels, particularly in cerebral artery [14].

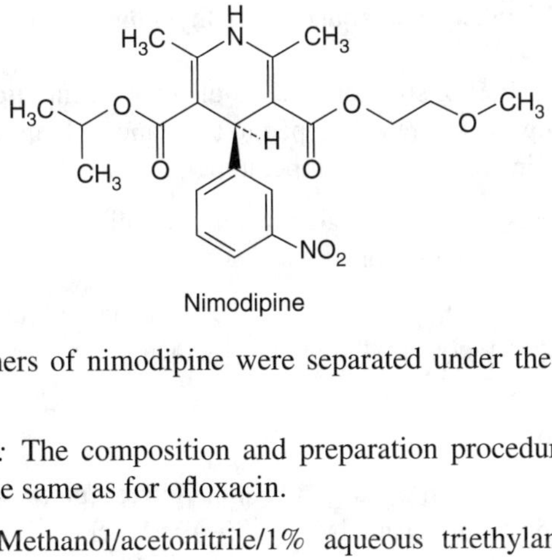

Nimodipine

The enantiomers of nimodipine were separated under the following conditions [11].

Stationary phase: The composition and preparation procedure is described in Section 14.2.1, the same as for ofloxacin.

Mobile phase: Methanol/acetonitrile/1% aqueous triethylammonium acetate (1.5:0.2:0.1, v/v/v).

Nimodipine ($1\ \mu l$) solution in ethanol at a concentration 1 mg/ml was applied to the TLC plate 10 mm from the edge. Chromatograms were developed to a distance 50 mm in a glass chamber $40 \times 40 \times 80$ mm in size at room temperature.

Detection: The spots on chromatograms were visualized under UV light at $\lambda = 365$ nm or by use of iodine vapor.

Under described conditions the R_F values of nimodipine enantiomers were 0.93 and 0.97, respectively, and selectivity factor was $\alpha = 2.43$.

14.2.3 ARYLOXYPROPANOLAMINE DERIVATIVES

A series of compounds based on aryloxypropanolamine structure is a diversified group of drugs widely used in therapy. Special attention should be paid to derivatives revealing α- and β-adrenoceptor activity.

14.2.3.1 Isoprenaline

β_1- and β_2-adrenergic receptor agonists, does not act on α receptor. It is mainly used in the treatment of asthma especially when caused by allergens [14].

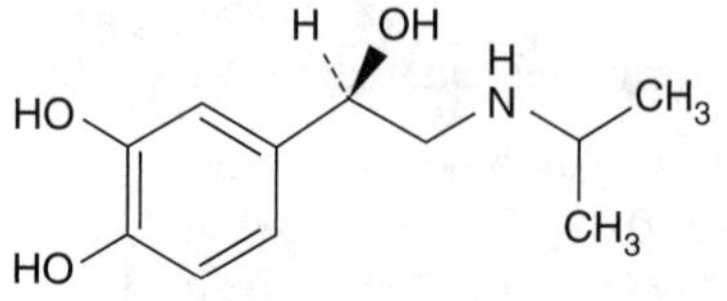

Isoprenaline

The separation of isoprenaline enantiomers by TLC has been developed by Zhu et al. [11].

Stationary phase: The composition and preparation procedure is described in Section 14.2.1, the same as for ofloxacin.

Mobile phase: Methanol/1% aqueous triethylammonium acetate at pH 4.1 (2:1, v/v).

On TLC plates, 1 μl of solution in ethanol at a concentration of 1 mg/ml was applied. Chromatograms were developed to a distance 50 mm in a glass chamber 40 × 40 × 80 mm in size at room temperature.

Detection: The spots on chromatograms were visualized under UV light at $\lambda = 365$ nm or by use of iodine vapor.

Under described conditions the R_F values of isoprenaline enantiomers were 0.26 and 0.39, respectively, and the selectivity factor was $\alpha = 1.82$.

14.2.3.2 Carvedilol

In addition to blocking both β_1- and β_2-adrenergic receptors, carvedilol also displays α_1-adrenergic antagonism. Isomer $S(-)$ is blocking mainly β-adrenergic receptors, while both enantiomers are blocking α_1-adrenergic receptors. Carvedilol is indicated in the treatment of hypertension and coronary disease [16].

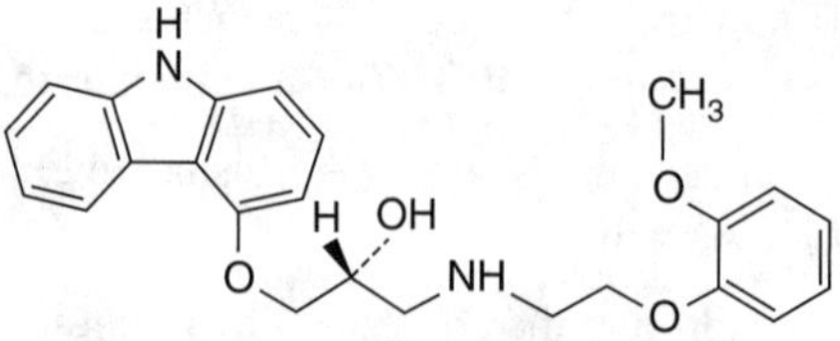

Carvedilol

The separation of carvedilol enantiomers was carried out under conditions described below [11].

Stationary phase: The composition and preparation procedure is described in Section 14.2.1, the same as for ofloxacin.

Mobile phase: Methanol/1% aqueous triethylammonium acetate at pH 4.1 (2:2, v/v).

On TLC plates, 1 μl of solution in methanol at a concentration of 1 mg/ml was applied. Chromatograms were developed to a distance 50 mm in a glass chamber 40 × 40 × 80 mm in size at room temperature.

Detection: The spots on chromatograms were visualized under UV light at λ = 365 nm or by use of iodine vapor.

Under described conditions the R_F values of carvedilol enantiomers were 0.34 and 0.54, respectively, and the selectivity factor was α = 2.28.

14.2.3.3 Metoprolol

Cardioselective β_1-receptor blocker without intrinsic sympathomimetic activity [14].

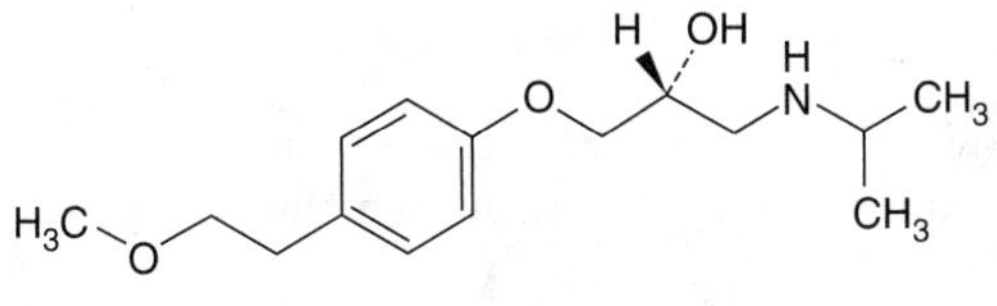

Metoprolol

Enantiomers of metoprolol were determined employing D-(−)-tartaric acid as chiral selector added to the mobile phase [14]. The results of experiments revealed that good resolution of enantiomers was achieved with 11.6 mmol/l D-(−)-tartaric acid in the mobile phase and in the impregnation solution at 25 ± 2°C [17].

Sample preparation: Standard solution of (±)-metoprolol tartarate at a concentration 20 mg/ml was prepared in methanol. The solution was diluted with methanol to concentrations of 5 and 10 mg/ml directly before application.

Stationary phase: Silica gel plates 60 F$_{254S}$ with concentrating zone (Merck) impregnated with mobile phase containing D-(−)-tartaric acid at ambient temperature for 90 min.

Mobile phase: Ethanol/water (70:30, v/v) with D-(−)-tartaric acid at a concentration 11.6 mmol/l as a chiral selector.

On chromatographic plates, 1 μl of samples was applied by applicator using disposable capillaries. The distance of the spots from the edge of the plates was 15 mm and the distance between spots was 10 mm. Chromatograms were developed to a distance of 60 mm for 40–50 min by ascending development in a chromatographic chamber saturated with the mobile phase for 30 min at ambient temperature. Chromatograms were then dried in air and registered densitometrically.

Detection: Chromatograms were scanned at $\lambda = 230$ nm with a Camag II TLC scanner, observed under UV light at $\lambda = 254$ nm and visualized in an iodine chamber.

Under established conditions the R_F values of S-($-$) and R-($+$)-metoprolol were 0.65 and 0.50.

Enantiomers of metoprolol were determined in urine after the extraction with toluene at pH 9.9 and derivatization with S-($+$)-benoxaprofen chloride [18].

Sample preparation: To 1 ml of urine in a stoppered test tube 1.0 ml of 0.5 mol/l carbonate buffer at pH 9.9 was added and shaken with 6.0 ml of toluene for 10 min. After centrifugation for 15 min, 4.0 ml of organic phase were transferred into another tube and evaporated to dryness in vacuum. The residue was dissolved in 1.0 ml of methylene chloride, then 100 mg of anhydrous sodium carbonate and 100 μl of derivatizing agent solution were added. The samples were left to react overnight at room temperature. Then 50 μl of methanol was added to stop the reaction and the solution was evaporated as described above. The residue was dissolved in 500 μl of cyclohexane by shaking vigorously.

Stationary phase: TLC plates Kieselgel 60 (Merck).

Mobile phase: Toluene/acetone (100:10, v/v) in ammonia atmosphere produced by two open 50 ml beakers filled with ammonia (33%) and standing in the chromatographic chamber at room temperature.

On chromatographic plates, 20 μl of solution in cyclohexane were applied with sample applicator in the form of 5-mm-wide bands. Chromatograms were developed in the freshly prepared mobile phase to a distance of 19.5 cm and then air-dried.

Detection: Chromatograms were scanned with KM 3 scanner (Carl Zeiss) at $\lambda_{ex} = 313$, $\lambda_{em} = 365$ nm.

Under established conditions the R_F values of R-($+$)-metoprolol and S-($-$)-metoprolol were 0.24 and 0.28, respectively.

The resolution of ($\pm$)-metoprolol into their enantiomers was achieved by TLC on silica gel plates impregnated with optically pure L-lysine (0.5%) and L-arginine (0.5%) as the chiral selectors. Different combinations of acetonitrile and methanol were used as the mobile phase. Spots were detected using iodine vapor. The detection limit for metoprolol was 0.26 μg [19].

Sample preparation: The solutions of racemic mixtures of ($\pm$)-metoprolol at a concentration 0.01 mol/l and their pure isomers were prepared in 70% ethanol.

Stationary phase: Thin-layer plates were prepared by spreading a slurry of silica gel G (30 g) in distilled water (60 ml containing 0.3 g of L-lysine or L-arginine) with a Stahl-type applicator. A few drops of ammonia were added to a slurry to maintain the pH above the isoelectric points of the amino acids. The plates were dried overnight at 60°C.

Mobile phase: Acetonitrile/methanol (15:4, v/v), but (15:5, v/v) when L-lysine was an impregnating reagent (15:3, v/v) and (16:4, v/v) when L-arginine was an impregnating reagent.

On chromatographic plates, 10 μl of racemic mixtures and solutions of pure enantiomers were applied side by side. Chromatograms were developed to a distance of 13 cm at 22 ± 2°C for 40 min in paper-lined rectangular glass chamber pre-equilibrated with the mobile phase for 10–15 min. The developed plates were dried at 60°C for 15 min.

Detection: The spots were visualized in an iodine chamber.

Under established conditions, the R_F values of (−) and (+) enantiomers were 0.15 and 0.25 when L-lysine was a chiral selector; 0.12 and 0.23 when L-arginine was a chiral selector.

Enantiomers of metoprolol were separated on LiChrosorb Diol HP-TLC plates with dichloromethane and *N*-benzoxycarbonyl-glycyl-L-proline as chiral selectors added to the mobile phase. The experiments revealed that saturation of the chromatographic system with water has a beneficial effect on chromatographic separation and reproducibility [20].

Sample preparation: Metoprolol solutions were prepared in 95% ethanol.

Stationary phase: LiChrosorb Diol F$_{254}$ HP-TLC plates.

Chromatographic plates 10×10 cm in size were stored at constant temperature (25°C) and relative humidity (80%) before use. Before sample application, the plates were dipped twice in the mobile phase for 1 min and allowed to dry for 5 min in between.

Mobile phase: *N*-benzoxycarbonyl-glycyl-L-proline in dichloromethane at a concentration 5 mmol/l.

On chromatographic plates, 0.5 or 1 μl of samples were applied by means of a sample applicator.

Detection: Scanning was performed with a Shimadzu TLC scanner at $\lambda = 280$ and 300 nm.

14.2.3.4 Propranolol

A nonselective β-blocker mainly used in the treatment of hypertension [14].

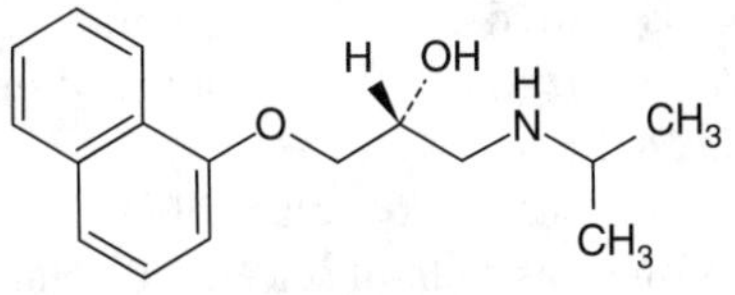

Propranolol

The enantiomers of propranolol were separated on molecularly imprinted polymer of R-(+)-propranolol, but tailing spots were obtained [21].

Sample preparation: Sample solution was prepared in methanol at a concentration 2 mg/ml.

Stationary phase: Polymerizing mixture consisted of 0.56 mmol R-(+)-propranolol, 3 mmol methacrylic acid, 0.14 mmol 2,2'-azo-*bis*(butyronitrile), 20 mmol ethylene glycol dimethacrylate, and 15 ml porogen $CHCl_3$. The mixture was degassed under vacuum and sonicated for 5 min. Polymerization was performed in a water bath at 60°C for 6 h under nitrogen. The resulting bulk polymer was ground in a mortar to a fine powder and sieved through a 100 μm sieve. After sieving, the polymer particles were soaked in a 9:1 mixture of acetonitrile and acetic acid for 24 h, then carefully washed with acetonitrile and dried overnight under vacuum. Micro glass plates (76 × 26 mm) were coated with 0.2 mm layers of a mixture of polymer (1 g), $CaSO_4$ (1 g), and water (2 ml). The plate was dried in air before use.

Mobile phase: Acetic acid in acetonitrile at a concentration 5%.

On chromatographic plates, 1 μl of sample solution containing 2 μg of determined substance was applied by means of Hamilton syringe. The plates were developed to a distance of 5–6 cm for 5–7 min in a presaturated chamber containing 10–15 ml of mobile phase at ambient temperature.

Detection: Developed plates were dried with a hair dryer, sprayed with a solution of anisaldehyde and then heated at 120°C until spot color developed.

Under established conditions the R_F values of S- and R-enantiomers were 0.78 and 0.32, respectively, and the selectivity factor was $\alpha = 2.4$.

A method for rapid determination of the enantiomers of propranolol after extraction from urine samples at pH 9.9 using toluene was developed. Enantiomers were determined after derivatization with S-(+)-benoxaprofen chloride [18]. The established determination conditions are described in Section 14.2.3.3.

Under established conditions the R_F values of R- and S-enantiomers were 0.32 and 0.39, respectively.

The separation of (±)-propranolol was achieved by direct method using silica gel plates impregnated with optically pure L-lysine (0.5%) and L-arginine (0.5%) as the chiral selectors. Different combinations of acetonitrile–methanol were used as the mobile phase. Spots were detected using iodine vapor. The detection limit for (±)-propranolol was 2.6 μg [19]. The established determination conditions are described in Section 14.2.3.3.

Under established conditions the R_F values of enantiomers (−) and (+) were 0.04 and 0.15 when L-lysine was a chiral selector, 0.15 and 0.39 when L-arginine was a chiral selector.

Enantiomers of propranolol were also separated on LiChrosorb Diol F_{254} HP-TLC plates with dichloromethane and N-benzoxycarbonyl-glycyl-L-proline as chiral selectors added to the mobile phase [20].

14.2.3.5 Oxprenolol

A nonselective β-blocker with some intrinsic sympathomimetic activity and a cell-membrane-stabilizing agent.

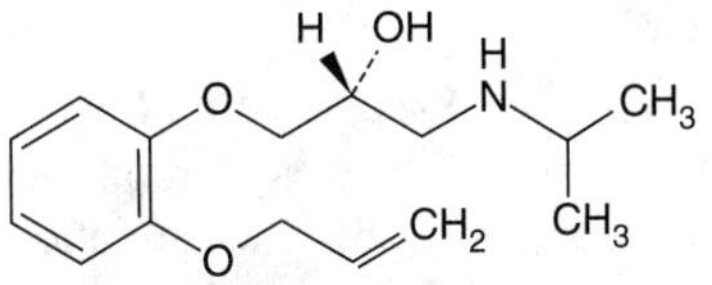

Oxprenolol

The enantiomers of oxprenolol were determined after the extraction from urine samples at pH 9.9 using toluene. The enantiomers were separated after derivatization with S-(+)-benoxaprofen chloride [18]. The established determination conditions are described in Section 4.2.3.3.

Under established conditions, the R_F values of R- and S-enantiomers were 0.32 and 0.38, respectively.

14.2.3.6 Atenolol

Cardioselective β-blocker without intrinsic sympathomimetic activity, not acting on the cell membrane [14].

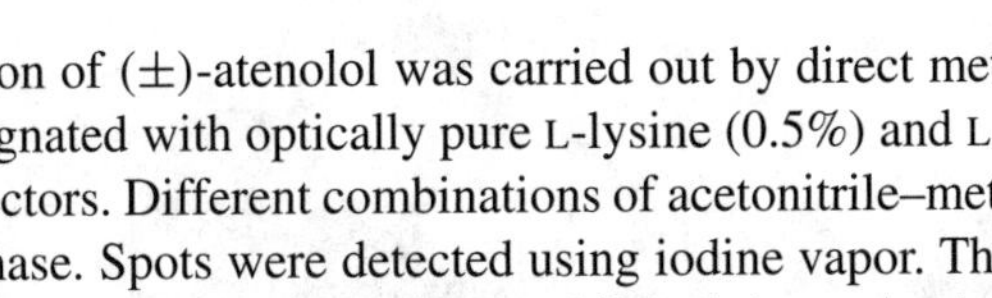

Atenolol

The separation of ($\pm$)-atenolol was carried out by direct method using silica gel plates impregnated with optically pure L-lysine (0.5%) and L-arginine (0.5%) as the chiral selectors. Different combinations of acetonitrile–methanol were used as the mobile phase. Spots were detected using iodine vapor. The detection limit for ($\pm$)-atenolol was 2.6 μg [19]. The established determination conditions are described in Section 14.2.3.3.

Under established conditions the R_F values of ($-$) and (+) enantiomers were 0.03 and 0.1 when L-lysine was a chiral selector, 0.22 and 0.30 when L-arginine was a chiral selector.

14.2.3.7 Alprenolol

A nonselective β-blocker with strong intrinsic sympathomimetic activity, to some extent a cell membrane stabilizing agent [14].

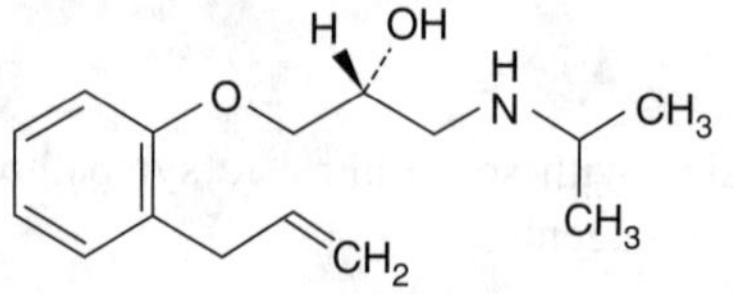

Alprenolol

Enantiomers of alprenolol were separated on LiChrosorb Diol F_{254} HP-TLC plates with dichloromethane and N-benzoxycarbonyl-glycyl-L-proline as chiral selectors added to the mobile phase [20]. The saturation of the chromatographic system with water has a beneficial effect on chromatographic separation and reproducibility.

Sample preparation: Alprenolol solutions were prepared in 95% ethanol.

Stationary phase: LiChrosorb Diol F_{254} HP-TLC plates.

Chromatographic plates 10×10 cm in size were stored at constant temperature (25°C) and relative humidity (80%) before use. Before sample application, the plates were dipped twice in mobile phase for 1 min and allowed to dry for 5 min in between.

Mobile phase: N-benzoxycarbonyl-glycyl-L-proline in dichloromethane at a concentration 5 mmol/l.

On chromatographic plates, 0.5 or 1 μl of samples were applied by means of a sample applicator.

Detection: Scanning was performed with a Shimadzu TLC scanner at $\lambda = 280$ and 300 nm.

14.2.4 PHENOTHIAZINE DERIVATIVES

14.2.4.1 Promethazine

Antihistamine and antiemetic medication; it has strong anticholinergic and sedative effects [14].

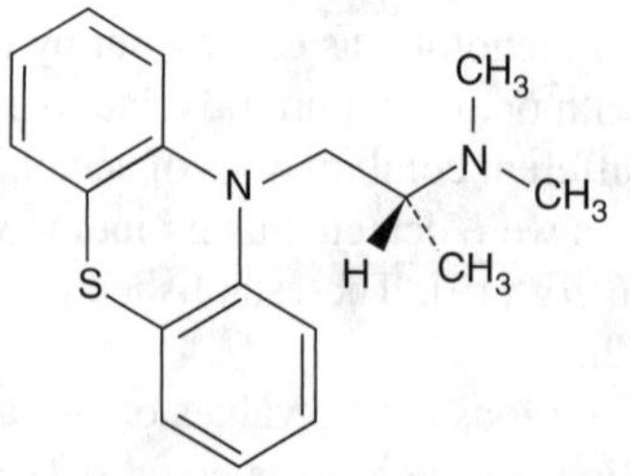

Promethazine

The separation of promethazine enantiomers was developed by Zhu et al. [11] under the following conditions.

Stationary phase: The composition and preparation procedure is described in Section 14.2.1, the same as for ofloxacin.

Mobile phase: Methanol/acetonitrile/1% aqueous triethylammonium acetate (2:0.2:0.1, v/v/v).

Promethazine solution (1 μl) in ethanol at a concentration 1 mg/ml was applied to the prepared plate 10 mm from the edge. Chromatograms were developed to a distance 50 mm in a glass chamber $40 \times 40 \times 80$ mm in size at room temperature.

Detection: The spots were identified by UV illumination at $\lambda = 365$ nm or by use of iodine vapor.

Under established conditions good separation of promethazine enantiomers was achieved, R_F values were 0.80 and 0.89 and selectivity factor was $\alpha = 2.02$.

14.2.5 GLUCOCORTICOSTEROIDS

14.2.5.1 Budesonide

Budesonide belongs to a group of drugs widely used in the treatment of chronic and asthmatic bronchitis due to its strong anti-inflammatory action, high selectivity, and few side effects [14].

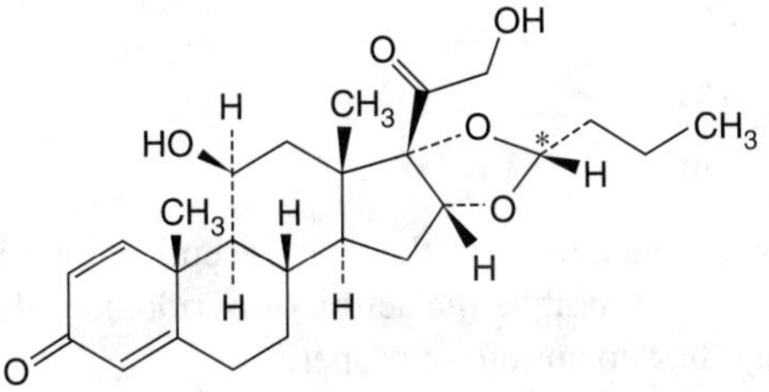

Budesonide

Budesonide is present in many pharmaceuticals in the form of two epimers 22R(+) and 22S(−), which are characterized by different strength and pharmacokinetic properties [22,23]. The anti-inflammatory properties of $R(+)$ isomer are nearly three times as strong as of $S(−)$. It also shows greater volume of distribution and plasma clearance. Budesonide $R(+)$ undergoes biotransformation more quickly than $S(−)$ isomer, that is why its systemic action is weaker [23].

Epimers of budesonide were separated by TLC under different conditions [24,25].

Stationary phase: Precoated HP-TLC NH$_2$ F$_{254S}$ (Merck), 10×20 cm.

Mobile phase: Chloroform/diethyl ether/methyl ethyl ketone (4:3:3, v/v/v).

Sample solutions prepared in chloroform at a concentration 0.001% were applied on the plates in the form of bands 1 cm wide. The development of chromatograms was performed by two methods: Conventional in a normal unsaturated chamber, chromatograms were developed to a distance of 9 cm and overpressurized TLC in a pressurized ultramicro chamber; chromatograms were developed to a distance of 16 cm.

Detection: Densitometrically at $\lambda = 254$ nm.

The epimers of budesonide were separated in both types of chromatographic chambers. The obtained R_F values for epimers of budesonide were 0.21 for $22R(+)$ and 0.24 for $22S(-)$ by conventional TLC, 0.29 for $22R(+)$ and 0.34 for $22S(-)$ by overpressurized TLC.

Epimers of budesonide were separated and determined quantitatively in chosen drugs under conditions described below [25].

Stationary phase: HP-TLC-Alufolien Cellulose plates (Merck) activated in 80°C for 60 min.

Mobile phase: 1% Aqueous solution of β-CD/methanol (15:1, v/v).

On chromatographic plates, 15 μl of standard solutions and sample solutions in methanol at a concentration 20 μg/ml were applied in the form of band 10 mm wide. Chromatograms were developed up to 13 cm and dried in air.

Detection: Densitometrically at $\lambda = 245$ nm.

Successful separation of isomers $R(+)$ and $S(-)$ was obtained and the obtained R_F values were 0.84 $R(+)$ and 0.93 $S(-)$.

14.2.6 OTHERS

14.2.6.1 Penicillamine

A pharmaceutical of the chelator class, the pharmaceutical form is D-penicillamine, L-penicillamine is toxic (it inhibits the action of pyridoxine). It is a metabolite of penicillin, although it has no antibiotic properties.

Penicillamine

Chromatographic separation of D- and L-penicillamine was carried out after derivatization with substituted benzaldehydes and heterocyclic aldehydes [26].

Sample preparation: 0.1 mmol of penicillamine was dissolved in 0.1 ml of water, then 0.5 ml of aldehyde solution of 0.1 mmol/l concentration in methanol was added. After sealing, the mixture was kept at 60–65°C for 8 h.

Stationary phase: TLC plates Chiralplate (Macherey-Nagel).

Mobile phase: Methanol/water/acetonitrile (50:50:200, v/v/v).

On chromatographic plates, 2 μl of sample solution were applied and dried with cold air.

Detection: The pretreatment of the plates, the chromatographic development, and visualization of the spots were performed by the method of Günther et al. [27] at 25°C.

The obtained R_F values after the reaction of penicillamine with 5-nitrofuraldehyde were 0.72 and 0.82 for D- and L-penicillamine, respectively.

14.2.6.2 Ascorbic Acid and Dehydroascorbic Acid

L-Enantiomer of ascorbic acid (AA) is known as vitamin C, the oxidation of AA leads to its conversion into L-dehydroascorbic acid (DHAA). The process is reversible, what is important when AA takes part in oxidation–reduction reactions such as hydroxylation of dopamine to noradrenaline and reduction of folic acid to tetrahydrofolic acid [28].

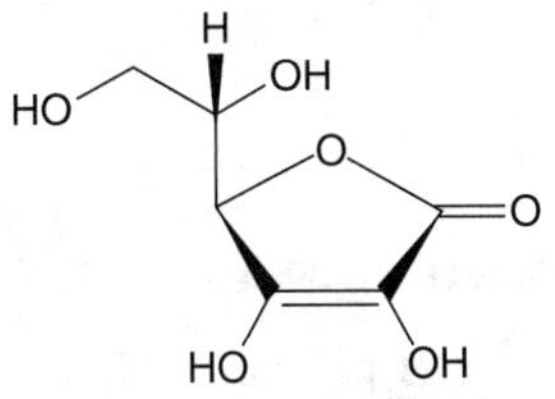

Ascorbic acid

Chromatographic separation was carried out on silica gel plates and cellulose plates impregnated with sodium borate (B). The experiments revealed that borate complexes of AA and DHAA could be separated from each other. TLC was also attempted using direct (D), reverse-phase (RP), and reverse-phase sodium borate (RP-B) TLC plates with and without metaphosphoric acid (MPA). A good separation of the three AA isomers and DHAA isomers was achieved on D and RP cellulose plates. Using an RP-silica-MPA plate, the three AA isomers as well as the three DHAA isomers were fairly separated from their own group members. The separation of six stereoisomers, three AA, and three DHAA was achieved using D-cellulose-MPA plate.

Stationary phase: Silica gel and cellulose TLC plates.

Silica gel was coated on inert flexible poly(ethyleneterephthalate) to 100 μm thickness. Polyacrylic was used as a binder in these plates. Cellulose plates were coated to a 160 μm thickness without binder on inert flexible ester base. Impregnation with B was accomplished by uniformly spraying a 3% B solution on silica gel or cellulose plates. The plates were air-dried and then heated to 100°C for 20 min. For RP-TLC the dried silica gel, cellulose, or borate plates were uniformly impregnated with silicone oil by allowing a 5% solution in ether to ascend the plate in the developing chamber. MPA plates were prepared by uniformly spraying a 2% solution on silica or cellulose plates, which were then air dried and heated at 100°C for 20 min. MPA plates were either sprayed with B or made RP. All of the above plates were stored in a desiccator until required.

Mobile phase: Acetonitrile/acetone/water/acetic acid (80:5:15:2, v/v/v/v), acetonitrile/butylnitrile/water/acetic acid (66:33:15:2, v/v/v/v).

Solutions of AA and DHAA at a concentration of 1 mg/ml were applied to the chromatographic plates in the amount of 10 μl and developed in a saturated chamber to a distance of 15 cm.

Detection: After drying, the plates were sprayed with 2% 2,4-DNP in ethanol containing 0.3 mol/l HCl.

The R_F values obtained for the three AA stereoisomers after the separation using D-cellulose-MPA plates were 0.42, 0.40, and 0.50 and for the three DHAA stereoisomers were 0.55, 0.52, and 0.47.

14.3 DRUGS OF NATURAL ORIGIN

14.3.1 PLANAR SEPARATIONS OF CHIRAL NATURAL COMPOUNDS

It is worth mentioning that natural products, due to their enantiopurity, play an important technical function in the development of suitable methods for chiral separations. Cellulose and amylose, which are easily accessible chiral polymers, after derivatization, are used as chiral stationary phases for both HPLC and TLC [29–31]. Microbial degradation of starch yielded CDs, β-CD-bonded silica gel provide the most popular chiral stationary phases.

Alkaloids, for example, quinine, are widely applied as catalytic reagents in chiral organic syntheses and also in the production of HPLC stationary phases for chiral separations. Cinchona alkaloids also induce the enantioselection of chiral fluoro-compounds analyzed by ^{19}F NMR spectroscopy [32]. (−)-Brucine and (−)-berberine mixed with silica gel were used for preparation of thin-layer plates to resolve racemic amino acids. Other optically active pure enantiomers of natural compounds are applied for impregnation of TLC plates: (+)-tartaric acid, L-aspartic acid [33]. (−)-Menthol is used for preparation of diastereomeric derivatives in indirect enantiomeric separation of, for example, 2- and 3-hydroxy acids [34].

Compared with column liquid chromatography, the number of enantiomeric separations by TLC is small [3,31]. Siouffi et al. [31] reviewed the most recent achievements in chiral separations by planar chromatography since 2001. According to this review, the number of references on chiral TLC separations does not exceed 200 in comparison with over 100,000 examples of HPLC resolutions. This disproportion is also true for natural compounds.

14.3.2 FLAVONOIDS

Flavonoids are one of the largest groups of phenolic natural compounds. They often occur in plants as O-glycosides. Epidemiological studies indicate that daily flavonoid intake in vegetables and fruits has a protective effect against cancer and coronary heart disease.

Chiral separation of flavonoids gains more attention owing to their utility as indicators of proof of origin for vegetable food products. Diastereomers of six flavanone-7-O-glycosides (naringin, prunin, hesperidin, neohesperidin, eriocitrin, and narirutin) were separated by chiral CE and this method was also applied to quality assessment of lemon juice [35]. Reviews of chromatographic, capillary electrophoretic, and capillary electrochromatographic techniques relevant to flavonoids were published recently [36,37].

The flavan-3-ols (+)-catechin and (−)-epicatechin are among these important dietary constituents owing to their potent antioxidant and anti-inflammatory properties [38]. These compounds usually occur together in plants. Nowadays, separation of catechins most often is achieved with such methods as HPLC or CE. For example, a micellar electrokinetic chromatography method was developed for the separation of (+)-catechin and (−)-epicatechin in *Theobroma cacao* beans. The method also proved suitable for the enantioseparation of (±)-catechin [39].

Planar chromatography (paper and TLC) has been employed for the analysis of these isomers as well.

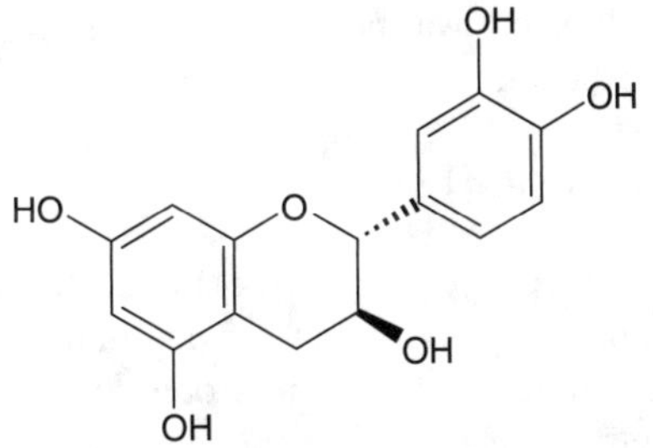

(+)-Catechin

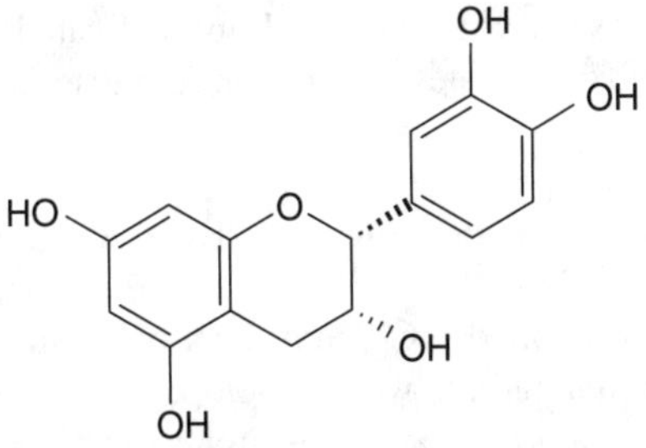

(−)-Epicatechin

Krzek et al. [40] separated (+)-catechin from (−)-epicatechin as follows:

Stationary phase: Aluminum-backed silica gel (Kieselgel 60) TLC plates (Merck).

Mobile phase: Butyl acetate/formic acid/water (90:5:5, v/v).

20 μl volumes of 0.1% methanol solutions of (+)-catechin and (−)-epicatechin were applied on the plate as 15 mm bands. The chromatograms were developed in a saturated chamber at room temperature to a distance of 15 cm.

Detection: Densitometry at $\lambda = 510$ nm.

Prior to analysis, the plates were sprayed with 1% vanilin in concentrated sulfuric acid and dried in a stream of warm air to obtain red spots with R_F values 0.54 for (+)-catechin and 0.43 for (−)-epicatechin. By this method, (+)-catechin was quantified in a commercial preparation (Catergen).

In another work [41], (+)-catechin and (−)-epicatechin were simultaneously determined in stem samples (with and without red heartwood) from *Fagus sylvatica* L. in the following conditions:

Stationary phase: Aluminum-backed silica gel F_{254} plates (Merck #1.05554), cut into 20 × 10 cm pieces and not prewashed before development.

Mobile phase: Diisopropyl ether/formic acid (9:1, v/v).

Samples (30 and 20 μl) were applied as 4 mm bands by means of a Camag Linomat 5 device. Development was performed in an unsaturated chamber (20 × 20 cm twin-trough) to a distance of 10 cm.

Detection: Densitometry at $\lambda = 513$ nm.

Both compounds were detected as red spots after derivatization by spraying with a fresh reagent prepared by dissolving 3 g vanilin in 100 ml absolute ethanol and adding 0.5 ml concentrated sulfuric acid dropwise.

TLC and densitometry were also used for qualitative and quantitative analysis of catechin epimers in rose-hip extracts from 15 *Rosa* L. species [38].

Stationary phase: Glass plates coated with silica gel 60 F_{254} (Merck) cut into 10 × 10 cm pieces, washed with methanol and activated at 110°C for 1 h for quantitative analysis and glass plates coated with cellulose and silica gel 60 (Merck) for qualitative analysis.

Mobile phase: Ethyl acetate/water/formic acid/acetic acid (25:4:0.6:0.4, v/v; upper phase).

Samples (20 μl) were applied as 6–8 mm bands by means of a Camag automatic TLC Sampler III. Chromatograms were developed "face down" in a horizontal Teflon DS chamber and saturated before use for 15 min, at room temperature, to a distance of 9 cm.

Detection: Densitometry at $\lambda = 254$ nm (reflectance mode).

The plates were sprayed with *bis*-diazotized sulfanilamide.

Lepri et al. [42] investigated the applicability of RP-TLC for the direct resolution of 19 structurally related racemic flavanones. Racemates of the following were examined: flavanone, 5-methoxyflavanone, 6-methoxyflavanone, 4'-methoxyflavanone, 6-hydroxyflavanone, 7-hydroxyflavanone, 2'-hydroxyflavanone, 4'-hydroxyflavanone, pinocembrin, pinocembrin-7-methyl ether, naringenin, isosakuranetin, sakuranetin, naringenin-7-glucoside, naringin, hesperetin, taxifolin, eriodictyol, and homoeriodictyol.

Stationary phase: (a) Homemade microcrystalline cellulose triacetate MCTA plates (a slurry of MCTA HPLC size [Fluka] with silica gel 60 GF$_{254}$ Merck as binder was prepared by mixing the latter 3 g with distilled water for 5 min then adding MCTA and ethanol and shaking for 5 min; layers of thickness 250 μm were used) and (b) silica plates C18-50/UV$_{254}$ (Macherey-Nagel) cut into 10 × 10 cm pieces.

Mobile phase: For stationary phase (a): ethanol/water (70:30, v/v) or methanol/water (80:20, v/v).

For stationary phase (b): 0.15 M β-CD aqueous solution with urea (32%) and sodium chloride (2%)/acetonitrile (80:20, v/v) or 0.05 M sodium bicarbonate and 0.05 M sodium carbonate solution containing 6% bovine serum albumin (BSA) and 12% 2-propanol or 0.1 M sodium carbonate solution containing 6% BSA and 6% 2-propanol.

Ethanol solutions of racemates (0.5–1.5 μl) were applied with a Hamilton syringe on the plates that were developed (ascending mode) in a Desaga thermostatted chamber saturated for 1 h at 25°C.

Detection: Densitometry in the reflection mode at $\lambda = 254$ nm.

MCTA plates with silica gel GF$_{254}$ as binder proved effective for the resolution of polysubstituted, nonglycosidic compounds. Mobile phases such as ethanol/water (70:30, v/v) and methanol/water (80:20, v/v) gave best results, and the resolution was improved with successive developments. Partial resolution was obtained for flavanone, 6-methoxy-, and 6-hydroxyflavanone in the mobile phase ethanol/water (80:20, v/v). Chromatographic separation of racemates on achiral stationary phase with β-CD as chiral selector added to the mobile phase was achieved in the case of flavanone and its 2'-hydroxy-, 4'-hydroxy-, and 4'-methoxy-derivatives possessing no hydroxy or methoxy substituents in the position 5, 6, or 7 of the fused aromatic ring. Separations of tested flavanones with BSA as a chiral selector added to the mobile phase gave satisfactory results in alkaline mobile phases for polysubstituted flavanones containing a methoxyl in position 3' or 4', for example, homoeriodictyol ($\alpha = 1.95$), isosakuranetin ($\alpha = 1.54$), and hesperetin ($\alpha = 1.62$).

14.3.3 ALKALOIDS

Alkaloids, the nitrogen-containing chiral secondary metabolites, can be divided into nonheterocyclic or heterocyclic alkaloids. The latter are called *typical alkaloids* and, with respect to the skeletal ring structure, may be further classified as, for example, quinoline, indole, pyridine, or tropane alkaloids. Alkaloids constitute a very diverse group not only in terms of structure, but also pharmacological activity. Many of them are important medicinal agents, such as analgesic morphine, antiarrhythmic ajmaline, and quinidine or anticancer vincristine and paclitaxel.

Techniques that are utilized in a great number of chiral alkaloid separations include HPLC and CE. For example, *Cinchona* alkaloids are most often analyzed by HPLC in the RP mode using UV detection, and this continues to be the method of choice [43]. Enantioselective separation of semisynthetic ergot alkaloids was performed by RP-HPLC with vancomycin and teicoplanin used as chiral selectors [44]. Four enantiomeric pairs of vincamine group alkaloids were separated by HPLC on a chiral stationary phase [45]. Enantioseparation of ephedrines was also performed by HPLC using β-CDs [46]. Stereoselective HPLC separation of atropine was achieved on a prepacked α_1-acid glycoprotein column [47]. Capillary zone electrophoresis was used for chiral resolution of lobeline [48]. Three different CE methods using derivatized β-CDs were developed for the separation of ephedrine and pseudoepherine stereoisomers [49]. Racemic mixture of tropane alkaloid ($\pm$)-hyoscyamine (= atropine) was also resolved by this method [50,51]. Two-dimensional gas chromatography (GC × GC) was employed for the separation of ephedrine-type alkaloids and their enantiomers in complex herbal preparations [52].

Reports on chiral TLC resolution of alkaloids are scarce.

14.3.3.1 Quinine

Quinine is the principal alkaloid of cinchona bark. It is a well-known agent in the treatment of chloroquine-resistant malaria; its use, however, has been curtailed due to serious side effects. The compound is also a common bitter flavoring in drinks. Its enantiomer quinidine is an antiarrhythmic drug.

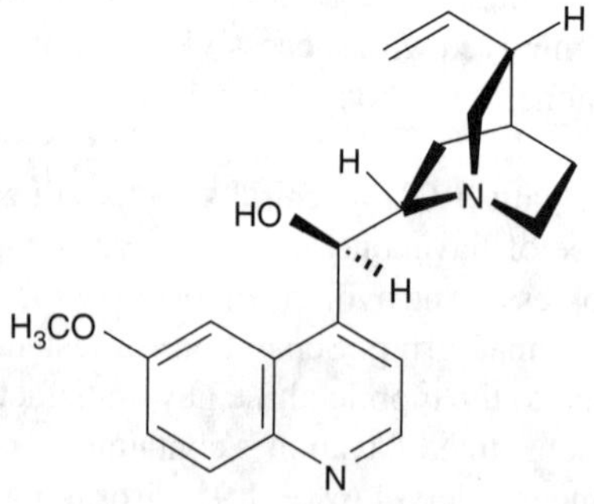

(−)-Quinine

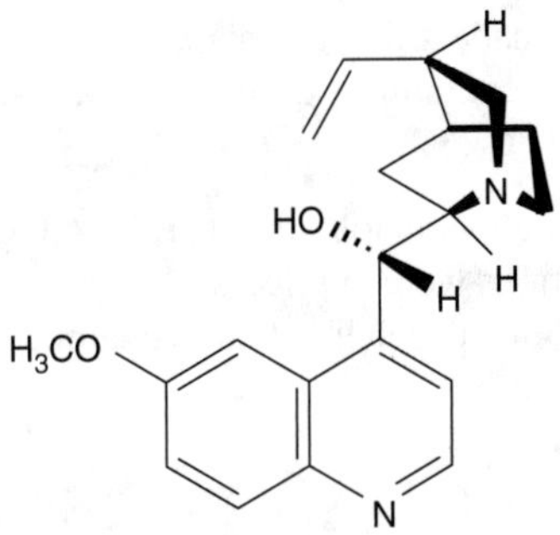

(+)-Quinidine

Quinine is separated from quinidine in normal phase TLC in mobile phases composed of chloroform, acetone, and ethyl acetate to which alcohols such as methanol, propanol, or butanol are added as polarity adjusters and ammonia or diethylamine to reduce spot tailing [43]. Common solvent systems used for the separation of cinchona alkaloids include chloroform/diethylamine (90:10, v/v) or toluene/ethyl acetate/diethylamine (70:20:10, v/v/v). After spraying the chromatogram with sulfuric acid, these alkaloids show a blue fluorescence in UV-365 nm. Treatment afterwards with iodoplatinate reagent gives a violet coloration in the visible [53].

14.3.3.2 Hyoscyamine

(±)-Hyoscyamine, known as atropine, is a tropane alkaloid with long tradition in medicinal use, for example, in ophthalmic solutions, as an antispasmodic and other. It is found in several plant species of the following genera: *Atropa, Datura, Hyoscyamus, Duboisia, Mandragora.* (−)-Hysocyamine is much more active than (+)-hyoscyamine.

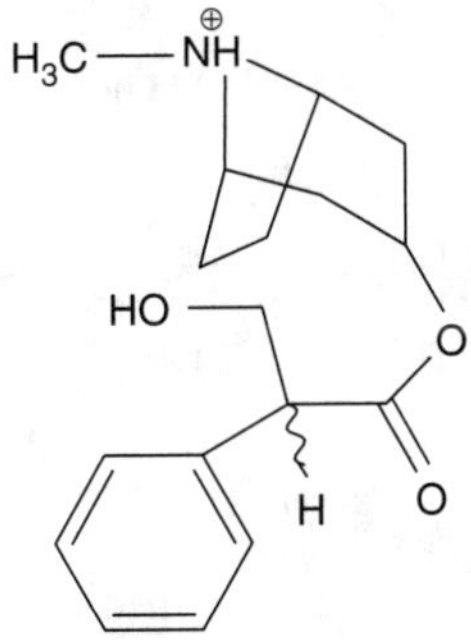

(+)-Hyoscyamine

Bhushan et al. [54] separated enantiomers of hyoscyamine on TLC plates impregnated with L-histidine.

Stationary phase: Homemade silica gel plates (a slurry of silica gel G [Merck] in double-distilled water containing optically pure L-histidine 0.5%; pH 7–8, activation overnight at 60°C).

Mobile phase: Acetonitrile/methanol/water (11.3:5.4:0.6, v/v/v).

Ethanol sample solutions were applied to the plates at a 10 μl level. Chromatograms were developed in a paper-lined glass chamber pre-equilibrated with the solvent system for 20–25 min.

Detection: Iodine vapor.

The authors have also investigated the effect of concentration of the impregnating reagent, pH, and changes in temperature on the resolution of enantiomers. The best results were obtained at pH between 7 and 8, at the temperature 18 ± 2°C, and at 0.5% of the chiral selector. The method allowed resolution of 6 μg of atropine into its enantiomers (+)-hyoscyamine hR_F $(=100 \times R_F) = 4.6$; (−)-hyoscyamine $hR_F = 9.2$.

14.3.3.3 Ephedrine

Ephedrine, isolated from *Ephedra sinica*, is the major active constituent of this plant species. It is used in the treatment of asthma, nasal congestion, and as a prophylactic to avoid hypotension associated with spinal anesthesia for cesarean section [55]. Herbal products containing *Ephedra* are used as antiobesity agents and energy boosters despite reports on serious side-effects, such as hypertension, seizures, myocardial infarction [56]. The 1S,2R-(+)-enantiomer of ephedrine is not present in the plant.

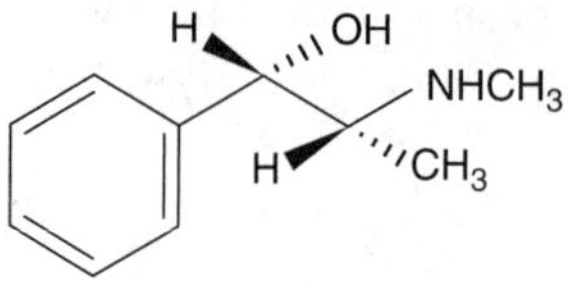

(−)-Ephedrine

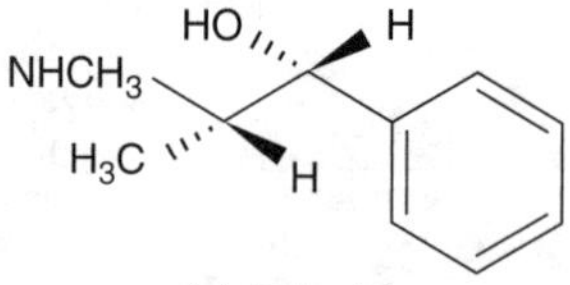

(+)-Ephedrine

Bhushan et al. [54] separated (±)-ephedrine on TLC plates impregnated with L-tartaric acid.

Stationary phase: Homemade silica gel plates (a slurry of silica gel G [Merck] in double-distilled water containing optically pure L-tartaric acid 0.5%; pH 7–8, activation overnight at 60°C).

Mobile phase: Acetonitrile/methanol/water (12:5:0.5, v/v/v).

Ethanol sample solutions were applied to the plates at a 10 μl level. Chromatograms were developed in a paper-lined glass chamber pre-equilibrated with the solvent system for 20–25 min.

Detection: Iodine vapor.

The method allowed resolution of as little as 2 μg of ($\pm$)-ephedrine into its enantiomers (+)-ephedrine hR_F = 9.8; (−)-ephedrine hR_F = 17.6.

14.3.4 Fatty Acids

The enantiomers of DL-α-hydroxypalmitic acid, DL-12-hydroxystearic acid, and DL-12-hydroxyoleic acid were resolved by TLC [57].

Stationary phase: Silica gel 60 F_{254} plates (Merck).

Mobile phase: Acetone/hexane/hydrochloric acid (3:5:2, v/v) containing 1% L-alanine for DL-α-hydroxypalmitic acid; development distance 5.5 cm; α = 1.63.

Hexane/acetonitrile/acetic acid/hydrochloric acid (5:2:1:2, v/v) with 2% L-alanine for DL-12-hydroxystearic acid; development distance 6 cm; α = 1.46.

Detection: Immersion of plate in 2% aqueous sodium hydroxide solution.

Enantiomers of DL-12-hydroxyoleic acid were separated on Chiralplate (Macherey-Nagel) with acetone/water (6:4, v/v) as the mobile phase; development distance 6 cm; α = 1.50; detection: iodine vapor.

REFERENCES

1. Bojarski, J., Chromatographic resolution of enantiomers, *Wiadomości Chemiczne*, 47, 279, 1993.
2. Bojarski, J., Aboul-Enein, H.Y., and Ghanem, A., What's new in chromatographic enantioseparations, *Curr. Anal. Chem.*, 1, 59–77, 2005.
3. Aboul-Enein, H.Y. et al., Application of thin-layer chromatography in enantiomeric chiral analysis — an overview, *Biomed. Chromatogr.*, 13, 531–537, 1999.
4. Morissey, I. et al., Mechanism of differential activities of ofloxacin, *Antimicrob. Agents Chemother.*, 40, 1775, 1996.
5. Chu, D.T.W., Nordeen, C.W., and Hardy, D.J., Synthesis, antimicrobial actives, and pharmacological properties of enantiomers of temafloxacin hydrochloride, *J. Med. Chem.*, 34, 160, 1991.
6. Foster, R.T. et al., Stereospecific high-performance liquid chromatographic assay of lomefloxacin in human plasma, *J. Pharm. Biomed. Anal.*, 13, 1243, 1995.
7. Machida, M. et al., Pharmacokinetics of gatifloxacin, a new quinolone and its enantiomers, *Nihon Kagahu Ryoho Gakkai Zasshi*, 47, 124, 1999.
8. Humphrey, G.H. et al., Pharmacokinetics of clinafloxacin enantiomers in human, *J. Clin. Pharmacol.*, 39, 1143, 1999.

9. Grellet, J., Ba, B., and Saux, M.C., High-performance liquid chromatography separation of fluoroquinolone enantiomeres: A review, *J. Biochem. Biophys. Methods*, 54, 221, 2002.

10. Bryskier, A. and Chantot, J.F., Classification and structure–activity relationship of fluoroquinolones, *Drugs*, 49(Suppl. 2), 16, 1995.

11. Zhu, Q. et al., β-Cyclodextrin-bonded chiral stationary phase for thin-layer chromatographic separation of enantiomers, *J. Planar Chromatogr.*, 14, 137, 2001.

12. Xu, L. et al., Preparation of cellulose Tris(benzoate)s for TLC and their chromatographic properties, *Fenxi Cesi Xuebao* (*Chinese J. Instrumental Anal.*), 22, 1, 2003.

13. Xu, L. et al., *Zongshan Daxue Xuebao, Ziran Kexueban Guangzhou*, 41, 115–117, 2002.

14. Podlewski, J. and Chwalibogowska-Podlewska, A., *Drugs for modern therapy. (in Polish)* Split Trading Warszawa, 2000.

15. Mielcarek, J., Normal-phase TLC separation of enantiomers of 1.4-didydropyridine derivatives, *Drug Develop. Ind. Pharm.*, 27, 175, 2001.

16. Zejc, A. and Gorczyca, M., *Chemistry of drugs. (in Polish)*, PZWL, Warszawa, 2002.

17. Lučić, B. et al., Direct separation of the enantiomers of ($\pm$)-metoprolol tartrate on impregnated TLC plates with D-($-$)-tartaric acid as a chiral selector, *J. Planar Chromatogr.*, 18, 294, 2005.

18. Pflugmann, G., Spahn, H., and Mutschler, E., Rapid determination of the enantiomers of metoprolol, oxprenolol and propranolol in urine, *J. Chromatogr.*, 416, 331, 1987.

19. Bhushan, R. and Thuku Thiongo, G., Direct enantioseparation of some β-adrenergic blocking agents using impregnated thin-layer chromatography, *J. Chromatogr. B*, 708, 330, 1998.

20. Tivert, A.M. and Backman, A., Separation of the enantiomers of β-blocking drugs by TLC with a chiral mobile phase additive, *J. Planar Chromatogr.*, 6, 216, 1993.

21. Suedee, R. et al., Thin-layer chromatographic separation of chiral drugs on molecularly imprinted chiral stationary phases, *J. Planar Chromatogr.*, 14, 194, 2001.

22. *European Pharmacopea*, 4th edn, Council of Europe, Strasbourg, 2002.

23. Zając, M. and Pawełczyk, E., *Chemistry of drugs. (in Polish)*, Akademia Medyczna, Poznań, 2000.

24. Ferenczi-Fodor, K., Kovacs, J., and Szepesi, G., Separation and determination of isomers on amino-bonded silica by conventional and overpressurized thin-layer chromatography, *J. Chromatogr.*, 392, 464, 1987.

25. Krzek, J. et al., Determination of budesonide $R(+)$ and $S(-)$ isomers in pharmaceuticals by thin-layer chromatography with UV densitometric detection, *Chromatographia*, 56, 759, 2002.

26. Kovacs-Hadady, K. and Kiss, I.T., Attempts for the chromatographic separation of D- and L-penicillamine enantiomers, *Chromatographia*, 24, 677, 1987.

27. Günther, K., Martens, J., and Schickedanz, M., Resolution of optical isomers by thin layer chromatography. Enantiomeric purity of D-penicillamine, *Arch. Pharm.* (Weinheim), 319, 461, 1986.

28. Roomi, M.W. and Tsao, C.S., Thin-layer chromatographic separation of isomers of ascorbic acid and dehydroascorbic acid as sodium borate complexes on silica gel and cellulose plates, *J. Agric. Food Chem.*, 46, 1406, 1998.

29. Kubota, T., Yamamoto, C., and Okamoto, Y., Tris(cyclohexylcarbamate)s of cellulose and amylose as potential chiral stationary phases for high-performance liquid chromatography and thin-layer chromatography, *J. Am. Chem. Soc.*, 122, 4056, 2000.

30. Mack, M. and Hauck, H.E., Separation of enantiomers in thin-layer chromatography, *Chromatographia*, 26, 197, 1988.

31. Siouffi, A.M., Piras, P., and Roussel, C., Some aspects of chiral separations in planar chromatography compared with HPLC, *J. Planar Chromatogr.*, 18, 5, 2005.

32. Abid, M. and Török, B., Cinchona alkaloid induced chiral discrimination for the determination of the enantiomeric composition of α-trifluoro-methylated-hydroxyl compounds by ^{19}F NMR spectroscopy, *Tetrahedron Asymm.*, 16, 1547, 2005.

33. Bhushan, R. and Martens, J., Direct resolution of enantiomers by impregnated TLC, *Biomed. Chromatogr.*, 11, 280, 1997.

34. Kim, K.R. et al., Configurational analysis of chiral acids as O-trifluoroacetylated(−)-menthyl esters by achiral dual-capillary column gas chromatography, *J. Chromatogr. A*, 891, 257, 2000.

35. Gel-Moreto, N., Streich, R., and Galensa, R., Chiral separation of six diastereomeric flavanone-7-O-glycosides by capillary electrophoresis and analysis of lemon juice, *J. Chromatogr. A*, 925, 279, 2001.

36. Molnar-Perl, I. and Füzfai, Zs., Chromatographic, capillary electrophoretic and capillary electrochromatographic techniques in the analysis of flavonoids, *J. Chromatogr. A*, 1073, 201, 2005.

37. Rijke, E. et al., Analytical separation and detection methods for flavonoids, *J. Chromatogr. A*, 1112, 31, 2006.

38. Nowak, R. and Hawrył, M., Application of densitometry to the determination of catechin in rose-hip extracts, *J. Planar Chromatogr.*, 18, 217, 2005.

39. Gotti, R. et al., Analysis of catechins in *Theobroma cacao* beans by cyclodextrin-modified micellar electrokinetic chromatography, *J. Chromatogr. A*, 1112, 345, 2006.

40. Krzek, J. and Janeczko, Z., Spectrophotometric and chromatographic-densitometric determination of (+)-catechin content in Catergen preparation, *Chemia Analityczna*, 30, 465, 1985.

41. Hofmann, T., Albert, L., and Rétfalvi, T., Quantitative TLC analysis of (+)-catechin and (−)-epicatechin from *Fagus sylvatica* L. with and without red heartwood, *J. Planar Chromatogr.*, 17, 350, 2004.

42. Lepri, L. et al., Reversed-phase planar chromatography of racemic flavanones, *J. Liq. Chrom. Rel. Technol.*, 22, 105, 1999.

43. McCalley, D.V., Analysis of the Cinchona alkaloids by high-performance liquid chromatography and other separation techniques, *J. Chromatogr. A*, 967, 1, 2002.

44. Tesařová, E., Záruba, K., and Flieger, M., Enantioseparation of semi-synthetic ergot alkaloids on vancomycin and teicoplanin stationary phases, *J. Chromatogr. A*, 844, 137, 1999.

45. Caccamese, S. and Prinzipato, G., Separation of the four pairs of enantiomers of vincamine alkaloids by enantioselective high-performance liquid chromatography, *J. Chromatogr. A*, 893, 47, 2000.

46. Herráez-Hernández, R. and Campins-Falcó, P., Chiral separation of ephedrines by liquid chromatography using β-cyclodextrins, *Anal. Chim. Acta*, 434, 315, 2001.

47. Breton, D. et al., Chiral separation of atropine by high-performance liquid chromatography, *J. Chromatogr. A*, 1088, 104, 2005.

48. Lin, X. et al., Chiral separation of lobeline and benzhexol by capillary electrophoresis using the reaction mixture of β-cyclodextrin, phosphorous-pentoxide and L-glutamic acid as chiral selector, *Anal. Chim. Acta*, 431, 41, 2001.

49. Phinney, K.W., Ihara, T., and Sander, L.C., Determination of ephedrine alkaloid stereoisomers in dietary supplements by capillary electrophoresis, *J. Chromatogr. A*, 1077, 90, 2005.

50. Mateus, L. et al., Enantioseparation of atropine by capillary electrophoresis using sulfated β-cyclodextrin: Application to a plant extract, *J. Chromatogr. A*, 868, 285, 2000.

51. Tahara, S. et al., Enantiomeric separation of atropine in Scopolia extract and Scopolia rhizome by capillary electrophoresis using cyclodextrins as chiral selectors, *J. Chromatogr. A*, 848, 465, 1999.

52. Wang, M. et al., Enantiomeric separation and quantification of ephedrine-type alkaloids in herbal materials by comprehensive two-dimensional gas chromatography, *J. Chromatogr. A*, 1112, 361, 2006.

53. Wagner, H., Bladt, S., and Zgainski, E.M., *Plant Drug Analysis*, Springer-Verlag, Berlin, 1984.

54. Bhushan, R., Martens, J., and Arora, M., Direct resolution of ($\pm$)-ephedrine and atropine into their enantiomers by impregnated TLC, *Biomed. Chromatogr.*, 15, 151, 2001.

55. Simon, L. et al., Dose of prophylactic intravenous ephedrine during spinal anesthesia for cesarean section, *J. Clin. Anesthesia*, 13, 366, 2001.

56. Vansal, S.S. and Feller, D.R., Direct effects of ephedrine isomers on human β-adrenergic receptor subtypes, *Biochem. Pharmacol.*, 58, 807, 1999.

57. Szulik, J. and Sowa, A., Separation of selected enantiomers of fatty hydroxy acids by TLC, *Acta Chromatographica*, 11, 233, 2001.

15 Chiral Separations Using Marfey's Reagent

Ravi Bhushan and Hans Brückner

CONTENTS

15.1 INTRODUCTION

The separation of enantiomers has always been regarded as one of the most challenging problems for chemists/scientists since the times of Pasteur (1848), van't Hoff and Le Bell (1874) when the idea of stereochemistry and stereoselectivity started coming in. The establishment of reliable analytical methods for determination of enantiomers has always been increasing in the fields of chromatography, asymmetric synthesis, mechanistic studies, studies of structure–function relationship of proteins, pharmacology, medicine, extraterrestrial chemistry, life sciences, and so forth.

Proteins are the most important constituents of all living systems. About 20 genetically encoded amino acids are the building blocks of proteins. All protein amino acids with the exception of glycine are chiral in nature. Multicellular organisms usually have L-amino acids. Racemization (epimerization) of optically active amino acids in dilute acid or base or at neutral pH may take place even in metabolically stable proteins of living mammals; as a consequence the protein–function relationship may be altered. Racemization during peptide synthesis, for example, is an unwanted process and too common. It is often assumed that racemization may not occur or it need not be examined. The production of optically pure amino acids analogs is a challenging task and also requires accurate analytical procedures to determine the optical purity during the course of asymmetric synthesis. It has not been recognized until recently that D-enantiomers of amino acids also do occur in the free and protein (peptide)-bound form in plants and animals, including man.

Chromatography has progressed from the time of classical column in 1906 to the capillary electrochromatography (CEC) in 2000. High-performance liquid chromatography (HPLC) has passed through many developmental phases, decreasing separation time due to reduction in particle size from >100 to 3 μm, improvements in pumping systems facilitating gradient elution, introduction of computer-controlled equipment with an integrated data system, and microcolumn or narrowbore columns requiring small sample volumes, analyte weights, and so forth. Separation of a variety of molecules, including diastereomers, on C8 or C18 RP columns and direct enantiomeric resolution on bonded phase packings such as polysaccharides, cyclodextrins, and macrocyclic antibiotics have been in practice during the past few decades. These chiral stationary phases (CSPs) eventually led to the decline of ligand-exchange chromatography. It is worth mentioning that gas chromatography (GC) proved to be one of the first methods resolving completely achiral derivatives (e.g., volatile analytes by esterification and acylation) of DL-amino acids on chiral stationary phase, which led to the development of Chirasil-Val as a GC CSP [1]; possible degradation of the analytes, conversion of Asn and Gln into Asp and Glu, and derivatization problems with basic amino acids His and Arg were considered to be the main weaknesses.

In contrast to HPLC, stationary phases for thin-layer chromatography (TLC) have not improved substantially with time. However, the reasons for using TLC include parallel separation of samples, high-throughput screening, static and sequential detection for identification, and integrity of the total sample. Moreover, TLC promises future prospects for improved separation performance.

15.2 RESOLUTION OF ENANTIOMERS (OR DIASTEREOMERS)

There have been two basic approaches for the chromatographic resolution of enantiomers: a direct and an indirect method. The resolution of a pair of enantiomers by reacting them with an optically pure chiral reagent, that is, the formation of diastereomers followed by their separation by chromatography in an achiral environment, is considered as an indirect approach and has been the most common means of achieving the resolution. The advantages include the commercial availability of a large number of chiral derivatizing reagents (CDRs) and a greater choice of chromatographic conditions. The enantiomer molecule and the CDR must possess an easily derivatizable and compatible functional group. The reaction should be fast and complete and proceed without steric constrains; otherwise, a variation in the formation rate of diastereomers may cause a kinetic resolution. Introduction of a chromophore enhances detection (say, for HPLC resolution). Resolution via diastereomer formation is usually improved when bulky groups are attached to the stereogenic center and when the stereogenic centers of both the reagent and the analyte are in close proximity in the resulting diastereomer. Various types of columns, mobile phases, and CDRs for the resolution of enantiomers of amino acids and their derivatives have been summarized [2].

The direct approach requires no chemical derivatization prior to separation process. Resolution is possible through reversible diastereomeric association between the chromatographic chiral environment and the solute enantiomers. The enantiomers may interact during the course of chromatographic process with a CSP or a chiral selector added to the mobile phase (CMPA) or a chiral selector mixed with/immobilized (especially in TLC) on the stationary phase [3,4]. Some important binding types present in enantioselective sorption process include coordination to transition metals (ligand exchange), charge transfer interaction, ion exchange, and inclusion phenomena (host–guest complex). Use of CSP in GC required achiral derivatization (and protection of side chain functionality) to produce volatile analytes, for example, esterification and acylation of amino acids; possible thermal degradation, racemization, and problems arising from derivatization chemistry have been the main weaknesses. Direct HPLC resolution of enantiomers of amino acids using various CSPs and CMPAs has been summarized by Bhushan and Joshi [2]. Direct methods have certain critical disadvantages, particularly in HPLC, for example, protein stationary phases are not durable over time and pH and also have low sample capacity. Besides, the correct elution order is difficult to be predicted because of the complexity of interactions with the protein [5]. Stationary phases with crown ethers and cyclodextrins, involving host–guest type complexation, often result in poor band shape and have slow kinetics on a chromatographic time scale.

On the other hand, TLC still seems to be more convenient, simple, less expensive, and efficient enough to ensure even difficult separation of pairs of analytes though its performance may sometimes be substantially lower than that of the instrumental modes of chromatography. Reviews on direct TLC resolution of enantiomers of amino acids and their derivatives including various other classes of compounds have appeared [3,6–9].

15.3 MARFEY'S REAGENT

The separation of diastereomeric pair via the indirect technique is sometimes simpler to perform and often has better resolution than with a direct method, because chromatographic conditions are much easily optimized. Marfey's Reagent (MR) has such attributes.

The MR can safely be considered as a chiral variant of the Sanger's reagent (2,4-DNFB). The Sanger's reagent (1945) was unique, because it provided corresponding DNP derivatives of amino acids that were identifiable by chromatography to establish amino acid sequence in peptides. However, there was no focus at that time toward resolution of enantiomers, may be because that was not required for sequence determination and more importantly there was no concern about the optical purity of proteins. Nevertheless, the DNP-amino acids are chiral molecules except for glycine. Notably, Zahn [10] in 1951 introduced a bifunctional variation of Sanger's reagent, namely 1,5-difluoro-2,4-dinitrobenzene (DFDNB), for the cross-linking of proteins.

This reagent was used by Peter Marfey in 1984 [11] to prepare a chiral variant of Sanger's reagent by substitution of one of the two fluorine atoms in DFDNB by L-alanine amide (H-Ala-NH$_2$). The resulting CDR was later called as the MR and is commonly abbreviated FDNP-L-Ala-NH$_2$ (FDAA). This reagent was used for enantiomeric separation of five amino acids by HPLC. The MR takes advantage of the remaining reactive aromatic fluorine that undergoes nucleophilic substitution with the free amino group on L- and D-amino acids (in the mixture), peptide, or target molecule and that of the stereogenic center in its alanine group (the L-form) to create diastereomers. It thus provides a structural feature to replace L-Ala-NH$_2$ by suitable chiral moieties such as those of other amino acids, and so forth. and have the analogs to examine efficiency and reactivity of the reagent.

Bhushan and Brückner [12] have discussed mechanistic aspects of resolution of amino acid diastereomers and the application of MR as a derivatizing agent in the form of different chiral variants, along with application to resolution of complex mixtures of DL-amino acids, amines and nonproteinogenic amino acids, peptides/amino acids from micro-organisms, and evaluation of racemizing characteristics. Earlier, B'Hymer et al. [13] discussed uses of MR mainly to resolution of amino acids. MR has largely been used for indirect resolution of enantiomers of amino acids. There are few reports dealing with TLC application of MR.

15.3.1 General Protocols for Synthesis

The cost of synthesis in terms of the cost of only reactants may be 50 times less than that of the commercial reagent. The general approach to synthesize the reagent (if not obtained commercially) and derivatization is that reported by Marfey [11]. However, certain modifications have been reported from time to time. Some of these are described below as a ready reference for the readers.

15.3.1.1 Synthesis of FDNP-L-Ala-NH$_2$

L-Ala-NH$_2$ · HCl (472 mg, 3.81 mmol) is dissolved in NaOH (1 M, 3.9 ml) and immediately acetone is added (60 ml). After adding anhydrous MgSO$_4$ (about 10 g) the contents are stirred at room temperature for about 3 h. MgSO$_4$ is removed by filtration and washed twice with a little acetone. DFDNB (668 mg, 3.27 mmol) is dissolved in acetone (15 ml) and to it is added the solution of L-Ala-NH$_2$ (in acetone) dropwise under magnetic stirring. The contents are then stirred for 30 min. Equal volume of water is added when golden yellow scales are formed. These are filtered, washed first with little 2:1 water–acetone mixture, then with water, and finally dried in air in the dark. The yield is 0.5 g (56%, mol. wt 272), m.p. 224–226°C. Another crop of crystals is obtained from the mother liquor upon removal of more acetone under vacuum. TLC in ethyl acetate as solvent and HPLC using 50 mM TEAP/acetonitrile (pH 3.0, from 10 to 50% acetonitrile during 1 h, flow rate 2 ml/min, at 340 nm) confirm the purity. The UV in 25 mM TEAP, pH 3.0 and 50% acetonitrile shows the maxima at 264, 338 nm, and shoulder at 380 nm. The reaction for synthesis is shown in Figure 15.1a.

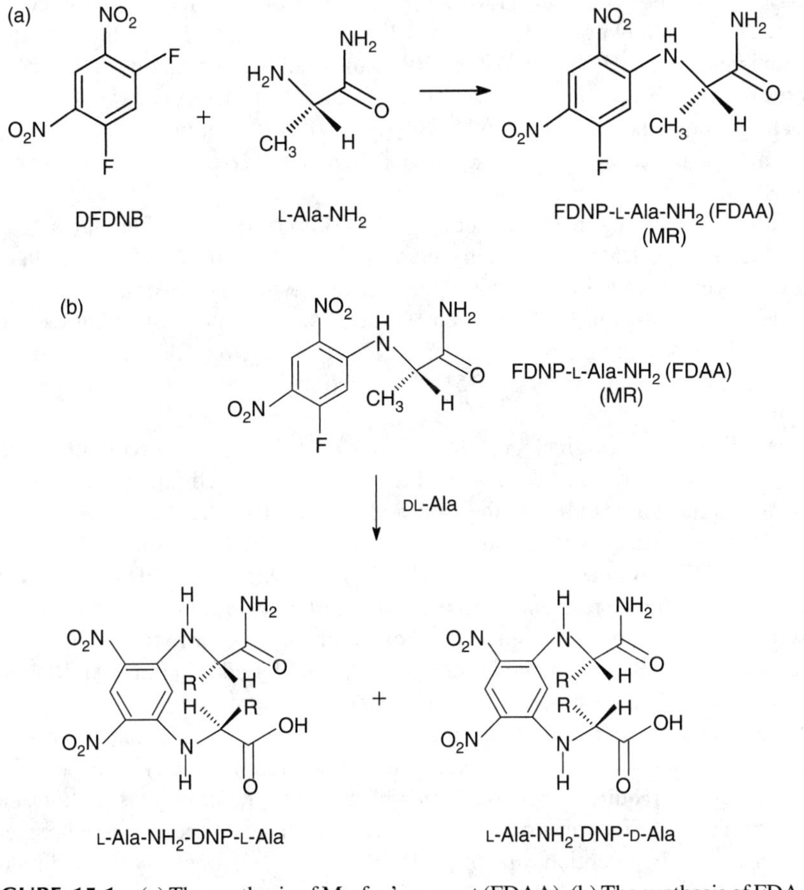

FIGURE 15.1 (a) The synthesis of Marfey's reagent (FDAA). (b) The synthesis of FDAA derivatives of amino acids.

Using modifications of the above protocol, Fujii et al. [14] prepared FDNP-L-Val-NH$_2$ (FDNPVA), FDNP-L-Phe-NH$_2$, FDNP-L-Ile-NH$_2$, FDNP-L-Leu-NH$_2$ (FDNPLA), and FDNP-D-Ala-NH$_2$, while Brückner and Keller-Hoehl [15] prepared FDNP-L-Ala-NH$_2$, FDNP-L-Val-NH$_2$, FDNP-L-Phe-NH$_2$, and FDNP-L-Pro-NH$_2$ with minor changes in reaction conditions, for example, heating the reaction mixture at 40–50°C for 1–2 h in some cases.

15.3.1.2　Synthesis of L- and D-diastereomers of amino acids

Derivatization of an amino acid with FDAA produces a diastereomer referred to as 2,4-dinitrophenyl-5-L-alanine amide amino acid or simply DNPA-amino acid. Aqueous solutions (50 mM) of amino acids (D- and L-isomers) are used as starting materials for synthesis. Solution of each of the amino acid (50 μl, 2.5 μmol) is placed in separate 2 ml plastic tubes. To each is added 100 μl of 1% acetone solution of FDAA (1 mg, 3.6 μmol), the molar ratio of FDAA to amino acid is 1.4:1, followed by NaHCO$_3$ (1 M, 20 μl, 20 μmol). The contents are mixed and heated over a hot plate at 30–40°C for 1 h with frequent mixing. After cooling to room temperature, HCl (2 M, 10 μl, 20 μmol) is added to each reaction mixture. After mixing, the contents are dried in a vacuum desiccator over NaOH pallets. Each residue is then dissolved in DMSO (0.5 ml). A 1:1 dilution of these is made (2.5 mM) and 5 μl sample of each is pooled and analyzed, for example, injected for HPLC.

It has been recognized that addition of DMSO or the use of mixtures of triethyl amine and DMSO drastically increased the reaction rate of derivatization and are recommended to overcome the low reactivities in the case of sterically hindered reagents. Further, the gradient elution with mixtures of sodium acetate buffer of pH 4.0 and methanol or aqueous TFA and acetonitrile are also suitable in order to accelerate analyses [15]. The reaction for the synthesis is shown in Figure 15.1b.

Application of original protocol of Marfey [11] was unsuccessful for the resolution of enantiomers of ring- and α-methyl substituted phenylalanines and phenylalanine amides [16]. In the modified procedure of derivatization, the analyte (1 mg) was dissolved in water (1 ml). An aliquot (100 μl) was mixed with a solution of FDAA in acetone (100 μl, 1.6% w/v, molar ratio of FDAA to analyte about 15:1). Higher reactant concentration and longer reaction times were also used. The reaction was allowed to stand overnight at 40°C or for 6 h at 50°C. The reaction was found to remain slow. It was stopped by addition of 2 M HCl and diluted twofold directly with the mobile phase. The application of a higher temperature and a longer reaction time may promote racemization of enantiomers that have hydrogen in α-position. Derivatization of peptides (e.g., oxytocin) with free α-amino group requires longer reaction time than that of amino acids [17]; instead of 90 min the reaction time for peptides is 12–24 h and five times excess of reagent is necessary. The retention time of peptides increases significantly depending on the size of the peptide fragment.

15.3.2 DETECTION

Marfey [11] carried out the analysis of five amino acids by HPLC, elution was done with linear gradient of acetonitrile in 50 mM TEAP buffer, pH 3.0, from 10 to 50% acetonitrile during 1 h, flow rate 2 ml/min, at 340 nm. L-Diastereomers were eluted from the RP column before D-diastereomers and all the ten diastereomers had a very similar absorption spectra characterized by a λ_{max} at 340 nm. The λ_{max} values are slightly different for diastereomers of different amino acids and vary slightly with the nature of solvent used. The spectra are stable if the solutions are kept in dark; otherwise, a gradual change occurs as a result of a photochemical decomposition of the absorbing chromophore. The hydrolyzed reagent appears as sharp peak and is separated from all the diastereomers. Detection at 340 nm makes Marfey derivatives insensitive to most buffer systems and most mobile phase impurities. The only detection interference is the excess MR itself. The chromatographic conditions can be adjusted to avoid coelution with any of the desired analyte peaks.

15.3.3 ADVANTAGES

There is little doubt that of the various methods available for the indirect HPLC resolution of enantiomers of amino acids, use of MR has been most successful. The reagent meets satisfactorily the characteristic features, as enumerated above, as a derivatizing agent in different situations. Marfey's method has been widely used for structural characterization of peptides, confirmation of racemization in peptide synthesis, and detection of small quantities of D-amino acids. Many other reagents used for prederivatization of free amino acids suffer inherent problems, for example, inability to react with all proteinogenic amino acids, unstable derivatives, poor detectability of certain amino acid derivatives, or lack of quantitative yield of the reaction.

MR provides a very simple and effective analytical method. Hydrolysis of peptides or proteins (may be microwave assisted) followed by derivatization of the resulting amino acids with the chiral MR adds a highly absorbing chromophore that converts the amino acids into UV-active diastereomers. This allows separation of D- and L-amino acids as diastereomers in the nanomole range on a nonchiral reversed phase standard column with the inherent rapidity of determination in HPLC. In addition, these derivatives of amino acids can be detected in both, simple UV as well as more selective mass spectrometric devices.

The method is considered to have an advantage over the method of Manning and Moore [18] in that it does not produce oligomeric products that can be formed when an L-amino acid N-carboxy anhydride (NCA) reacts with a mixture of L- and D-amino acids. Major advantages of MR over other precolumn derivatizations include (i) possibility to carry out chromatography on any multipurpose HPLC instrument without column heating, (ii) simultaneous detection of proline in a single chromatographic run, and (iii) stable amino acid derivatives.

The Marfey's-derivatives of D- and L-amino acids can be identified by coinjection of standard derivatized D- and L-amino acids; the reciprocity principle of chromatography makes it possible, using both enantioisomeric reagents, to determine the opposite stereoisomer's retention time without measuring for the corresponding authentic sample.

The other advantage of MR is the possibility to increase its hydrophobicity (thus increasing α and R_s of the derivatives), by replacing the chiral selector Ala-NH$_2$ of the reagent with other amino acids such as Leu-NH$_2$ or Val-NH$_2$ [19–22]. This method can simultaneously identify amino acids with the correct absolute configuration under gradient elution conditions and is highly sensitive. Since D- and L-amino acids are separated by HPLC as diastereomers of MR and the L-amino acid derivative is usually eluted before the corresponding D-isomer, the method is useful to determine absolute configuration. These characteristics make the method quite flexible for resolution and quantification of such optical isomers.

15.3.4 DISADVANTAGES

Amino acids such as tyrosine and histidine or containing two amino groups such as ornithine and lysine can form both mono- and di-substituted Marfey's derivatives; thus, doubling the number of peaks in a chromatogram for these amino acids. Using MR in excess formation of mono-substituted derivative can be minimized. Reaction of hydroxyl group of tyrosine with MR produces both mono- and di-substituted derivatives [21]; increasing the strength of base during derivatization reaction increases the yield of di-substituted tyrosine and minimizes the appearance of the peak for mono-substituted derivative.

Enantiomeric purity of the reagent is very important when the reagent is produced synthetically; enantiomerically pure L-alanine amide must be used and no racemization must have occurred during the synthesis. A small quantity of D-isomer of MR can cause a false detection of the other enantiomer in the chromatographic analysis of the diastereomer. Purity of the final reagent is required to be determined by a direct chromatographic procedure. Determination of absolute configuration is difficult for a peptide containing unusual amino acids without having their standard samples.

15.4 CHIRAL VARIANTS AND STRUCTURAL ANALOGS OF MR

Resolution of 19 pairs of DL-proteinogenic AAs, containing neutral, acidic, basic, or aromatic side chains, by HPLC, was made possible with three chiral variants of MR prepared by the reaction of DFDNB with Val-NH$_2$, Phe-NH$_2$, and Pro-NH$_2$. Differences in Δt_R (retention times) of these diastereomers were compared with those obtained by derivatization of the same set of 19 DL-AAs with the original MR [20]. However, the FDNP-Val-NH$_2$ gave the largest Δt_R values.

Some of the major variants prepared and studied include (a) FDNP-Val-CONHR with the group R as, -H, -t-Bu, -CH(CH$_3$)C$_6$H$_5$, -C$_6$H$_5$, and

-*p*-nitrophenyl; (b) FDNP-Val-CO*OR* with the group *OR* as, -OH, -OCH$_3$, and -O-*t*-Bu; (c) FDNP-Ala-Ala-NH$_2$; and (d) FDNP-PEA and FDNP-Valol (-Val or -Ala bonded to FDNP through α-NH-). Thus, these were the substituted FDNP-Val amide, FDNP-Val esters, FDNP-amines, and FDNP-dipeptide amide. The amino acids analyzed were DL-Val, -Glu, -Ser, -Lys (mono), and -Lys (di). These variants were used to determine the structural parameters responsible for larger Δt_R values of AA diastereomers and to optimize HPLC separation of diastereomers formed by reaction of these FDNP reagents with certain AAs [21]. FDNP-Val-NH$_2$ has been reported to provide retention times varying between 4 and 20 min in comparison to FDAA for the determination of configuration and stereochemical purity of cysteine residues in peptides [17].

Based on considerations of structural analogy and reactivity Brückner and Strecker [23] carried out nucleophilic substitution of the three halogens (chlorine or fluorine) of trihalo-*s*-triazines (2,4,6-trihalo-1,3,5-triazines), that is, cyanuric chloride or cyanuric fluoride, respectively, by (i) reaction with either methanol, 2-naphthol, 1-methoxy naphthalene, or 4-aminoazobenzene, providing UV-absorbing chromogenic dihalo-*s*-triazines; (ii) reaction with L-alanine amide yielding chiral monohalo-*s*-triazines; and (iii) reaction (of the monohalo-*s*-triazine) with selected D- or L-amino acids to form diastereomeric derivatives, which were separated by RP (C18) HPLC using mixtures of water, acetonitrile, and TFA as eluents. Thus, the dinitro fluoro moiety of, for example, FDAA was replaced by a *s*-triazine moiety having a suitable chromophore. These chiral monohalo-*s*-triazines were found capable of resolving certain DL-amino acids. The resolution was lower in some cases in comparison with those obtained by reaction of the same DL-amino acids with other chiral variants, described before [20,21]. Analysis of results suggested that the nitro groups in the dinitro fluoro moiety (say, FDAA) was contributing to better resolution, while the increasing bulkiness of the substituent *R* in the monohalo-*s*-triazine led to decrease in resolution. Nevertheless, these variants provided a general approach for the design and construction of tailor made reagents suitable for precolumn derivatization and liquid chromatographic separation of resulting diastereomers of the amino acid enantiomers. The approach was extended to other CDRs by Brückner and Leitenberger [24].

When FDAA was replaced by L-FDLA (1-fluoro-2,4-dinitrophenyl-5-L-leucine amide), an enhanced sensitivity, hydrophobicity, and thermal stability (in comparison to L-FDAA) were observed. Electrospray ionization and frit-fast atom bombardment (Frit-FAB) were applied as the interface [25,26]. It also required changing the original mobile phase of acetonitrile–phosphate buffer to acetonitrile–TFA as the former was not volatile. Derivatized amino acids with FDLA showed almost the same retention behavior as that with FDAA.

Harada et al. [27] prepared a chiral anisotropic reagent, 1-fluoro-2,4-dinitrophenyl-5-(*R*, *S*)-phenylethyl amine [(*R*, *S*)-FDPEA], for the determination of absolute configuration of α-carbon of primary amino compounds. The reagent has advantages in terms of reactivity, reliability, and effectiveness and has successfully been applied to a peptide (microgenin) produced by cyanobacterium. The reagent was prepared by Marfey's method [11] while the derivatives were prepared by treatment of primary amino compounds with FDPEA under slightly basic conditions followed by TLC preparative separation. The amino compounds analyzed include L-isoleucinol, D-, and L-phenylalaninols.

15.5 MECHANISM FOR RESOLUTION OF DIASTEREOMERS

Use of MR provides an indirect approach for separation of diastereomers prepared prior to chromatography. The separation of FDAA derivatives of amino acids has been presented in literature based on H-bonding, different conformations, and difference in the hydrophobicity of the two diastereomers based on side chains in the α-amino acids. These are briefly described below.

15.5.1 HYDROGEN BONDING

Marfey [11], Brückner and Keller-Hoehl [20], and Brückner and Gah [21] discussed independently that the resolution of L- and D-derivatives is essentially due to intramolecular H-bonding. Marfey's conclusions were based on the structural features and elution behavior of five amino acid derivatives, while those of Brückner et al. were based on structural features correlated with separation behavior of a large number of derivatives including several chiral variants (as mentioned above) and construction of molecular models. These are briefly described in the following.

Marfey [11] attributed the reason for the L-diastereomer eluting before the D-isomer to a stronger intramolecular H-bonding in D-isomer than in L-isomer. He suggested that the carboxyl group can H-bond to either an *ortho* situated nitro group producing a nine-member ring or, more likely, the carbonyl oxygen of the *meta*-situated L-Ala–NH$_2$ forming a 12-membered ring. Stronger H bonding in a D-diastereomer would produce a more hydrophobic molecule, which would be expected to interact more strongly with the RP column and thus have a stronger retention time than a L-diastereomer. Nature of the amino acid side chain is also responsible for the differences in elution times of the diastereomers; the ionizable side chains of Asp and Glu decrease the separation, while neutral and hydrophobic side chains increase it. The construction of (space filling Corey–Pauling–Koltun) molecular models of the L–L and D–L diastereomers (the first letter refers to the configuration of AA to be analyzed and the second to that of the reagent) showed [15,21] that:

(i) In the L–L, the carboxyl group of the analyte is located extremely close to the carboxamide of the reagent, thus facilitating the formation of an intramolecular H-bridge (Figure 15.2). Formation of an H-bridge between the carboxyl and the carboxamide group is not possible to such an extent in the case of D–L diastereomer. The non H-bonded free carboxyl group of the D-AA in D–L diastereomer results in a gap and less symmetrical form of the molecule in the D–L diastereomer causing stronger interaction with the alkyl chains of the

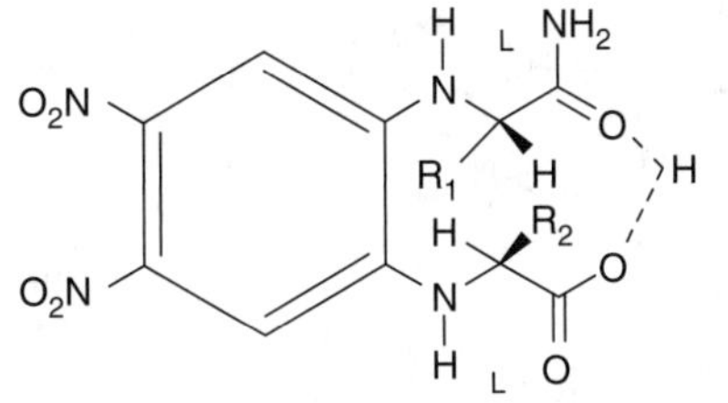

FIGURE 15.2   L–L Diastereomer showing H-bonding.

reversed phase and thus greater retention time in comparison with the L–L diastereomer. These observations are in agreement with Marfey's explanation [11] that H-bond is the most important feature, which brings differences in the free energies of diastereomers owing to its formation/nonformation and for obtaining large Δt_R values [20].

(ii) In the L–L and D–L diastereomers of Asp and Glu acids (to be analyzed) the respective β- and γ-carboxyl groups probably do not exhibit a different behavior as exhibited by mono carboxyl amino acids. The relatively higher hydrophilicities in these cases lead to shorter retention times and lower Δt_R values as compared to neutral side chain AAs.

(iii) AAs in diastereomers have high steric hindrance due to side chains and there is almost no rotational freedom, for example, this is lower for valine having a C^α-isopropyl chain as compared to Ala having C^α-methyl side chain; these steric factors further facilitate resolution of diastereomers. The models also suggested that the proline ring in the respective diastereomer was perpendicular to the benzene ring leading to steric hindrance in the reagent FDNP-Pro-NH$_2$.

(iv) All reagents with the formulae FDNP-L-AA-NHR, FDNP-L-AA-OR, and FDNP-L-Ala-Ala-NH$_2$ have resolution capacity as diastereomers formed by reaction with DL amino acids, as in all cases an intramolecular H-bridge can be formed for the L–L diastereomer. This is not the case with reagents having structures FDNP-NHR; the moiety R alters the electronegativity of the carboxy group in —CONHR and —COOR, and also the hydrophobicities of diastereomers. The larger Δt_R values are obtained by diastereomers formed by reaction of H-L-Val-OH with FDNP-L-Val-OH since in the resulting L–L diastereomer the —COOH groups of amino acids can form H-bridges to each other.

15.5.2   Conformation

The UV spectra of the FDAA derivatives of the amino acids (under study) suggested [26] that a stable conformation, including intramolecular H-bonding, between the two nitro groups in the benzene ring and both amino groups of the amino acid and L-alaninamide, is formed as a planar molecule of a three-ring system, similar to

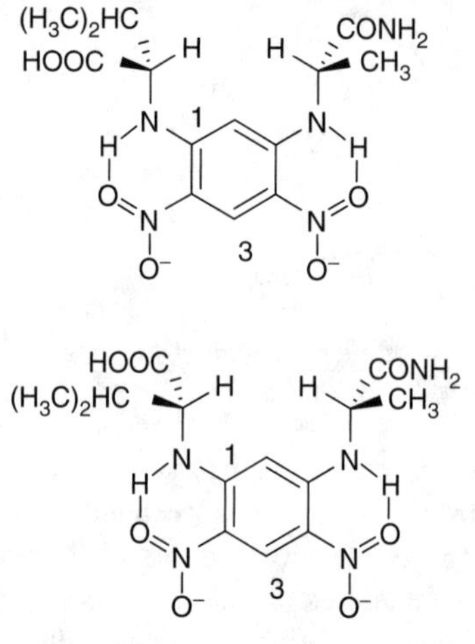

FIGURE 15.3 Plausible conformations of the L–L and L–D diastereomers of L- and D-valine with FDAA. (Reprinted with permission from Fujii et al., 1997; Copyright (1997) American Chemical Society.)

anthracene (Figure 15.3). Further, the NMR spectra of L- and D-valine derivatized with FDAA indicated that both α-protons were spatially situated near H-6 of the benzene ring in both L- and D-amino acid derivatives. Thus the resulting conformations of the L- and D-valine derivatives (Figure 15.3), in which each substituent (except for the amino groups of Val and L-alaninamide) was oriented perpendicular to the planar molecule of the dinitrobenzene, were stable and predominant in solution. Therefore, in the FDAA derivative of D-valine the more hydrophobic substituents of valine and L-alaninamide (the isopropyl group and methyl group, respectively) were in a *cis* (Z)-type arrangement to the plane of the dinitrobenzene, whereas the FDAA derivative of L-valine had the opposite arrangement (*trans*, [E] type). The resulting conformations of the L- and D-amino acid derivatives were stable.

The resolution of the L- and D-amino acid derivatives was due to difference in their hydrophobicity, which is derived from the *cis*- or *trans*-type arrangement of two more hydrophobic substituents at both α-carbons of an amino acid and L-alanine amide; thus, the *cis* (Z)-type arrangement of FDAA derivative interacts more strongly with ODS silica gel and has a longer retention time than that of the *trans* (E)-type arrangement. Therefore, the L-amino acid derivative is usually eluted first from the column.

The experimental observation that the retention time of amino acids derivatized with FDAA were dependent on the hydrophobicity of amino acids further confirms the proposed mechanism, for example, the FDAA derivative of an amino acid, which has a larger difference in hydrophobicity between the α-carboxyl group and the side chain, has a longer retention time and a better resolution. Separation behavior of FDAA derivatives of amino acid methyl esters and amino compounds without the α-carboxyl group, such as 1-phenylethylamine, alanilol, and valinol, showed that the retention time of methyl esters became longer than those of the parent amino acids, that is, the resolution power of amino acid methyl ester decreased in comparison with that of the parent amino acid.

The mechanism was further supported by the separation behavior of certain amino acids derivatized with chiral variants of FDAA in which L-Ala-NH$_2$ was replaced with L-Val-NH$_2$, L-Phe-NH$_2$, L-Ile-NH$_2$, and L-Leu-NH$_2$ since the retention time became longer, and the resolution became better with the increase of their length of alkyl side chains in the amino acid amides [14]. The amino acid derivatized with the reagent of D-alaninamides showed the completely opposite elution order. These results indicated that the α-carboxyl group of an amino acid was not always essential for the resolution and the separation behavior could be explained without consideration of intramolecular H-bonding (between the carboxyl of analyte and the carboxamide of the reagent).

The conformation for the derivative of the primary amino compounds formed with FDPEA was quite similar (Figure 15.3) to those of the FDLA derivatives of amino acids and primary amino compounds [27], it was investigated by UV and NMR, and is based on intramolecular H-bonding between the nitro groups and α-amino groups of the primary amino compounds and is suggestive of the separation mechanism.

15.5.3 DIFFERENCE IN THE HYDROPHOBICITY

Nevertheless, the mechanism discussed by Marfey [11], Brückner and Keller-Hoehl [20], and Brückner and Gah [21] based on different conformations and intramolecular H-bonding, correlates the separation behavior to differences in the hydrophobicity of the two diastereomers where the D-diastereomer produces a more hydrophobic molecule, which would be expected to interact more strongly with the RP column and thus has a stronger retention time.

The main difference in the proposed separation mechanisms can be looked in terms of the structure of the plausible conformations (Figures 15.2 and 15.3). Looking to the structure drawn by Marfey [11] and Brückner and Keller-Hoehl [20] it is apparent that a 12-member ring due to H-bond formation between carbonyl oxygen of the *meta*-situated L-Ala-NH$_2$ with the H atom of the carboxyl group of the L- or D-amino acid is equally likely as there is a free rotation between the C$^{\alpha}$- and amino−N of the amino acid moiety that allows equal chances for the−COOH group to come on either side (and this rotation is not affecting the configuration).

On the other hand, the figure 15.3 shows that a H-bond between oxygen atom of the nitro group in position 2 of the benzene ring with amino −H of the amino acid (Val, in this case) and another H-bond formation between nitro group at position 4 in the benzene ring and the amino −H of the alanine amide moiety (of the reagent) is possible, that is, forming a six-member ring on either side. A six-member ring, nearly planar, on both sides provides a more stable state of the complex (in comparison with a 9- or 12-member ring). Such a structure is possible for both L- and D-configurations of the amino acid. Thus, the two structures shown in Figure 15.3 behave as diastereomers due to the difference in steric arrangement of the groups at the stereogenic center of the α-amino acid (to be analyzed). Further, the hydrophobic methyl (of alanine amide) and isopropyl (of valine) groups in D-Val-DNPA are *cis* to the plane of dinitrobenzene and thus interact more strongly with ODS silica gel and have a longer retention time. The FDAA derivatives of D-amino acids have *cis*-type arrangement because in most of the α-amino acids the side chain is more hydrophobic than the carboxylic group. Therefore, FDAA derivative of the L-enantiomer is usually eluted before the corresponding D-isomer. Participation of the nitro groups at hydrogen bonds is supported by the fact that *s*-triazine CDRs, which are considered to be structurally related to MR [23], in general are less effective for the indirect chiral separation of amino acids [28].

Thus, in both the cases (Figures 15.2 and 15.3), H-bond plays a role in the overall stability of the diastereomeric complex except that the site of H-bond is different. The three-point rule [29] proposed for resolution of enantiomers considers H-bond as one of the important factors along with $\pi-\pi$ interactions and steric repulsions between the CSP and one of the enantiomeric forms to distinguish between the two enantiomeric forms. In the application of MR the stationary phase is achiral, but the MR being chiral is responsible for diastereomeric formation and the differential interaction of the diastereomers with the ODS causes separation.

15.6 TLC APPLICATIONS OF MARFEY'S REAGENT

15.6.1 ENANTIOMERIC RESOLUTION VIA FDAA DIASTEREOMERS

First TLC application of MR has been reported by Ruterbories and Nurok [30] FDAA derivatives were prepared for 22 DL-amino acids and for 18 L-amino acids. These were chromatographed either on silica gel TLC plates using the binary solvent system, acetic acid-*tert*-butyl methyl ether, or on C18-bonded plates using the binary solvent system methanol/0.3 M sodium acetate. The difference of R_f values for a given solute pair is shown as ΔR_f, while the maximum value of ΔR_f is referred to as $(\Delta R_f)_{max}$. Among the 22 DL-amino acids separated, there were DL-ethionine and DL-citrulline along with 20 proteinogenic amino acids on C18 silica layers, whereas only 14 could be separated on silica gel layer. It was not possible to separate all diastereomers in single run due to overlapping of spots. The $(\Delta R_f)_{max}$ values and mobile phase compositions for normal and reverse phase resolution are shown in Tables 15.1 and 15.2, respectively. The L-isomer was found to have the higher R_f in reverse phase thin-layer chromatography (RP TLC) as was the case in RP HPLC [11]. The behavior was attributed to H-bonding as per Marfey's original model [11].

There was a good agreement between computer prediction and actual experimental separations. Computer-simulated and experimental chromatograms of isoleucine, leucine, phenylalanine, and arginine, were compared and, are shown in Figure 15.4.

Resolution of enantiomers of amino acids through MR has been found to be simple and rapid on RP TLC plates without using impregnated plate or a chiral mobile phase additive. Enantiomers of glutamate and aspartate

TABLE 15.1
$(\Delta R_f)_{max}$ for Diastereomeric FDAA Amino Acid Derivatives on Silica Gel Layers

No.	Parent amino acid	$(\Delta R_f)_{max}$	χ^a (mole fraction of methanol)
1	DL-Alanine	0.06	0.26
2	DL-Arginine	0.00	(No resolution)
3	DL-Asparagine	0.00	(No resolution)
4	DL-Aspartic acid	0.00	(No resolution)
5	DL-Citrulline	0.14	0.75
6	DL-Cystine	0.00	(No resolution)
7	DL-Ethionine	0.09	0.20
8	DL-Glutamic acid	0.00	(No resolution)
9	DL-Histidine	0.00	(No resolution)
10	DL-Isoleucine	0.12	0.18
11	DL-Leucine	0.10	0.19
12	DL-Lysine	0.09	0.40
13	DL-Methionine	0.08	0.23
14	DL-Norvaline	0.11	0.22
15	DL-Norleucine	0.14	0.21
16	DL-Phenylalanine	0.07	0.22
17	DL-Proline	0.06	0.58
18	DL-Serine	0.00	(No resolution)
19	DL-Threonine	0.00	(No resolution)
20	DL-Tryptophan	0.09	0.22
21	DL-Tyrosine	0.05	0.24
22	DL-Valine	0.10	0.22

Mobile phase: acetic acid-*tert*-butyl methyl ether.

[a] Mole fraction of acetic acid-*tert*-butyl methyl ether at which $(\Delta R_f)_{max}$ occurred.

Source: From Ruterbories, K.J. and Nurok, D., *Anal. Chem.*, 59, 2735, 1987.

were separated most effectively with solvent consisting of 25% acetonitrile in triethylamine–phosphate buffer (50 mM, pH 5.5). Separation of L- and D-serine was achieved with 30% acetonitrile solvent. The enantiomers of threonine, proline, and alanine were separated with 35% acetonitrile solvent, and those of methionine, valine, phenylalanine, and leucine with 40% of acetonitrile solvent. The possibility of using TLC for quantitative determination of amino acid enantiomers was shown by the quantitative recovery of D- and L-alanine from the TLC plate in the range of 0.56–4.48 nmol. The yellow spots were scrapped off the plate after chromatography, and extracted with methanol/water (1:1, v/v). The absorbance of the extracts was measured at 340 nm [31]. Since FDAA is light sensitive, the derivatives are also required to be protected from exposure to light during all procedures. The R_f values and the L/D ratio of R_f is given in Table 15.3.

TABLE 15.2

$(\Delta R_f)_{max}$ **for Diastereomeric FDAA Amino Acid Derivatives on Bonded C18 Layers**

No.	Parent amino acid	$(\Delta R_f)_{max}$	X^a
1	DL-Alanine	0.17	0.29
2	DL-Arginine	0.06	0.24
3	DL-Asparagine	0.13	0.19
4	DL-Aspartic acid	0.22	0.07
5	DL-Citrulline	0.12	0.23
6	DL-Cystine	0.08	0.35
7	DL-Ethionine	0.20	0.40
8	DL-Glutamic acid	0.11	0.12
9	DL-Histidine	0.08	0.23
10	DL-Isoleucine	0.21	0.42
11	DL-Leucine	0.20	0.43
12	DL-Lysine	0.15	0.43
13	DL-Methionine	0.20	0.36
14	DL-Norvaline	0.20	0.38
15	DL-Norleucine	0.20	0.43
16	DL-Phenylalanine	0.18	0.41
17	DL-Proline	0.13	0.30
18	DL-Serine	0.11	0.19
19	DL-Threonine	0.21	0.24
20	DL-Tryptophan	0.15	0.36
21	DL-Tyrosine	0.22	0.53
22	DL-Valine	0.21	0.37

Mobile phase: methanol–sodium acetate buffer (0.03 M, pH 4).

[a] Mole fraction of methanol–sodium acetate buffer (0.03 M, pH 4) at which $(\Delta R_f)_{max}$ occurred.

Source: From Ruterbories, K.J. and Nurok, D., *Anal. Chem.*, 59, 2735, 1987.

Chromatograms showing the best resolution for certain FDAA amino acids are shown in Figure 15.5.

15.6.2 SEPARATION OF FDAA DERIVATIVES BY 2D TLC

The presence of free D-alanine, D-proline, and D-serine was demonstrated in mammalian tissue, using a mutant mouse strain lacking D-amino acid oxidase [32] by derivatizing the free amino acids, from kidney and serum, with FDAA. The diastereomers were separated by two-dimensional TLC (2D TLC) and finally reverse phase chromatography was used to obtain resolution of D- and L-isomers, that is, no enantiomeric resolution was achieved on TLC plates. The TLC pattern of reference neutral amino acids derivatized with FDAA is shown in Figure 15.6.

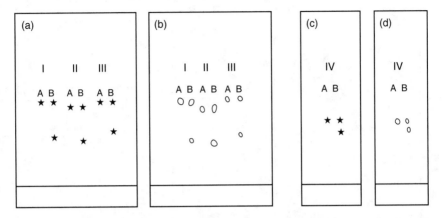

FIGURE 15.4 Computer-simulated (a) and experimental (b) chromatograms for FDAA derivatives of isoleucine (I), leucine (II), phenylalanine (III), and arginine (IV) (c and d). IA, IIA, IIIA, and IVA are for the derivatives of L-amino acids while IB, IIB, IIIB, and IVB are for the derivatives of DL-amino acids. (From Ruterbories, K.J. and Nurok, D., *Anal. Chem.*, 59, 2735, 1987.)

TABLE 15.3
Separation of FDAA Amino Acids on a RP TLC Plate

Racemate	Solvent (acetonitrile %)[a]	R_f L	D	L/D
Ser	25	0.201	0.165	1.220
Glu	25	0.150	0.114	1.319
Asp	25	0.176	0.133	1.377
Ser	30	0.367	0.320	1.147
Glu	30	0.317	0.271	1.149
Ser	35	0.437	0.394	1.109
Glu	35	0.417	0.375	1.112
Asp	35	0.438	0.403	1.087
Thr	35	0.395	0.328	1.204
Pro	35	0.276	0.224	1.232
Ala	35	0.270	0.203	1.330
Ser	40	0.475	0.449	1.058
Glu	40	0.450	0.420	1.071
Met	40	0.326	0.239	1.364
Val	40	0.294	0.200	1.470
Phe	40	0.235	0.143	1.643
Leu	40	0.216	0.144	1.500

[a] Mobile phase: acetonitrile–triethylamine–phosphate buffer (50 mM, pH 5.5). Time: 20 min; temperature: 25°C; solvent front: 75 mm.

Source: From Nagata, Y., Iida, T., and Sakai, M., *J. Mol. Cat. B Enzym.*, 12, 105, 2001.

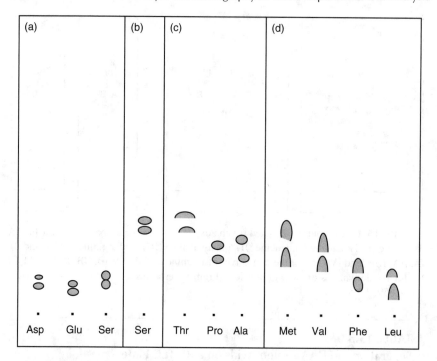

FIGURE 15.5 Separation of optical isomers of FDAA amino acids on reversed phase TLC plates. Mobile phase: acetonitrile/50 mM triethylamine/phosphate buffer (pH 5.5); 25:75 (a); 30:70 (b); 35:65 (c); or 40:60 (v/v) for (d). The upper and the lower spots are L- and D-isomers, respectively. Time: 20 min; temperature: 25°C; solvent front: 75 mm. (From Nagata, Y., Iida, T., and Sakai, M., *J. Mol. Cat. B Enzym.*, 12, 105, 2001)

TLC was carried on Whatman K6 precoated 0.25 mm layers of silica gel using *n*-butanol/acetic acid/water (3:1:1, v/v) and phenol/water (3:1, v/v) for first and second dimension, respectively. The yellow FDAA derivatives were scraped off the TLC plate and were extracted with methanol/water (1:1, v/v) for HPLC analysis.

The method was further extended for analysis of D- and L-amino acids in mouse kidney in a similar manner. HPLC methods, for the resolution of amino acid enantiomers that utilized a chiral stationary phase or a chiral mobile phase additive were considered expensive and less satisfactory due to the high retention results in broad peaks. The separation of FDAA derivatives followed by HPLC, as described above, was found to be much more successful as there was neither needed a post column derivatization nor fluorimetric analysis and moreover a subnanomolar sensitivity was attained [33]. D- and L-enantiomers of glutamate, aspartate, asparagine, serine, threonine, alanine, proline, tyrosine, valine, methionine, isoleucine, leucine, phenylalanine, and histidine were derivatized with FDAA and the diastereomers were separated by 2D TLC. Each was separated except for the two spots comprising Tyr and Val, and Ile, Leu, and Phe. Only histidine was separated further into D- and L-diastereomers by TLC. The excess hydrolyzed FDAA moved to the front

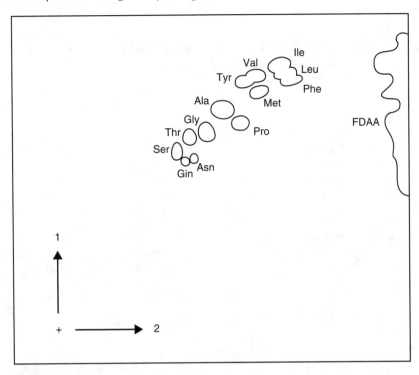

FIGURE 15.6 Two-dimensional TLC of FDAA-derivatized neutral amino acids. (+), Starting point, 1 for first dimension, and 2 for second dimension. Mobile phase in first dimension: *n*-butanol/acetic acid/water (3:1:1, v/v). Mobile phase in second dimension: Phenol/water (70:30, v/v). (From Nagata, Y., Yamamoto, K., Shimojo, T., Konno, R., Yasumura, Y., and Akino, T., *Biochim. Biophys. Acta*, 1115, 208, 1992.)

of the second solvent, resulting in complete separation of FDAA from the derivatized amino acids. This was of advantage for the HPLC analysis, because the elution time of hydrolyzed FDAA was close to that of some FDAA-amino acids. Heuser and Meads [34] separated 14 chiral amino alcohols and amines by HPTLC on RP-18 WF$_{254}$ plates after derivatization with MR.

A modified Marfey's method with TLC scanning was reported to monitor the enantiomeric purity of synthesized (D)-phenyl[3-^{14}C]alanine [4]; that is, >99% was measured [35].

15.6.3 Resolution of FDPA Derivatives of Proteinaceous Amino Acids by Normal and RP TLC

Racemic and nonracemic amino acids were derivatized with chiral reagent 1-fluoro-2,4-dinitrophenyl-5-L-phenylalanine amide (FDPA) according to Marfey's method [11]. Twelve proteinaceous α-amino acids were derivatized with FDPA, a chiral variant of MR. The resulting diastereomers were separated

TABLE 15.4

The hR_f Values of FDPA Diastereomers of Certain Proteinaceous Amino Acids Resolved on Normal Phase TLC

No.	Amino acids	hR_f		
		L	D	D/L
1	Alanine	37.5	62.5	1.67
2	Phenylalanine	45.0	75.0	1.67
3	Leucine	50.0	82.5	1.65
4	Isoleucine	47.5	80.0	1.68
5	Valine	40.0	70.0	1.75
6	Cysteine	37.5	67.5	1.80
7	Serine	32.5	60.0	1.85
8	Methionine	40.0	72.5	1.81
9	Asparagine	27.5	40.0	1.45
10	Glutamic acid	25.0	45.0	1.80
11	Aspartic acid	37.5	65.0	1.73
12	Threonine	40.0	62.5	1.56

Source: From Bhushan, R., Brückner, H., and Kumar, V., Unpublished data.

on normal phase and reverse phase thin layer chromatographic plates. The derivatives were well resolved with binary mobile phase combinations. In normal phase TLC best resolution was achieved with phenol/water (70:30, v/v), while in RP TLC the best resolution was achieved with combination of acetonitrile/triethylamine/phosphate buffer (50 mM, pH 5.5). The successful combinations for various FDPA diastereomers in reverse phase were (i) acetonitrile–TEAP buffer (50 mM, pH 5.5, 50:50, v/v) for alanine, phenylalanine, leucine, and isoleucine; (ii) acetonitrile–TEAP buffer (50 mM, pH 5.5, 30:70, v/v) for serine, valine, cysteine, and methionine; and (iii) acetonitrile–TEAP buffer (50 mM, pH 5.5, 40:60, v/v) for asparagine, glutamic acid, aspartic acid, and threonine. The effects of buffer concentration, pH, and concentration of organic modifier on resolution have been studied. This indirect method allowed the resolution of DL-amino acids in nanomole range. The hR_f values are shown in Tables 15.4 and 15.5 for normal and reverse phase modes, respectively [36].

15.6.4 RESOLUTION OF CERTAIN NONPROTEINACEOUS AMINO ACIDS BY NORMAL AND RP TLC

Some nonprotein α-amino acids were derivatized with chiral variant of MR FDPA, according to modified Marfey's method [11]. The resulting diastereomers were separated by normal and reverse phase TLC in nanomole range. In normal phase TLC, the amino acids used were phenylglycine, cysteic acid, isovaline,

TABLE 15.5
The hR_f Values of FDPA Diastereomers of Certain Proteinaceous Amino Acids Resolved on RP TLC

No.	Amino acids	% Acetonitrile	hR_f		
			L	D	L/D
1	Alanine	50	40.0	50.0	1.25
2	Phenylalanine	50	33.8	41.3	1.22
3	Leucine	50	31.3	38.8	1.24
4	Isoleucine	50	30.0	37.5	1.25
5	Valine	30	09.2	18.5	2.01
6	Cysteine	30	12.3	21.5	1.75
7	Serine	50	45.1	55.2	1.22
8	Methionine	30	13.8	23.1	1.67
9	Asparagine	40	24.6	35.4	1.44
10	Glutamic acid	40	29.2	33.8	1.12
11	Aspartic acid	40	30.7	36.9	1.20
12	Threonine	40	24.6	35.4	1.44

Source: From Bhushan, R., Brückner, H., and Kumar, V., Unpublished data.

TABLE 15.6
The hR_f Values of FDPA Diastereomers of Certain Nonproteinaceous Amino Acids Resolved on Normal Phase TLC

No.	Amino acids	hR_f		
		L	D	D/L
1	Phenylglycine	38	72	1.894
2	Cysteic acid	11	29	2.636
3	Isovaline	39	72	1.846
4	2-Amino-*n*-butyric acid	34	69	2.029
5	Norvaline	38	74	1.947
6	2-Aminoadipic acid	19	43	2.263
7	2-Aminooctanoid acid	47	80	1.702

Mobile phase: phenol/water (3:1, v/v). Solvent front: 9 cm.

Source: From Bhushan, R., Brückner, H., and Kumar, V., Unpublished data.

2-aminobutyric acid, norvaline, 2-aminoadipic acid, and 2-aminooctanoic acid; the best resolution was obtained with solvent combination of phenol/water (3:1, v/v). Only pipecolinic acid failed to resolve with this mobile phase combination. The hR_f are given in Table 15.6.

TABLE 15.7
The hR_f Values of FDPA Diastereomers of Certain Nonproteinaceous Amino Acids Resolved on RP TLC

No.	Amino acids	CH$_3$CN (%)[a]	hR_f		
			D	L	D/L
1	2-Amino-*n*-butyric acid	50	33	39	1.181
2	2-Aminooctanoic acid	50	19	26	1.368
3	Norvaline	50	29	37	1.276
4	Cysteic acid	45	37	40	1.081
5	2-Aminoadipic acid	45	19	27	1.421
6	Isovaline	30	18	20	1.111
7	Pipecolinic acid	30	18	24	1.333
8	Phenylglycine	30	20	23	1.150

[a] Mobile phase: acetonitrile with triethylamine–phosphate buffer (50 mM, pH 5.5, v/v). Solvent front: 9 cm.

Source: From Bhushan, R., Brückner, H., and Kumar, V., Unpublished data.

In the reverse phase TLC the amino acids used were phenylglycine, cysteic acid, isovaline, 2-aminobutyric acid, norvaline, 2-aminoadipic acid, 2-amino octanoic acid, and pipecolinic acid. Various solvent combinations such as methanol–TEAP buffer and acetonitrile–TEAP buffer were tried for resolution when combinations of acetonitrile with triethylamine–phosphate buffer (50 mM, pH 5.5) were found successful. The hR_f values are given in Table 15.7. As compared with normal phase TLC, hR_f values and resolution were significantly low, but spots were more compact.

It was observed that in normal phase TLC, the hR_f values increased with increase in bulkiness of side chains, but in reverse phase HPLC these values decreased. This can be attributed due to increase in hydrophobic character of side chains. The diastereomer of L-enantiomer has lesser affinity for the C18 silica gel than the corresponding D-enantiomer in reverse phase TLC due to their more hydrophilic character. Therefore, derivative of L-enantiomer moved faster than those of the corresponding D-enantiomer. Observed elution order in normal phase TLC was opposite to that in reverse phase [36]. Since FDPA and its derivatives are light sensitive, all procedures were protected from light exposure and the derivatives were kept in dark at 0–4°C.

15.7 CONCLUSION

Structural features of MR provide a flexibility and possibility to increase its hydrophobicity by replacing the chiral selector Ala-NH$_2$ of the reagent with other suitable moieties and make the method quite flexible for resolution and quantification of DL-amino acids, in particular, in different situations. The separation mechanism involves the formation of H-bridge to provide a stable conformation, which is responsible for elution behavior, and the α-amino group is essential for the resolution of both diastereomers.

Derivatization with MR permitted a very sensitive method for detection of racemization (enantiomerization or epimerization) of amino acids, in particular, and in several instances revealed better information than shown by CPA/MALDI-MS. Though majority of studies using MR are with HPLC, there are relatively few reports on its TLC applications. Nevertheless, the reagent provides a versatile approach for indirect separation of DL-amino acids and certain other compounds capable of derivatizing with MR.

ACKNOWLEDGMENTS

The authors are thankful to Alexander von Humboldt-Stiftung, Bonn, Germany for the award of a fellowship, and to the Indian Institute of Technology, Roorkee, for granting leave of absence to Ravi Bhushan.

REFERENCES

1. Frank, H., Nicholson, G.J., and Bayer, E., Rapid gas chromatographic separation of amino acid enantiomers with a novel chiral stationary phase, *J. Chromatogr. Sci.*, 15, 174, 1977.
2. Bhushan, R. and Joshi, S., Resolution of enantiomers of amino acids by HPLC, *Biomed. Chromatogr.*, 7, 235, 1993.
3. Bhushan, R. and Martens, J., Direct resolution of enantiomers by impregnated TLC, *Biomed. Chromatogr.*, 11, 280, 1997.
4. Bhushan, R. and Martens, J., Amino acids and their derivatives. In: *Handbook of Thin Layer Chromatography*, Sherma, J. and Fried, B., Eds., Marcel Dekker Inc., New York, pp. 373–415, 2003.
5. Pirkle, W.H. and Pochapsky, T.C., Considerations of chiral recognition relevant to the liquid chromatographic separation of enantiomers, *Chem. Rev.*, 89, 347, 1989.
6. Martens, J. and Bhushan, R., Enantiomerentrennung von — Aminosäuren mittels Dünnschichtchromatographie, *Chem.-Ztg.*, 112, 367, 1988.
7. Martens, J. and Bhushan, R., TLC enantiomeric separation of amino acids, *Int. J. Peptide Protein Res.*, 34, 433, 1989.
8. Martens, J. and Bhushan, R., Importance of enantiomeric purity and its control by TLC, *J. Pharm. Biomed. Anal.*, 8, 259, 1990.
9. Martens, J. and Bhushan, R., Direct resolution of enantiomers of ibuprofen by liquid chromatography, *Biomed. Chromatogr.*, 12, 309, 1998.
10. Zahn, H., The reaction of 1,3-difluoro-4,6-dinitrobenzene with fiber proteins and amino acids, Kolloid Zeitschrift, 121, 39–45, 1951.
11. Marfey, P., Determination of D-amino acids. II. Use of a bifunctional reagent, 1,5-difluoro-2,4-dinitrobenzene, *Carlsberg Res. Commun.*, 49, 591, 1984.
12. Bhushan, R. and Brückner, H., Marfey's reagent for chiral amino acid analysis: A review, *Amino Acids*, 27, 231, 2004.

13. B'Hymer, C., Montes-Bayon, M., and Caruso, J.A., Marfey's reagent: Past, present, and future uses of 1-fluoro-2,4-dinitrophenyl-5-L-alanine amide, *J. Sep. Sci.*, 26, 7, 2003.

14. Fujii, K., Ikai, Y., Mayumi, T., Oka, H., Suzuki, M., and Harada, K., A non empirical method using LC/MS for determination of the absolute configuration of constituent amino acids in a peptide: Elucidation of limitations of Marfey's method and of its separation mechanism, *Anal. Chem.*, 69, 3346, 1997.

15. Brückner, H. and Keller-Hoehl, C., HPLC separation of DL-amino acids derivatized with N^2-(5-fluoro-2,4-dinitrophenyl)-L-amino acid amides, *Chromatographia*, 30, 621, 1990.

16. Péter, A., Olajos, E., Casimir, R., Tourwe, D., Broxterman, Q.B., Kaptein, B., and Armstrong, D.W., High performance liquid chromatographic separation of the enantiomers of unusual α-amino acids analogues, *J. Chromatogr. A*, 871, 105, 2000.

17. Szabó, S., Szokan, Gy., Khlafulla, A.M., Almas, M., Kiss, C., Rill, A., and Schön, I., Configuration and racemization determination of cysteine residues in peptides by chiral derivatization and HPLC: Application to oxytocin peptides, *J. Peptide Sci.*, 7, 316, 2001.

18. Manning, J.M. and Moore, S., Determination of D- and L-amino acids by ion exchange chromatography as L-D and L-L dipeptides, *J. Biol. Chem.*, 243, 5591, 1968.

19. Harada, K., Mayumi, T., Shimada, T., Suzuki, M., Kondo, F., and Watanabe, M.F., Occurrence of four depsipeptides, aeruginopeptins with microcystins from toxic cyanobacteria, *Tetrahedron Lett.*, 34, 6091, 1993.

20. Brückner, H. and Keller-Hoehl, C., HPLC separation of DL-amino acids derivatized with N^2-(5-fluoro-2,4-dinitrophenyl)-L-amino acid amides, *Chromatographia*, 30, 621, 1990.

21. Brückner, H. and Gah, C., HPLC separation of DL-amino acids with chiral variants of Sanger's reagent, *J. Chromatogr.*, 555, 81, 1991.

22. Brückner, H., Wittner, R., and Godel, H., Fully automated HPLC separation of DL-amino acids derivatized with O-phthaldialdehyde together with N-isobutyryl-cysteine: Application to food samples, *Chromatographia*, 32, 383, 1991.

23. Brückner, H. and Strecker, B., Use of chiral monohalo-s-triazine reagents for the liquid chromatographic resolution of DL-amino acids, *J. Chromatogr.*, 627, 97, 1992.

24. Brückner, H. and Leitenberger, M., LC separation of derivatized DL-amino acid by aminopropyl silica-bonded Marfey's reagent and analog, *Chromatographia*, 42, 683, 1990.

25. Harada, K., Fujii, K., Mayumi, T., Hibino, Y., and Suzuki, M., A method using LC/MS for determination of absolute configuration of constituent amino acids in peptide — Advanced Marfey's method, *Tetrahedron Lett.*, 36, 1515, 1995.

26. Fujii, K., Ikai, Y., Oka, H., Suzuki, M., and Harada, K., A non empirical method using LC/MS for determination of the absolute configuration of constituent amino acids in a peptide: Combination of Marfey's method with mass spectrometry and its practical application, *Anal. Chem.*, 69, 5146, 1997.

27. Harada, K., Shimizu, Y., and Fujii, K., A chiral anisotropic reagent for determination of the absolute configuration of a primary amino compound, *Tetrahedron Lett.*, 39, 6245, 1998.

28. Brückner, H. and Wachsmann, M., Design of chiral monochloro-*s*-triazine reagents for the liquid chromatographic separation of amino acid enantiomers, *J. Chromatogr. A*, 998, 73, 2003.

29. Dalgliesh, C.E., The optical resolution of aromatic amino acids on paper chromatograms, *J. Chem. Soc.*, 137, 3940, 1952.

30. Ruterbories, K.J. and Nurok, D., Thin-layer chromatographic separation of diastereomeric amino acid derivatives prepared with Marfey's reagent, *Anal. Chem.*, 59, 2735, 1987.

31. Nagata, Y., Iida, T., and Sakai, M., Enantiomeric resolution of amino acids by thin-layer chromatography, *J. Mol. Cat. B Enzym.*, 12, 105, 2001.

32. Nagata, Y., Yamamoto, K., Shimojo, T., Konno, R., Yasumura, Y., and Akino, T., The presence of free D-alanine, D-proline and D-serine in mice, *Biochim. Biophys. Acta*, 1115, 208, 1992.

33. Nagata, Y., Yamamoto, K., and Shimojo, T., Determination of D- and L-amino acids in mouse kidney by high-performance liquid chromatography, *J. Chromatogr.*, 575, 147, 1992.

34. Heuser, D. and Meads, P., HPTLC separation of chiral amino alcohols and amines after derivatization with Marfey's reagent, *J. Planar Chromatogr. Modern TLC*, 6, 324, 1993.

35. Koltai, E., Alexin, A., Rutkai, Gy., and Toth-Sarudy, E., Synthesis of optically pure (D)-phenyl [3-^{14}C] alanine, *J. Label. Compd. Radiopharm.*, 41, 977, 1998.

36. Bhushan, R., Brückner, H., Gupta, D., and Kumar, V., *J. Planar Chromatogr.*, (2007) communicated.

Index